23rd IAPRI World Conference on Packaging (IAPRI Bangkok 2022)

Packaging for World Crisis: Survival and Sustainability

Bangkok, Thailand
12-16 June 2022

ISBN: 978-1-7138-9645-6

Printed from e-media with permission by:

Curran Associates, Inc.
57 Morehouse Lane
Red Hook, NY 12571

Some format issues inherent in the e-media version may also appear in this print version.

Printed with permission by Curran Associates, Inc. (2024)

For permission requests, please contact Thai Packaging Centre (TPC)
at the address below.

Thai Packaging Centre (TPC)
Thailand Institute of Scientific and Technological Research (TISTR)
196 Phaholyothin Rd.
Ladyao Chatuchuk Bangkok 10900

Phone: 0 2579 1121
Fax: 0 2579 7573

Additional copies of this publication are available from:

Curran Associates, Inc.
57 Morehouse Lane
Red Hook, NY 12571 USA
Phone: 845-758-0400
Fax: 845-758-2633
Email: curran@proceedings.com
Web: www.proceedings.com

iapri
Bangkok 2022

23rd World Packaging Conference
12-16 June 2022

Conference Proceedings
The 23rd IAPRI World Conference on Packaging
(IAPRI Bangkok 2022)

CONTENTS

CONTENTS

CONTENTS

ORGANIZING COMMITEE

CONSULTANT

Mrs. Chutima Eamchotehawalit

CHAIR

Miss Pattra Maneesin

COMMITTEE

Miss Rachaneepen Pensit
Miss Sirirat Thavonrat
Mr. Bundit Fungsin
Miss Siriporn Larpkiattaworn
Miss Thanomchit Wanawatthanakul
Miss Anchana Pattanasupong
Mrs. Alissara Kuprasit
Miss Patama Lewlertmongkol
Miss Boonnanida Soda
Mrs. Sompong Pornchainoppakorn
Miss Rattana Ngamvisitsil
Mrs. Siriwan Tungsangprateep
Mrs. Bussakorn Praditniyakul
Mrs. Wisanee Nuamekin
Miss Nantanee Chaimongkalayon
Miss Pornteera Rattanarat
Miss Gansuda Wangchanachai
Mr. Thanyakorn Areerachakul

CONFERENCE SECRETARY

Mrs. Ratchaneewan Kulchan

CONFERENCE CO-SECRETARY

Mrs. Jeerawan Sutthilux
Miss Yanisa Sobsatian

SCIENTIFIC COMMITEE

Alexander Bardenshtein	Danish Technological Institute	Denmark
Changfeng Ge	Rochester Institute of Technology	USA
Cristina Guzmán	Universidad de Monterrey	Mexico
Damien Erre	ESIReims	France
Daniel Hellström	Lund	Sweden
Eric Martine	HEIG-VD	Switzerland
Frank Welle	Fraunhofer IVV	Germany
Gordon Robertson	University of Queensland	Australia
Gregory Batt	Clemson University	USA
Ingo Bueren	International Packaging Institute, Schaffhausen	Switzerland
Javier Zabaleta	Itene	Spain
Jay Singh	Cal Poly State University	USA
Katsuhiko Saito	Kobe University	Japan
Kiyomi Sakamoto	Shonan University of Medical Science	Japan
Mika Vähä-Nissi	VTT	Finland
Paul Singh	Packaging Forensic Associates	USA
Renee Wever	Linköping University	Sweden
Rafael Gavara	IATA-CSIC	Spain
Randy Beaudry	Michigan State University	USA
Ratchaneewan Kulchan	TISTR	Thailand
Roland Klooster	University of Twente	Netherlands
Selcuk Yildirim	ZHAW	Switzerland
Silvia Dantas	CETEA	Brazil
Soren Ostergaard	Danish Technological Institute	Denmark
Dong Sun Lee	Kyungnam University	South Korea
Siriwan Tungsangprateep	TISTR	Thailand
Supachai Pisuchpen	Prince of Songkla University	Thailand
Thomas Goedecke	BAM	Germany
Vincent Rouillard	Victoria University	Australia
Wanee Chinsirikul	NSTDA	Thailand
Yves Wyser	Nestlé Research Centre	Switzerland
Zhi-Wei Wang	Jinan University	China

The 23rd IAPRI World Conference on Packaging

CONFERENCE PROGRAMME

12 – 15 June 2022

Sunday June 12th — AVANI Hotel Sukhumvit

Time	Program	
12.00 - 13.00	Lunch at Greenhouse Restaurant, AVANI Hotel Sukhumvit	
13.00 - 14.00	**Distribution Packaging CoP**	Chair: Vincent Rouillard
14.00 - 15.00	**Packaging & Consumers CoP**	Chair: Virpi Kohonen
15.00 - 16.00	**Active & Intelligent Packaging CoP**	Chair: Selcuk Yildirim
16.00 - 17.00	**Sustainable Packaging CoP**	Chair: Carlos Diaz-Acosta
18.00 - 20.00	Welcome Reception at Cashmere Room, AVANI Hotel Sukhumvit	

Monday June 13th — Bangkok International Trade & Exhibition Centre (BITEC), Bangkok, Thailand

Time	Program	
8.00 - 9.00	Registration	
9.15 - 9.45	**Opening Ceremony** • Welcoming Remark by *Prof. Dr. Jay Singh, President IAPRI* • Welcoming Remark by *Dr. Chutima Eamchotchawalit, Governor of Thailand Institutes of Scientific and Technological Research (TISTR)* • Opening Speech by *Prof. Dr. Sirirurg Songsivilai, Permanent Secretary of Ministry of Higher Education, Science, Research and Innovation Thailand*	Room: Nile 1-4
9.45 - 10.00	Coffee Break	
10.00 - 10.45	Keynote Lecture on **"Trends and Innovation on Sustainable Packaging"** • *Prof. Pierre Pienaar, President, World Packaging Organization (WPO)*	
10.45 - 11.30	Keynote Lecture on **"Impact of Chain of Custody on Packaging Verification and Traceability"** • *Mr. Takehiro KANEKO, Manager, Transport Packaging Div., Japan Packaging Institute (JPI), Japan*	
11.30 - 12.30	Panel Discussion on **"Biodegradable & Compostable, Research, Development, Standard and Testing"** • *Prof. Dr. Suwabun Chirachanchai, The Petroleum and Petrochemical College, Chulalongkorn University* • *Mr. Viboon Pungprasert, Bio Plastics Commercial Manager, PTT Global Chemical PCL* • *Dr. Pichet Rojanapitayakorn, Senior Researcher, SCG Chemicals Public Company Limited* • *Dr. Pattra Maneesin, Deputy Governor Industrial Services Group, TISTR*	Moderator: Asst. Prof. Dr. Supachai Pisuchpen, Faculty of Agro-Industry, Prince of Songkla University
12.30 - 13.30	Lunch	Room: Bhiraj Hall 1
13.30 - 13.55	**Oral Presentation** (20 mins each plus 5 mins question) **Session: 1. Packaging sustainability** **2. Distribution and transport packaging** **3. Active & intelligent packaging** **4. Packaging materials**	Session Chair: 1. Dr. Roland ten Klooster 2. Dr. Supoj Pratheepthinthong 3. Dr. Sathukarn Kabcum 4. Dr. Siriporn Larpkiattaworn

Time	Program	
13.30 - 13.55	1.1 Investigating the Environmental Benefits of Increasing the Pallet Top Deck Stiffness in Unit Load System - *Saewhan Kim, Virginia Tech, United States* 2.1 Response of a Packaging System to a Transient Event: Real-time Replication VS SRS Simulation - *Damien Erre, Université de Reims, France* 3.1 Retardation of Lipid Oxidation of Sacha Inchi (Plukenetia volubilis L.) Oil Using Bio-based UV-Shielding Metal Oxide-Incorporated Active Packaging Films - *Busarin Chongcharoenyanon, Kasetsart University, Thailand* 4.1 Determination and Pre-treatment of Hazardous Substances in Food Can - *Shuai Wu, Hunan University of Technology, China*	Session Room: 1. Amber 1 2. Amber 2 3. Amber 3 4. Silk 4
14.00 - 14.25	1.2 Physical Properties Improvement of In Situ Modified Poly(lactide) Film for Fresh Produce Packaging - *Piyawanee Jariyasakoolroj, Kasetsart University, Thailand* 2.2 Impact of the Horizontal Acceleration Phenomenon Duration on the Tilt or Toppling of the Load - *Baptiste Boutrige, Esireims, France* 3.2 Development of a Colorimetric Total Volatile Base-Nitrogen Sensor for Potential Real-Time Freshness Monitoring of Pre-packaged Food - *Darylle Jerome Ortiz, Department of Science and Technology, Philippines* 4.2 Determination of Stretch Film Properties with Hysteresis for Freight Transport Applications - *Elora Leguebe, Esireims, France*	
14.30 - 14.55	1.3 Sorting of Plastic Packaging on Base of 3D Relief Code - *Roland ten Klooster, University of Twente, Netherlands* 2.3 Structural Damage Analysis of the Contaminated Insects Subjected to Drop Shock - *Saki Matsumoto, National Agriculture and Food Research Organization, Japan* (Virtual Presentation) 3.3 Antimicrobial Activity of Biodegradable Lignin Nanoparticles in Polyvinyl Acetate Based Film Against Major Spoilage Microorganisms in Mango Fruit - *Angel Jr Talongon Basbasan, Kasetsart University, Thailand* 4.3 Evaluating the Relationship Between the Quality of 3D Printing and Rheological Properties of PLA and PLAPHA - *Ehsan Behzadfar, Ryerson University, Canada*	
15.00-15.15	Poster session	
15.15-15.30	Coffee break	
15.30 - 15.55	**Session: 5. Packaging sustainability** **6. Distribution and transport packaging** **7. Packaging for food & agriculture** **8. Packaging materials** 5.1 Designing Reusable Packaging: Tool Development, Validation, and Implications - *Bjorn de Koeijer, University of Twente, Netherlands* 6.1 Effect of the Sustained Forces During Transportation on the Creep Behavior of Stretch Film Affecting Unit Load Stability - *Manuel Alfredo Garcia-Romeu, Polytechnic University of Valencia, Spain* (Virtual Presentation) 7.1 Multi-accelerant Approach for Shelf-life Determination of Oxygen-sensitive Foods in Polymeric Packaging Materials - *Anbuhkani Muniandy, Purdue University, United States* 8.1 Compatibilization of Polyethylene Polypropylene Blends via Ionic Liquid to Enhance Performance for Packaging Applications - *Carlos A. Diaz, Rochester Institute of Technology, United States*	Session Chair: 5. Dr. Jay Singh 6. Dr. Supoj Pratheepthinthong 7. Dr. Bundit Fungsin 8. Dr. Rachaneepen Pensit Session Room: 5. Amber 1 6. Amber 2 7. Amber 3 8. Silk 4
16.00 - 16.25	5.2 Investigation of the Consumer Perceptions of e-Commerce Packaging - *Eduardo Molina, Virginia Tech, United States* 6.2 A Review and Update of the Corrugated Fiberboard Cushioning - *Changfeng Ge, Rochester Institute of Technology, United States* (Virtual Presentation) 7.2 Buckling Strength Estimation in Ventilated Corrugated Paperboard Packages - *Amin Joodaky, Michigan State University, United States* 8.2 The Effect of Alcoholic Solutions on the Thermo-mechanical Properties of Poly(lactic acid) - *Rafael A. Auras, Michigan State University, United States* (Virtual Presentation)	

CONFERENCE PROGRAMME

Time	Program	
16.30 - 16.55	5.3 Life Cycle Assessment of Returnable Mailers used for Apparel Electronic Commerce: A Case Study in Canada - *Jonghun Park, Toronto Metropolitan University, Canada* 6.3 Measurement of the Vertical Vibrations Experienced by Pallets During Material Handling - *Clark Sabattus, Virginia Tech, United States* (Virtual Presentation) 7.3 Poly (Butylene Adipate-co-Terephthalate)/Thermoplastic Starch Blended Phosphate via Blown Film Extrusion as Functional Biodegradable Food Packaging - *Phanwipa Wongphan, Kasetsart University, Thailand* 8.3 Performance of Recyclable Thermoforming Sheets for Sous Vide Potatoes - *Marit Kvalvåg Pettersen, NOFIMA AS, Norway* (Virtual Presentation)	
17.00 - 17.25	7.4 Application of Biodegradable Film as Modified Atmosphere Packaging for Red Chili (Capsicum annum cv. Jinda) - *Dane Archibald G. Balanon, Kasetsart University, Thailand* 8.4 Evaluation of Melt-Processed Polyhydroxyalkanoates and Zinc Oxide Nanocomposites as Packaging Materials - *Chris Vanheusden, Hasselt University, Belgium* (Virtual Presentation)	
18.00 - 20.00	Official Dinner	Room: Bhiraj Hall 1

Tuesday June 14th

Bangkok International Trade & Exhibition Centre (BITEC), Bangkok, Thailand

Time	Program	
8.00 - 9.00	Registration	
9.00 - 9.45	Keynote Lecture on **"Impact of Consumer Trends on the Global Packaging Industry"** - *Prof. Dr. Jay Singh, President IAPRI, Professor of Packaging Program Director, Cal Poly State University, San Luis Obispo, California, USA.*	Room: Nile 1-4
9.45 - 10.30	Keynote Lecture on **"The Future of Fruit & Vegetable Packaging"** - *Prof. Dr. Randolph M. Beaudry, Professor of Department of Horticulture, Michigan State University*	
10.30 - 10.45	Poster Session	
9.45 - 11.00	Coffee Break	
11.00 - 11.25	**Oral Presentation** (20 mins each plus 5 mins question) **Session: 9. Active & intelligent packaging** **10. Distribution and transport packaging** **11. Medical, cosmetic & pharmaceutical packaging** **12. Packaging for hazardous & dangerous goods** 9.1 UV-shielding Properties of Poly(lactic acid) (PLA)/Poly(butylene adipate-co-terephthalate) (PBAT) Films Incorporating Metal Oxides - *Uruchaya Sonchaeng, Kasetsart University, Thailand* 11.1 Usability Study of Aluminum Pouch of PTP (Press Through Package) Among Patients and Pharmacist - *Kiyomi Sadamoto, Shonan university of medical sciences, Japan* (Virtual Presentation) 12.1 Development of an Analysis and Testing Concept for the Evaluation of Impact Targets in the Mechanical Safety Testing of Dangerous Goods Packagings - *Nikolaos Lengas, Bundesanstalt für Materialforschung und –prüfung (BAM), Germany* (Virtual Presentation)	Session Chair: 9. Dr. Nopparat Viriyakitpattana 10. Dr. Jay Singh 11. Dr. Siriwan Tungsangprateep 12. Dr. Ratchaneewan Kulchan Session Room: 9. Amber 1 10. Amber 2 11. Amber 3 12. Silk 4

Time	Program	
11.30 - 11.55	9.2 Antimicrobial Paper Coated with Lignin Nanoparticle for Bread Packaging Applications - *Nyein Chan Moe, Kasetsart University, Thailand* 10.2 Analysis of Deckboard Stiffness Effect and Location Effect on Plastic Pails on Wooden Pallets - *Mary Paz Alvarez Valverde, Center for Packaging and Unit Load Design, United States* 11.2 Predicting Different Effects of Over-the-Counter Medication Label on Patients' Attention by Using Machine Learning Approaches - *Shiva Esfahanian, Michigan State University, United States* (Virtual Presentation) 12.2 Flow Properties of Powdery or Granular Filling Substances of Dangerous Goods Packagings - Comparison of the Measurement of the Angle of Repose and the Determination of the Hausner Ratio - *Eva Schlick-Hasper, Bundesanstalt für Materialforschung und –prüfung (BAM), Germany* (Virtual Presentation)	
12.00 - 12.25	9.3 Development of Antifungal Breathable Film Incorporated with Lignin Nanoparticles for Mango Packaging and Distribution - *Thanakrit Ananphadung, Kasetsart University, Thailand* 10.3 Development of a Friction-Driven Finite Element Model to Simulate the Load Bridging Effect of Unit Loads Stored in Warehouse Racks - *Eduardo Molina, Virginia Tech, United States*	
12.30 - 13.30	Lunch	Room: Bhiraj Hall 1
13.30 - 13.55	**Session: 13. Active & intelligent packaging** **14. Packaging Design, ergonomics & human factors** **15. Packaging for food & agriculture** **16. Packaging materials** 13.1 Releasing Active Food Packaging: Can Natural Extracts be a Safety Concern? - *Ana Sanches Silva, National Institute of Agrarian and Veterinary Research, Portugal* 14.1 Future-driven Packaging Design: A Foresight Method to Aid in Designing Solutions for Future Challenges - *Thomas Goudsblom, University of Twente, Netherlands* 15.1 Enhanced Preservation Functionality of Bioplastic Materials for Sustainable Food Packaging - *Nathdanai Harnkarnsujarit, Kasetsart University, Thailand* 16.1 Microencapsulation and Thermal Enhancement of Fish Oil Based Phase Change Material Through Epoxidation Process - *Yeyen Laorenza, Kasetsart University, Thailand*	Session Chair: 13. Dr. Rujira Deewatthanawong 14. Sathukarn Kabcum 15. Dr. Siriwan Tungsangprateep 16. Dr. Nopparat Viriyakitpattana Session Room: 13. Amber 1 14. Amber 2 15. Amber 3 16. Silk 4
14.00 - 14.25	13.2 Development and Characterization of Ethyl Maltol Incorporated Starch-Based Films for Active Biodegradable Food Packaging - *Khwanchat Promhuad, Kasetsart University, Thailand* 14.2 Tool Design on Sustainable Consumer Behaviour for Food Packaging Companies - *Tao L.F. Heslenfeld, University of Twente, Netherlands* 15.2 Roles of Titanium Dioxide, Zinc Oxide and Silicon Dioxide on Morphology, Properties and Migration of Thermoplastic Cassava Starch and Poly(Butylene Adipate-co-Terephthalate) Nanocomposite Films - *Danaya Phothisarattana, Kasetsart University, Thailand* 16.2 Factors Influencing Mechanical Damage on Tinplate Coatings - *Mieke van den Berg, University of Twente, Netherlands* (Virtual Presentation)	
14.30 - 14.55	13.3 Industrial By-products as Raw Materials for Active Films: The Approach of the ValICET Project - *Raquel Sendon, University of Santiago de Compostela, Spain* 15.3 Investigating the Effects of Packaging Design Factors on Customer Purchasing Decisions for Goat Manure Fertilizer - *Narucha Tanaiutchawoot, Suranaree University of Technology, Thailand* 16.3 Effects of Film Tension and Contamination on Seal Strength and Seal Integrity of PP/LDPE + Talc Based Flexible Food Packaging Films - *Ilknur Ilhan, University of Twente, Netherlands* (Virtual Presentation)	

Time	Program	
15.00 - 15.25	15.4 Measuring the Hexane Vapor Transmission Rate of barrier paper materials - *Yves Wyser, Société des produits Nestlé / Switzerland,* (Virtual Presentation)	
15.30 -15.45	Poster Session	
15.45 - 16.00	Coffee Break	
16.00 - 17.30	Panel Discussion on **"Global Packaging Trends in Post-Covid Era"** • Pierre Pienaar, World Packaging Organization (WPO) • Chakravarthi, World Packaging Organization (WPO) • Nerida Kelton, World Packaging Organization (WPO) • Hiroko Akieda, World Packaging Organization (WPO) • Henky Wibawa, World Packaging Organization (WPO) • Soha Atallah, World Packaging Organization (WPO) (Moderator)	Room: Nile 1-4
18.00 - 20.00	Cultural Dinner at Chambray Grand Ballroom, AVANI Hotel Sukhumvit	

Wednesday June 15th **Bangkok International Trade & Exhibition Centre (BITEC), Bangkok, Thailand**

Time	Program	
8.30 - 9.00	Registration	
9.00 - 9.35	Keynote Lecture on **"Crucial role of Active & Intelligent Packaging in Market"** • *Dr. Ratchaneewan Kulchan, Director of Packaging Development Laboratory, Thai Packaging Centre, TISTR*	Room: Nile 1-4
9.35 - 10.15	Keynote Lecture on **"Hermetic Storage Technology Reducing: Post-harvest Losses of Dried Agricultural Commodities"** • *Dr. Melanie Blanca-Ocreto, Manager for Customer Support Department, Global Support and Production, GrainPro Philipines, Inc.*	
10.15 - 10.30	Coffee Break	
10.30 - 10.55	**Oral Presentation** (20 mins each plus 5 mins question) **Session: 17. Distribution and transport packaging** **18. Active & intelligent packaging** **19. Logistics & supply chain** 17.1 On Friction Measurements and Certification of Slip-Preventing Materials for Road Freight Usages - *Julien Lepine, Universite Laval, Canada* (Virtual Presentation) 18.1 Pyrogallol incorporated Thermoplastic Starch Based Films as Oxygen Scavenging Active Bioplastic Packaging - *Juthathip Promsorn, Kasetsart University, Thailand* 19.1 The Role of Digital Specification Data Management in Streamlining Packaged-Product Supply Chains - *Adam Armstrong, California Polytechnic State University, United States* (Virtual Presentation)	Session Chair: 17. Dr. Supoj Pratheepthinthong 18. Dr. Sathukarn Kabcum 19. Dr. Roland ten Klooster Session Room: 17. Amber 1 18. Amber 2 19. Amber 3
11.00 - 11.25	17.2 Measurement and Analysis of Low-acceleration and Long-duration Longitudinal Events Using Delivery Van - *Péter Böröcz, Széchenyi István University, Hungary* (Virtual Presentation) 18.2 Pilot-scale Production of Bioactive EVOH-based Films Containing Methyl Anthranilate: Processing, Functional Properties and Antifungal Activity Evaluation - *Alejandro Aragón-Gutiérrez, ITENE, Spain* (Virtual Presentation) 19.2 Impacts of the Local E-Commerce - *Cristina Fernanda Guzman Siller, Universidad de Monterrey, Mexico* (Virtual Presentation)	

CONFERENCE PROGRAMME

Time	Program	
11.30 - 12.00	**Closing Ceremony**	Room: Nile 1-4
12.00 - 13.00	**Lunch**	Room: Bhiraj Hall 1
13.00 - 14.00	**General Meeting**	Room: Silk 4
14.00 - 16.30	**Visit ProPak Asia 2022 / Thai Packaging Centre**	
18.00 - 20.00	**Cruise Dinner - Chao Phraya Princess**	

iapri
Bangkok 2022

23rd World Packaging Conference

12-16 June 2022

KEYNOTE SPEAKERS AND PANELISTS

Prof. Pierre Pienaar

Keynote Lecture on **"Trends and Innovation on Sustainable Packaging"**

Pierre Pienaar's interest in packaging started in 1984 after having studied pharmacy. Soon after joining a large pharmaceutical company, Pierre became concerned about a number of packaging-related issues in the pharmaceutical world.

This combined field has taken him to the far corners of the world, always involving one or other aspects of research in packaging. Pierre has attended international congresses and conferences where he has delivered and continues to present papers and continues to judge national and international packaging competitions. He continues to write articles for numerous packaging magazines around the world.

Pierre has a Master of Science Degree (Packaging Engineering/ Technology) from Brunel University, UK. He also has a Master of Manufacturing and Production Degree from University of Hertfordshire, UK. He is a registered Certified Packaging Professional in over 60 countries.

He has been National President of the South African Institute of Packaging (IPSA), an honorary life member of (IPSA), a Fellow of the Australian Institute of Packaging (AIP), Professional member of the Australian Food, Industry and Science Technology as well as the Refrigerated Warehouse and Transport Association of Australia.

He is a packaging engineer in his own global packaging consulting business, PackTech Solutions Pty Ltd and has extensive experience in the pharmaceutical and food and beverage packaging industries with over 35 years of experience in the field of packaging science and engineering as well as its related subject matter. He is the current National Education Director, Past National President of the AIP and past Vice President of the World Packaging Organisation (WPO), responsible for global packaging education.

He has lectured in the technology and science of packaging at various universities and institutions around the world for the past thirty years. He currently lectures the Masters students in Food Innovation and Packaging at University of Melbourne (Australia), as well as: Bond University (Australia), University of Newcastle (Australia), University of New South Wales (Australia), Monash University (Australia), Queensland University of Technology (Australia), University of Southern Queensland (Australia). Beijing Institute of Graphic Communication (China), Sichuan University (China), Jiangnan University (China), Tra Vinh University (Vietnam), Gadjah Mada University (Indonesia), Rochester Institute of Technology (USA) and Wien University of Applied Sciences (Austria).

He is regularly called to be an expert witness in court cases, due to his extensive experience in packaging science, packaging engineering and the technology thereof.

He holds a double professorship in Packaging Engineering and is the current President of the World Packaging Organisation (WPO).

Additional: Should you want to know more about Pierre or his experience, please view this 3 minute company clip: https://bizvideomedia.wistia.com/medias/9mtggpkr81 or visit www.packsol.com.au or LinkedIn PierrePienaar@LinkedIn or Google Pierre Pienaar Packaging.

Mr. Takehiro KANEKO

Keynote Lecture on **"Impact of Chain of Custody on Packaging Verification and Traceability"**

Date of Birth
Aug. 17, 1956

Professional associations
Japan Packaging Institute

Position
Manager, Transport Packaging Div.
Japan Packaging Research Institute

Contact information
4-1-1, Tsukiji, Chuo-Ku, Tokyo 104-0045 Japan
Phone: +81-3-3543-1189 FAX: +81-3-3543-8970

Education
University - Graduated

Licenses and certificates
Certified Packaging Professional

Professional experience

1. Marantz Japan Inc. Apr. 1979 – Aug. 1990
   Quality Assurance and Control for new products in design stage.
2. Sony Corporation Sept. 1990 – Nov. 2014
   - Internal governance and technical support of packaging engineering for business groups. (Standardization, compliance, education, etc.)
   - ISO standardization
     Convenor of ISO/TC122/SC4/WG4, and ISO/TC122/SC3/WG7, and expert of related SC or WG in ISO/TC122 "packaging".
   - Develop JIS (Japan Industrial Standard) in the field of packaging, entrusted from JPI as WG member.
3. Japan Packaging Institute Jan. 2015 – Present
   - Develop ISO standards as a secretariat of ISO/TC122.
     Secretariat of ISO/TC122/SC3/WG10, ISO/TC122/SC3/WG12, and expert of related SC or WG in ISO/TC122 "packaging".
   - Expert of ISO/TC308
   - Coordinate domestic WGs for developing ISO standards.
   - Technical support for member company of JPI in the field of transport packaging.

Prof. Dr. Jagjit (Jay) Singh

Keynote Lecture on **"Impact of Consumer Trends on the Global Packaging Industry"**

Industrial Technology & Packaging,
Orfalea College of Business,
Cal Poly State University, San Luis Obispo, CA 93407
Phone: (805) 756 2129 **Fax:** (805) 756 6111
Email: jasingh@calpoly.edu

PROFESSIONAL EMPLOYMENT

- 09/2010 – Present: Professor, Industrial Technology, Orfalea College of Business, Cal Poly State University, San Luis Obispo, CA. Director, Packaging Program
- 09/2007 – 08/2010: Associate Professor, Industrial Technology, Orfalea College of Business, Cal Poly State University, San Luis Obispo, CA. Director, Packaging Program
- 08/2003 – 09/2007: Assistant Professor, Industrial Technology, Orfalea College of Business, Cal Poly State University, San Luis Obispo, CA. Director, Packaging Program
- 01/2003 – 08/2003: Instructor/Research Associate, School of Packaging, Michigan State University, East Lansing, MI
- 01/1995 – 12/2002: Instructor/Research Assistant, School of Packaging, Michigan State University, East Lansing, MI

EDUCATION

- Ph.D., Packaging Science, 2002. Michigan State University, East Lansing, MI. Dissertation Title: Package Performance for Liquid Hazardous Materials in High Altitude Shipments
- Master of Science, Packaging Science, 1998. Michigan State University, East Lansing, MI. Thesis Title: Evaluating Performance of Internal Packaging for Damage to Glassware
- Bachelor of Engineering, Electrical Engineering, 1992. University of Pune, India

ACADEMIC & PROFESSIONAL HONORS

1. R. David LeButt Award (2019): Packaging Education Award. International Safe Transit Association (ISTA)
2. Distinguished Scholarship Award (2017): Recognition of achievement in scholarship and creative activity across theentire range of disciplines represented at Cal Poly.
   The award honors work conducted primarily at Cal Poly andcelebrates both exemplary specific accomplishments and outstanding bodies of achievement
3. Guest Professor at Jinan University, China (2017 - present): Invited to visit Jinan University to teach, lecture, and/orperform research.
4. President (2016 - present): International Association of Packaging Research Institutes, IAPRI (an internationalmembership association to promote packaging as a global science to important target groups within internationalorganizations and research foundations)
5. Seminar Director (2013 – present) freshPACKmoves Seminar, Monterey, CA. Focused on packaging innovation, foodsafety & traceability, sustainable packaging solutions, and cold- chain logistics for fresh perishables. Designed foraccess to first-hand intelligence, data-driven research, and top-notch speakers.

Prof. Dr. Jagjit (Jay) Singh

6. Conference Chair (2012) - 18th IAPRI World Packaging Conference, June 17-21, 2012, Cal Poly State University, SanLuis Obispo. 170 leading national and international packaging researchers (industry and academia) from 24 countrieswere availed of 93 scientific presentations on 11 topics areas ranging from packaging for food & agriculture tohazardous & dangerous goods.
7. Global Board of Directors (2012 – 2016) – International Association of Packaging Research \ Institutes (an internationalmembership association to promote packaging as a global science to important target groups within internationalorganizations and research foundations)
8. Global Board of Directors (2012 – 2016) – International Safe Transit Association (an organization focused on thespecific concerns of transport packaging with the mission to develop and deliver standards, educational programs,and tools for the economic, social, and environmental optimization of packaging systems)
9. Editor-in-Chief (January 2009 – April 2010) - Journal of Applied Packaging Research • Publisher: DEStech Publications,Inc. (ISSN: 1557-7244) • Subject: international forum for the dissemination of refereed research papers, reviewarticles, tutorials, and news about innovative and emerging technologies for the packaging industry.
10. Most Outstanding Faculty, Packaging and Logistics, Orfalea College of Business, Cal Poly State University, 2008-2014
11. Most Promising Scientific Work - Characterization and Compression Properties of Injection Carbon-NanotubeComposites • 23rd International Association of Packaging Research Institutes' (IAPRI) Symposium, Windsor, UK, June2007
12. Division 1 Chair, committee D10.18 of ASTM (2009 – 2010)
13. Outstanding Industrial Technology Professor Award (2006)- Region 6, National Association of Industrial Technology(NAIT)
14. Best Presentation by a Young Scientist - A Comparison of RFID (Radio Frequency Identification) Tag Performance forFresh Produce Containers, 22nd IAPRI World Symposium on Packaging, Campinas, Brazil, June 2005

Prof. Dr. Randolph M. Beaudry

Keynote Lecture on **"The Future of Fruit & Vegetable Packaging"**

Current Appointment

50% Extension	Postharvest Specialist for Fruits and Vegetables
50% Research	Postharvest physiology of quality maintenance

Education And Experience

B.S.	Horticulture	The Ohio State University	1981
M.S.	Horticulture	University of Georgia	1985
Ph.D.	Horticulture	University of Georgia	1989
Assistant Professor	Horticulture	Michigan State University	3/1989
Associate Professor	Horticulture	Michigan State University	7/1995
Professor	Horticulture	Michigan State University	7/2001
Acting Chairperson	Horticulture	Michigan State University	7/2007-7/2009

Specialization and Areas of Professional Experience

- Optimizing storage (CA, MAP) and harvesting techniques for preserving product quality
- Physical and physiological factors affecting the O2 tolerance of stored commodities
- Maturity assessment of horticultural commodities
- Development and evaluation of non-destructive quality assessment techniques
- Use of natural volatiles for prevention of decay of packaged produce
- Glycolytic regulation of carbon flux during storage of fruits and vegetables
- Aroma biochemistry and molecular regulation

Professional Societies

- American Society for Horticultural Science
- International Society for Horticultural Science
- American Society of Plant Biologists
- American Association for the Advancement of Science
- International Fresh-cut Produce Association

Student Advisement

Doctoral advisor for 8 students
Master of Science advisor for 15 students
Doctoral Committee member for 44 students
Master of Science committee member for 39 students
External committee member for 13 international students
Undergraduate research project advisor for 11 students
Undergraduate minority program advisor for 3 students
High School Honors Science Program advisor for 13 students

Prof. Dr. Randolph M. Beaudry

Honors And Awards

2002 USDA Group Honor Award for Excellence for 2001
2006 ASHS Fruit Publication Award for 2005
2010 Gold Award for Poster, International Horticulture Congress, Lisbon
2010 Distinguished Service Award, Michigan State Horticultural Society
2012 Fruit Man of the Year, Michigan Pomesters
2015 Researcher of the Year, International Fruit Tree Association
2015 Fellow, American Society of Horticultural Science.
2017 - 2019 Research Division Vice President, ASHS
2019 Fulbright-Nehru Distinguished Fellowship

Synergistic Activities

Teaching and Advising - I do not have a formal teaching appointment, however, I have taught several courses over the course of my career at MSU. I have served on over 130 committees of masters and PhD students in a number of departments in the colleges of Agriculture and Natural Resources and Natural Sciences. I have served on committees of PhD students from New Zealand, Holland, Canada and Australia.

Extension (170+ presentations to lay audiences, 70+ outreach articles)
I act as the state resource to Cooperative Extension Service field agents and growers in the areas of Apple harvest maturity evaluation, Apple storage disorders, Recommendations for air and CA storage of apple and blueberry fruit, Blueberry harvest and storage operations, Modified atmosphere packaging ,Effects of ReTain on fruit maturation and storability, Effects of 1-MCP on fruit ripening and storability, Cultivar affects apple slice quality

Research
The primary thrust of my research program is the optimization of storage strategies for the maintenance of postharvest quality of fruits and vegetables. Outputs since joining MSU include: Patents and IP disclosures - 15; Book chapters - 12; Review articles - 2; Refereed Publications 136 total; Seminars/Invited presentations to professional meetings - 82

DR. RATCHANEEWAN KULCHAN

Keynote Lecture on **"Crucial role of Active & Intelligent Packaging in Market"**

Position:

Director of Packaging Development Laboratory
Thai Packaging Centre, Thailand Institute of Scientific and Technological Research

Company's Address

Thai Packaging Centre (TPC)
Thailand Institute of Scientific and Technological Research (TISTR)
196 Phaholyothin Rd. Ladyao Chatuchuk Bangkok 10900
Tel. 0 2579 1121 **Fax:** 0 2579 7573
e-mail: ratchaneewan@tistr.or.th

Education

1993 B.Sc. Food Science and Technology, Rajamangala Institute of Technology, Thailand
2008 M.Sc. Packaging Technology, Kasetsart University, Thailand
2016 Ph.D. Packaging Technology, Kasetsart University, Thailand

Special

- Packaging Technology
- Active and Intelligent Packaging
- Fruit and Vegetable Packaging
- Shelf life Evaluation and Prediction

Research Experience

1996 Effect of Packaging Material on Quality of Orchid.
1997 Performance Improvement of corrugated Fibreboard Box for Transportation of Peach.
Effect of Wrapping Material and Ventilation Hole on Temperature Inside Orchid Packaging
2001 Dehydrated Papaya and Packaging for Dehydated Fruits.
2002 Development of Packaging for Durian Freeze Dried
2003 Development of Packaging for Fully Blended Juices and By-Products-from Lemon.
2004 Packaging Designed for OTOP

Award:

2004 Development of Packaging for Fully Blended Juices and By-Products-from Lemon get the Best Research from Thailand Institute of Scientific and Technological Research, Thailand.
2005 Designed Boxes of Ko-Kret, s Earthernware Get the Asia Star Award, Thailand.

Present Research :

- Smart Label to Indicate Food Freshness of Seafood
- Research and Development of Packaging for Products of Bastard Oleaster to Commercial

DR. MELANIE BLANCA-OCRETO

Keynote Lecture on **"Hermetic Storage Technology Reducing: Post-harvest Losses of Dried Agricultural Commodities"**

077 del Pilar St., Barangay V
Alfonso, Cavite, Philippines
Mobile Number/WhatsApp: +639178045475
Email: melanie@grainpro.com;
melanieocreto@gmail.com

Academic Background
Doctor of Philosophy in Agricultural Engineering (2016)
A cademic Achievement Award Recipient
Major in Crop Processing
Minor in Food Science
University of the Philippines, Los Baños
Laguna, Philippines
Dissertation: Development of Long-Term Storage Systems for Sweet Sorghum (*Sorghum bicolor* (L.) Moench) Stalks

Master of Engineering in Food Engineering & Bioprocess Technology (2009)
Asian Institute of Technology Pathumthani, Thailand
Thesis: Enzymatic Delignification of Lignocellulosic Rice Straw and Water Hyacinth for Biofuel Applications

Bachelor of Science in Agricultural Engineering (2003)
Cum Laude
University of Southern Mindanao Kabacan, 9407 Cotabato, Philippines
Research: *Design, Fabrication & Evaluation of Corn Jabber*

Professional Background
Registered Professional Agricultural Engineer
License Number 5532
Board of Agricultural Engineering
Professional Regulation Commission
Republic of the Philippines

Service Record
Assistant Professor III (August 2019 to present)
Department of Engineering Sciences
College of Engineering and Agro-Industrial Technology
University of the Philippines, Los Baños, Laguna
Courses Taught: *Mathematical Methods in Engineering, Statics of Rigid Bodies, FluidMechanics*

DR. MELANIE BLANCA-OCRETO

Assistant Manager for Technical Support Department (November 2016 to July 2019)
GrainPro Philippines, Inc.
Subic Bay Freeport
Duties and Responsibilities: *Providing trial protocols, monitoring and drafting final reports forfield trials; Developing and organizing modules, multimedia visual aids and serving as postharvest specialist during trainings, conferences and field installations; Drafting and updating product instruction manuals and user guides; Providing supports to sales, responding to queries and complaints, and attending meetings with clients; Assisting Research and Development during Alpha- and Beta-testing of new products*

Assistant Professor I (May 2015 to June 2016)
Instructor III (October 2012 to April 2015)
Instructor II (August 2011 to September 2012)
Instructor I (June 2009 to July 2011)
Substitute Instructor I (July 2004 to August 2007)
College of Engineering & Computing
University of Southern Mindanao
Kabacan, 9407 North Cotabato, Philippines
Courses Taught:
Master of Science in Agricultural and Biosystems Engineering
Food Process Engineering; Biomass and Energy

Bachelor of Science Degree Courses
Soil & Water Conservation Engineering; Irrigation and Drainage; Fluid Mechanics; Environmental Engineering; Hydrometeorology; Food Engineering, Refrigeration and Airconditioning; Engineering Mathematics

Other Functions:
(a) Adviser – General Engineering Society
(b) Conduct review classes for Licensure Exam for Agricultural Engineers

Instructor III (February 2004 to June 2004)
College of Agriculture, Mindanao State University – General Santos City
Courses Taught: *Farm Machinery, Machine Design, Basic Computer Applications; Statistics*

Apprentice (April 2002-May 2002)
Pioneer Hi-Bred Seeds Philippines Inc.
Polomolok, South Cotabato

Professional Dossier
Suwabun Chirachanchai, Ph.D.

Panel Discussion on **"Biodegradable & Compostable, Research, Development, Standard and Testing"**

The Petroleum and Petrochemical College,
Chulalongkorn University
E-mail: csuwabun@chula.ac.th

Professional Experience:

Sep 2016 – Aug 2020
Dean of The Petroleum and Petrochemical College, Chulalongkorn University

July 2012 – June 2016
President of Polymer Society of Thailand

Research Interests:

1. Polymer Modifications, Functional Polymers
2. Controlled Molecular Recognition and Inclusion Compounds
3. Biopolymers, Chitin-Chitosan
4. Nanomaterials, Polymer Composites
5. Polymer Electrolyte Membrane Fuel Cells
6. Biodegradable Plastics

Publications: (*: Corresponding Author) (8 publications in 3 years (2019-2022)) (total 123 publications)

1. Phunpee, S.; Ruktanonchai, U.R.; Chirachanchai, S.* "Brush-Structured Chitosan/ PolyHEMA with Thymine and Its Synergistic Effect on the Specific Interaction with ssDNA and Cellular Uptake" Langmuir, 38, 18, 5915–5923, 2022. (Impact Factor 2020: 3.882)
2. Wiwatsamphan, P.; Chirachanchai, S.* "Persistently reversible pH-/thermo-responsive chitosan/poly (N-isopropyl acrylamide) hydrogel through clickable crosslinked interpenetrating network" Polymer Degradation and Stability, 198, 109874, 2022. (Impact Factor 2020: 5.030)
3. Pokprasert, A.; Theato, P.; Chirachanchai, S.* "Proton donor/acceptor copolymer brushes on sulfonated poly(ether ether ketone) membrane: An approach to construct efficient proton transfer pathway in polymer electrolyte membrane fuel cell" Polymer, 240, 124523, 2022. (Impact Factor 2020: 4.430)
4. Kertsomboon, T.; Agarwal, S.; Chirachanchai, S.* "UCST-Type Copolymer through the Combination of Water-Soluble Polyacrylamide and Polycaprolactone-Like Polyester" Macromolecular Rapid Communications, 2000243, 2020. (Impact Factor 2019: 4.886)
5. Kertsomboon, T.; Chirachanchai, S.* "Amphiphilic biodegradable co-networks of Poly(butylene succinate)-Poly(ethylene glycol) chains for nano-gelation via Click chemistry and its potential dispersant for multi-walled carbon nanotubes" Polymer Degradation and Stability, 179, 109266, 2020. (Impact Factor 2019: 4.032)

Professional Dossier
Suwabun Chirachanchai, Ph.D.

6. Pokprasert, A.; Chirachanchai, S.* "Polymer electrolyte membrane with magnetic nanoparticles containing benzimidazole terminals: An approach to induce proton transfer species on membrane surface" International Journal of Hydrogen Energy, 45, 16, 9989-9999, 2020. (Impact Factor 2018: 4.084)
7. Yenpech, N.; Intasanta, V.; Tashiro, K.; Chirachanchai, S.* "Color and shape reversible, recoverable and repeatable mechanochromic shape memory polycaprolactone: a single material with dual functions" Polymer Chemistry, 11, 91-101, 2020. (Impact Factor 2018: 4.760)
8. Yenpech, N.; Intasanta, V.; Chirachanchai, S.* "Laser-triggered shape memory based on thermoplastic and thermoset matrices with silver nanoparticles" Polymer, 182, 121792, 2019. (Impact Factor 2018: 3.771)

Mr. Viboon Pungprasert

Panel Discussion on **"Biodegradable & Compostable, Research, Development, Standard and Testing"**

Bioplastics Commercial Manager
PTT Global Chemical, Plc.

Mr. Viboon Pungprasert has been in bioplastics industry more than 10 years. Start with experiences with NatureWorks, the largest PLA bioplastics producers in the world. He used to be General Manager of NatureWorks Asia Pacific, ltd which has office located in Bangkok. The main duty is to set up NatureWorks business in South East Asia, and persuade Thai government to promote Thailand as Bioplastic hubs for the region. Currently, Mr. Viboon is taking care of bioplastics business of PTT Global Chemical (GC) as GC is the main distributor of NatureWorks PLA, and PTTMCC BioPBS. His another position is the president of Thai Bioplastics Industry Association (TBIA) that consists of more than 50 companies in bioplastic industries in Thailand as member. TBIA has mission to bring Thailand as a leader in bioplastics in Asia as well as Asean region.

Dr. Pichet Rojanapitayakorn

Panel Discussion on "Biodegradable & Compostable, Research, Development, Standard and Testing"

Dr. Pichet Rojanapitayakorn has been working with SCG Chemicals for more than 15 years, having different roles within R&D, application development, open innovation research partnership and innovation management. He received his PhD in chemical engineering from Chulalongkorn university and joined university of Connecticut, USA to do postdoctoral research in the field of polymer blend/composite for biomedical applications from 2001 to 2004. He started his career at SCG Chemicals in 2004 from technical improvement of polyolefin production plant and developing new products mainly polyolefins either in polymerization process or post polymerization process. During his career path at SCG Chemicals, he has conducted research works in different kinds of R&D theme, for example, polymer blend & composite for new applications, polymer synthesis, polymer coating, inorganic development for plastic additives, recycling of plastic waste, bioplastics development for various applications. He also took part in collaboration with other research institutes and explored new technologies to serve open innovation strategy of SCG Chemicals. At present, he is the senior research, taking care of the research to support circular economy and sustainability direction. He has experienced working with bioplastics since 2005 and developed bioplastic compound for commercialization since 2012.

Dr.Pattra Maneesin

Panel Discussion on "Biodegradable & Compostable, Research, Development, Standard and Testing"

Thailand Institute of Scientific and Technological Research (TISTR),
35 Mu 3 Technopolis, Tambon Khlong Ha,
Amphoe Khlong Luang, Pathum thani 12120, Thailand
Phone : 66-2-577 9202
Fax : 66-2-577 9009
E-mail : pattra_m@tistr.or.th

Education and degrees held

- Ph.D. in Packaging (2001), Michigan State University, USA
- M.S. in Post Harvest and Food Process Engineering (1995), Asian Institute of Technology (AIT), Thailand
- B.S. Honors in Agricultural Industry (1989), King Mongkut's Institute of Technology (Ladkrabang), Thailand

Training

- Belgium Packaging Institute, Belgium: Training course on "Packaging", 1990
- Packaging Technology Center, Food Technology Institute, Brazil: Training course on "Glass, Metal and Flexible Packaging for Processed Foods", 1991
- Training and Productivity Authority of Fiji, Fiji: APO Training course on "Value Addition to Agricultural Products", 2009 - Taiwan Environmental Management Association (TEMA),
- Taiwan: APO Workshop on Eco-design

Position held

- Oct 2020 – Present Deputy Governor Industrial Services, Thailand Institute of Scientific and Technological Research (TISTR)
- 2018 – Sep 2020 Director, Thai Packaging Centre, Thailand Institute of Scientific and Technological Research (TISTR)
- 2017 – 2018 Director, Packaging Development Laboratory, Thai Packaging Centre, Thailand Institute of Scientific and Technological Research (TISTR)
- 2008 – 2017 Senior Research Officer, Thai Packaging Centre, Thailand Institute of Scientific and Technological Research (TISTR)
- 1989– 2008 Research Officer, Thai Packaging Centre, Thailand Institute of Scientific and Technological Research (TISTR)

Dr. Pattra Maneesin

Experiences

1. Secretary General of Asian Packaging Federation (APF), 2018 - present
2. Packaging expert for the International Trade Centre (ITC) mission on supply chain and value added products, Dhaka, Bangladesh, 2009
3. Packaging expert for the International Trade Centre (ITC) Proinvest – regional training programme on packaging Accra, Ghana and Bangkok, Thailand, 2011
4. Packaging consultant on "Post harvest packaging", Colombo, Sri Lanka, 2008-2009
5. Packaging consultant, Thai Packaging Association, 2005- present
6. Thai Industry Standard (TIS) committee on drafting and revising packaging standards
7. Special Lecturer for related topics in packaging area e.g.
   - Chamber of Commerce of Sri Lanka, Training course on Post-harvest packaging. Bangkok, 2008 and 2009.
   - Vietnamese Agricultural Department, Training course on Packaging for agricultural and value added products, Bangkok, 2008.
   - Agricultural Department of Myanmar, Training course on Packaging for agricultural community, Bangkok, 2008
   - National Food Institute, Training course on Container Integrity, Bangkok, 2002- present
   Professional area of specialization or interest
   - Food packaging, active & intelligent packaging e.g. antimicrobial packaging, odor removing packaging etc., modified atmosphere packaging (MAP) of fresh foods, post-harvest packaging, export packaging for fresh produces, shelf life extension of food products

Mr. Chakravarthi AVPS

Panel Discussion on **"Global Packaging Trends in Post-Covid Era"**

CEO, Managing Director of Ecobliss India
(a joint venture with Netherlands based MNC)
a global leader in innovative blister packaging solutions.

Winner of Worldstar awards (the highest in packaging excellence) for five years in a row with 100% success.

And the lone winner of an AmeriStar in pharmaceutical packaging from India.
Has more than three decades of experience in paper, printing, and packaging fields.
Served as chairman of Indian Institute of Packaging, Hyderabad and on the National Board for more than a decade.

Actively participated packaging related events worldwide for more than two decades by now.

Recently appointed as Ambassador of World Packaging Organisation.

Instrumental in developing various pharmaceutical compliance packs for US and regulatory markets!

Published articles in various magazines and participated in live programmes of public interest in TV shows.

Toured extensively worldwide, during the past two and half decades!

Spoke on various international platforms. Has been jury for different award competitions.
Delivered lectures and chaired various sessions on different topics in FICCI, CII, IIP, CPHI, APF, WPO, Iphex, FAPPCCI, WPC and so many other platforms.

Was honoured with a special recognition award at India Pharma Awards by UBM for his contribution to the pharma packaging industry during CPHI India in 2017.

Ms. Nerida Kelton

Panel Discussion on **"Global Packaging Trends in Post-Covid Era"**

Nerida is the Executive Director for the Australian Institute of Packaging (AIP), Vice President Sustainability & Save Food and the ANZ Board member for the World Packaging Organisation (WPO).

She has worked in the Packaging industry for over 23 years, is the lead for the Save Food Packaging Consortium project within the Fight Food Waste Cooperative Research Centre and was the packaging representative on the Department of Environment and Energy's National Food Waste Strategy Steering Committee.

Nerida is passionate about educating the industry on the important role that packaging plays in minimising Food Waste and how designing Save Food Packaging can make a difference. She is also one of the creators of the annual Christmas Foodbank hamper program which is designed to help those who are vulnerable and in need. The volunteer program has packed over 12,000 hampers to the value of over AU$1 million over the last 10 years.

Nerida is also committed to helping educate and train packaging professionals in the importance of sustainable and circular packaging design and recognising best practice in this area.

Nerida Kelton is a qualified public relations practitioner with a Bachelor of Arts Degree (Double Major in Japanese), a Bachelor of Business in Public Relations and a Postgraduate Degree in Public Relations and Communications.

Nerida is also a full member of the Public Relations Institute of Australia (PRIA), a Member of the International Packaging Press Organisation (IPPO), a full member of the Australian Institute of Packaging and is the youngest and first female to have ever received Life Membership to the Supply Chain & Logistics Association of Australia for her significant contribution to the association and the wider industry. Her articles are published in an extensive range of publications across Australasia on a regular basis.

With over 27 years' experience in Association & Event Management and Public Relations Nerida has spent the majority of her career working with non-profit industry-based educational organisations and has extensive experience in working with volunteer driven associations.

Ms. Hiroko Akieda

Panel Discussion on **"Global Packaging Trends in Post-Covid Era'**

Ms. Hiroko Akieda joined the Japan Packaging Institute (JPI) seven years ago and appointed section manager of the International Business Department in 2016, a role that makes the most of her multilingual skills and worldwide travel experience.

Established in 1963, JPI enjoys a long and close relationship with the World Packaging Organisation, which was founded in Tokyo in 1968. Ms. Akieda feels honoured to have become the first WPO global ambassador to represent Japan and one of just four global ambassadors recruited for a position that was newly created by current WPO President Pierre Piennar in 2018.

She believes her ambassadorial responsibilities will complement her global activities for JPI whose mission statement contains the message: "No border for packaging" (or "Packaging without borders"?). For JPI, Ms. Akieda is in charge of international promotion, strengthening global ties with other national bodies, institutes and universities, and engaging the interest of press and media, to develop the Japanese packaging industry and raise its profile. She is part of the JPI exhibition team behind TOKYOPACK, a biennial international packaging show that provides a platform for Japan's cutting-edge packaging technology and reinforces Japan's reputation across the world for exquisite design and high production values. Another of her duties is arranging inspection tours of Japanese packaging companies for overseas visitors. These are a chance for the world's packaging professionals to exchange information and update their technical knowledge.

JPI acts as the Secretariat of ISO/TC122-packaging and is entrusted by JISC-Japanese Industrial Standards Committee to develop global standards. Ms. Akieda has been deeply involved in its activities as ISO/ TC122 Committee Manager Support Team member and also as Secretary of ISO/TC122/WG9 – Accessible design for packaging, contributing to the ISO/TC122 goal of disseminating knowledge and advancing packaging technology around the world.

Ms. Akieda was awarded a Bachelor of Education (BEd) degree by Yokohama National University and she also studied Spanish and international business at University of Salamanca in Spain and University of Lima in Peru.

Mr. Henky Wibawa

Panel Discussion on **"Global Packaging Trends in Post-Covid Era'**

Henky Wibawa serves as Indonesian Packaging Consultant and Executive Director of Indonesian Packaging Federation based in Jakarta. He has more than 30 years of experience in market and business development and research across the packaging manufacturing industries: Argha Karya PI, VAW Europack, Alcan Flexipack, Amcor Flexibles Indonesia. Then since beginning Year 2014 actively through his various engagements with the government, academia and industry, he sees many challenges and opportunities for the packaging industry. As Asia-Pacific and particularly Indonesian consumer goods are growing, and focusing on packed foods that are healthy, natural, and produced sustainably, packaging manufacturers need to innovate new materials, processes and business models very quickly to fulfil these demands. This also implies that the stakeholders need to ensure that the whole value chain, from the farms to consumer is creating less carbon footprint. The rise of digital technologies and artificial intelligence is allowing technology companies and entrepreneurs to disrupt the way FMCG manufacturers operate and engage with consumers, which forces packaging companies to innovate faster.

Henky studied and graduated as Electrical Engineer from Ruhr-Universitaet Bochum, Germany

Ms. Soha Atallah

Panel Discussion on **"Global Packaging Trends in Post-Covid Era'**

Ms. Soha Atallah has an extensive experience of 18 years in industrial development, packaging, marketing and food safety in Lebanon and the Arab Region.

She has established herself as a proactive player in the Lebanese industry, founding LibanPack, the Lebanese Packaging Centre, and raising awareness on packaging development through consulting and through the organization of the yearly competition to university students StarPack.

During her work experience she provided packaging and labelling design services to more than 200 industrial companies in Lebanon and other Arab Countries. Ms. Atallah has been working for more than 11 years with UNIDO as an international expert on packaging and marketing. Ms. Atallah has an M.A from LAU and recently she did and executive course on Strategic Marketing Management at Harvard Business School.

CONFREENCE PROCEEDINGS :

ACTIVE AND INTELLIGENT PACKAGING

• ORAL PRESENTATION

• GENERAL STREAM

AG-O01 Retardation of Lipid Oxidation of Sacha Inchi (Plukenetia volubilis L.) Oil Using Bio-based UV-Shielding Metal Oxide-Incorporated Active Packaging Films

AG-O02 Development of a Colorimetric Total Volatile Base-Nitrogen Sensor for Potential Real-Time Freshness Monitoring of Pre-packaged Food

AG-O03 Antimicrobial Activity of Biodegradable Lignin Nanoparticles in Polyvinyl Alcohol-Based Film Against Major Spoilage Microorganisms of Mango Fruit

AG-O04 UV-shielding Properties of Poly(lactic acid) (PLA)/Poly(butylene adipate-co-terephthalate) (PBAT) Films Incorporating Metal Oxides

AG-O05 Antimicrobial Paper Coated with Lignin Nanoparticles for Bread Packaging Applications

AG-O06 Development of Antifungal Film Incorporated with Lignin Nanoparticles for Mango Packaging and Distribution

AG-O07 Releasing Active Food Packaging: Can Natural Extracts be a Safety Concern?

AG-O08 Development and Characterization of Ethyl Maltol Incorporated Starch-Based Films for Active Biodegradable Food Packaging

AG-O09 Industrial By-Products as Raw Materials for Active Films: The Approach of the ValICET Project

AG-O10 Pyrogallol Incorporated Thermoplastic Starch Based Films as Oxygen Scavenging Active Bioplastic Packaging

AG-O11 Pilot-scale Production of Bioactive EVOH-Based Films Containing Methyl Anthranilate: Processing, Functional Properties and Antifungal Activity Evaluation

• POSTER PRESENTATION

AG-P01 Development of Electrospun Chitosan Nanofibers for Antimicrobial Food Packaging

AG-O01

Retardation of Lipid Oxidation of Sacha Inchi (*Plukenetia volubilis* L.) Oil Using Bio-based UV-Shielding Metal Oxide-Incorporated Active Packaging Films

Busarin Chongcharoenyanon[1,3], Uruchaya Sonchaeng[1,3], Piyawanee Jariyasakoolroj[1,3], Ubonratana Siripatrawan[2,3], Panuwat Suppakul[1,3]

[1]*Department of Packaging and Materials Technology, Kasetsart University, Bangkok, 10900, Thailand*
[2]*Department of Food Technology, Faculty of Science, Chulalongkorn University, Bangkok, 10330, Thailand*
[3]*Research Unit on Smart Packaging Technology for Foods, Kasetsart University Research and Development Institute (KURDI), Kasetsart University, Bangkok, 10900, Thailand*

***Correspondence to:** Department of Packaging and Materials Technology, Kasetsart University, Bangkok, 10900, Thailand. E-mail panuwat.s@ku.ac.th*

ABSTRACT: Bio-based composite films from poly(lactic acid) (PLA) and poly(butylene adipate-co-terephthalate) (PBAT) were incorporated with zinc oxide (ZnO) and titanium dioxide (TiO_2) as UV-shielding active packaging films. Sacha inchi (Plukenetia volubilis L.) oil is susceptible to lipid oxidation. Photooxidation of sacha inchi oil produces singlet oxygen, which reacts with polyunsaturated fatty acids (PUFAs) and produces fatty acid hydroperoxides. Their decomposed products produce undesirable odors and flavors. The oil was poured into film pouches and stored at room temperature for 4 weeks. Retardation of lipid oxidation was indicated by peroxide value (PV) and percent of free fatty acid (%FFA) of the packed oils. For all film treatments, the PV range tended to increase with time, and the boundary between two regions was identified in terms of the time and PV that characterized the induction period (IP). Changes in the IP of sacha inchi oil with different film treatments were negligible. But after this point, the results also showed that PV increased linearly with storage time and increased in acceleration after the second week. The protection with PLA/PBAT/TiO_2 film and PLA/PBAT/ZnO film were associated with gradual increases in PV and the increase was dramatic for the control treatment. The PLA/PBAT/TiO_2 film showed promise for UV-shielding packaging applications.

Keyword: Metal-oxide, Oxidative stability, Polybutylene adipate-co-terephthalate (PBAT), Polylactic acid (PLA), Sacha inchi oil

1. INTRODUCTION

Plastic is an ideal material for active packaging. However, the packaging waste management and depletion of petroleum; raw material for conventional plastics are the main environmental problems. This issue urges the development and commercialization of bio-based plastics as a sustainable material to solve problems [1]. Poly(lactic acid) (PLA) and poly(butylene adipate-co-terephthalate) (PBAT) are bio-based material. The blending of PLA and PBAT improves the mechanical properties of the composite film such as impact strength and ductility compared to PLA solely [2, 3]. Furthermore, a large amount of benzene rings in PBAT molecular chains also provides the possibility to improve the UV resistance of PLA [4]. In the application of food packaging, UV light is one of the factors that deteriorate food products, especially edible oil and lipid-contained foods. The addition of metal oxides such as titanium dioxide (TiO_2) and zinc oxide (ZnO) to the PLA can improve the UV shielding properties of the film. Only 2% of ZnO can absorb 61% of the UV-A and shield 57% of the UV-B [5]. However, the UV-shielding properties of PLA/PBAT incorporated with metal oxides have not been studied. The addition of metal oxides to plastic films can block UV rays which are suitable for oils packaging.

Sacha inchi (*Plukenetia volubilis* L.) oil provides important health benefits to humans. It contains a large amount of polyunsaturated fatty acids (PUFAs); linolenic acid (ω-3) and linoleic acid (ω-6). PUFAs control blood pressure and reduces triglycerides which protect against cardiovascular disease [6, 7]. However, PUFAs render lipid oxidation. The iodine value (IV), degree of unsaturation of a lipid, of sacha inchi oil is 193.1 which is relatively high compared to other oils such as coconut oil (IV = 10), canola oil (IV = 94-120), soybean oil (IV = 117 to 143) and fish oil (IV = 185) [8]. Photooxidation of sacha inchi oil produces singlet oxygen, which reacts with polyunsaturated fatty acids (PUFAs) and produces fatty acid hydroperoxides. Their decomposed products produce undesirable odors and flavors. Active packaging with UV-shielding properties can retard the lipid oxidation and prolong the shelf-life of sacha inchi oil. A UV-shielding active packaging can reduce the addition of active compounds into food.

Peroxide value (PV) and free fatty acids are one of the most frequently determined quality parameters during oil production, storage, and marketing. The results of PV estimation give a clear indication of lipid autoxidation [9]. Liu *et al.* (2014) [4] assessed the change in the quality of sacha inchi oil over 1 year by PV and acid value (AV). The oil samples were kept in colorless and green glass bottles. The PV increased until the 10th month of storage and began to decline. High temperature and light affect the increase of the PV during the early storage. However, the PV decreased due to secondary oxidation which changes peroxide to other substances. The formation of free fatty acids is an important measure of food rancidity, which leads to many effects such as hydrolyzed and oxidation rancidity [4]. % FFA is an indicator of lipid rancidity. The quantity of free fatty acids is an important factor for classifying sacha inchi oil into commercial grades [10]. A promising approach to maintaining the integrity of an oxidation-sensitive oil is the UV-shielding active packaging films.

The objective of this research was to investigate the efficiency of zinc oxide (ZnO) and titanium dioxide (TiO_2) as UV-shielding additives in PLA/PBAT film for retardation of lipid oxidation of high-value edible oil; sacha inchi oil.

2. MATERIALS AND METHODS

Bio-based active film preparation

PLA Ingeo™ Biopolymer 4043D (NatureWorks LLC, Minnetonka, Minnesota, United States) and PBAT ecoflex® F Blend C1200 (BASF SE, Ludwigshafen, Germany) resins were mixed with a fixed blending ratio of 85/15, using a counter-rotating twin-screw extruder (Labtech Engineering LTE 20-40, Samutprakarn, Thailand). The processing temperature range of 150 - 170 °C and screw speed of 150-200 rpm was set to obtain PLA/PBAT 85/15 compound. The addition of TiO_2 or ZnO (Chemipan Corporation Co., Ltd., Bangkok, Thailand).was systematically varied as listed in Table 1. All compounds were pelletized and dried at 50 °C for 1 day before being blown to be film by a single-screw extruder (Labtech Engineering LE-25-30/C, Samutprakarn, Thailand), equipped with a blown film unit (Labtech Engineering LF-400 Samutprakarn, Thailand). The blown films were produced in the temperature range of 160 - 175 °C with an annular die temperature of 160 °C and a nip-roll speed of 2.7 m/min.

Table 1 Blending ratios of PLA/PBAT/TiO_2 and PLA/PBAT/ZnO compounds and films

No.	Sample	PLA (%wt)	PBAT (%wt)	ZnO (%wt)	TiO_2 (%wt)
1	PLA/PBAT	85	15	-	-
2	PLA/PBAT/ZnO 1%	84.15	14.85	1	-
3	PLA/PBAT/TiO_2 1%	84.15	14.85	-	1

Sacha inchi (*Plukenetia volubilis* L.) oil extraction and storage

Sacha inchi beans (Piangdoi, Chiangmai, Thailand) were extracted with an electric oil extraction machine at 100°C then the oil was centrifuged by a refrigerated centrifuge (Sigma Laboratory Centrifuges, Darmstadt, Germany) at 3200 rpm/20 °C for 10 minutes. The supernatant oil was packed in the bio-based film pouch of 14*17 cm. Each pouch contained 80 ml of Sacha inchi oil. The pouches were sealed with a heat sealer. The oil samples were stored at room temperature under natural light for 1, 2, 3, 7, 15, and 30 days.

Peroxide Value (PV)

The PV was analyzed according to the AOAC Official Method 965.33 (2019) Peroxide Value of Oils and Fats, Titration Method. Chloroform, ethanol and acetic acid were purchased from RCI Labscan Limited (Bangkok, Thailand). Sodium thiosulfate and potassium iodide were purchased from Kemaus (Australia).

Percent of Free Fatty Acid (%FFA)

The %FFA was analyzed according to the ISO 660:1996 Animal and vegetable fats and oils – Determination of acid value and acidity. Ethanol was purchased from RCI Labscan Limited (Bangkok, Thailand). Sodium hydroxide and phenolphthalein were purchased from Kemaus (Australia).

Statistical analysis

The results of the PV and FFA values were subjected to a one-way analysis of variance using the SPSS Statistics software version 28.0.0.0 (190). Statistical significance of the difference between mean values was determined using Duncan's new multiple range test at $p < 0.05$.

3. RESULTS AND DISCUSSION

Peroxide Value

Sacha inchi oil was analyzed for PV according to the AOAC method (2019) 965.33 Peroxide Value of Oils and Fats. The results were shown in Figure 1. The PV of the oil before packing in the biobased film was 0.05 mEq/kg fat. The mean PV of the oil sample stored for 1, 2 and 3 days were 0.95, 0.93 and 1.83 mEq/kg fat, respectively. The PV was slightly higher to 3.05 mEq/kg fat on day 7. The value rapidly went up to 10.09 and 26.05 mEq/kg fat on day 15 and 30, respectively. The oil was packed in 3 types of film and showed some differences in PV on day 3, 15 and 30. The effect of metal oxides on UV-shielding bioplastic PLA/PBAT film was clearly noticed on day 15. The PV of the oil in PLA/PBAT/ZnO film (8.13 mEq /kg Fat) and PLA/PBAT/TiO_2 (8.80 mEq /kg Fat) film were lower compared to PLA/PBAT film (13.33 mEq /kg Fat). The PV range tended to increase with time, and the boundary between the two regions was identified in terms of the time and PV that characterized the induction period (IP). Changes in the IP of sacha inchi oil with different film treatments were negligible. But after this point, protection with PLA/PBAT/TiO_2 and PLA/PBAT/ZnO were associated with gradual and significant increases in PV, respectively, and the increase was dramatic for the control treatment.

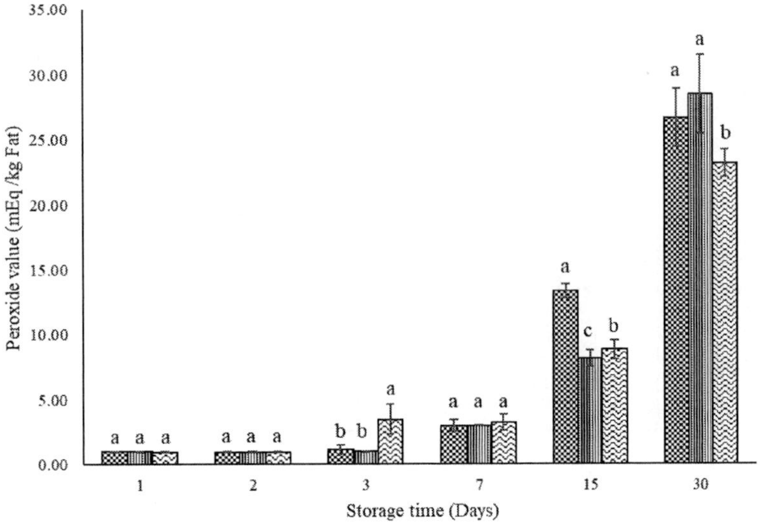

Figure 1 Peroxide value of sacha inchi oil in biobased film.

▩ ; PLA/PBAT film, ▥ ; PLA/PBAT/ZnO film, ▨ ; PLA/PBAT/TiO_2 film

Free Fatty Acid

Sacha inchi oil preserved in metal oxide-incorporated active packaging films were analyzed for percent of free fatty acid (%FFA) followed ISO 660:1996 method Animal and vegetable fats and oils – Determination of acid value. and acidity. The results of the analysis were shown in Figure 2. The averaged %FFA of the oil before packing was 0.21% and after storage on days 1 and 2 were 0.20% and 0.22%, respectively. At 3 days, %FFA decreased to 0.17% and then increased on day 7 to 0.30%. The differences in %FFA among the oil packed in different films were noticed on day 7. PLA/PBAT/ZnO film gave the highest %FFA compared to PLA/PBAT film and PLA/PBAT/TiO_2. The decrease in %FFA gradually occurred on day 15 and 30. The reduction of %FFA was affected by the transformation of chemical substances in oxidation reactions [7].

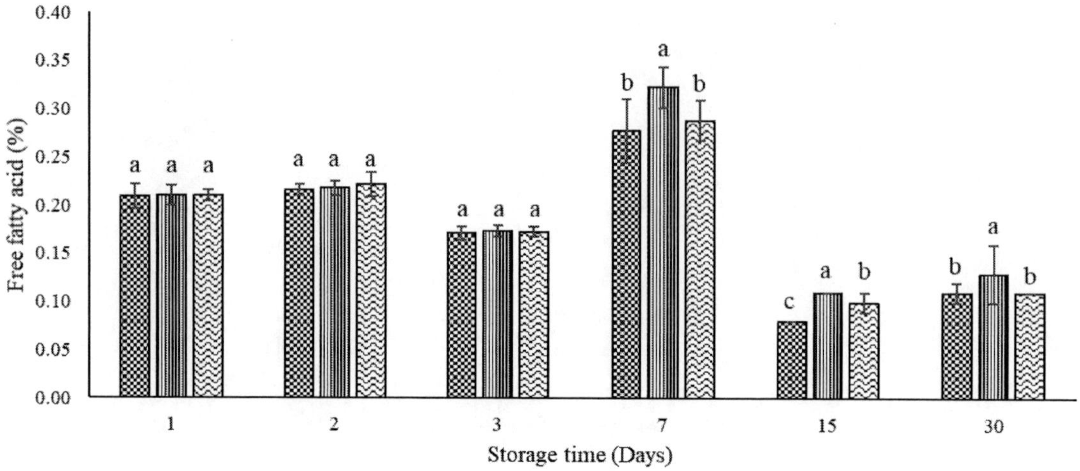

Figure 2 Percent of free fatty acid of sacha inchi oil in biobased film.

; PLA/PBAT film, ; PLA/PBAT/ZnO film, ; PLA/PBAT/TiO_2 film

4. CONCLUSIONS

The effects of metal oxide incorporation on UV-shielding properties of bio-based active packaging films were investigated. Sachi inchi oil was selected due to its high value from the health benefits and chemical structure of polyunsaturated fatty acids which were prone to lipid oxidation. The oil was packed in PLA/PBAT incorporated with ZnO or TiO_2 film pouches and then stored at room temperature for 1, 2, 3, 7, 15 and 30 days. The properties of the oil, PV and %FFA, were analyzed. The PV range tended to increase with time. The %FFA increased during day 1-7 but decreased on day 15 to 30 due to chemical changes in the oxidation reaction. The different properties of the oil packed in different films were noticed. PLA/PBAT/TiO_2 showed promise for UV-shielding packaging applications.

5. ACKNOWLEDGEMENT

The Kasetsart University Research and Development Institute (KURDI), Bangkok, Thailand provided financial support for this research, FF(KU) 25.64.

6. REFERENCES

[1] Al-Itry, R., Lamnawar, K. and Maazouz, A. (2014). Rheological, morphological, and interfacial properties of compatibilized PLA/PBAT blends. Rheol Acta, 53, pp. 501-517, DOI: 10.1007/s00397-014-0774

[2] Gutiérrez, L.F., Rosada, L.M. and Jiménez, A. (2011). Chemical composition of Sacha Inchi (Plukenetia volubilis L.) seeds and characteristics of their lipid fraction. Grasas y Aceites, 62(1), DOI:10.3989/gya044510.

[3] Iwata, T. (2015). Biodegradable and bio-based polymers: Future prospects of eco-friendly plastics. Angew. Chem. Int. Ed, 54, pp. 3210-3215.

[4] Liu, Q., Xu, YK., Zhang, P., Na, Z., Tang, T. and Shi, YX. (2014). Chemical composition and oxidative evolution of Sacha Inchi (Plukentia volubilis L.) oil from Xishuangbanna (China). Grasas Aceites, 65(1), DOI: 10.3989/gya.075713.

[5] Lizundia, E., Marie, C.P., Alain, G., Jose, L.V., and Sandra, D. (2019). Impact of ZnO Nanoparticle Morphology on Relaxation and Transport Properties of PLA Nanocomposites. Polymer Testing, 75, pp. 175-84, DOI: 10.1016/j.polymertesting.2019.02.009.

[6] Li, W., Sun, C., Li, Ch., Xu, Y., Tan, H. and Zhang, Y. (2021). Preparation of effective ultraviolet shielding poly (lactic acid)/poly (butylene adipate-co-terephthalate) degradable composite film using co-precipitation and hot-pressing method. International Journal of Biological Macromolecules, 193, pp. 540-547, DOI: 10.1016/j.ijbiomac.2021.09.097.

[7] Maurer, N.E., Sakoda, B.H., Chagman, G.P. and Rodriguez-Saona, LE. (2012). Characterization and authentication of a novel vegetable source of omega-3 fatty acids, sacha inchi (Plukenetia volubilis L.) oil. Food Chem, 132(2), pp. 1173-80, DOI: 10.1016/j.foodchem.2012.02.143.

[8] Maroid, A.A., Saeed, M.E. and Hussein, I. (2017). Vangueria madagascariensis (Rubiaceae) as New Oil Source. Unconventional Oilseeds and Oil Sources, 28, pp. 167-180.

[9] Sharma, S., Jaiswal, A.K., Duffy, B. and Jaiswal, S. (2020). Ferulic acid incorporated active films based on poly(lactide) /poly(butyleneadipate-co-terephthalate) blend for food packaging. Food Packaging and Shelf Life, 24, 100491, DOI: 10.1016/j.fpsl.2020.100491.

[10] Ramos-Escudero, F., González-Miret, M.L., Viñas-Ospino, A. et al. (2019). Quality, stability, carotenoids and chromatic parameters of commercial Sacha inchi oil originating from Peruvian cultivars. J Food Sci Technol 56, 4901-4910, DOI: 10.1007/s13197-019-03960-x

AG-O02

Development of a Colorimetric Total Volatile Base-Nitrogen Sensor for Potential Real-Time Freshness Monitoring of Pre-packaged Food

Darylle Jerome Ortiz[1]*, Ray Anne Grace Garalde[1], Mary Joy Paico[1]

[1]*Industrial Technology Development Institute, Department of Science and Technology, Taguig City, Metro Manila, 1631, Philippines*

***Correspondence to:** Industrial Technology Development Institute, Department of Science and Technology, Taguig City, Metro Manila, 1631, Philippines. daryllejerome.ortiz@gmail.com*

ABSTRACT: The interest on sensors for real-time freshness monitoring of pre-packaged food continuously rises in response to increasing demand for safer products. This study aimed to develop a colorimetric sensor that is chemically responsive to total volatile base-nitrogen (TVB-N) for potential monitoring of food freshness. The sensor was made from a combination of natural food dye of sweet potato and black carrot concentrate which was impregnated on filter paper by centrifugation method. The sensor was then encased with polyethylene film to address its hydrophilicity. Optimum dye concentration and centrifugation time parameters for the development of sensor was identified at 5% w/w and 5 minutes at 3000 rpm, respectively, achieving stable color values (CIELAB) for the prototype. Exposure to TVB-N simulants (30 and 40 mg-N/100 ml ammonia solution) at storage temperatures 4°C, 12°C, and 30°C was carried out to determine the sensor's chromatic response. Total color difference (TCD) of color values and digital images were used as index of response over time. The simulation resulted to color change from bright red to dark red to violet color on sensor after 8 days. Color transition to dark red color was observed on sensor stored at 4°C and 12°C when subjected to 30 and 40 mg-N/100 ml ammonia solution, whereas, violet color was detected only at 30°C on both simulant concentrations. The results demonstrated the sensing performance of the developed colorimetric sensor at different ammonia concentration thus, indicating its potential to detect TVB-N as spoilage metabolite of food. The response of the sensor to actual pre-packaged product still needs further investigation.

Keyword: Intelligent packaging; TVB-N; freshness indicator; natural food dye; chromatic response

1. INTRODUCTION

Freshness, as an index of product acceptability and determinant of quality, has to be preserved to provide consumer satisfaction. However, instances of spoilage happen even within the declared shelf life of food products leading to food lost and waste (FLW) or worse, foodborne disease outbreaks (FBDOs) once accidentally consumed. These significant challenges are faced by both developed and developing countries even in this modern era. FBDOs affects an an estimate of 30% of people in industrialized countries each year,[1] whereas, twenty-five (25%) of all food losses per year worldwide along the supply chain (equivalent to around 1.3 billion tons per year)[2], accounts to microbial deterioration.[1] In the Philippine scenario, FBDOs brought 13,577 morbidity[3] between 2005-2018 and FLW accounts to 9.3 million tons per year[4].

Packaging plays an important role in food supply chain. At the end of processing, it is responsible on the subsequent stages that a product will run through. However, spoilage of food still exists due to microbial contamination even with the total observance of proper packaging of products and sometimes incidentally promoted by improper temperature storage. This paves way for research on optimizing food packaging with additional functionalities more than its traditional role—containment, protection from outside environment, freshness keeping, and facilitation of transport and distribution—that is, to provide real-time information to consumers.[5]

Freshness indicator (FI) (also known as food quality indicator/FQI)is a type of Intelligent Packaging (IP) technology which provides a concise, direct, and scientifically-sound information on the state of freshness of food.[6] This is meant to be a simple device with complex structure and/or method fabrication that performs intelligent function without requiring complex operation for use.[6-7] This type of indicator is placed inside the food package and is part of the packaging label to indicate as visual indicator for real-time quality of food.[8] FIs usually senses volatiles that acts as stimuli for its response. Food products, commonly muscled-based foods, give off volatile compounds during its degradation such as trimethylamine, ammonia and dimethylamine which are considered as total volatile basic nitrogen (TVB-N). These are results of destructive activities of microorganism during spoilage and is one the the most important freshness indicator on monitoring food quality and safety.[9]

Natural pigments from black carrot and sweet potato in the form of anthocyanins are widely known to express pH-responsive colors[10] These crops are determined to contains high amounts of acylated anthocyanins—identified to be more stable on wider pH range and elevated temperature[11], and express wide color range at higher pH[12] compared to fruit anthocyanin. Many researches used black carrot and different sweet potato as source of anthocyanin due to its gaining popularity. Moradi et al.[13] incoporated black carrot anthocyanins in a bacterial cellulose nano to develop a novel pH indicator for monitoring of fish freshness. Wangmo et al.[14] developed a novel sensor based on bead-counting of purple sweet potato anthocyanin forming a purple sweet potato (PSP) tapioca pearls that are sensitive to ammonia and pH changes and was used in freshness monitoring of shrimp.

This present work was aimed to develop a colorimetric sensor through impregnation of natural dye pigments from black carrot and sweet potato through centrifugation. More so, this aimed determine the developed sensor's chromatic response to TVB-N as spoilage metabolite of food to evaluate its possible applicability for pre-packaged food.

2. MATERIALS AND METHODS

2.1. Determination of optimum natural dye concentration

A commercial natural dye concentrate from sweet potato and black carrot (IMCD Philippines Corp., Philppines) was used as source of color pigment for the developed colorimetric sensor. Different concentrations (w/w) of natural dye solution at 100% (control), 0.3%, 1%, and 5% were prepared by mixing liquid colorant concentrate with distilled water. Each solution were stirred for 30 min using magnetic stirrer. Pigments from solution of was immobilized in a filter paper strips (Whatman 42, 2.5 cm x 2.5 cm). A falcon tube with strip was dispensed with 25 ml of natural dye solution and centrifugated for 30 min. Impregnated strips were then dried for approximately 24 h using oven set at 40°C.

Two sensor strips (2) per treatment were stored in a dark environment and were measured for color values (CIELab) after drying (Day 0) and on Day 7 using a chromameter (Konica Minolta Color Reader CR-20, Japan). Color values were recorded and evaluated for significant difference after storage.

2.2. Determination of optimum centrifugation time

Identified optimum concentration was subjected to various centrifugation time. Colorimetric sensors were fabricated following the steps described in 2.1 except on centrifuge duration. Centrifugation was done for 30 min (control), 20 min, 15 min, 10 min, and 5 min. Test for stability of produced sensors was then carried out by storing four (4) sensor strips in a dark environment and measured for color values on Day 0 and Day 7. Obtained color values were evaluated used to calculate for the total color difference (TCD) of each strip after storage using the equation below:

$$\Delta E = \sqrt{(L*)^2 + (a*)^2 + (b*)^2} \qquad (1)$$

2.3. Simulation test for TVB-N exposure

Developed colorimetric sensor was tested for chromatic response by exposure to TVB-N simulants. Sensor strips were fabricated using the method optimized in 2.1 and 2.2. Sensors were encased using PE film (20 μm) to address its hydrophilicity. TVB-N simulants, on the other hand, were prepared by diluting ammonia (25%; Merck, Germany) with distilled water producing 30 and 40 mg-N/100 mL concentrations.

TVB-N exposure was done by using a glass container (240 ml internal volume) with metal cap as simulation chamber. The colorimetric sensors were attached under the metal cap. The simulation chambers were filled with approximately 20 ml simulant and closed with cap with sensor strips. The test was carried out at 30°C, 12°C, and 4°C for 8 days.

Three (3) sensor strips from each simulant concentrations at different temperatures were monitored for color values per day. Total color difference (Equation 1) were calculated using the color values obtained from every sampling interval. Regression of TCD per day was calculated to determine the rate of color change (TCD/day) of sensor strips.

3. RESULTS AND DISCUSSION

3.1. Optimum natural dye concentration

Optimum dye concentration was determined based on the intensity and stability (decrease) of color of the produced sensors in terms of color values (CEILab). Figure 1 illustrates the colorimetric sensors developed using different natural dye concentrations. The fabricated sensor strips using pure liquid natural dye showed an extremely dark red color and was found to be damp after 24 hours of drying. This can be attributed to the high total soluble solids (TSS) content of the utilized liquid natural dye, making the strips hygroscopic. More so, color bleedin was observed on these sensors during the storage test which signifies that the strips are not completely dried.

Sensor strips fabricated using lower concentrations of natural dye (0.3%, 1%, 3%, and 5% w/w) developed a progressive color from light pink to red. An increasing color intensity of sensor strips was observed with the increasing amount of natural dye in each solution. Also, these sensors achieved complete drying as compared to those fabricated with pure natural dye solution.

The results also revealed that colorimetric sensors developed using 3% and 5% natural dye solutions have the most stable color in terms of TCD after after a week of storage in dark environment. TCD values for each sensor strips were at 1.73 and 2.03, respectively, signifying a perceptible color change but though close observation. Higher TCD values were observed on sensor strips developed using 100%, 0.3%, and 1% concentrations which suggests a significant color change in terms of L*, a*, and b* values thus, a more detectable color change. Among the natural dye solution, 5% was identified as the optimum concentration considering its stability and visual quality in terms of red color intensity.

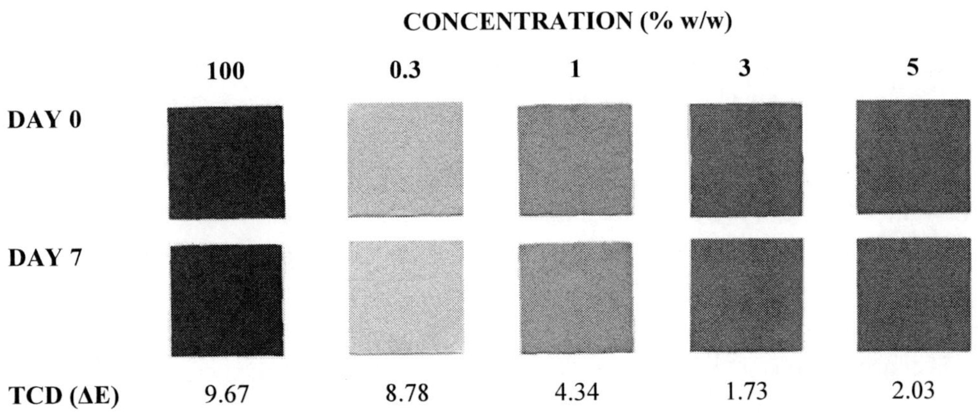

Figure 1. Change in color of colorimetric sensors developed using different concentrations of natural dye solution.

3.2. Optimum centrifugation time

The fabrication method was further optimized by testing for shortest possible duration of the centrifugation step. Table 1 lists the TCD values for each sensor strips developed at various centrifugation time using 5% natural dye solution. Based on the results of TCD calculation, reduced centrifugation time still resulted to a non-detectable change in color. Also, statistical analysis of results revealed that centrifugation time has no significant effect on TCD values of sensors after storage for 7 days. Therefore, reducing the duration of pigment impregnation process for up to 5 min achieves the same color intensity and stability of the sensor compare to control.

Table 1. Effect of centrifugation time to TCD compared to control sample.

CENTRIFUGE TIME (min)	TOTAL COLOR DIFFERENCE (ΔE) (Day 0 vs Day 7)
30 (Control)	2.1 ± 0.3^a
20	1.9 ± 0.2^a
15	2.0 ± 0.2^a
10	1.8 ± 0.2^a
5	1.9 ± 0.2^a

TCD (ΔE) scores interpretation: $0 < \Delta E < 1$:No noticed difference; $1 < \Delta E < 2$: Experienced observer can notice difference; $2 < \Delta E < 3.5$: Unexperienced observer also notices difference; $3.5 < \Delta E < 5$: Clear difference between colors; $5 < \Delta E$: Two color are obviously different.[15]

Values sharing the same letters are not significantly different at $p<0.05$.

3.3. Chromatic response of colorimetric sensor exposed to TVB-N simulants

The colorimetric sensors produced using the developed fabrication method were tested for its chromatic response to TVB-N simulants contaning different concentrations of ammonia. The test was carried out to determine its applicability on pre-packaged meat product which can be related to its sensitivity to TVB-N as spoilage metabolite for this product type. TCD was identified as index of color change and calculated using the obtained values of from color measurements.

The TCD values of colorimetric sensors during exposure to varying ammonia concentrations was illustrated in Figure 2. Based on the results, drastic increase in TCD was observed on sensors exposed to 30 and 40 mg-N/100 ml ammonia solutions at 30°C. These sensors exhibited scores above the limit (TCD > 3.5) from Day 2 up to end od storage test. These observations can be related to the rapid color change of sensors from bright red to violet based on results of visual color transition after 8 days (Figure 3). In contrast, slow increase in TCD was observed on sensors exposed to both amomonia concentrations at 4°C and 12°C. Based on calculations, sensors exposed on the said conditions breached the limit for from Day 6 to Day 8. The visual color change revealed the change of these sensors from bright red to dark red after 8 days of storage for both ammonia concentrations.

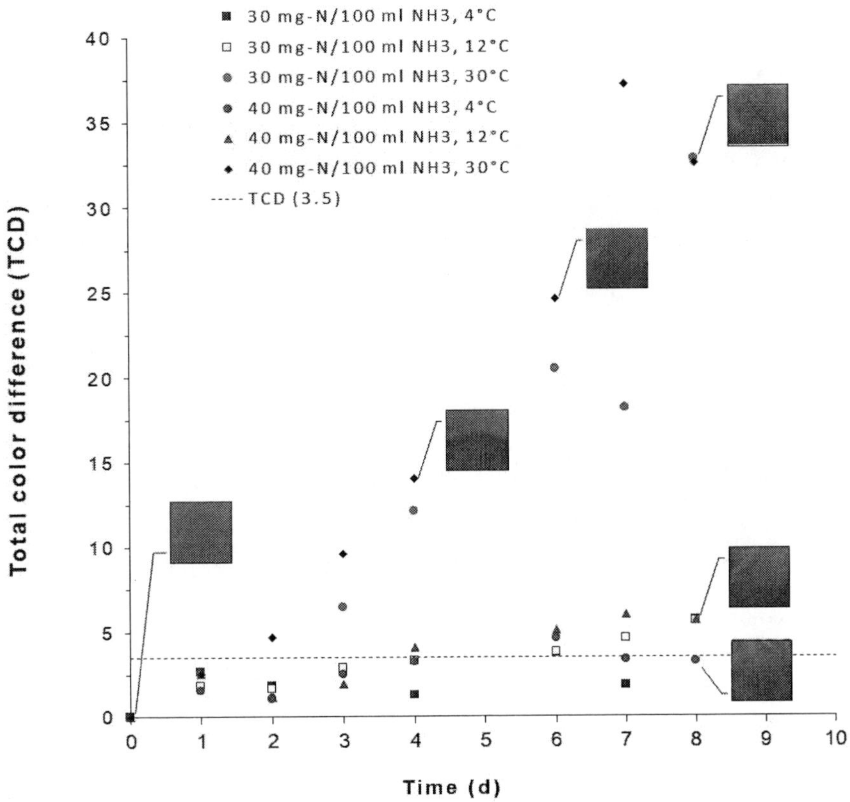

Figure 2. TCD of colorimetric sensor during TVB-N simulation test for 8 days at different temperature storage.

A 30 mg-N/100 ml NH_3

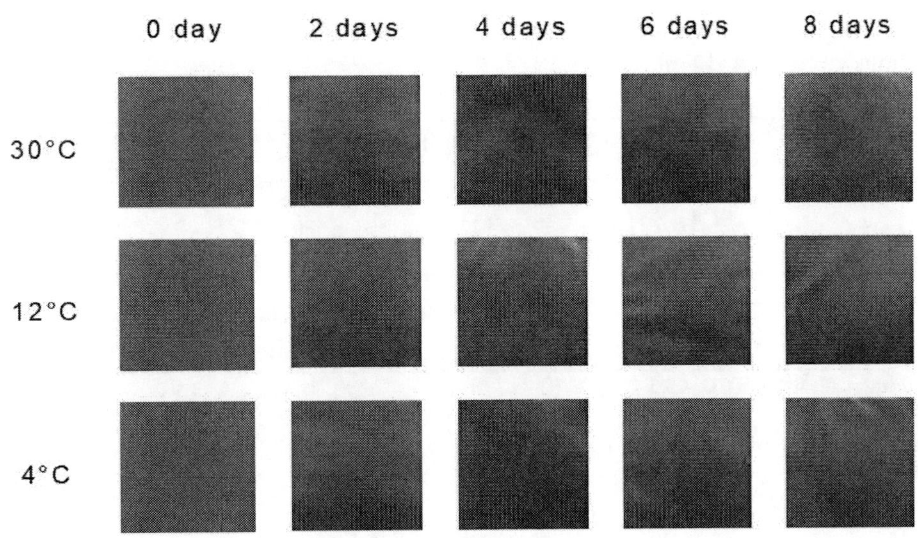

B 40 mg-N/100 ml NH_3

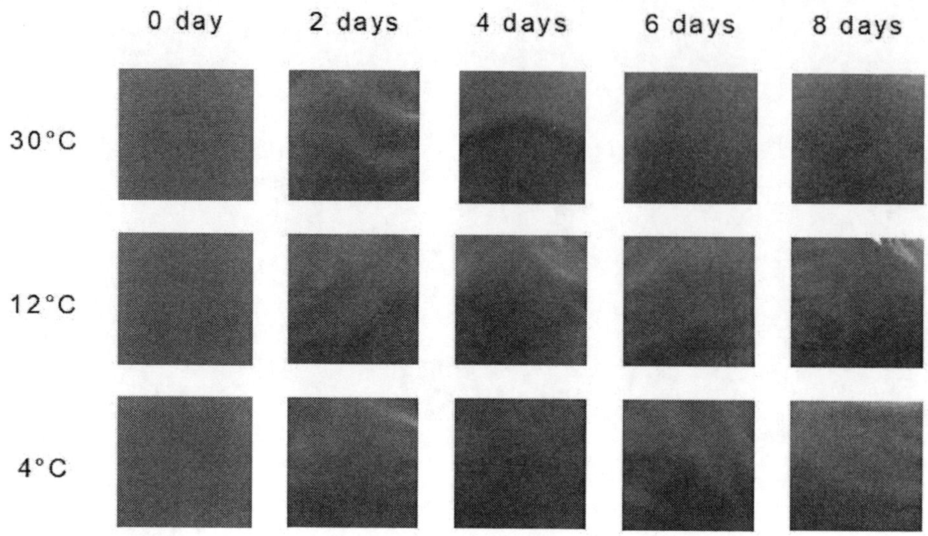

Figure 3. Visual color transition of colorimetric sensor during TVB-N simulation test for 8 days at different temperature storage.

In addition, the rate of color change based on TCD per day (TCD/day) was calculated using regression analysis to further express the effect of different ammonia concentrations on the color transition of the sensor (Table 2). Increasing TCD/day values was observed on sensors with increasing storage temperature. Sensors exposed to both simulant concentrations at 30°C was identified to have higher TCD/day values compared to those stored at 12°C hence faster color change, while, TCD/day of sensors at 4°C cannot be calculated since color change was observed to be not recognizable after storage test (TCD < 3.5) which suggests slow color transition of sensors as exposed to TVB-N simulants at lower temperature.

Table 2. TCD per day (TCD/day) of colorimetric sensor after exposure to 30 and 40 mg-N/100 mL concentrations stored at 4, 12 and 30 °C.

Simulant Concentration (mg-N/100g ml NH_3)		**TCD/day** 30 °C	12 °C	4 °C
40	Trial 1	4.28	0.80	**
	Trial 2	4.72	0.70	**
	Average	4.50	0.75	-
30	Trial 1	3.27	0.71	**
	Trial 2	3.66	0.50	**
	Average	3.47	0.61	-

**Not further calculated. TCD < 3.5.

4. CONCLUSIONS

A colorimetric sensor was successfully developed using a combination of natural dye concentrate from sweet potato and black carrot. Optimizing the dye concentration and centrifugation time at 5% w/w and 5 min, respectively, resulted to stable color and desirable red color intensity of sensor hence, decrease on said parameters do not affect the quality of the developed sensor. TCD values, visual color transition, and rate of color change (TCD/day) revealed the chromatic response of the sensors from bright red to dark red when stored at 4°C and 12°C, and bright red to violet at 30°C, after exposing to varying concentrations of ammonia solution. These results suggest the sensing performance of the developed sensor for the detection of present TVB-N in headspace. Hence, the developed colorimetric sensor can potentially be used for freshness monitoring of pre-packaged food during storage.

5. ACKNOWLEDGEMENT

The authors would like to acknowledge the Industrial Technology Development Institute of the Department of Science and Technology (ITDI-DOST) - Philippines for funding this research work. Also, to our mentor, Ms. Daisy E. Tañafranca, Division Chief of the Packaging Technology Division (PTD) of ITDI-DOST for her guidance and support on this study.

6. REFERENCES

[1] Bondi, M., Messi, P., Halami, P., Papadopoulou, C., and de Niederhausern, S. (2014, Jul 6). Emerging microbial concerns in food safety and new control measures. Hindawi Publishing Corporation BioMed Research International Vol. 2014 Article ID 251512. http://dx.doi.org/10.1155/2014/251512.

[2] Gustavsson, J., Cederberg, C., and Sonesson, U. (2011 May 16-17). Global Food Losses and Food Waste: Extent, Causes and Prevention; Study Conducted for the International Congress Save Food! at Interpack 2011, Düsseldorf, Germany; Food and Agriculture Organization of the United Nations: Rome, Italy.

[3] Azanza, M.P., Membrebe, B.Ñ., Sanchez, R.G., Estibo, E.M., Dollete, U.G., Feliciano, R., & Garcia, N.K. (2018). Philippine Journal of Science 148 (2). pp. 317-336. ISSN 0031-7683.

[4] United Nations Environment Programme. (2021). Food Waste Index Report 2021. https://www.unep.org/resources/report/unep-food-waste-index-report-2021. ISBN No.:978-92-807-3868-1.

[5] Wilbey, A. (2013). Emerging food packaging technologies. International Journal of Dairy Technology, 137-152.

[6] Shao, P., Liu, L., Yu, J., Lin, Y., Gao, H., and Chen, H. (2021). An overview of intelligent freshness indicator packaging for food quality and safety monitoring. Trens in Food Science & Technology 118; 285-296.

[7] Sohail, M., Sun, D. W., & Zhu, Z. W. (2018). Recent developments in intelligent packaging for enhancing food quality and safety. Critical Reviews in Food Science and Nutrition, 58(15), 2650-2662.

[8] Pacquit, A., Crowley, K., & Diamond, D. (2008). Smart packaging technologies for fish and seafood products. In J. Kerry & P. Butler (Eds.), Smart packaging technologies for fast moving consumer goods (pp. 75-98). England: John Wiley & Sons, Ltd.

[9] Moosavi-Nasab, M. Khoshnoudi-Nia, S., Azimifar, Z., and Kamyab, S. (2021). Evaluation of the total volatile basic nitrogen (TVB-N) content in fish fillets using hyperspectral imaging coupled with deep learning neural network and meta-analysis. Scientific Reports. https://doi.org/10.1038/s41598-021-84659-y.

[10] Hamedi, S. and Koosha, M. (2020 Nov). Designing a pH-responsive drug delivery system for the release of black-carrot anthocyanins loaded in halloysite nanotubes for cancer treatment. Applied Clay Science Vol 197. https://doi.org/10.1016/j.clay.2020.105770.

[11] Kirca, A., Ozkan, M., and Cermeroglu, B. (2007). Efect of temeperature, solid content and pH on the satability of black carrot anthocyanins. Food Chem., 10, pp. 212-218. https://doi.org/10.1016/j.foodchem.2006.01.019.

[12] Giusti, M.M. and Wrolstad, R.E. (2003). Acylated anthocayanins from edible sources and their applicaions in food systems. Biochemical Engineering Journal 14 (3), pp. 217-225.

[13] Moradi, M. Tajik, H., Almasi, H., Forough, M., and Ezati, P. (2019). A novel pH-sensing indicator based on bacterial cellulose nanofibers and black carrot anthocyanins for monitoring fish freshness. Carbohydrate Polymers Vol 222. https://doi.org/10.1016/j.carbpol.2019.115030.

[14] Wangmo, L. Suratsawadee, A., Ratvijitvech, T., Siripinyanond, A. (2021). A novel sensor based on bead-counting of purple sweet potato tapioca pearl for freshness monitorin of shrimp. Food Chemistry Vol 368. https://doi.org/10.1016/j.foodchem.2021.130863.

[15] Mokrzycki, W.S. and Tatol, M. (2012). Colour difference ΔE – A survey. Machine Graphics and Vision 20(4), pp. 383-411. https://www.researchgate.net/publication/236023905_Color_difference_Delta_E_-_A_survey.

AG-O03

Antimicrobial Activity of Biodegradable Lignin Nanoparticles in Polyvinyl Alcohol-Based Film Against Major Spoilage Microorganisms of Mango Fruit

Angel Jr Basbasan[1]*, Bongkot Hararak[2], Charinee Winotapun[2], Kanchana Boonruang[3], Pattarin Leelaphiwat[1], Wanee Chinsirikul[4], Vanee Chonhenchob[1]

[1]*Department of Packaging and Materials Technology, Faculty of Agro-Industry, Kasetsart University, Bangkok 10900 Thailand*
[2]*National Metal and Materials Technology Center, National Science and Technology Development Agency, Pathum Thani 12120, Thailand*
[3]*Department of Horticulture, Faculty of Agriculture, Kasetsart University, Bangkok 10900, Thailand*
[4]*National Nanotechnology Center, National Science and Technology Development Agency, Pathum Thani 12120, Thailand*

***Correspondence to:** Department of Packaging and Materials Technology, Faculty of Agro-Industry, Kasetsart University, Bangkok 10900 Thailand. E-mail: amboy25@gmail.com*

ABSTRACT: Lignin is a natural polymer existing in plants and a by-product of the paper-and-pulp and bioethanol production industries. It offers a sustainable way to inhibit the growth of microorganisms due to its inherent antimicrobial property. Coupled with nanotechnology, antimicrobial activity is substantially enhanced. Limited information is available about lignin nanoparticles (LNPs) for food packaging applications. Hence, this study investigated the antimicrobial activity of LNPs against major spoilage-causing microorganisms of mango fruit using potato dextrose agar as the growing medium. The difference in antimicrobial activity between raw lignin and LNPs both with 0.2 % (w/v) concentration against *Colletotrichum gloeosporioides* and *Lasiodiplodia theobromae* that causes anthracnose and stem-end-rot in mango, respectively was determined. Compared with raw lignin, 0.2 % (w/v) LNPs showed at least 9 times antimicrobial strength against *C. gloeosporioides* but moderate against *L. theobromae*. Increasing LNPs concentration to 0.5 % (w/v) increased the antimicrobial activity to 72.02 and 60.79 % against *C. gloeosporioides* and *L. theobromae*, respectively. However, increasing further to 2 and 5 % (w/v) gave no significant ($p \leq 0.05$) difference in inhibition. To investigate the synergistic effect, thymol with concentrations of 0.5 and 1 % (w/v) was added into the 0.5 % LNPs solution. Results showed that 0.5 % LNPs with thymol at a concentration as low as 0.5 % exhibited fungicidal effects against both *C. gloeosporioides* and *L. theobromae*. Further, the antimicrobial activity of 0.5 % (w/v) LNPs with 0.5 % (w/v) thymol incorporated in PVA-based film (PVA/LNP/Thy) was investigated. Growth inhibition of both *C. gloeosporioides* and *L. theobromae* was observed in PVA/LNP/Thy but not in the neat PVA film. Results obtained in this study suggest that the antimicrobial activity of 0.5 % (w/v) LNPs with 0.5 % (w/v) thymol has the potential to prolong the storage life of packed mango fruit by inhibiting the growth of major spoilage causing microorganisms *C. gloeosporioides* and *L. theobromae*.

Keyword: Lignin nanoparticles, Antimicrobial packaging, Fungicidal, Polyvinyl alcohol, Mango

1. INTRODUCTION

Biodegradable polymers are considered a sustainable and eco-efficient alternative packaging material for products derived from fossil sources [1] e.g., conventional plastic films (CPF). However, compared to CPF, biodegradable plastic films (BPF) in general are inferior in performance e.g., less strength and low to moderate barrier properties. The development of BPF for food packaging applications with enhanced properties without compromising biodegradability led to the incorporation of natural and active compounds such as lignin. Lignin is the most abundant aromatic renewable biopolymer having a complex structure and possessing functional groups responsible for active functions [2]. The advantages of lignin lie in its wide natural availability and economic cost aside from being environment-friendly. The utilization of lignin to enhance properties of BPF for food packaging purposes is a growing concern of interest driven by: (a) lignin possesses natural antioxidant and antimicrobial properties, (b) the aim of developing a packaging material that is at least with the same level of strength and barrier properties as CPF, and (c) valorization of lignin offers a sustainable way of utilization of renewable resources. Lignin is a major by-product of the paper-making industry on which it is extracted during cellulose isolation processes to have fibers with more than 98 % cellulose content [3] and from the second generation (2G) production of bioethanol from lignocellulosic biomass [4] e.g., rice straw, sugar cane bagasse, corn stover, etc.

Relatively large particle size of lignin – 10 to 100 μm, limits its application, as the properties of the resulting composites are significantly dependent on particle size and interfacial bonding of the polymeric matrix [5]. To widen the application of lignin where nanoparticles are desired, methods for the preparation of LNPs have been established such as polymerization, acid precipitation, solvent exchange, ultrasonication, and alkaline precipitation among others [6]. In the area of food packaging, several published works have shown the beneficial effects of the presence of LNPs in biodegradable materials. Incorporation of LNPs enhanced mechanical behavior and thermal stability of wheat gluten bionanocomposite [7]; increased absorption capacity of chitosan composite [8]; improved UV barrier property [9] and increased the impact strength and thermal stability of PLA based composites [10]. In terms of antimicrobial properties, most of the previous research have tested LNPs against bacteria such as *E. coli*, *S. aureus*, and *Salmonella* species. However, insufficient information can be found about the antimicrobial activity of LNPs against yeast, molds, and other types of fungi which also cause major deterioration in fresh and processed food products. Therefore, the present study generally aimed to determine the antimicrobial activity of LNPs against *C. gloeosporioides* and *L. theobromae*, two fungi specie that causes major deterioration in mango i.e., anthracnose and stem-end-rot in mango, respectively [11].

The present study used poly (vinyl alcohol) or PVA as the polymeric matrix to maintain the full biodegradability of the resulting composite. Four phases of experiments were employed, namely: (1) Antimicrobial activity difference between raw lignin and LNPs; (2) Antimicrobial activity of LNPs with varied concentrations; (3) Antimicrobial activity of 0.5 % (w/v) LNPs with thymol; and (4) Antimicrobial activity of 0.5 % (w/v) LNPs with 0.5 % (w/v) thymol incorporated into PVA-based film.

2. MATERIALS AND METHODS

2.1 Materials

Softwood kraft lignin (BioPivaTM100) powder was purchased from UPM Biochemicals, Finland. PVA Mowiol® with an average Mw of ~205,000 g/mol was purchased from Sigma-Aldrich (Thailand). Pure culture *C. gloeosporioides* was obtained from the Plant Protection Research and Development Office of the Department of Agriculture, Bangkok, Thailand while pure culture of *L. theobromae* was isolated from Nam Dok Mai mango following the method of Rafiullah [12].

2.2 Methods

2.2.1 Difference in the antimicrobial activity between raw lignin and LNPs

2.2.1.a Preparation of raw lignin-deionized water suspension. Raw lignin powder was mixed with deionized water to have a 0.2 % (w/v) concentration. The mixture was homogenized for five minutes using IKA T18 digital Ultra Turrax homogenizer, set at 9000 rpm.

2.2.1.b Preparation of LNPs-deionized water suspension. Anti-solvent precipitation assisted by ultrasonication was employed to prepare LNPs-deionized water suspension. Raw lignin was purified first by fractionation using acetone. The obtained acetone-soluble lignin was dried at 80 °C in a vacuum oven and ground into powders. Acetone-soluble lignin was dissolved in acetone at room temperature. After lignin was fully dissolved, deionized water was gradually added under ultrasonication (VCX 500 sonicator, 500-watt, 20 kHz, and 13 mm probe) for 5 minutes. The volume of deionized water was adjusted to have a final concentration of 0.2 % (w/v) LNPs.

2.2.1.c Determination of antimicrobial activity. Antimicrobial activity was carried out through poisoned-food method as described by Balouiri [13], with slight modification. Mycelial plug having a diameter of eight-millimeter which was taken from the edge of actively growing pure fungi culture was placed on the center of the newly prepared PDA containing raw lignin or LNPs, on which the final concentration in the PDA was 0.2 % (w/v). Pure PDA in Petri dishes served as the control. All prepared samples were stored in an ambient dark place at 30 ° C ± 2. Mycelial growth was measured using a digital Vernier caliper by taking the average of the diameters in perpendicular direction. The antimicrobial activity was quantified by taking the growth inhibition in percentage using the formula [14]:

$$\text{Growth inhibition (\%)} = 100 - \frac{\text{Ø with lignin}}{\text{Ø control}} \times 100 \qquad (1)$$

2.2.2 Antimicrobial activity of LNPs with varied concentrations

In this phase of the study, the antimicrobial activity of LNPs with 0.2, 0.5, 2.0, and 5.0 % (w/v) was determined, on which pure PDA served as the control. The procedures described in 2.2.1.b were followed but the amount of LNPs and deionized water were adjusted to obtain the desired LNPs varied concentrations. Similarly, procedures presented in 2.2.1.c were followed for the determination of antimicrobial activity.

2.2.3 Antimicrobial activity of 0.5 % (w/v) LNPs and thymol

Preparation of LNPs-thymol-deionized water colloidal suspension. Thymol was dissolved in a previously prepared LNPs-acetone solution using a magnetic stirrer, for 2 minutes. A 100 ml deionized water was gradually added into the solution intermittently within 6 minutes under ultrasonication (VCX 500 sonicator, 500-watt, 20 kHz, and 13 mm probe). Afterward, the mixture was sonicated further for 6 more minutes. The ratio of LNPs and thymol in the colloidal suspension were as follows:

Control	-	Pure PDA
Treatment 1	-	0.5 g LNPs: 0 g thymol (0.5:0)
Treatment 2	-	0.5 g LNPs: 0.5 g thymol (1:1)
Treatment 3	-	0.5 g LNPs: 1 g thymol (1:2)

The antimicrobial activity was determined following the procedures described in 2.2.1.c.

2.2.4 Antimicrobial activity of 0.5 % (w/v) LNPs with 0.5 % (w/v) thymol incorporated into PVA-based film

2.2.4.a Preparation of PVA-based films. PVA-based film samples were prepared through casting method. Neat PVA solution was prepared by dissolving eight grams of PVA powder in a 200 ml deionized water. For PVA/0.5%LNPs solution, eight-grams PVA powder was dissolved in a 200 ml previously prepared 0.5% LNPs-deionized water suspension. For PVA/LNPs/thymol solution, four ml of the previously prepared solutions of 0.5%LNPs/0.5%thymol was added into the neat PVA solution to obtain PVA/0.5%LNPs/0.5%thymol. Each of the prepared solutions was mixed using a magnetic stirrer for three hours. To form the film, each of the solutions was poured into a 0.3 cm thick casting glass mold and dried inside an oven set at 40 °C for 72 hours. To summarize, the following were the films prepared:

Control	-	Neat PVA film
Treatment 1	-	PVA/0.5%LNPs
Treatment 2	-	PVA/0.5%LNPs/0.5%thymol

2.2.4.b Determination of antimicrobial activity of PVA-based films. Vapor diffusion method according to Boonruang [15] was adopted to determine the antimicrobial activity of the prepared film samples against *C. gloeosporioides* or *L. theobromae*, with slight modification. Petri dishes containing pure PDA were inoculated with an eight-mm diameter mycelial plug taken from the margin of actively growing mycelia of pure culture of the target fungi. A 6 x 6 cm of a film sample was fixed inside the cover of the petri dish; Petri dishes were then sealed immediately with parafilm. Storage condition for prepared samples and the antimicrobial activity of the films was quantified as described in 2.2.1.c.

3. RESULTS AND DISCUSSION

3.1 Difference in the antimicrobial activity between raw lignin and LNPs

Both raw lignin and LNPs showed antimicrobial activity against both *C. gloeosporioides* and *L. theobromae*. Based on scanning electron analysis (SEM), the particle size of raw lignin was D (0.9) 57.52 μm while LNPs were ~100 nm. The increased total surface area to volume ratio of LNPs has a significant increase in its antimicrobial activity. Quantifying the difference, obtained results showed that LNPs are nine times stronger in inhibiting the growth of *C. gloeosporioides* compared to raw lignin. Moreover, the antimicrobial activity of LNPs were observed for up to 10 days while 8 days for raw lignin. On the other hand, although there was a slight difference of 5 to 11 % between LNPs and raw lignin, the inhibition was stronger against *L. theobromae* up to 6 days but the growth inhibition was observed in a shorter duration compared with *C. gloeosporioides*.

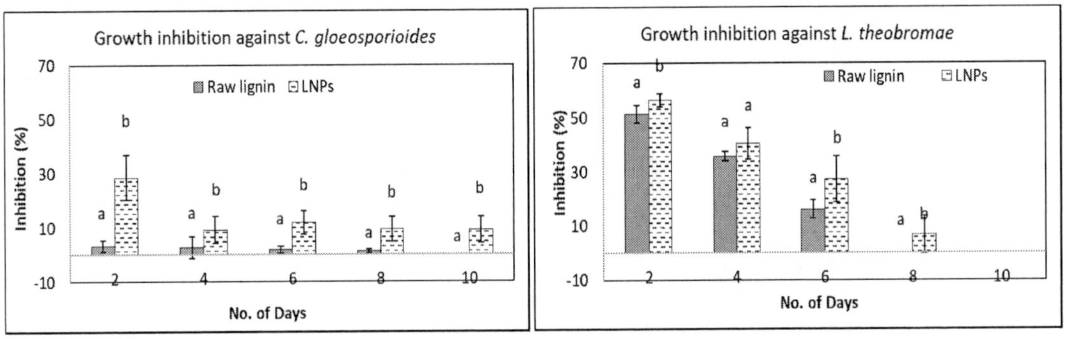

Figure 1 Antimicrobial activity of raw lignin and LNPs against *C. gloeosporioides* and *L. theobromae*

3.2 Antimicrobial activity of LNPs with varied concentrations

Increasing the LNPs concentration from 0.2 to 0.5 % (w/v) showed a significant increase of 72.02 and 60.79 % in the antimicrobial activity against both *C. gloeosporioides* and *L. theobromae*, respectively, as shown in Figure 2. Interestingly, increasing further the concentration to 2 and 5 % (w/v) showed no significant increase in inhibiting the growth of the target fungi. Figure 3 shows the mycelial growth of *C. gloeosporioides* on PDA with LNPs having 0.5, 2.0, and 5.0 % (w/v) concentrations. Based on the employed statistical analysis by DMRT, the mycelial growth has no significant difference at $\alpha = 0.05$. This could be the result of the agglomeration of LNPs which occurred in both 2.0 and 5.0 % (w/v) concentrations, but not in 0.5 % (w/v) concentration. The agglomeration of LNPs (visible at the backside of the Petri dishes containing the growth medium) can hide some of the surface area of the nanoparticles affecting the full/complete antimicrobial activity at higher concentrations. A similar result was obtained by Domínguez-Robles [16], showing that the adhesion of *S. aureus* to PBS/lignin composite containing 2.5, 5, 10, and 15 % (w/w) lignin showed a similar degree of adherent bacteria at $\alpha = 0.05$. As discussed by Zhang [17], key factors such as high surface energy, large number of hydrogen bonds, and Van der Waals forces between nanoparticles make the LNPs highly prone to agglomeration. Moreover, agglomeration of LNPs can also be the effect of concentration or amount of LNPs added, type of treatment agent, and dispersing agent. Anti-solvent precipitation method which was used in this study to prepare LNPs has the disadvantage of non-completely dispersion of LNPs in the solvent.

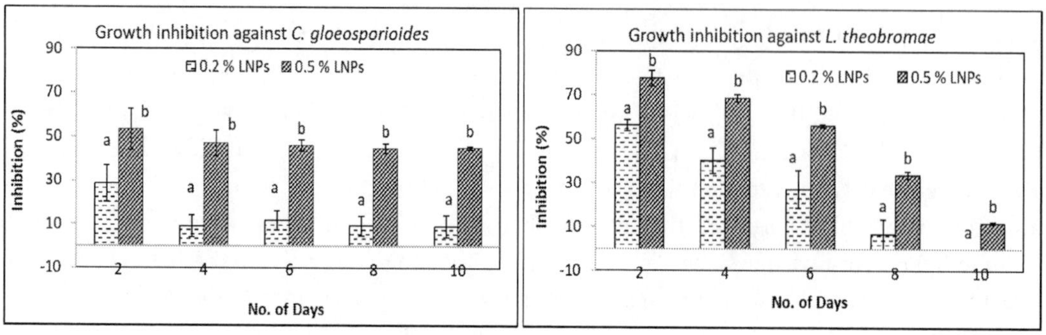

Figure 2 Antimicrobial activity of 0.2 and 0.5 % (w/v) LNPs against *C. gloeosporioides* and *L. theobromae*

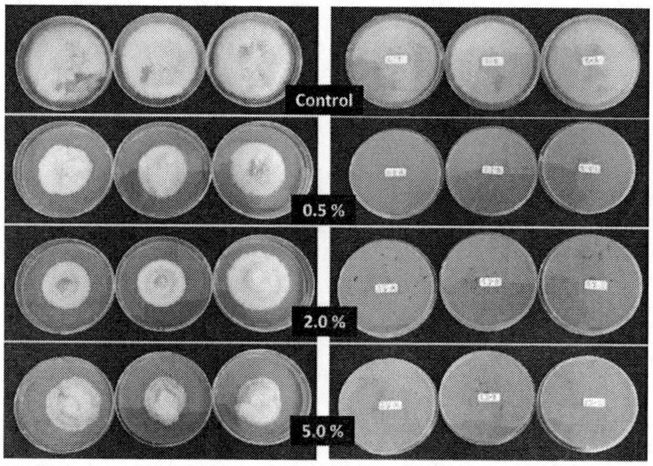

Figure 3 Mycelial growth of *C. gloeosporioides* (left); agglomeration of LNPs visible on the backside of the Petri dish (right), after 12 days of incubation

3.3 Antimicrobial activity of 0.5 % (w/v) LNPs and thymol

Thymol is known as a strong antimicrobial compound but its volatility affects the remaining content in a material after processing, hence affecting the antimicrobial effect. At this phase of the study, only a very low amount of thymol was used and was mixed with LNPs to protect it from possible degradation during sterilization of the PDA growth medium at 121 ° C for 15 minutes. The results as shown in Figure 4, thymol as low as 0.5 % (w/v) mixed with LNPs showed a fungicidal effect against both the fungi specie. With this fungicidal effect, it was implied that LNPs were able to protect the thymol during the sterilization process of the PDA growth media, as one of the beneficial properties of lignin is its thermal stability.

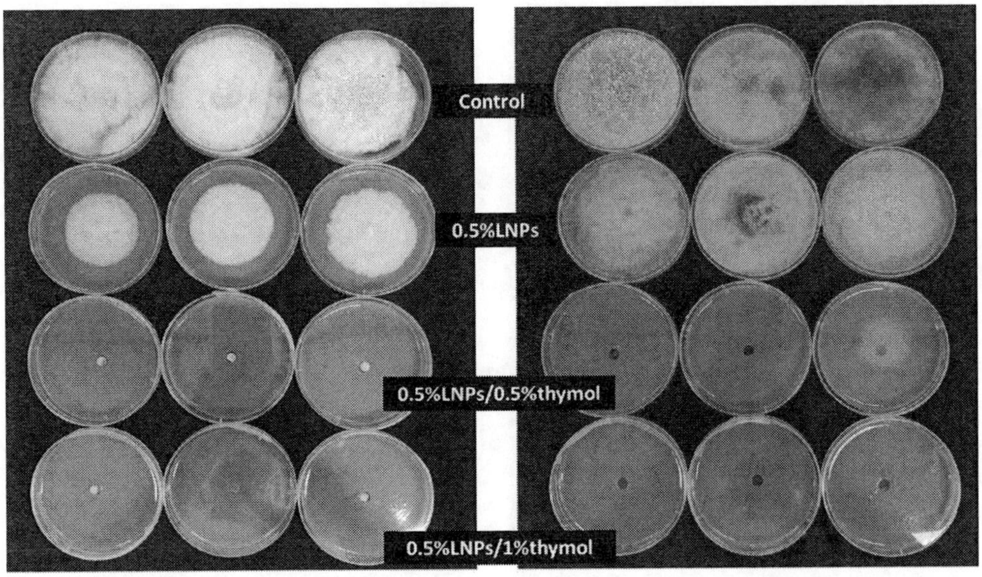

Figure 4 Effect of low amount of thymol nanoencapsulated in LNPs on the mycelial growth of *C. gloeosporioides* (left) and *L. theobromae* (right) after 12 days of incubation

3.4 Antimicrobial activity of 0.5 % (w/v) LNPs with 0.5 % (w/v) thymol incorporated into PVA-based film (PVA/0.5LNPs/0.5thymol)

LNPs with the concentration of 0.5 % (w/v) in PVA-based film showed a very low to no growth inhibition against *C. gloeosporioides* and *L. theobromae*, with 0.55 and 1.43 %, respectively after 6 days of incubation. As compared with the results in 3.1 and 3.2 where LNPs had a direct contact effect on the fungi studied, in this phase of the study, the functional group in LNPs responsible for the antimicrobial activity cannot diffuse out easily in six days, as an effect of the complex structural composition of LNPs and the state of LNPs as a solid compound. Meanwhile, inhibition growth of 70.77 % and 71.06 % was observed against *C. gloeosporioides* and *L. theobromae*, respectively as an antimicrobial activity of 0.5 % LNPs and 0.5 % thymol incorporated into the PVA-based film. In 3.3, fungicidal effect was observed while fungistatic effect was observed in this phase of the study. This could be the effect of occurred nanoencapsulation of thymol by the LNPs, as a result, only a portion lower than 0.5 % thymol was released within 6 days, which is not sufficient enough to totally inhibit the growth of both the fungi specie studied.

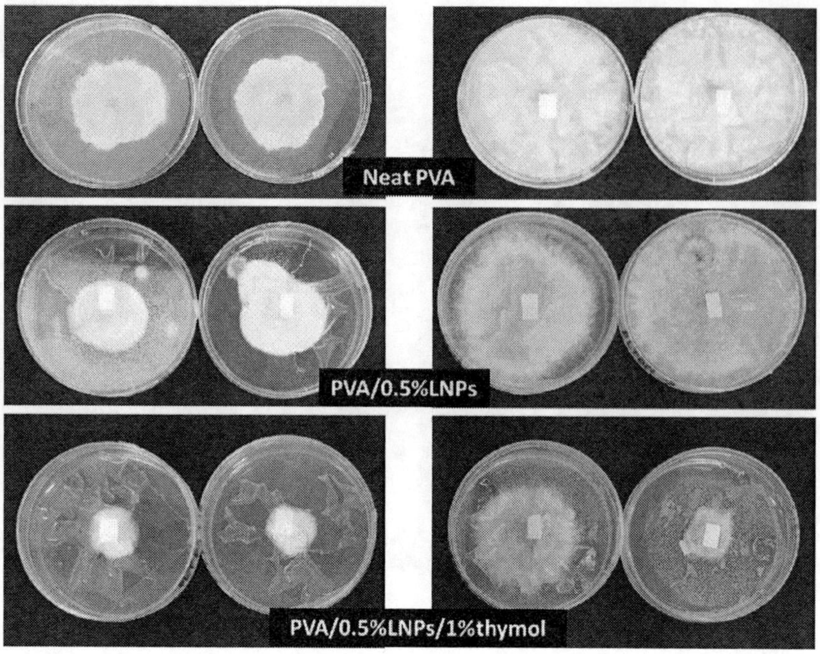

Figure 5 Antimicrobial activity 0.5 % LNPs and 0.5 % thymol incorporated in PVA-based film against *C. gloeosporioides* (left) and *L. theobromae* (right), after 6 days of incubation

4. CONCLUSIONS

Compared to raw lignin, lignin nanoparticles can provide stronger antimicrobial activity in inhibiting the growth of *C. gloeosporioides* and *L. theobromae*, which cause anthracnose and stem-end-rot in mango, respectively. While LNPs can inhibit the growth of the fungi in in-direct contact using poisoned food method, its antimicrobial activity was not observed in the PVA-based film due to its highly complex molecular structure and its solid state, as a result, diffusion of functional groups responsible for the antimicrobial activity is difficult to take place. On the other hand, LNPs are effective as nanocapsules for thymol, a known volatile antimicrobial compound. The synergistic effect of 0.5 % (w/v) of thymol and 0.5 % (w/v) LNPs can inhibit the growth of *C. gloeosporioides* and *L. theobromae* if it is in-direct contact although the inhibition decreases to 70.77 % and 71.06 % if it is incorporated into PVA-based film, nevertheless, growth inhibition can still be considered high.

5. REFERENCES

[1] Urbina, L., Eceiza, A., Gabilondo, N., Corcuera, M. Á. and Retegi, A. (2019). Valorization of apple waste for active packaging: multicomponent polyhydroxyalkanoate coated nanopapers with improved hydrophobicity and antioxidant capacity. *Food Packaging and Shelf Life*, 21.

[2] Sipponen, M. H., Lange, H., Crestini, C., Henn, A. and Osterberg, M. (2019). Lignin for Nano- and Microscaled Carrier Systems: Applications, Trends, and Challenges. *ChemSusChem*, 12, 10, 2039-2054.

[3] Fernández-Rodríguez, J., Gordobil, O., Robles, E., González-Alriols, M. and Labidi, J. (2017). Lignin valorization from side-streams produced during agricultural waste pulping and total chlorine free bleaching. *Journal of Cleaner Production*, 142, 2609-2617.

[4] Zhao, J., Quan, C. and Fan, S. (2013). Role of Lignin in Bio-Ethanol Production from Lignocellulosic Biomass. *Journal of Biobased Materials and Bioenergy*, 7, 5, 533-540.

[5] Gao, W. and Fatehi, P. (2019). Lignin for polymer and nanoparticle production: Current status and challenges. *The Canadian Journal of Chemical Engineering*, 97, 11, 2827-2842.

[6] Chauhan, P. S. (2020). Lignin nanoparticles: Eco-friendly and versatile tool for new era. *Bioresource Technology Reports*, 9.

[7] Yang, W., Kenny, J. M. and Puglia, D. (2015). Structure and properties of biodegradable wheat gluten bionanocomposites containing lignin nanoparticles. *Industrial Crops and Products*, 74, 348-356.

[8] Sohni, S., Hashim, R., Nidaullah, H., Lamaming, J. and Sulaiman, O. (2019). Chitosan/nano-lignin based composite as a new sorbent for enhanced removal of dye pollution from aqueous solutions. *Int J Biol Macromol*, 132, 1304-1317.

[9] Yang, W., Fortunati, E., Dominici, F., Giovanale, G., Mazzaglia, A., Balestra, G. M., Kenny, J. M. and Puglia, D. (2016). Effect of cellulose and lignin on disintegration, antimicrobial and antioxidant properties of PLA active films. *Int J Biol Macromol*, 89, 360-368.

[10] Spiridon, I., Leluk, K., Resmerita, A. M. and Darie, R. N. (2015). Evaluation of PLA–lignin bioplastics properties before and after accelerated weathering. *Composites Part B: Engineering*, 69, 342-349.

[11] Konsue, W., Dethoup, T. and Limtong, S. (2020). Biological Control of Fruit Rot and Anthracnose of Postharvest Mango by Antagonistic Yeasts from Economic Crops Leaves. *Microorganisms*, 8, 3.

[12] Rafiullah Khan, M., Suwanamornlert, P., Sangchote, S. and Chonhenchob, V. (2020). Antifungal activity of propyl disulphide from neem against Lasiodiplodia theobromae and Neofusicoccum parvum causing stem end rot in mango. *J Appl Microbiol*, 129, 5, 1364-1373.

[13] Balouiri, M., Sadiki, M. and Ibnsouda, S. K. J. J. o. p. a. (2016). Methods for in vitro evaluating antimicrobial activity: A review, 6, 2, 71-79.

[14] Balaguer, M. P., Lopez-Carballo, G., Catala, R., Gavara, R. and Hernandez-Munoz, P. J. I. J. o. F. M. (2013). Antifungal properties of gliadin films incorporating cinnamaldehyde and application in active food packaging of bread and cheese spread foodstuffs, 166, 3, 369-377.

[15] Boonruang, K., Kerddonfag, N., Chinsirikul, W., Mitcham, E. J. and Chonhenchob, V. (2017). Antifungal effect of poly(lactic acid) films containing thymol and R-(-)-carvone against anthracnose pathogens isolated from avocado and citrus. *Food Control*, 78, 85-93.

[16] Dominguez-Robles, J., Larraneta, E., Fong, M. L., Martin, N. K., Irwin, N. J., Mutje, P., Tarres, Q. and Delgado-Aguilar, M. (2020). Lignin/poly(butylene succinate) composites with antioxidant and antibacterial properties for potential biomedical applications. *Int J Biol Macromol*, 145, 92-99.

[17] Zhang, Z., Terrasson, V. and Guénin, E. (2021). Lignin nanoparticles and their nanocomposites. *Nanomaterials*, 11, 5, 1336.

AG-O04

UV-shielding Properties of Poly(lactic acid) (PLA)/Poly(butylene adipate-co-terephthalate) (PBAT) Films Incorporating Metal Oxides

Uruchaya Sonchaeng[1], Busarin Chongcharoenyanon[1], Piyawanee Jariyasakoolroj[1], Ubonrat Siripatrawan[2], Panuwat Suppakul[1]*

[1]Department of Packaging and Materials Technology, Faculty of Agro-Industry, Kasetsart University, Bangkok, 10900, Thailand
[2]Department of Food Technology, Faculty of Science, Chulalongkorn University, Bangkok, 10330, Thailand

***Correspondence to:** Panuwat Suppakul. Department of Packaging and Materials Technology, Faculty of Agro-Industry, Kasetsart University, Bangkok, 10900, Thailand.*
E-mail: panuwat.s@ku.ac.th

ABSTRACT: Poly(lactic acid) (PLA)/poly(butylene adipate-co-terephthalate) (PBAT) films with blending ratio of 85:15 wt% were investigated for potential UV-shielding packaging applications. Two UV-absorbing metal oxides, namely, zinc oxide (ZnO) and titanium dioxide (TiO_2), were incorporated into the PLA/PBAT films. PLA/PBAT films with ZnO 1 and 3 wt% and TiO_2 1 and 3 wt% could be fabricated with good reproducibility using a blown film extrusion process. In the UV region (below 400 nm), the light transmittance values of PLA/PBAT/ZnO 1% and PLA/PBAT/ZnO 3% films were lower than 20% and 3%, respectively, while the percent light transmittance values of PLA/PBAT/TiO_2 1% and PLA/PBAT/TiO_2 3% films were lower than 3% and 1%, respectively. The visual transparency, color, and light transmittance of the films were in good agreement, i.e., the higher the opacity, the higher the L* values in the CIELAB color space, and the lower the transmittance. The opacity and the L* values of the films were in the following order: PLA/PBAT < PLA/PBAT/ZnO 1% < PLA/PBAT/ZnO 3% < PLA/PBAT/TiO_2 1% < PLA/PBAT/TiO_2 3%. The ATR-FTIR (Attenuated Total Reflection-Fourier Transform Infrared) spectra did not show a noticeable difference or band shifting when comparing the PLA/PBAT films with and without the metal oxides. Both ZnO and TiO_2 contributed to improvement in the UV-shielding properties of the PLA/PBAT films. However, using the films as a food contact material may require further investigation on safety.

Keyword: Smart packaging, UV-shielding, PLA, PBAT, Metal oxides

1. INTRODUCTION

UV (Ultraviolet) light is a form of electromagnetic radiation in the wavelength range below 400 nm. UV light can be the cause of food deterioration, especially for foods or food components that are sensitive to photodegradation or photooxidation. Different strategies have been developed to prevent or slow down such deterioration to improve the shelf-life of packaged food products. "Smart packaging" is one of those strategies. Smart packaging is packaging innovation that uses technology to provide additional packaging functions that are more than containing the products. The terms "active packaging" and "intelligent packaging" are often used, sometimes interchangeably, with smart packaging. Concerning food packaging, these two terms can be distinguished as follows: active packaging means packaging that acts on the environment surrounding food to increase the shelf-life and intelligent packaging means packaging that does not act directly on food with any action other than monitoring the condition of the packaged product [1]. By these definitions, packaging with UV-shielding properties can be categorized as active packaging. Examples of intelligent packaging are time-temperature indicators, chemical sensors, and barcode/QR code labels.

In this research project, blend films from poly(lactic acid) (PLA) and poly(butylene adipate-*co*-terephthalate) (PBAT) were incorporated with UV absorbing metal oxides, namely, zinc oxide (ZnO) and titanium dioxide (TiO_2). Both PLA and PBAT are biodegradable and the blends of PLA/PBAT are biodegradable products [2,3]. At proper blending ratios, the properties of PLA and PBAT can complement each other [4–7]. There are reports of polymers incorporating ZnO and TiO_2 showing potential as UV-shielding materials [8,9]. However, to the authors' knowledge, no data are available on the UV-shielding effects of ZnO and TiO_2 in PLA/PBAT films. Thus, this research aimed to investigate the UV-shielding properties of the PLA/PBAT films incorporating ZnO and TiO_2.

2. MATERIALS AND METHODS

2.1 Materials

Commercial PLA (Ingeo™ Biopolymer 4043D) and PBAT (ecoflex® F Blend C1200) resins were the products of NatureWorks LLC (Minnetonka, MN, USA) and BASF Corporation (Ludwigshafen, Germany), respectively. Rutile titanium dioxide (TiO_2 with 95% purity) and zinc oxide (ZnO with 99.75% purity, particle size less than 44 μm) were purchased from Chemipan Corporation Co., Ltd. (Bangkok, Thailand). All components were used as received without further purification.

2.2 Film production

The PLA and PBAT resins were mixed with a fixed blending ratio of PLA:PBAT 85:15, using a 20-40 counter-rotating twin-screw extruder (Labtech Engineering LTE, Samutprakarn, Thailand) with a processing temperature range of 150–170°C and screw speed of 150–200 rpm. The addition of TiO_2 or ZnO was systematically varied as listed in Table 1. All compounds were pelletized and dried at 50°C for 1 d, before film blowing by a Labtech Engineering single-screw extruder LE-25-

30/C, equipped with a Labtech Engineering LF-400 blown film unit. The blown films were produced in the temperature range of 160–175°C with an annular die temperature of 160°C and a nip-roll speed of 2.7 m/min. The thickness of each film was measured at least in triplicate using a digital micrometer (Mitutoyo, Tokyo, Japan).

Table 1 Blending ratios of PLA/PBAT incorporating ZnO and TiO_2

Sample	PLA (%wt)	PBAT (%wt)	ZnO (%wt)	TiO_2 (%wt)
PLA/PBAT	85.00	15.00	-	-
PLA/PBAT/ZnO 1%	84.15	14.85	1.00	-
PLA/PBAT/ZnO 3%	82.45	14.55	3.00	-
PLA/PBAT/ZnO 5%	80.75	14.25	5.00	-
PLA/PBAT/TiO_2 1%	84.15	14.85	-	1.00
PLA/PBAT/TiO_2 3%	82.45	14.55	-	3.00
PLA/PBAT/TiO_2 5%	80.75	14.25	-	5.00
PLA/PBAT/ZnO + TiO_2 1%	84.15	14.85	0.50	0.50
PLA/PBAT/ZnO + TiO_2 3%	82.45	14.55	1.50	1.50
PLA/PBAT/ZnO + TiO_2 5%	80.75	14.25	2.50	2.50

2.3 Film properties

The light transmittance of the films was measured using an Evolution 350 UV-Vis Spectrophotometer (Thermo Fisher Scientific, Waltham, MA, USA) in the wavelength range of 200–800 nm. The color of the films in the CIELAB color space was measured by an UltraScan VIS spectrophotometer (HunterLab, Reston, VA, USA). Fourier Transform Infrared (FTIR) analysis coupled with an Attenuated Total Reflectance (ATR) was performed using a Bruker Tensor 27 Spectrometer (Bruker, San Jose, CA, USA) at wavenumbers of 500–4000 cm^{-1}, a resolution of 4 cm^{-1} and 32 scans.

2.4 Statistical analysis

Statistical analysis was conducted using SAS OnDemand for Academics software (SAS Institute Inc., North Carolina, USA).

3. RESULTS AND DISCUSSION

3.1 Film production

For PLA/PBAT incorporating ZnO and TiO_2 with different blending ratios (listed in Table 1), some of the resulting blown films were hard and brittle and some did not show good reproducibility. Only the finished flexible blown films with good reproducibility and consistency were used for further experiments. These films were PLA/PBAT, PLA/PBAT/ZnO 1%, PLA/PBAT/ZnO 3%, PLA/PBAT/TiO_2 1%, and PLA/PBAT/TiO_2 3%, as summarized in Table 2 together with their thicknesses. The rest of the samples were discarded.

Table 2 PLA/PBAT, PLA/PBAT/ZnO, and PLA/PBAT/TiO_2 blown films with good reproducibility and consistency

Samples	Thickness* (mm)
PLA/PBAT (Control)	0.054±0.004 [a]
PLA/PBAT/ZnO 1%	0.048±0.009 [ab]
PLA/PBAT/ZnO 3%	0.045±0.003 [b]
PLA/PBAT/TiO_2 1%	0.052±0.004 [a]
PLA/PBAT/TiO_2 3%	0.049±0.003 [ab]

* Values are shown as mean ± standard deviation with different letters indicating significant differences between groups (Tukey's HSD, $P<0.05$).

3.2 Film properties

Figure 1a) shows the light transmittance of the film samples. The films incorporating ZnO and TiO_2 have lower transmittance compared to the transmittance of the control PLA/PBAT film. The transmittances of all the films below 300 nm are approximately zero. Comparing the same amount of ZnO and TiO_2 (i.e., ZnO 1% versus TiO_2 1% and ZnO 3% versus TiO_2 3%), the films containing TiO_2 have lower transmittance than the films containing ZnO. Furthermore, the higher the amount of ZnO or TiO_2, the lower the transmittance. The films behave the same way in the UV region (below 400 nm). As shown in Figure 1b), the UV-light transmittance values of the PLA/PBAT/ZnO 1% and PLA/PBAT/ZnO 3% films are lower than 20% and 3%, respectively. It can be seen from the inset in Figure 1b) that the UV-light transmittance values of the PLA/PBAT/TiO_2 1% and PLA/PBAT/TiO_2 3% films are lower than 3% and 1%, respectively. With effective UV-shielding properties of the PLA/PBAT/ZnO and the PLA/PBAT/TiO_2 films, the films can be a good choice for packaging of UV-light sensitive foods and beverages, such as milk and dairy products, meat and meat products, vegetable oils, wine, and dried nuts [10].

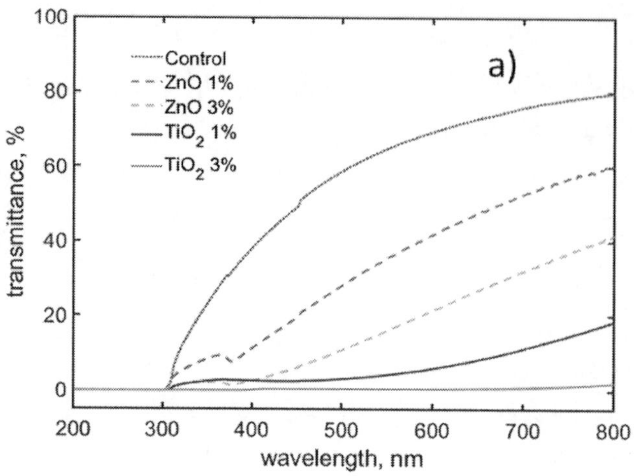

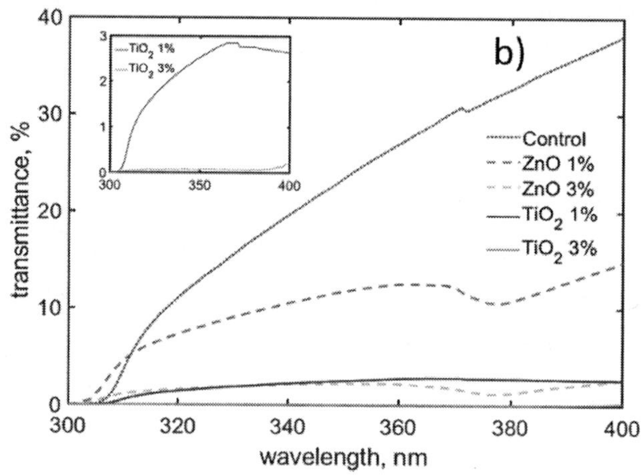

Figure 1 Light transmittance of the PLA/PBAT film (Control) and PLA/PBAT film incorporating ZnO and TiO_2 in a) the entire range and b) the UV range (below 400 nm). The inset shows enlarged transmittance of PLA/PBAT/TiO_2 films.

From the visual transparency of the PLA/PBAT films containing TiO_2 and ZnO shown in Figure 2, the film opacity increases as the amount of ZnO and TiO_2 increases, and the films with TiO_2 have higher opacity than the films with ZnO. The color measurement in CIELAB color space is shown in Figure 3. The a* values (red/green coordinate) among all the films are consistent. The b* values (yellow/blue coordinate) increase, showing the shift from blue toward yellow, as the amounts of metal oxides increase. The increasing L* values (light/dark coordinate) show the shift toward lighter color as the amounts of metal oxides increase. The opacity and the L* values of the films are in the following order: PLA/PBAT < PLA/PBAT/ZnO 1% < PLA/PBAT/ZnO 3% < PLA/PBAT/TiO_2 1% < PLA/PBAT/TiO_2 3%. The visual transparency, color, and light transmittance of the films are in good agreement, i.e., the higher the opacity, the higher the L* values, and the lower the transmittance. Regarding the films' appearances and colors, the films may not be suitable for products that require transparent packaging. The transparency of the films can be improved by reducing the particle sizes of the metal oxides [11,12].

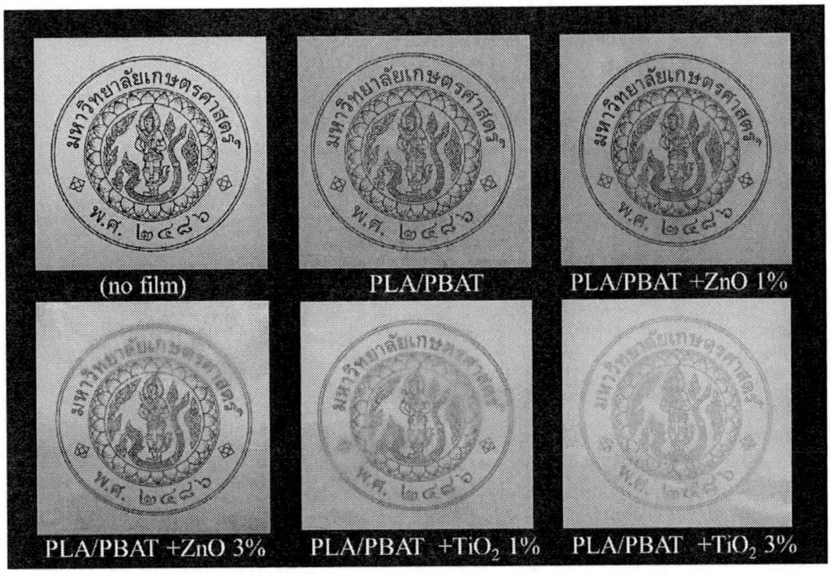

Figure 2 Visual transparency of the film samples

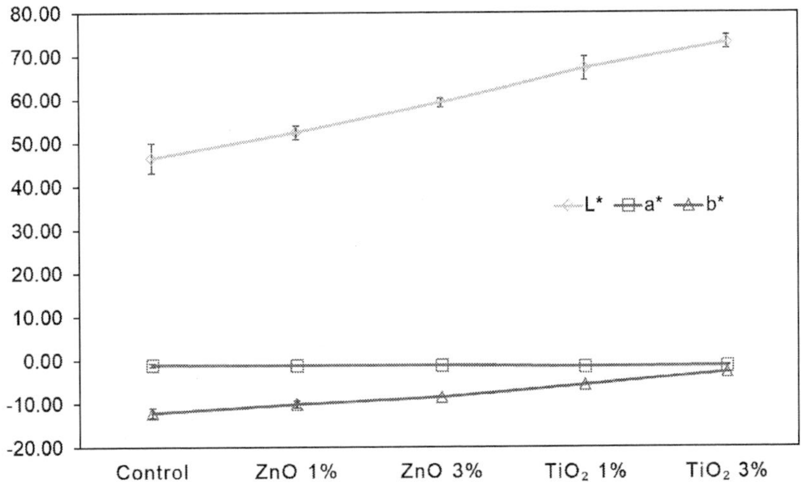

Figure 3 CIELAB color results.

The ATR-FTIR spectra of the films are shown in Figure 4. The peak positions of the films reflect the respective PLA and PBAT peaks [3]. The addition of ZnO or TiO_2 to PLA/PBAT film does not show a noticeable difference or band shifting when compared with the PLA/PBAT spectrum, implying no chemical interactions between the PLA/PBAT matrix and the metal oxides.

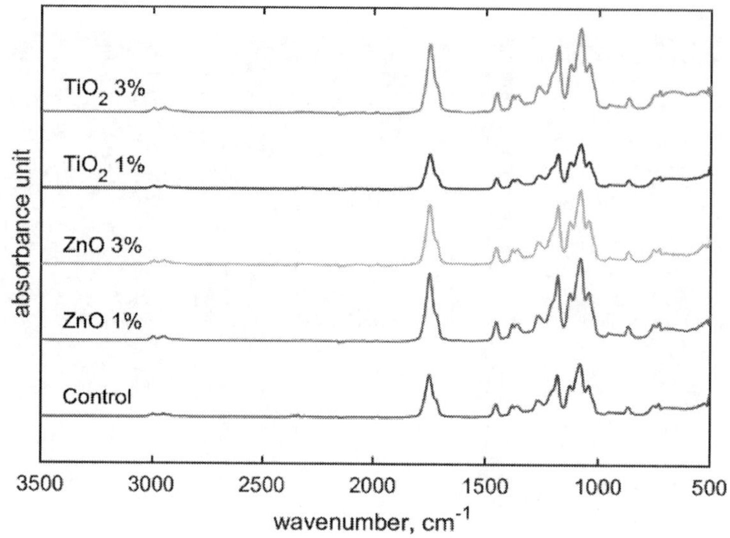

Figure 4 ATR-FTIR spectra of PLA/PBAT (Control), PLA/PBAT/ZnO, and PLA/PBA/TiO_2 films.

4. CONCLUSION AND REMARK

Both ZnO and TiO_2 contribute to improvement in the UV-shielding properties of the PLA/PBAT 85:15 films. Incorporating TiO_2 to PLA/PBAT results in films with higher opacity and better UV-shielding than adding ZnO at the same amount. However, the high opacity may limit the use of the TiO_2 incorporated films for products that require transparent packaging. It is important to note that both ZnO and TiO_2 are currently listed as a substance generally recognized as safe (GRAS) according to the US Food and Drug Administration (FDA). Nevertheless, the European Food Safety Authority (EFSA) announced in 2021 that TiO_2 was no longer considered a safe food additive [13]. Thus, using the films with ZnO and TiO_2 as a food contact material may require further investigation on safety.

5. ACKNOWLEDGEMENT

This research was supported by Kasetsart University Research and Development Institute (KURDI) FF(KU) 25.64.

6. REFERENCES

[1] Drago E, Campardelli R, Pettinato M, Perego P. Innovations in Smart Packaging Concepts for Food: An Extensive Review. Foods 2020;9:1628. https://doi.org/10.3390/foods9111628.

[2] Fu Y, Wu G, Bian X, Zeng J, Weng Y. Biodegradation Behavior of Poly(Butylene Adipate-Co-Terephthalate) (PBAT), Poly(Lactic Acid) (PLA), and Their Blend in Freshwater with Sediment. Molecules 2020;25:3946. https://doi.org/10.3390/molecules25173946.

[3] Weng YX, Jin YJ, Meng QY, Wang L, Zhang M, Wang YZ. Biodegradation behavior of poly(butylene adipate-co-terephthalate) (PBAT), poly(lactic acid) (PLA), and their blend under soil conditions. Polym Test 2013;32:918–26. https://doi.org/10.1016/j.polymertesting.2013.05.001.

[4] Choi B, Yoo S, Park S. Carbon Footprint of Packaging Films Made from LDPE, PLA, and PLA/PBAT Blends in South Korea. Sustainability 2018;10:2369. https://doi.org/10.3390/su10072369.

[5] Sancheti I, Katkar P, Thorat S, Radhakrishnan S, Kulkarni MB. PBAT/PLA/HNT filled blends for active food packaging. Mater Res Express 2020;6:125378. https://doi.org/10.1088/2053-1591/ab7c1f.

[6] Moustafa H, El Kissi N, Abou-Kandil AI, Abdel-Aziz MS, Dufresne A. PLA/PBAT Bionanocomposites with Antimicrobial Natural Rosin for Green Packaging. ACS Appl Mater Interfaces 2017;9:20132–41. https://doi.org/10.1021/acsami.7b05557.

[7] Liewchirakorn P, Aht-Ong D, Chinsirikul W. Practical Approach in Developing Desirable Peel-Seal and Clear Lidding Films Based on Poly(Lactic Acid) and Poly(Butylene Adipate-Co-Terephthalate) Blends. Packag Technol Sci 2018;31:296–309. https://doi.org/10.1002/pts.2321.

[8] Lizundia E, Penayo MC, Guinault A, Vilas JL, Domenek S. Impact of ZnO nanoparticle morphology on relaxation and transport properties of PLA nanocomposites. Polym Test 2019;75:175–84. https://doi.org/10.1016/j.polymertesting.2019.02.009.

[9] Yang T-C, Noguchi T, Isshiki M, Wu J-H. Effect of titanium dioxide on chemical and molecular changes in PVC sidings during QUV accelerated weathering. Polym Degrad Stab 2014;104:33–9. https://doi.org/10.1016/j.polymdegradstab.2014.03.023.

[10] Kwon S, Orsuwan A, Bumbudsanpharoke N, Yoon C, Choi J, Ko S. A Short Review of Light Barrier Materials for Food and Beverage Packaging. KOREAN J Packag Sci Technol 2018;24:141–8. https://doi.org/10.20909/kopast.2018.24.3.141.

[11] Zeljko M, Ocelić Bulatović V, Špada V, Blagojević SL. Environmentally Friendly UV-Protective Polyacrylate/TiO2 Nanocoatings. Polymers (Basel) 2021;13:2609. https://doi.org/10.3390/polym13162609.

[12] Sakamoto M, Okuda H, Futamata H, Sakai A, Iida M. Influence of Particle Size of Titanium Dioxide on UV-ray Shielding Property. KONA Powder Part J 1996;14:183–9. https://doi.org/10.14356/kona.1996025.

[13] Younes M, Aquilina G, Castle L, Engel K, Fowler P, Frutos Fernandez MJ, et al. Safety assessment of titanium dioxide (E171) as a food additive. EFSA J 2021;19. https://doi.org/10.2903/j.efsa.2021.6585.

AG-O05

Antimicrobial Paper Coated with Lignin Nanoparticles for Bread Packaging Applications

Nyein Chan Moe[1], Charinee Winotapun[2], Bongkot Hararak[2], Panitee Suwanamornlert[3], Pattarin Leelaphiwat[1], Vanee Chonhenchob[1]*

[1]Department of Packaging and Materials Technology, Faculty of Agro-Industry, Kasetsart University, Bangkok 10900, Thailand
[2]National Metal and Materials Technology Center, National Science and Technology Development Agency, Pathumthani 12120, Thailand
[3]College of Integrative Medicine, Dhurakij Pundit University, Bangkok 10210, Thailand

***Correspondence to:** Department of Packaging and Materials Technology, Faculty of Agro-Industry, Kasetsart University, Bangkok 10900, Thailand. E-mail: vanee.c@ku.ac.th*

ABSTRACT: Antimicrobial packaging has received much attention in the past decades for both research and commercialization. As lignin is abundantly available, sustainable and inexpensive, and possesses promising antimicrobial activities, it has become an attractive natural material for food packaging applications. This study was aimed to investigate lignin nanoparticles (LNPs) as an antimicrobial agent for paper coating for bread. Different concentrations (0, 0.5, 1.0, and 2.0 wt%) of LNPs were tested against major spoilage microorganisms in bread; *Aspergillus niger* and *Penicillium spp.* Interestingly, the highest antimicrobial activities against both microorganisms were 2 wt% LNPs. Then, 2 wt% LNPs and chitosan (CTS) solution was prepared for paper coating. Paper commonly used to make bread tray was coated with 2 wt% LNPs/chitosan solution. Mechanical and antimicrobial properties of the coated paper were determined. The 2 wt% LNPs/chitosan coating resulted in the greatest antimicrobial effect against mycelial growth. The 2 wt% LNPs/chitosan coated paper also exhibited the highest hydrophobicity. Tensile strength of the 2 wt% LNPs/chitosan coated paper was highest among others. Finally, bread samples were placed in the paper tray coated with 2 % LNPs/chitosan compared with commonly available paper tray and plastic bag for bread packaging. LDPE and 1% LNPs incorporated LDPE film were used as secondary package on application of bread. The results showed that the coated paper trays could delay fungal growth and extended the shelf life of bread.

Keyword: Antimicrobial, Lignin nanoparticles, Coating, Bread, Shelf life

1. INTRODUCTION

The incorporation of antimicrobial agents into packaging film is a promising way to control the growth of microorganisms in packaged foods. Lignin, the most abundant polymer next to cellulose, naturally possesses aromatic (phenolic) compound. Lignin as a heterogeneous biomacromolecule [1] is made up of an integrated network block of aromatic units derived from the radical coupling of its phenylpropanoid monomers [2] for active functions. Polyphenols in lignin provide not only antioxidant but also antimicrobial activity. Lignin effectively inhibited not only bacteria but also fungi such as *Aspergillus niger* [3]. Aside from possessing antimicrobial properties, lignin is low cost and widely available, making it an attractive alternative for chemical preservatives. Million tons of lignin are produced from the paper-making industry per year. Moreover, lignin has less or no detrimental effects on the environment since lignin is derived from plants.

The important characteristics of lignin nanoparticles to define its potential application are the chemical composition, size, shape, surface charge, hydrophobicity, and the presence of functional groups [2]. According to Chen [3], lignin nanoparticles can enhance good antimicrobial behaviors. Aside from naturally having antimicrobial and antioxidant properties, LNPs could affect positively the thermal stability and tensile strength of the film [2]. The inherent antimicrobial property of lignin could be used to prolong the shelf life of perishable food products such as bread. The demand for bread has gradually increased worldwide and the market is expected to grow annually. However, bread is prone to spoilage; without the use of preservatives, it has a short shelf life of 3 to 4 days, resulting in limited marketability and significant economic losses [4]. The high moisture content and water activity of bread favor the mold growth [5] mostly caused by fungi such as *Aspergillus niger* and *Penicillium* spp. which cause about 60% of spoilage in bakery products during storage [6]. Hence, this present study aimed to prolong the shelf life of bread by using lignin nanoparticles (LNPs) as an antifungal agent to inhibit the growth of *Aspergillus niger* and *Penicillium* spp, two of the most common fungi that cause deterioration in bread. The LNPs were coated on paper and used as tray in the packaging system for the bread. Moreover, neat LDPE and LDPE incorporated with 1 % LNPs were used as the secondary packaging material. Results of this study could be explored further to extend shelf life, not only for bread but also for other perishable food products.

2. MATERIALS AND METHODS

2.1 Materials

Lignin nanoparticles suspension and chitosan flakes with 90 % deacetylation degree were provided by National Metal and Materials Technology Center (MTEC), Thailand. Lactic acid (85 %) was purchased from KEMAUS, Australia. Potato dextrose agar (PDA) and peptone water were purchased from Sigma-Aldrich (St Louis, MO, USA). Commercial paperboard substrate with a grammage of 250 g/m^2 was purchased from a bakery shop in Bangkok, Thailand. LDPE films (neat and incorporated with 1 % LNPs) were provided by MTEC, Thailand.

2.2 Methods

2.2.1 Preparation of coating solution and coating on paper

Two grams of chitosan was dissolved in 100 ml 1 % (v/v) lactic acid using a magnetic stirrer for 24 hours at ambient condition. The chitosan-lactic acid solution and the previously prepared LNPs suspensions with concentrations of 0.5, 1.0, and 2.0 wt% were sterilized at 121 °C for 15 minutes. Afterwards, chitosan and LNPs suspension were homogenized for 30 minutes using a hot plate stirrer set at 60 °C and 300 rpm. The coating solution was coded as 0.5%LNPs/CTS, 1%LNPs/CTS, and 2%LNPs/CTS.

Three ml of the coating solutions were used to coat the paper by using a rod coater (PI-1210 Film Coater, Japan) with a coating speed of 50 mm/min, at ambient condition. Coated papers were dried in an oven (UNE 200; Memmert, Schwabach, Germany) at 80 °C for 1 minute.

2.2.2 Antifungal activity of coating solutions

Poisoned food method was used to determine the antifungal activity of the coating solutions. Potato dextrose agar (PDA) powder was mixed with LNPs/CTS solutions. In addition, the inhibition of pure chitosan (2 wt%) and pure LNPs (0.5 wt%) was studied for comparison. The new mixture was then sterilized, cooled down, poured into Petri dishes, and let it solidified.

An 8 mm mycelial plug taken from the edge of an actively growing pure culture of *Aspergillus niger* or *Penicillium* spp was placed at the center of the Petri dish containing the solidified LNPs/CTS solutions with PDA. Pure PDA served as the control. Petri dishes were then sealed immediately using parafilm. All prepared samples were incubated in an ambient condition. The antifungal activity was quantified by taking the average of the perpendicular diameters of the mycelial growth using Vernier caliper, and applied in the following equation:

$$\text{Antifungal activity (\%)} = ((Dc-Dt)/Dc) \times 100 \qquad (1)$$

where Dc, diameter of mycelial growth in control plate
Dt, diameter of mycelial growth with antifungal agent

2.2.3 Characterization of the coated paper

Prior to the characterization of the coated paper samples, all the coated papers were pre-conditioned for one day in a desiccator (27 °C ± 1; 65 ± 2% RH) as indicated in ISO 187.

2.2.3.a Surface hydrophobicity. The moisture sensitivity of the surface of the coated paper was determined using a water contact angle meter (Dataphysics OCA15EC, Dataphysics Instruments GmbH, Germany). Briefly, 3 μl was dropped on a sample size of 1 x 10 cm fixed in a sample holder.

2.2.3.b Mechanical properties. Tensile strength, Young's modulus, and elongation of coated papers with specimen size of 15 x 100 mm were determined using Instron Universal Testing Machine (Model 5965, Instron, USA) in accordance with ASTM D828.

2.2.4 Packaging testing on bread

2.2.4.a Sample preparation. The coating solution that showed the highest antifungal activity in 2.2.2 was used in the packaging testing of bread. Further, neat LDPE and LDPE incorporated with 1 % LNPs were used to packaged bread samples. As primary packaging, paper coated with the antifungal solution was used as a tray to contain bread samples. In brief, the following were the treatments imposed:

Control	-	Neat LDPE
Treatment 1	-	Neat LDPE with LNPs/CTS coated paper (LDPE-LNPs/CTS coated paper)
Treatment 2	-	LDPE incorporated with 1% LNPs with LNPs/CTS coated paper (LNPs/LDPE-LNPs/CTS coated paper)

A slice of bread was packed following the treatments mentioned above. Three samples were prepared for each treatment. All samples were kept under ambient condition.

2.2.4.b Determination of bread quality. The quality of the bread was monitored through (1) In-package gas composition, (2) moisture content (MC) and water activity (a_w), and (3) mold growth through visual inspection and yeast and mold count analysis. In-package gas composition (O_2 and CO_2) was determined every day for up to 10 days by using a Headspace Gas Analyzer (Dansensor, Model- CheckMate 3, Denmark). Moisture content was determined using a moisture analyzer set at 105 °C. On the other hand, water activity was determined using a water activity analyzer (Testo 650, Testo Inc., USA). Both MC and a_w were monitored at 3, 5, 7, and 9 days. Visual inspection of mold growth was done also at 3, 5, 7, and 9 days, and was supported by yeast and mold count analysis.

2.2.5 Statistical analysis

Analysis of variance (ANOVA) among samples was calculated by using SPSS and Duncan's new multiple range test was performed at a significance level of p □ 0.05.

3. RESULTS AND DISCUSSION

3.1 Antifungal activity of coating solutions

The antifungal activity of coating solutions showed differently against *Aspergillus niger* and *Penicillium* spp. The stronger antifungal activity against *Aspergillus niger* was observed in coating solutions with LNPs in which a higher concentration of LNPs shows stronger antifungal activity. Meanwhile, the inhibition of pure chitosan and pure LNPs showed less inhibition. Against *Penicillium* spp, 1 % LNPs/CTS showed the strongest inhibition up to 5 days, after that 2 % LNPs/CTS showed the greatest inhibition up to 7 days. Similarly, pure chitosan and pure LNPs showed less inhibition against *Penicillium* spp. The results implied that the presence of LNPs in the coating solution contributes to the increased antifungal activity, with the increasing concentration up to 2 %.

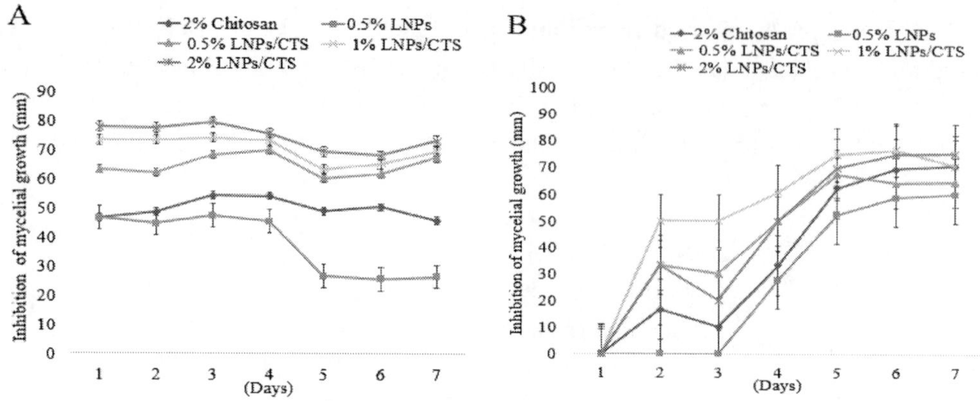

Figure 1 Antifungal activity of different coating solutions against (a) *Aspergillus niger* and (b) *Penicillium* spp

3.2 Characterization of the coated paper

3.2.1 Surface hydrophobicity

One of the beneficial properties of lignin is its hydrophobicity. Compared with pure 2 wt% chitosan, 0.5 wt% LNPs can significantly increase the surface hydrophobicity of the coated paper. Moreover, the effect of mixing chitosan and LNPs together increased further the surface hydrophobicity of the coated paper on which higher concentration of LNPs up to 2 wt% showed higher hydrophobicity.

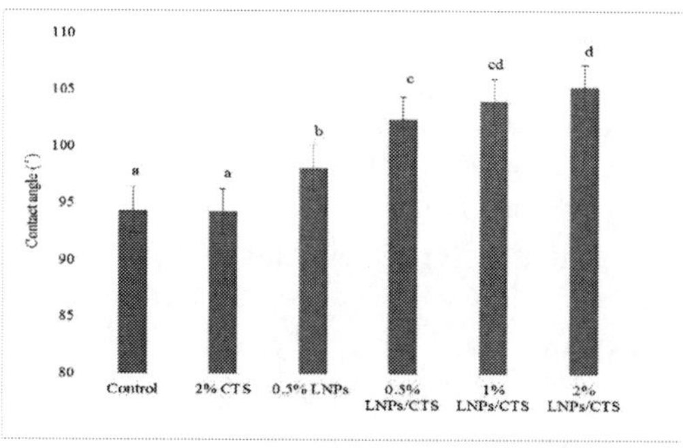

Figure 2 Surface hydrophobicity of coated paper as affected by LNPs concentration

3.2.2 Mechanical properties

The blend of chitosan and LNPs significantly increased the tensile strength and Young's modulus of the coated paper. It is evident because pure chitosan and LNPs are almost similar to the uncoated paper (control). Paper samples coated with 2% LNPs/CTS showed the highest tensile strength and Young's modulus of 59.14 MPa and 2705.46 MPa, respectively. This could be the effect of benzene rings present in the lignin, which provide rigidity.

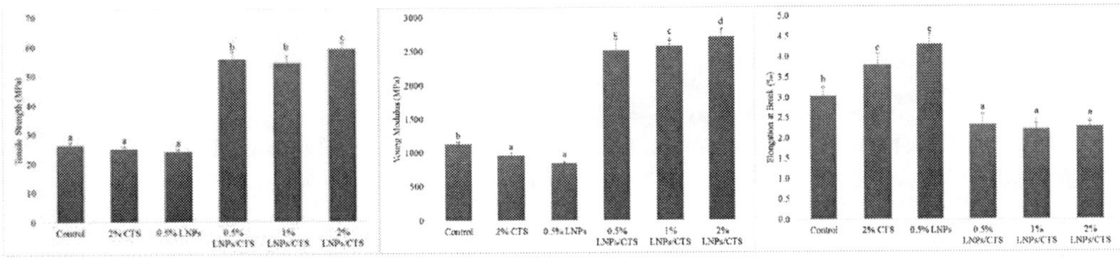

Figure 3 Properties of Tensile Strength, Young Modulus and Elongation of LNPs coated paper

3.3 Packaging testing on bread

3.3.1 In-package gas composition

In-package gas composition can be associated with the microbial count as microbes consumed O_2 and generated CO_2 during their growth. It was observed that O_2 concentration started to decrease on day 7 for both neat LDPE and neat LDPE with LNPs/CTS coated paper while O_2 concentration in LDPE incorporated with 1% LNPs with LNPs/CTS coated paper decreased on day 8. This is an indicator that there is a higher count of microbes present in the first two treatments. Similarly, the increasing concentration of CO_2 implied the increased number of microbes present in the samples. Among the treatments, LDPE incorporated with 1% LNPs with LNPs/CTS coated paper showed the lowest percentage of CO_2. With this result, it can be stated the presence of LNPs both in the film and on the coated paper slows down the proliferation of molds in the bread.

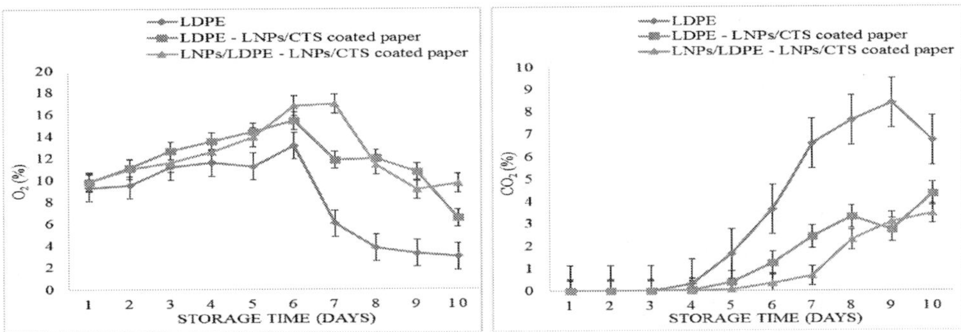

Figure 4 In-package gas composition

3.3.2 Moisture content (MC) and water activity (a_w)

A slightly decreased MC of all bread samples was observed. This is due to the good moisture barrier property of LDPE. For a_w, no significant difference was observed among treatments. But then, it was notable that a_w of bread samples under treatment 2 decreased from day 0 to day 5, but a_w of both control and treatment 1 continuously increased after day 3. Also, it should be noted that the initial a_w of the bread is high because the bread used has no preservatives added.

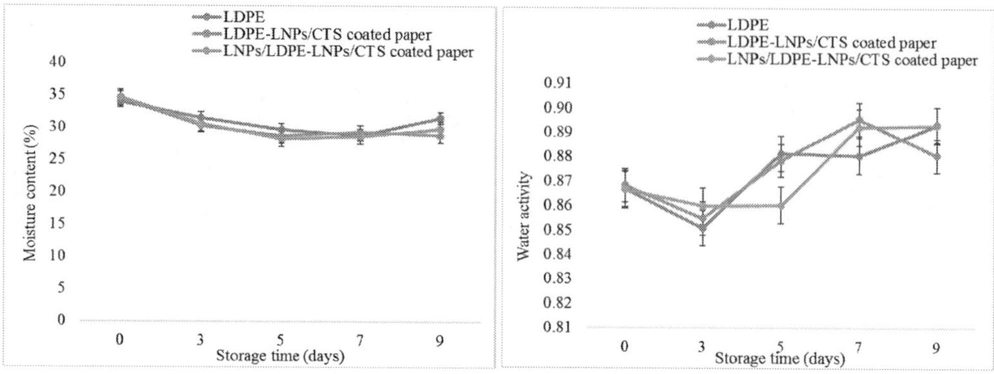

Figure 5 Moisture content and water activity of bread samples

3.3.3 Mold growth

Visually, mold growth was observed in bread under treatment 1 for 4 days longer compared with control which molds were observed on day 5. Moreover, there was no mold growth observed in bread under treatment 2 up to day 9, as shown in Figure 6. Based on this observation, the shelf life of bread is significantly extended with the application of LNPs in the LDPE film and in the coating solution.

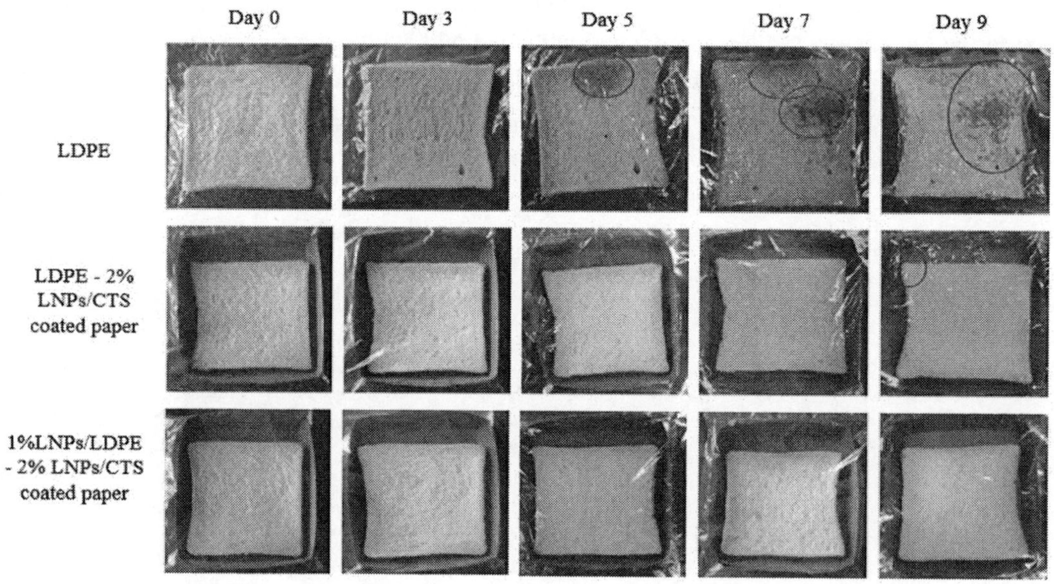

Figure 6 Visual appearance and mold growth on bread. Circles in red indicate the area of mold growth.

The visual appearance and the mold growth of the bread samples can be correlated with the yeast and mold count presented in Table 1. On day 9, control samples have the highest yeast and mold count, while bread samples under treatments 1 and 2, have a far lower count of yeast and mold.

Table 1. Yeast and mold count of bread samples

Films	Yeast & Molds (Log cfu/g)				
	Day 0	Day 3	Day 5	Day 7	Day 9
LDPE	ND	3.5	3.6	3.6	4.9
LDPE-2% LNPs coated paper	ND	3.4	3.5	3.6	3.8
1% LNPs/LDPE-2% LNPs coated paper	ND	3.4	3.5	3.5	3.7

ND: Not detectable

4. CONCLUSIONS

Lignin nanoparticles is a promising antifungal agent against the common cause of bread spoilage: *Aspergillus niger* and *Penicillium* spp. This study revealed that using paper coated with LNPs and applied in the packaging system could help to extend the shelf life of the bread. Moreover, LNPs can improve the properties of coated paper. Overall, LNPs showed a beneficial effect in the active packaging of bread due to its natural antifungal property.

5. REFERENCES

[1] Ajao, O., Jeaidi, J., Benali, M., Restrepo, A. M., El Mehdi, N., & Boumghar, Y. (2018). Quantification and variability analysis of lignin optical properties for colour-dependent industrial applications. Molecules, 23(2), 377. DOI:10.3390/molecules23020377

[2] Schneider, W. D. H., Dillon, A. J. P., & Camassola, M. (2021). Lignin nanoparticles enter the scene: A promising versatile green tool for multiple applications. Biotechnology Advances, 47, 107685. DOI: 10.1016/j.biotechadv.2020.107685

[3] Chen, S., Chen, H., Yang, S., & Fan, D. (2021). Developing an antifungal and high-strength soy protein-based adhesive modified by lignin-based polymer. Industrial Crops and Products, 170, 113795. DOI: 10.1016/j.indcrop.2021.113795

[4] Suwanamornlert, P., Kerddonfag, N., Sane, A., Chinsirikul, W., Zhou, W., & Chonhenchob, V. (2020). Poly (lactic acid)/poly (butylene-succinate-co-adipate) (PLA/PBSA) blend films containing thymol as alternative to synthetic preservatives for active packaging of bread. Food Packaging and Shelf Life, 25, 100515. DOI: 10.1016/j.fpsl.2020.100515

[5] Nionelli, L., Wang, Y., Pontonio, E., Immonen, M., Rizzello, C. G., Maina, H. N., ... & Coda, R. (2020). Antifungal effect of bioprocessed surplus bread as ingredient for bread-making: Identification of active compounds and impact on shelf-life. Food Control, 118, 107437. DOI: 10.1016/j.foodcont.2020.107437

[6] Noshirvani, N., Ghanbarzadeh, B., Mokarram, R. R., & Hashemi, M. (2017). Novel active packaging based on carboxymethyl cellulose-chitosan-ZnO NPs nanocomposite for increasing the shelf life of bread. Food Packaging and Shelf Life, 11, 106-114. DOI: 10.1016/j.fpsl.2017.01.010

AG-O06

Development of Antifungal Film Incorporated with Lignin Nanoparticles for Mango Packaging and Distribution

Thanakrit Ananphadung[1], Bongkot Hararak[2], Charinee Winotapun[2], Kanchana Boonruang[3], Pattarin Leelaphiwat[1], Wannee Chinsirikul[4], Vanee Chonhenchob[1]*

[1]Department of Packaging and Materials Technology, Kasetsart University, Bangkok, 10900, Thailand
[2]National Metal and Materials Technology Center, National Science and Technology Development Agency, Pathum Thani, 12120, Thailand
[3]Department of Horticulture, Kasetsart University, Bangkok, 10900, Thailand
[4]National Nanotechnology Center, National Science and Technology Development Agency, Pathum Thani, 12120, Thailand

***Correspondence to:** Department of Packaging and Materials Technology, Kasetsart University, Chatuchak, Bangkok, 10900, Thailand.*
E-mail address: vanee.c@ku.ac.th (V. Chonhenchob)

ABSTRACT: Lignin nanoparticles (LNPs) has received worldwide attention in recent years due to its sustainability, renewability, and abundant presence in nature, combined with its antimicrobial and antioxidant activities and the attempt to explore high value-added lignin applications. There has been a limited number of studies describing the use of lignin in fresh produce applications. Hence, this study aimed to develop the breathable film incorporated with LNPs to take advantage of its promising antimicrobial effects on prolonging the shelf life of mango fruit during distribution and export. The antifungal activities were tested against *Colletotrichum gloeosporioides*, which is a major pathogen causing anthracnose disease leading to a major loss of mango fruit. The film containing LNPs had the highest antifungal activities in vitro as compared to the control film without LNPs. The fruit packed in LNPs film and placed in the box had the longest storage life, followed by that in control film and without film, respectively. The LNPs films could potentially be used to extend the storage life of mango fruit suitable for storage, distribution, and export.

Keyword: Lignin nanoparticles, Antifungal film, Postharvest, Essential oil

1. INTRODUCTION

Mango (*Mangifara indica.* L) is well known as an economically important fruit in Thailand. It is highly nutritious, full of bioactive compounds (e.g. carotenoids, vitamin C) with an attractive flavor and taste. The storage life of mango for export should be at least 3 weeks to have sufficient time for transport and distribution. Anthracnose is a major postharvest disease of many tropical and subtropical fruits, including mango, thus affecting the quality and shelf-life [1]. The infection usually occurs in the pre-harvest stage when the spores of the fungus or mycelia infect the fruit at the early development stage. Generally, preharvest infection remains or hidden throughout the fruit development stage until ripening [2] and it will be visible during storage and distribution, hence the occurrence of spoilage. The spoilage during storage and distribution or in the entire supply chain that can reach up to 50 % [4] is a major contributor to economic losses.

Lignin is the most abundant and renewable aromatic biopolymer on earth [5]. Aside from being renewable and biodegradable, it offers many unique functions that are suitable for packaging applications such as antimicrobial and antioxidant properties [6], and UV-shielding [7]. It has been reported that lignin composite films could be formed by melt-extrusion process [8]. Several studies developed packaging films incorporated with micro-scale lignin. These films were reported to possess radical scavenging activity and showed resistance to common pathogens [9], however, the mechanical properties tended to be affected [10,11]. Nevertheless, micro-scale lignin could be processed into lignin nanoparticles (LNPs) through available methods. Due to its high surface area and higher reactivity, LNPs can improve film properties.

Cinnamaldehyde (CNs) a safe, biodegradable, environment-friendly, and natural volatile essential oil has been widely studied for its antimicrobial properties. Generally, the volatile components in essential oils are a mixture of aromatic and aliphatic compounds commonly used in the food industry [13]. Recently, many researchers have reported that CNs is an effective antifungal compound against spoilage caused by anthracnose and other diseases in the postharvest of mango and other fruits [3,14,15,16]. Many studies have revealed that the quality of essential oils is based on their compositions, which are influenced by the extraction methods [17,18]. Hence, it is necessary to optimize the concentration of essential oils that are effective to preserve the quality of mango at postharvest.

The objectives of the present study were to develop biodegradable PBS-based films and determine the inhibition effect of LNPs and CNs on the growth of *Colletotrichum gloeosporioides*, in order to design packaging films that are suitable for mango fruits during storage, distribution, and export, hence prolonging the shelf life.

2. MATERIALS AND METHODS

2.1 Materials

Commercial biodegradable polybutylene succinate (PBS) was purchased from PPT Global Chemical Public Company Limited (Bangkok, Thailand). Softwood kraft lignin (BioPivaTM100) was purchased from UPM Biochemicals (Helsinki, Finland), and was transformed into nanoparticles by National Metal and Materials Technology Center (Pathumthani, Thailand). FCC grade (≥ 98%) *trans*-Cinnamaldehyde was purchased from Sigma-Aldrich (St Louis, USA). Pure culture of *C. gloeosporioides* isolated from mango was purchased from Plant Protection Research and Development Office, Bangkok, Thailand.

2.2 Methods

2.2.1 Preparation of films

Poly(butylene-succinate) (PBS) pellets were dried and mixed with dry LNPs powder at a weight ratio of 98/2. Pellets were physically blended using a double screw extruder (Haake Polylab OS, Germany). Temperature profile from the feeding zone to the die was 110/130/140/150/154/155/155/160 °C at a screw speed of 100 rpm. PBS/CNs masterbatch was prepared using an internal mixer (Haake Rheomex OS, Germany) of 310 cm^2 with a fill factor 0f 0.8, and rotor speed of 60 rpm, at 120 °C for 12 min. Temperature profile from feeding to mixer was 120/120/120. The 20 wt%. Cinnamaldehyde (CNs) was feed into the internal mixer once PBS was in the melt state. The bulk polymer blend was crushed into pellets. All masterbatches were mixed in the following film concentration: PBS/1%LNPs, PBS/10%CNs, PBS/1%LNPs,/7.5%CNs, and PBS/1%LNPs,/10%CNs. The mixture was introduced into the hopper of a blown film single screw extruder (Thermo Scientific™ HAAKE Rheomex for the HAAKE Polylab OS torque rheometer platform, German) connected with annular die to form a film. The temperature profile used for feeding, compression, metering and die zone was 150/155/160 and 160°C. Thickness of the films (30 ± 5 μm) was measured randomly at different points. Neat PBS film were prepared as control. Formulations of blown films is summarized as follows:

Table 1 Composition of PBS-LNPs, PBS-CNs and PBS-LNPs-CNs blends

Samples	PBS (wt%)	LNPs (wt%)	CNs (wt%)	Code
1	99	1	-	PBS/1%LNPs
2	91.5	1	7.5	PBS/1%LNPs/7.5%CNs
3	89	1	10	PBS/1%LNPs/10%CNs

2.2.2 Determination of antifungal activity of LNPs

Poisoned-food method as described by Balouiri [19] was followed. Mycelial plug with an 8 mm diameter was taken from the edge of actively growing pure culture of *C.gloeosporioides*, then placed on the center of prepared sterile potato dextrose agar (PDA) containing 1% LNPs. All prepared samples were incubated at 25 °C. Perpendicular diameters of the mycelium were monitored using a Vernier caliper every 3 days for 15 days. The antimicrobial activity was calculated using Eq. 1

$$\text{Inhibition percentage } (\%) = 100 - \frac{\mu_P}{\mu_V} \times 100 \qquad (1)$$

where μ_p is the maximum growth rate with film sample and μ_r is the maximum growth rate of control.

2.2.3 Determination of antifungal activity of the films

The antimicrobial activity of the films was determined by vapor contact assay as described by Soylu [20]. Preparation and incubation of samples and quantification of the antifungal activity described in 2.2.2 were followed except that the monitoring was done every 2 days for 6 days. Additionally, a 2.5 cm diameter cut of film sample was fixed inside the cover of the Petri dish, then immediately sealed with Parafilm. PBS films without any blended compound served as the control.

2.2.5 Statistical analysis

Analysis of variance (ANOVA) was performed at a significance level of $p \leq 0.05$ using SPSS software (SPSS Inc., Chicago, IL, USA).

3. RESULTS AND DISCUSSION

3.1 Antifungal activity of LNPs

The antifungal activity of LNPs was conducted *in vitro*. It was observed that there was a significant difference in the mycelial growth of *C.gloeosporioides* between control and with 1 % LNPs. The results revealed that LNPs with concentration as low as 1 wt% showed good antimicrobial activity against *C.gloeosporioides*. This can be equated to the large surface area of LNPs for reactivity. Also, the direct contact of *C.gloeosporioides* with the functional groups responsible for antifungal activity in LNPs makes it effective to slow down the growth of the mycelium. Translating the difference in mycelial growth, LNPs showed a growth inhibition as high as > 60 % on day 3. Starting day 6, the inhibition decreased but interestingly, inhibition was sustained at > 50 % up to 10 days. At the end of the observation period, day 15, growth inhibition was still observed with 30 %.

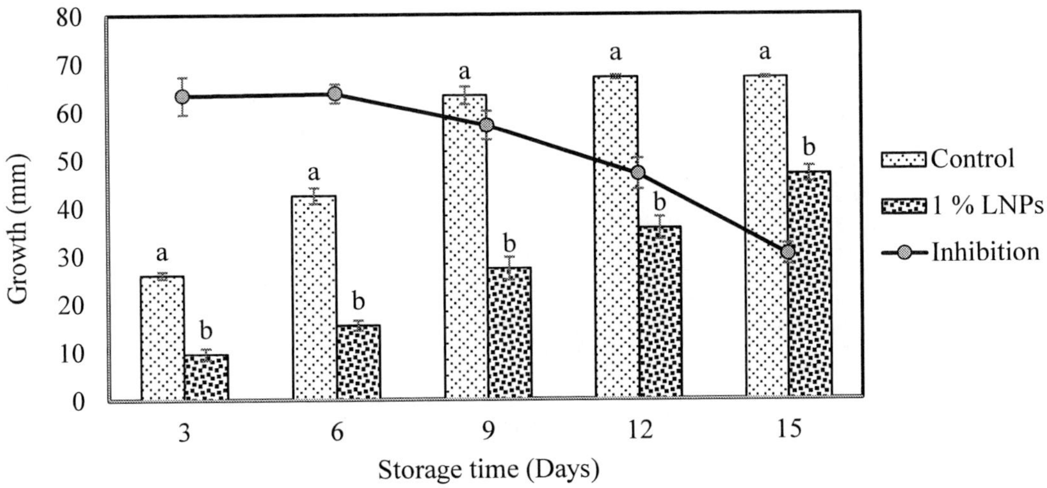

Figure 1 Antifungal activity of 1%LNPs against *C.gloeosporioides*

3.2 Antifungal activity of the films

The antifungal activity of film samples was monitored until 6 days. Compared to the results in 3.1, the antifungal activity of 1 % LNPs incorporated in PBS was low ranging from < 20 % on day 2 down to < 5 % on day 6. This is because LNPs is not a volatile compound, aside from its natural state which is solid. Also, lignin has a highly complex molecular structure on which its functional groups could not diffuse out easily and act the full function as an antifungal. The addition of cinnamaldehyde helped in the growth inhibition of *C.gloeosporioides*. It was expected that cinnamaldehyde alone (PBS/10%CNs) has high antimicrobial activity but cinnamaldehyde plus LNPs showed stronger inhibition. At day 2, PBS/1%LNPs/10%CNs showed the highest inhibition followed by PBS/1%LNPs/7.5%CNs. However, on days 4 and 6, lower concentration of CNs in the PBS composite e.g. 7.5 % showed a stronger inhibition compared to 10 % CNs in the PBS composite. The possible reason could be traced to the miscibility between cinnamaldehyde and PBS. It is interesting, however, that a lower concentration of cinnamaldehyde with LNPs (PBS/1%LNPs/10%CNs) showed stronger antifungal activity than the higher concentration of cinnamaldehyde alone (PBS/10%CNs). The presence of LNPs in the composite contributed to the antifungal activity. Also, the LNPs could serve as nanocapsule on which it protected the cinnamaldehyde from volatilization during the processing of film. Possibly, the remaining cinnamaldehyde in PBS/1%LNPs/10%CNs was higher than PBS/10%CNs, after film processing.

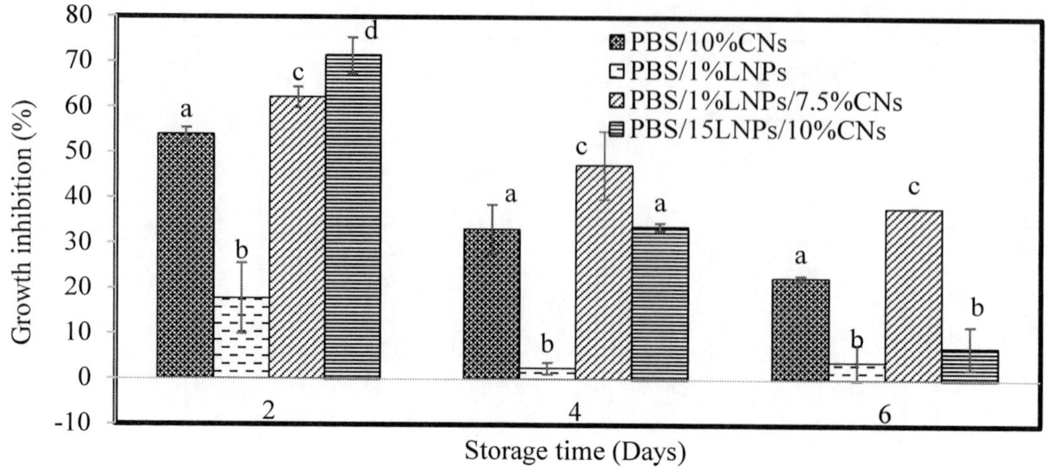

Figure 2. Antifungal activity of films against *C.gloeosporioides*

4. CONCLUSIONS

LNPs possesses strong antifungal property against *C.gloeosporioides* as shown in the *in vitro* testing using PDA as the growth medium. However, LNPs showed low antifungal activity when it was incorporated into PBS film. Aside from its natural state which is solid, it is not a volatile compound. Moreover, the highly complex molecular structure of lignin makes it difficult for the functional groups responsible for antimicrobial property to diffuse. Nevertheless, the result of this study showed that LNPs could enhance the antifungal property of a film incorporated with cinnamaldehyde. Hence, the antifungal film developed in this study can be used for the packaging of mango during storage and distribution.

5. ACKNOWLEDGEMENT

Authors would like to thank the National Metal and Materials Technology Center (MTEC) and the Department of Packaging and Materials Technology, Faculty of Agro Industry, Kasetsart University for their materials and financial support.

6. REFERENCES

[1] Boonruang K, Kerddonfag N, Chinsirikul W, Mitcham E.J. and Chonhenchob V. (2017). Antifungal effect of poly(lactic acid) films containing thymol and R-(-)-carvone against anthracnose pathogens isolated from avocado and citrus. Food Control, 78, pp. 85-93, DOI: 10.1016/j.foodcont.2017.02.032

[2] Dinh SQ, Chongwungse J, Pongam P. and Sangchote S. (2003). Fruit infection by Colletotrichum gloeosporioides and anthracnose resistance of some mango cultivas in Thailand. Australasian Plant Pathology, 32, pp. 533-538, DOI: 10.1071/AP03053

[3] Perumal AB, Nambiar RB, Sellamuthu P.S. and Emmanuel R.S. (2021). Use of modified atmosphere packaging combined with essential oils for prolonging post-harvest shelf life of mango (cv. Banganapalli and cv. Totapuri). LWT. 148. 111662. DOI: 10.1016/j.lwt.2021.111662.

[4] Janisiewicz W.J. and Korsten L. (2002). Biological control of postharvest diseases of fruits. Annu Rev Phytopathol, 2002, 40:411-4, DOI: 10.1146/annurev. Phyto.40.120401.130158

[5] Calvo-Flores FG, Dobado JA, Isac-García J. and Matín-Martínez FJ. (2015). Lignin and lignans as renewable raw materials: chemistry, technology and applications. John Wiley & Sons Ltd

[6] Yang W, Owczarek JS, Fortunati E, Kozanecki M, Mazzaglia A, Balestra GM, Kenny JM, Torre L. and Puglia D. (2016). Antioxidant and antibacterial lignin nanoparticles in polyvinyl alcohol/chitosan films for active packaging. Industrial Crops and Products, 94, DOI: 10.1016/j.indcrop.2016.09.061

[7] Zhang X, Liu W, Liu W. and Qiu X. (2020). High performance PVA/lignin nanocomposite films with excellent water vapor barrier and UV-shielding properties. International Journal of Biological Macromolecules, 142, pp. 551-558, DOI: 10.1016/j.ijbiomac.2019.09.129

[8] Xing Q, Buono P, Ruch D, Dubois P, Wu L. and Wang W. (2019). Biodegradable UV-Blocking Films through Core–Shell Lignin–Melanin Nanoparticles in Poly(butylene adipate-co-terephthalate). ACS Sustainable Chemistry & Engineering, 7,DOI: 10.1021/acssuschemeng.8b05755.

[9] Domínguez-Robles J, Larrañeta E, Fong ML, Martin NK, Irwin NJ, Mutjé P, Tarrés Q. and Delgado-Aguilar M. (2020). Lignin/poly(butylene succinate) composites with antioxidant and antibacterial properties for potential biomedical applications. International Journal of Biological Macromolecules, 145, pp. 92-99, DOI: 10.1016/j.ijbiomac.2019.12.146.

[10] Alexy P, Košíková B. and Podstránska G. (2000). The effect of blending lignin with polyethylene and polypropylene on physical properties. Polymer, 41:13, pp. 4901-4908, DOI: 10.1016/S0032-3861(99)00714-4.

[11] Xiao Z, Liu W, Liu W. and Qiu X. (2019). High performance PVA/lignin nanocomposite films with excellent water vapor barrier and UV-shielding properties. International Journal of Biological Macromolecules, 142, DOI: 10.1016/j.ijbiomac.2019.09.129.

[12] Chauhan P.S. (2020). Lignin nanoparticles: Eco-friendly and versatile tool for new era. Bioresource Technology Reports, 9, 100374, DOI: 10.1016/j.biteb.2019.100374.

[13] Bakkali F, Averbeck S, Averbeck D. and Idaomar M. (2008). Biological effects of essential oils – A review. Food and Chemical Toxicology, 46:2, pp. 446-475, DOI: 10.1016/j.fct.2007.09.106.

[14] Sivakumar D, Wijeratnam RSW, Wijesundera R.L.C. and Abeyesekere M. (2002).Control of postharvest diseases of rambutan using cinnamaldehyde. Crop Protection, 21:9, pp. 847-852, DOI:10.1016/S0261-2194(02)00051-0.

[15] Ummarat N. and Seraypheap K. (2021). Application of Essential Oils for Maintaining Postharvest Quality of 'Rongrien' Rambutan Fruit. Agriculture, 11, 1204, DOI: 10.3390/agriculture11121204.

[16] Jardim I, Denilson O, Geraldo S, Vicente C. and Paulo S. (2018). (E)-cinnamaldehyde from the essential oil of Cinnamomum cassia controls Meloidogyne incognita in soybean plants. Journal of Pest Science, 91, DOI: 10.1007/s10340-017-0850-3.

[17] Schaneberg B. and Ikhlas K. (2002). Comparison of Extraction Methods for Maker Compounds in the Essential Oil of Lemon Grass by GC. Journal of agricultural and food chemistry, 50, 1345-9, DOI: 10.1021/jf011078h.

[18] Reyes-Jurado F, Franco-Vega A, Ramírez-Corona N, Palou E. and Lo´pez-Malo A. Essential Oils: Antimicrobials Activities, Extraction Methods, and Their Modeling. Food Engineering Reviews, 7, DOI: 10.1007/s12393-014-9099-2.

[19] Balouiri M, Sadiki M. and Ibnsouda S.K. (2016). Methods for in vitro evaluating antimicrobial activity: A review, Journal of Pharmaceutical Analysis, 6:2, pp.71-79, DOI: 10.1016/j.jpha.2015.11.005.

[20] Soylu EM, Kurt Ş. and Soylu S. (2010). In vitro and in vivo antifungal activity of essential oils of various plants against tomato grey mould disease agent Botrytis cinerea. International journal of food microbiology, 143, 183-9, DOI: 10.1016/j.ijfoodmicro.2010.08.015.

AG-O07

Releasing Active Food Packaging: Can Natural Extracts be a Safety Concern?

Ana Sanches Silva[1,2,3]*, Letricia Pereira Barbosa[4], Raquel Sendón[4]

[1]National Institute for Agrarian and Veterinary Research (INIAV), I.P., 4485-655 Vairão, Vila do Conde, Portugal;
[2]Center for Study in Animal Science (CECA), ICETA, University of Porto, 4051-051 Porto, Portugal;
[3]University of Coimbra, Faculty of Pharmacy, 3000-548 Coimbra, Portugal;
[4]Analytical Chemistry, Nutrition and Food Science Department, Pharmacy Faculty, University of Santiago de Compostela, 15782 Santiago de Compostela, Spain;

***Correspondence to:** National Institute for Agrarian and Veterinary Research (INIAV), I.P., Rua dos Lagidos, Lugar da Madalena, 4485-655 Vairão, Vila do Conde, Portugal; E-mail: ana.silva@iniav.pt*

ABSTRACT: Active food packaging has originated a revolution in the packaging world, allowing a positive interaction with food. Since this concept appeared, it has already suffered progresses and nowadays releasing active packaging tends to be more sustainable and associate with natural additives and biopolymers. In this paper, the advancements of releasing active packaging intended for food purposes will be addressed. Examples of releasing active systems developed using plants extracts (e.g. thyme, rosemary, tea) and industrial fruits' by-products will be presented. Factors affecting the biological activity of natural extracts will be focused but also their potential contamination by plant toxins, mycotoxins, heavy metals or pesticides residues in order to draw conclusions about their safety and need of standardization. The release of active compounds from active food packaging to foods or food simulants and their effects in the oxidation status of packaged foods will be discussed.

Keyword: Active Food Packaging; Biological Activity; Contaminants; Natural Extracts; Releasing systems

1. INTRODUCTION

It is consensual that active food packaging represents an uprising for food industry. They allow an intended and positive interaction between the packaging and the packaged food to the final aim of extending food shelf-life [1].
One of the reasons why active food packaging is considered sustainable is related with the fact that allows to reduce food waste. Lately, there is a tendency to use natural compounds in releasing active food packaging. A considerable number of studies found in the literature use essential oils, extracts from plants such as rosemary, oregano, thyme, or even, fruit's by-products to obtain these natural extracts.

However, plants may be contaminated by toxins, produced by plants themselves, so-called plant toxins, or by fungi, i.e., mycotoxins. The preparation of extracts from plants can concentrate active compounds such as phenolics but, if these plants are contaminated, they can ultimately originate contaminated extracts. Pesticides residues are other group of analytes of interest to determine in natural extracts, especially in those prepared from fruits and vegetables.

In this communication, the progresses of releasing active packaging intended for food purposes will be addressed. Examples of releasing active systems developed using plants extracts and industrial fruits' by-products will be presented. Factors affecting the biological activity of natural extracts will be focused but also their potential contamination by mycotoxins or pesticides residues in order to infer conclusions about their safety and need of standardization. The release of active compounds from active food packaging to foods/food simulants and their effects in the oxidation status of packaged foods will be discussed.

2. RESULTS AND DISCUSSION

2.1 Natural extracts incorporated in packaging

In the last few years there is a tendency to substitute synthetic additives by natural additives both in food and in releasing active packaging. In part this is explained because some synthetic additives are associated with adverse side effects like allergies [2]. These natural extracts include those obtained by plants, fruits by-products and essential oils. Table 1 compiles some of the natural extracts that have been recently used in releasing active food packaging, the main compounds contributing for their activity and the assays used to evaluate their biological properties.

Table 1- Examples of natural extracts that have been used to prepare active packaging and the assays used to evaluate their biological properties.

Natural extract (Solvent)	Main compounds contributing for biological activity	Assays used to evaluate biological properties	Main conclusions	Ref.
Ethanolic Rosemary extract (Ethanol) and Ethanolic Thyme extract (Ethanol)	Rosemary: Rosmarinic acid, carnosic acid, carnosol	DPPH, β□Carotene bleaching assay, Total phenolic compounds content, Total flavonoids content, Identification of compounds by HPLC, antimicrobial activity was evaluated by diffusion disc against S. aureus ATCC 25923, L. monocytogenes ATCC 13932, E. coli ATCC 25922 and C. perfringens ATCC 13124.	The rosemary ethanolic extract presented highest antioxidant capacity and contained the greatest amount of carnosol, rosmarinic acid and carnosic acid. The chosen rosemary extract was also very easily obtained and produced with a low-toxicity extraction solvent. The rosemary extract also inhibited three of the four tested microorganisms, showing a great antimicrobial activity.	[3]
Apple and ginger (50:50, w/w) (Ethanol)	Phenolic compounds	DPPH, β□Carotene bleaching assay, Total phenolic compounds content, Total flavonoids content, Identification of the phenolic compounds by HPLC	Extracts obtained from ginger + apple by-products presented the highest antioxidant capacity and the highest phenolic content. Industrial fruits by-products contain phenolic compounds with excellent biological properties that justify their valorization.	[4]
Apple, carrot, beet and ginger (50:29:20:1, w/w) (Ethanol)	Phenolic compounds			
Green tea extract (Ethanol)	Phenolic compounds, mainly catechins	Identification of the phenolic compounds by HPLC- DAD and confirmation by HPLC-ESI-MS/MS	HPLC-DAD and HPLC-ESI-MS/MS analysis of the green tea extract have shown a large presence of antioxidant compounds belonging to the catechin group, particularly of flavan-3-ols groups, thus showing its potential antioxidant activity	[5]
Lemon (Ethanol)	Phenolic compounds	DPPH, β□Carotene bleaching assay, Total phenolic compounds content, Total flavonoids content, Identification of the phenolic compounds by HPLC	Phenolic compounds responsible for the high antioxidant activity of citrus by-products: hydroxycinnamic acids, flavonols glycosides, and flavanone glycosides. Eriocitrin was the main phenolic compound (17.49 mg/g) determined in lemon by-products, while hesperidin (4.9 mg/g) was the main phenolic in orange by-products. The antioxidant capacity of the apple byproduct was mainly due to hydroxycinnamic acids (e.g., rosmarinic acid) and other specific compounds such as epicatechin and phlorizin.	[6]
Orange (Ethanol)				
Apple (Ethanol)				

Commonly, these extracts are characterized in terms of their antioxidant and antimicrobial capacity. Besides the determination of individuals antioxidants, such as phenolic compounds, by high performance liquid chromatography, 1,1-diphenyl-2-picryl hydrazyl (DPPH) radical scavenging assay is one of the most frequently used methods to determine the antioxidant capacity of extracts (Table 1). In what concerns to antimicrobial capacity, disc diffusion test is one of the preferred methods.

Although natural extracts can exhibit good biological properties like antioxidant and antimicrobial capacity (Table 1), they may potentially be contaminated by plant toxins [7], mycotoxins [8], heavy metals [9] or even pesticides [10] similarly to what happens with its source material. However, research papers that prepare natural extracts intended to be incorporated in releasing active packaging does not address this issue, and they limit to evaluate the biological properties.

Kowalska [9] determined the level of contamination of 240 plant materials with heavy metals (Cd, Pb, As, and Hg), namely herbal raw materials, spices, tea, and coffee. This paper reported that lead levels exceeded the limits in 18% of the samples (24 herbs).

Caldeirão *et al.* [8] evaluated the presence of mycotoxins and trace elements in herbs and herbal infusions commercialized in Brazil. From 58 herb samples, 42 (72%) were contaminated, being beauvericin the most frequent mycotoxin (43% of the samples), followed by sterigmatocystin and HT-2 toxin, present in 37% and 24% of the samples, respectively.

Montiel-León *et al.* [10] studied fruits and vegetables from organic and conventional agriculture and found that 47% of the 133 analysed samples had levels above the limit of detection for at least one pesticide. Imidacloprid (0.08–29 μg/kg), acetamiprid (0.11–108 μg/ kg), and clothianidin (0.13–141 μg/kg) were the most frequent while atrazine was reported in about 30% of the lettuce samples.

There are many factors affecting the variability of the natural extracts obtained from plants. The composition of the extracts depends on the plant species, part of the plant used, edaphoclimatic conditions and extraction methodology. Moreover, the activity of the final extract obtained from a plant is affected by many factors as compiled in Figure 1. Due to all these variability factors it would be of great importance to find in the future means of standardization of these extracts.

Selected solvent (e.g.: water, ethanol, acetone)
Amount of plant material/solvent volume ratio
Number of extractions
Time of contact between solvent and plant material
Light protection
Storage temperature

Figure 1- Factors affecting the variability of the extracts related with the extraction method and storage conditions.

2.2 Effect of active packaging incorporating natural extracts in food orfood simulants

In the literature there are many examples of active packaging incorporating natural extracts (Table 2). From the analysis of Table 2 is possible to conclude that there are many combinations between natural extracts and polymers, especially biopolymers.

Table 2- Compiles recent studies that developed active packaging incorporating natural extracts.

Natural extract	Polymerix matrix	Method of incorporation in the polymer	Methods to determine effectiveness	Material tested to evaluate activity	Reference
Mango kernel extract	Soyprotein isolate or fish gelatin	Casting	DPPH, FRAP, ABTS,	Ethanolic film and extract	[11]
Green tea extract	PLA	Blown extrusion	Extract: DPPH, β□Carotene bleaching assay, Total phenolic compounds content, Total flavonoids content, Food: p-anisidine value, TBARS assay and hexanal monitoring	Extract and Food; smoked salmon	[12]

Natural extract	Polymerix matrix	Method of incorporation in the polymer	Methods to determine effectiveness	Material tested to evaluate activity	Reference
Cocoa bean shells	Poly(lactic acid)	Casting	DPPH		[13]
Herba Lophatheri extract	Chitosan	Casting	DPPH	Fim	[14]
Mulberry fruit extract	κ-carrageenan	Casting	DPPH	Film and Food (Milk)	[15]
Olive leaf extract	κ-carrageenan	Casting	DPPH, Folin-Ciocalteu	Film	[16]
Rosemary extract	Whey protein	Casting	Hexanal content, TBARS	Food (Salami)	[1]
Rosemary extract and chitosan	Poly(lactic acid)	Melt mixing and compression moulding	ABTS	Ethanolic extract of the film	[17]
Sage and lemon balm leaves	Poly(lactic acid)	Melt blending	ABTS, DPPH, in situ gas-phase hydroxyl radical generation, reducing power, Folin-Ciocalteu	Fim	[18]
Apple peel ethanolic extract	Chitosan and gelatin	Casting	DPPH, ABTS	Film	[19]
Curcumin	Carboxymethyl cellulose	Casting	DPPH, ABTS	Film	[20]
Mangosteen (*Garcinia mangostana* L.) rind	Chitosan	Casting	DPPH, peroxide value, TBARS	Film and food (Soybean oil)	[21]
Red pitaya (*Hylocereus polyrhizus*) peel extract	Starch/polyvinyl alcohol	Casting	DPPH	Film and Food (Shrimp)	[22]
Green tea extract	Whey protein	Casting	Extract: DPPH, β□Carotene bleaching assay, Total phenolic compounds content, Total flavonoids content, Food: Fatty acids profile, peroxide value, p-anisidine value, TBARS	Extract and Food (Salmon)	[23]
P. dioica Polysaccharide-Enriched Extract	Food-grade polypropylene (PP)	Coating using electrospinning	Microbial growth (total mesophile aerobic colony counts), colour stability, lipid oxidation, TBARS, and sensory analysis	Extract and Food (Minced chicken breasts)	[24]
Fucus vesiculosus extract	Whey protein	Casting	Extract: DPPH, β□Carotene bleaching assay, Total phenolic compounds content, Food: TBARS	Extract, Film and Food (chicken breasts)	[25]

DPPH- 1,1-diphenyl-2-picryl hydrazyl; TBARS- Thiobarbituric acid reactive substances assay.

Moreover, many factors considerably affect the migration of natural extracts incorporated in active packaging to food and therefore affect the final effectiveness of the releasing active packaging: (i) activity of the natural extract, (ii) type of polymer, (iii) % of the natural extract incorporated in the polymer, (iv) time and temperature of contact between packaging and food; (v) characteristics of food. Regarding the % of the natural extract incorporated in the polymer, the highest % of incorporation is not always the best in terms of effectiveness, especially concerning antioxidant extracts. For example, Martins *et al.* [12] found prooxidant effects when incorporating 2% (w/w) green tea extract in polylactic acid (PLA). The same effect was not found when incorporating 1% (w/w) ethanolic green tea in PLA.

4. CONCLUSIONS

Lately there is a tendency for using natural extracts in releasing active packaging, however it is important to have in consideration not only their biological properties but also their possible contamination by toxins, heavy metals or pesticide residues. Therefore, in order to standardize natural extracts, it is recommended to evaluate which active compounds should be monitored and their minimum levels to achieve efficacy. In addition, it is of great importance the monitorization of possible contaminants. Moreover, in order to obtain successful releasing active packaging, it is important to define in which polymers the natural extracts should be incorporated, at what levels, for which foods the new releasing active packaging is intended and the storage conditions (time and temperature of contact).

5. ACKNOWLEDGEMENT

ValICET (Valorise foods and Improve Competitiveness through Emerging Technologies applied to food by-products within the circular economy framework) project is supported by the PRIMA program (Partnership for research and innovation in the Mediterranean area, Section 2 Funded Projects 2020, Agri-Food Value Chain) and funded in Spain by the Spanish Ministry of Science and Innovation and the State Research Agency (AEI) (PCI2021-121965). A. Sanches Silva is grateful to the Foundation for Science and Technology (FCT) for funding the ValICET project in Portugal (PRIMA/0001/2020). The work was also supported by UIDB/00211/2020 with funding from FCT/MCTES.

6. REFERENCES

[1] Andrade MA, Ribeiro-Santos R, Guerra M, Sanches-Silva A (2019) Evaluation of the oxidative status of salami packaged with an active whey protein film. Foods 8:. https://doi.org/10.3390/foods8090387

[2] Sadighara, Parisa, Safta, Mehdi, Limam, Intissar, Ghanati, Kiandokht, Nazari, Zahra, Karami, Marzieh and Abedini A (2022) Association between food additives and prevalence of allergic reactions in children: a systematic review. Rev Environ Health

[3] Andrade MA, Ribeiro-Santos R, Costa Bonito MC, et al (2018) Characterization of rosemary and thyme extracts for incorporation into a whey protein based film. Lwt 92:497–508. https://doi.org/10.1016/j.lwt.2018.02.041

[4] Andrade MA, de Oliveira Torres LR, Silva AS, et al (2020) Industrial multi-fruits juices by-products: total antioxidant capacity and phenolics profile by LC–MS/MS to ascertain their reuse potential. Eur Food Res Technol 246:2271–2282. https://doi.org/10.1007/s00217-020-03571-3

[5] Vilarinho F, Stanzione M, Buonocore GG, et al (2021) Green tea extract and nanocellulose embedded into polylactic acid film: Properties and efficiency on retarding the lipid oxidation of a model fatty food. Food Packag Shelf Life 27:. https://doi.org/10.1016/j.fpsl.2020.100609

[6] Barbosa CH, Andrade MA, Séndon R, et al (2021) Industrial fruits by-products and their antioxidant profile: Can they be exploited for industrial food applications? Foods 10:1–16. https://doi.org/10.3390/foods10020272

[7] Izcara S, Casado N, Morante-Zarcero S, et al (2022) Miniaturized and modified QuEChERS method with mesostructured silica as clean-up sorbent for pyrrolizidine alkaloids determination in aromatic herbs. Food Chem 380:. https://doi.org/10.1016/j.foodchem.2022.132189

[8] Caldeirão L, Sousa J, Nunes LCG, et al (2021) Herbs and herbal infusions: Determination of natural contaminants (mycotoxins and trace elements) and evaluation of their exposure. Food Res Int 144:. https://doi.org/10.1016/j.foodres.2021.110322

[9] Kowalska G (2021) The safety assessment of toxic metals in commonly used herbs, spices, tea, and coffee in poland. Int J Environ Res Public Health 18:. https://doi.org/10.3390/ijerph18115779

[10] Montiel-León JM, Duy SV, Munoz G, et al (2019) Occurrence of pesticides in fruits and vegetables from organic and conventional agriculture by QuEChERS extraction liquid chromatography tandem mass spectrometry. Food Control 104:74–82. https://doi.org/10.1016/j.foodcont.2019.04.027

[11] Maryam Adilah ZA, Jamilah B, Nur Hanani ZA (2018) Functional and antioxidant properties of protein-based films incorporated with mango kernel extract for active packaging. Food Hydrocoll 74:207–218. https://doi.org/10.1016/j.foodhyd.2017.08.017

[12] Martins C, Vilarinho F, Sanches Silva A, et al (2018) Active polylactic acid film incorporated with green tea extract: Development, characterization and effectiveness. Ind Crops Prod 123:100–110. https://doi.org/10.1016/j.indcrop.2018.06.056

[13] Papadopoulou EL, Paul UC, Tran TN, et al (2019) Sustainable Active Food Packaging from Poly(lactic acid) and Cocoa Bean Shells. ACS Appl Mater Interfaces 11:31317–31327. https://doi.org/10.1021/acsami.9b09755

[14] Wang L, Guo H, Wang J, et al (2019) Effects of Herba Lophatheri extract on the physicochemical properties and biological activities of the chitosan film. Int J Biol Macromol 133:51–57. https://doi.org/10.1016/j.ijbiomac.2019.04.067

[15] Liu Y, Qin Y, Bai R, et al (2019) Preparation of pH-sensitive and antioxidant packaging films based on κ-carrageenan and mulberry polyphenolic extract. Int J Biol Macromol 134:993–1001. https://doi.org/10.1016/j.ijbiomac.2019.05.175

[16] da Rosa GS, Vanga SK, Gariepy Y, Raghavan V (2020) Development of Biodegradable Films with Improved Antioxidant Properties Based on the Addition of Carrageenan Containing Olive Leaf Extract for Food Packaging Applications. J Polym Environ 28:123–130. https://doi.org/10.1007/s10924-019-01589-7

[17] Vasile C, Stoleru E, Darie-Niţa RN, et al (2019) Biocompatible materials based on plasticized poly(lactic acid), chitosan and rosemary ethanolic extract I. effect of chitosan on the properties of plasticized poly(lactic acid) materials. Polymers (Basel) 11:. https://doi.org/10.3390/polym11060941

[18] Gavril GL, Wrona M, Bertella A, et al (2019) Influence of medicinal and aromatic plants into risk assessment of a new bioactive packaging based on polylactic acid (PLA). Food Chem Toxicol 132:. https://doi.org/10.1016/j.fct.2019.110662

[19] Riaz A, Lagnika C, Abdin M, et al (2020) Preparation and Characterization of Chitosan/Gelatin-Based Active Food Packaging Films Containing Apple Peel Nanoparticles. J Polym Environ 28:411–420. https://doi.org/10.1007/s10924-019-01619-4

[20] Roy S, Rhim JW (2020) Carboxymethyl cellulose-based antioxidant and antimicrobial active packaging film incorporated with curcumin and zinc oxide. Int J Biol Macromol 148:666–676. https://doi.org/10.1016/j.ijbiomac.2020.01.204

[21] Zhang X, Liu J, Yong H, et al (2020) Development of antioxidant and antimicrobial packaging films based on chitosan and mangosteen (Garcinia mangostana L.) rind powder. Int J Biol Macromol 145:1129–1139. https://doi.org/10.1016/j.ijbiomac.2019.10.038

[22] Qin Y, Liu Y, Zhang X, Liu J (2020) Development of active and intelligent packaging by incorporating betalains from red pitaya (Hylocereus polyrhizus) peel into starch/polyvinyl alcohol films. Food Hydrocoll 100:105410. https://doi.org/10.1016/j.foodhyd.2019.105410

[23] Castro FVR, Andrade MA, Silva AS, et al (2019) The contribution of a whey protein film incorporated with green tea extract to minimize the lipid oxidation of salmon (Salmo salar L.). Foods 8:1–16. https://doi.org/10.3390/foods8080327

[24] Reboleira J, Adão P, Guerreiro SFC, et al (2020) Poultry shelf-life enhancing potential of nanofibers and nanoparticles containing porphyra dioica extracts. Coatings 10:1–14. https://doi.org/10.3390/coatings10040315

[25] Andrade MA, Barbosa CH, Souza VGL, et al (2021) Novel active food packaging films based on whey protein incorporated with seaweed extract: Development, characterization, and application in fresh poultry meat. Coatings 11:1–12. https://doi.org/10.3390/coatings11020229

AG-O08

Development and Characterization of Ethyl Maltol Incorporated Starch-Based Films for Active Biodegradable Food Packaging

Khwanchat Promhuad, Nathdanai Harnkarnsujarit*

Department of Packaging and Materials Technology, Faculty of Agro-Industry, Kasetsart University, Thailand

***Correspondence to:** Department of Packaging and Materials Technology, Faculty of Agro-Industry, Kasetsart University, Thailand.*
E-mail: nathdanai.h@ku.ac.th

ABSTRACT: Active packaging has direct influences on the quality and safety of packaged food products by preventing chemical and biological induced quality changes. Ethyl maltol is a volatile flavoring compound with antimicrobial capacity, while its flavor is compatible with bakery products. The incorporation of ethyl maltol into the package possibly produced active packaging for extending the shelf-life of the bakery. The objectives of this research were to develop biopolymer films that incorporated ethyl maltol and investigate the effects of the active films on the shelf life of bakeries. The ethyl maltol was incorporated into starch films up to 10% and the morphology and properties were characterized. Results showed the increased intensity of the infrared absorption O-H stretching band as increased ethyl maltol which suggested an increased inter and intra-molecular hydrogen bonding between ethyl maltol and starch. The film with ethyl maltol had an increased contact angle, indicating enhanced surface hydrophobicity. The crystallinity of the films depended on the concentration of ethyl maltol. Only 10% ethyl maltol showed the recrystallization behavior and the films became opaque due to phase separation of the crystallites. Oxygen permeability (OP) and water vapor permeability (WVP) decreased with the addition of the ethyl maltol due to the formation of network interactions between starch and ethyl maltol. The films were effectively delayed fungal growth in packaged butter cake, corresponding with an increasing shelf-life of more than 2 - 3 times. The results indicated that ethyl maltol incorporated starch film showed promising features as active biodegradable food packaging material.

Keyword: Active packaging, bakery product, volatile compound, antimicrobial, food packaging

1. INTRODUCTION

Environmentally friendly materials made from natural and renewable resources have suddenly got attention [1-3]. Tapioca is a major crop in Thailand to produce starch. Starch has been used in the field of degradable polymers as an abundant and low-cost raw material. Blend films containing starch are used for prospective materials in the agriculture, medicinal, and packaging industries [4-7]. Edible films made from starch have the right physical properties which are odorless, tasteless, colorless, and impermeable to oxygen [8-9].

Recrystallization of starch and protein polymers causes bakery products to fast deteriorate due to moisture loss, and firmness loss [10]. This is known as 'staling'. Wheat starch, which contains amorphous networks of starch and gluten proteins, is a common ingredient in bakery products. These networks are in a thermodynamically non-equilibrium condition, and they tend to crystallize [11]. Furthermore, fungal growth is promoted by oxygen, the permeability of the packaging is crucial to the shelf-life of bakery products [12].

Packaging serves to protect food by preserving quality, safety, and shelf life. When volatile substances are incorporated into polymers, active packaging is created, which can improve the quality and safety of packaged meals. An important aspect that contributes to antimicrobial efficiency is the release and subsequent contact between antimicrobial compounds from the package and foods either by direct or indirect contact [13]. However, it is impossible to have direct contact between the packaging and the surface area of extremely porous goods like bakery products. As a reason, releasing volatile compounds into headspace is an alternate technique for improving antimicrobial efficiency in bakery and porous food products that come into touch with the food package in an indirect way [14].

Ethyl maltol is used as a flavor enhancer for food. Since 1967, ethyl maltol, a synthetic version, has been available. Many foods, including coffee, soybeans, cereals, bread, malt beverages, and chocolate milk, include this compound as a flavor enhancer [15-17]. The biological safety of ethyl maltol after acute, subacute, and chronic administration was investigated, and the findings indicated that ethyl maltol is safe [18].

The objective of this research was to create antifungal films from starch-containing ethyl maltol. The effects of ethyl maltol on the film's physical, mechanical, and barrier properties were studied. The antifungal efficiency and quality of packaged butter cake were also examined. The research hypothesized that interaction between ethyl maltol and starch would alter and control ethyl maltol release, affecting packaging qualities, food quality, and antifungal efficacy of the films. Furthermore, improved bioplastic packaging functionality will compensate for bioplastics' limited barrier properties and increase the quality and safety of packed foods.

2. MATERIALS AND METHODS

2.1 Film preparation

Acetylated tapioca starch (AS, DS 0.01 - 0.03, Siam Modified Starch Co., Ltd., Thailand) was mixed with glycerol (Asian Scientific Co., Ltd., Thailand) and ethyl maltol (EMA, Sigma-Aldrich, Switzerland) at 0%, 1%, and 10% w/w of total solid. Weight ratio for powder: glycerol was 100:35. The mixture was mixed using a stirrer at temperatures 90 °C for 60 minutes. Films dried at 50 °C for 13 hours using a hot air oven.

2.2 Fourier-transform infrared spectroscopy (FTIR)

Film sample (1 cm × 1 cm) was determined for functional group by using attenuated total reflection FTIR spectra using a Bruker Tensor 27 FT-IR Spectrometer (Bruker OPTIK GmbH, Germany) between 500-4000 cm^{-1} at a resolution of 4 cm^{-1} with 64 co- added scans per spectrum. Spectra were standardized using the spectrum of air.

2.3 Tensile properties

Tensile strength (TS) and elongation at break (EB) (ASTM D882-09) were determined using an Instron Universal Testing Machine (Model 5965, Instron, USA) with crosshead speed at 500 mm/min. The films (150 mm × 25 mm) were stored at 50 % RH 48 h before measurement.

2.4 Surface hydrophobicity

Surface wetting properties were measure. Distilled water (3 μL) was dropped on the films. Images were captured for contact angle using SCA 20 image analysis software (Dataphysics OCA15EC, Dataphysics Instrument GmbH, Germany). Values were averaged from ten drops of the sample.

2.5 Water vapor permeability (WVP)

Water vapor transmission rate (WVTR) was determined by the standard cup method (ASTM E96-80) at 25 ± 2 °C and 50 ± 2 % RH. Films were cut into 7 cm diameter circles and placed on aluminum cups containing dried silica gels and an O-ring sealed with paraffin wax. The samples were weighed and stored at a climate chamber (Binder KBF 720, GmbH, Germany) for 14 days. The WVP ($g \cdot mm \cdot Pa^{-1} \cdot m^{-2} \cdot day^{-1}$) was calculated using Equation 1.

$$WVP = (WVTR \text{ x } L)/ \Delta P \qquad (1)$$

where water vapor transmission rate (WVTR, $g \cdot m^{-2} \cdot day^{-1}$) derives from the slope of the linear weight increase, L (mm) is the thickness and ΔP is water vapor pressure difference (Pa) between each side of the films.

2.6 Oxygen permeability

Oxygen transmission rate (OTR) was evaluated following ASTM D3985-81 using an oxygen permeation analyzer (model 8500, Illinois Instrument, USA). Duplicate films were cut into circles with a diameter of 14 cm and conditioned at 25 °C and 50% RH for 48 h. Equation 2 was used to compute the oxygen permeability (OP; $cm^3 \cdot mm \cdot m^{-2} \cdot day^{-1} \cdot atm^{-1}$).

$$OP = (OTR \times L)/\Delta P \qquad (2)$$

where ΔP (atm) is the difference in partial pressure between each side of the films and L is the thickness (mm). The OP data were averaged from three specimens.

2.7 Packaging of bakery

Butter cake was obtained from a local bakery shop (Bangkok, Thailand) and stored at room temperature. The surface of the butter cake was directly contacted, and the tray was sealed with the films. The samples were stored in PP bags at room temperature. The properties of butter cake were tested throughout storage.

2.8 Statistical analysis

Duncan's multiple range test was used to calculate analysis of variance (ANOVA) among samples using IBM SPSS Statistics for Windows (version 23, SPSS Inc., USA) at a 95% confidence level.

3. RESULTS AND DISCUSSION

3.1 FTIR

The strongest absorption band at 1018 cm^{-1} was linked to $-CH_2$ bending of polysaccharides in IR spectra in the fingerprint region (700 to 1700 cm^{-1}) (Figure 1). Strong IR absorption between 1000 and 1300 cm^{-1} was due to C=O stretching [19]. The vibration of C-O-H deformation in amorphous starch was linked to the absorbance at 1017 cm^{-1}. The coupling of C-O, C-C, and O-H stretching, as well as bending and asymmetric stretching of the C-O-C glycosidic linkage, were attributed to bands at 1080 and 1150 cm^{-1} [20]. The addition of 10% ethyl maltol clearly increased intensity between 1000 and 1300 cm^{-1}. Ethyl maltol also changed the intensity of the O-H stretching band associated with free, intermolecular, and intramolecular hydroxyl groups, ranging from 3000 to 3700 cm^{-1} [21-22].

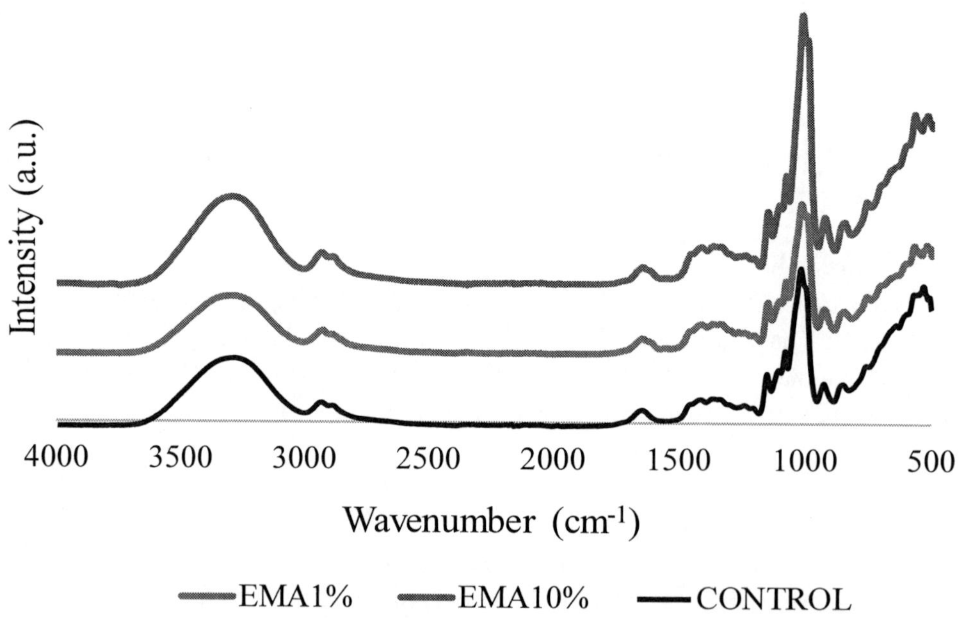

Figure 1 FTIR analysis as absorption spectra of starch-based films containing ethyl maltol (EMA) at different concentrations (0, 1, and 10%).

3.2 Mechanical properties

The addition of ethyl maltol to films had an insignificant effect on the TS and EB. When ethyl maltol was added, the starch network developed structural changes and polymer chain movements were facilitated under stress, improving film flexibility and network structure is very stable, with strong molecular orientation [23].

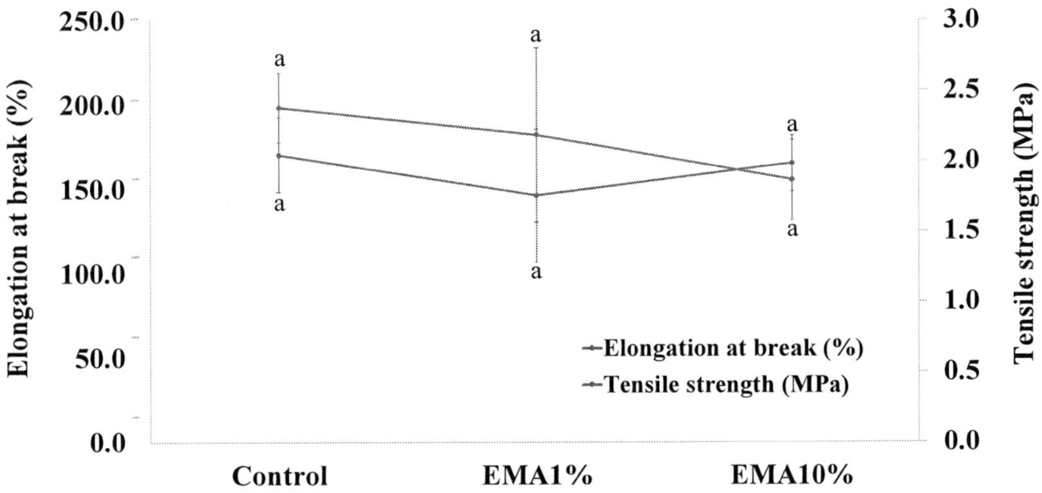

Figure 2 Mechanical properties as tensile strength and elongation at break of starch-based films containing ethyl maltol (EMA) at different concentrations (0, 1, and 10%).

3.3 Surface hydrophobicity

Surface hydrophobicity was evaluated by means of contact angle (Figure 3). An increase in ethyl maltol content gave a significant increase in the contact angle of the films. These results indicate that the wettability of the films decreased with an increase in the ethyl maltol content. This phenomenon was attributed to the low wettability of ethyl maltol which was associated with the hydrophobic acetyl groups present in ethyl maltol [24]. This finding suggests that ethyl maltol decreased the wettability of the surface.

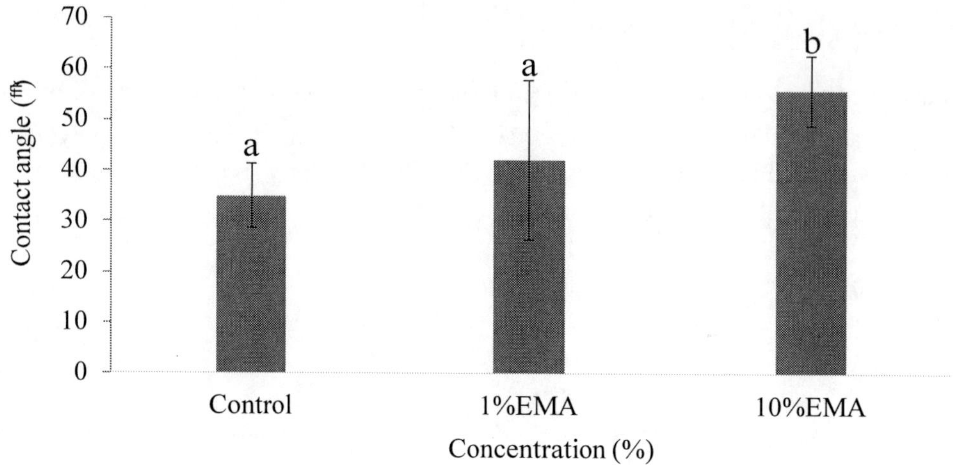

Figure 3 Contact angle (CA) of starch-based films containing ethyl maltol (EMA) at different concentrations (0, 1, and 10%).

3.4 Barrier properties

Water vapor permeability (WVP) and oxygen permeability (OP) of films are shown in Figure 4. Increased ethyl maltol to 10% highly increased WVP due to non-homogeneity of the network which allowed for greater water vapor diffusion. This impact could be attributable to ethyl maltol's higher polarity and capacity to sustain moisture content. The high affinity of glycerol for water, which favors water molecule diffusion, accounts for the substantial WVP of these films [25]. Ethyl maltol is a tiny hydrophilic molecule. It can readily be introduced between starch molecules to form hydrogen bonds with hydroxyl groups. Ethyl maltol increased the hydrophilicity of the matrices, which reduced the OP values [24]. Several factors can affect the OTR of plasticized films, including the plasticizer's physical state and molecular weight, changes in the film bulk and surface structure, chemical interactions between the plasticizer and oxygen, and the ability of small molecules to fill voids in the polymer matrix [26]. Starch films are considered to be good oxygen barriers because of their tight packing, highly organized hydrogen-bonded network structure, and poor solubility [27].

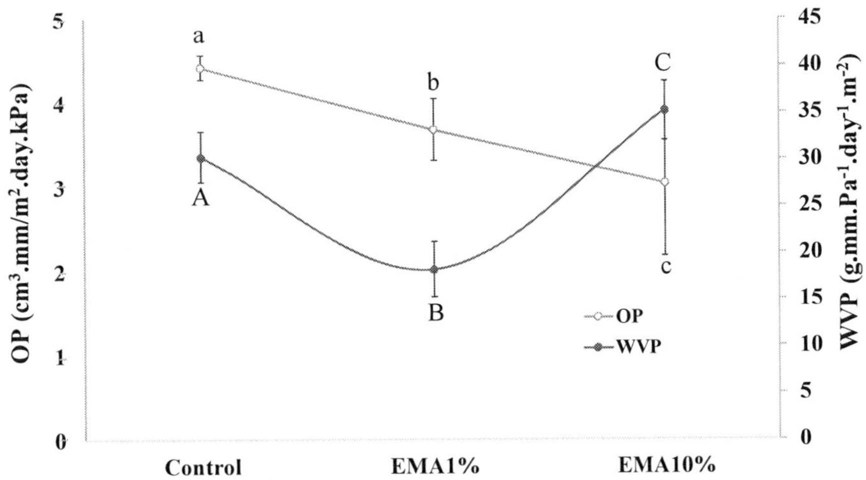

Figure 4 barrier properties Water vapor permeability (WVP), oxygen permeability (OP) as of starch-based films containing ethyl maltol (EMA) at different concentrations (0, 1, and 10%).

3.5 Packaging of bakery

Mold expansion was effectively inhibited by ethyl maltol films (Figure 5). After 4 days of storage at room temperature, the film with ethyl maltol and commercial film displayed the mold growth. Conversely, films containing 1% and 10% ethyl maltol showed the mold growth after 9 and 13 days, respectively. Release of ethyl maltol efficiently inhibited microorganisms, further storage resulted in an antimicrobial function.

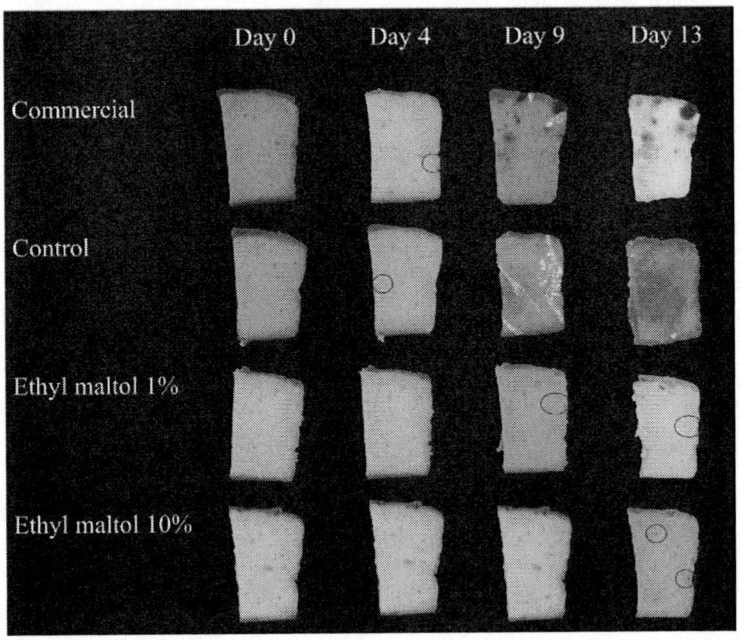

Figure 4 Appearance of packaged butter cake stored in starch-based films containing ethyl maltol (EMA) at different concentrations (0, 1 and 10%).

4. CONCLUSIONS

The starch-based film with 10% ethyl maltol decreased oxygen permeability (OP) and increased water vapor permeability (WVP) and surface hydrophobicity. Ethyl maltol decreased the wettability of the matrices. However, adding ethyl maltol to films had an insignificant effect on the TS and EB levels. Moreover, the film contained with ethyl maltol showed inhibitory effects against fungal growth in butter cake. Ethyl maltol possibly improved the antimicrobial function of active packaging which preserved the quality of packaged bakery.

5. ACKNOWLEDGEMENT

This research was supported by Kasetsart University Research and Development Institute (KURDI), FF(KU)25.64.

6. REFERENCES

[1] Ma X. F. and Yu J. G. (2004). The plasticizers containing amide groups for thermoplastic starch. Carbohydrate Polymers, 57, 197-203. DOI: 10.1016/j.carbpol.2004.04.012

[2] Nakamura E. M, Cordi L, Almeida G. S. G, Duran N. and Mei L. H. I. (2005). Study and development of LDPE/starch partially biodegradable compounds. Journal of Materials Processing Technology, 162-163, 236-241. DOI: 10.1016/j.jmatprotec.2005.02.007

[3] Yu L, Dean K. and Li L. (2006). Polymer blends and composites from renewable resources. Progress in Polymer Science, 31, 576-602. DOI: 10.1016/j.progpolymsci.2006.03.002

[4] Funke U, Bergthaller W. and Lindhauer M. G. (1998). Processing and characterization of biodegradable products based on starch. Polymer Degradation and Stability, 59, 293-296. DOI: 10.1016/S0141-3910(97)00163-8

[5] Hulleman Stephan H. D, Janssen Frank H. P. and Feil H. (1998). The role of water during plasticization of native starches. Polymer, 39, 2043-2048. DOI: 10.1016/S0032-3861(97)00301-7

[6] Lu Y. S, Tighzerta L, Dole P. and Erre D. (2005). Preparation and properties of starch thermoplastics modified with waterborne polyurethane from renewable resources. Polymer, 46, 9863-9870. DOI: 10.1016/j.polymer.2005.08.026

[7] Wongphan, P., Khowthong, M., Supatrawiporn, T., & Harnkarnsujarit, N. (2022). Novel edible starch films incorporating papain for meat tenderization. Food Packaging and Shelf Life, 31, 100787. DOI: 10.1016/j.fpsl.2021.100787

[8] Sadeghizadeh-Yazdi J, Habibi M, Kamali A. A. and Banaei M. (2019). Application of edible and biodegradable starch-based films in food packaging: A systematic review and meta-analysis. Current research in nutrition and food science journal, 7(3), 624-637. DOI: 10.12944/CRNFSJ.7.3.03

[9] Phinainitisatra, T., & Harnkarnsujarit, N. (2021). Development of starch□based peelable coating for edible packaging. International Journal of Food Science & Technology, 56(1), 321-329. DOI: 10.1111/ijfs.14646

[10] Bumbudsanpharoke, N., Wongphan, P., Promhuad, K., Leelaphiwat, P., & Harnkarnsujarit, N. (2022). Morphology and permeability of bio-based poly (butylene adipate-co-terephthalate)(PBAT), poly (butylene succinate)(PBS) and linear low-density polyethylene (LLDPE) blend films control shelf-life of packaged bread. Food Control, 132, 108541. DOI: 10.1016/j.foodcont.2021.108541

[11] Gray J. A. and Bemiller J. N. (2003). Bread staling: molecular basis and control. Comprehensive reviews in food science and food safety, 2(1), 1-21. DOI: 10.1111/j.1541-4337.2003.tb00011.x

[12] Rahman M, Islam R, Hasan S, Zzaman W, Rana M. R, Ahmed S. and Sunny A. R. (2022). A Comprehensive Review on Bio-Preservation of Bread: An Approach to Adopt Wholesome Strategies. Foods, 11(3), 319. DOI: 10.3390/foods11030319

[13] Chatkitanan, T., & Harnkarnsujarit, N. (2021). Effects of nitrite incorporated active films on quality of pork. Meat Science, 172, 108367. DOI: 10.1016/j.meatsci.2020.108367

[14] Chawla R, Sivakumar S. and Kaur H. (2021). Antimicrobial edible films in food packaging: Current scenario and recent nanotechnological advancements-a review. Carbohydrate Polymer Technologies and Applications, 2, 100024. DOI: 10.1016/j.carpta.2020.100024

[15] Heath H. B. (1978). Flavor Technology, Profiles, Products and Applications. The AVI Publ. Co. Inc: Wesport, Conecticut.

[16] Hui Y. H. (1991). Encyclopedia of food science and technology. New York: Wiley.

[17] LeBlanc D. T. and Akers, H. A. (1989). Maltol and ethyl maltol: From the larch. Tree to successful food additive. Food Technology, 43, 78–84.

[18] Gralla E. J, Stebbins R. B, Coleman G. L. and Delahunt, C. S. (1969). Toxicity studies with ethyl maltol. Toxicology and Applied Pharmacology, 15, 604–613. DOI: 10.1016/0041-008X(69)90062-3

[19] Nagar M, Sharanagat V. S, Kumar Y, Singh L. and Mani S. (2019). Influence of xanthan and agar-agar on thermo-functional, morphological, pasting and rheological properties of elephant foot yam (Amorphophallus paeoniifolius) starch. International journal of biological macromolecules, 136, 831-838. DOI: 10.1016/j.ijbiomac.2019.06.133

[20] Liu Q, Charlet G, Yelle S. and Arul J. (2002). Phase transition in potato starch–water system I. Starch gelatinization at high moisture level. Food Research International, 35(4), 397-407. DOI: 10.1016/S0963-9969(01)00134-X

[21] Phothisarattana, D., Wongphan, P., Promhuad, K., Promsorn, J., & Harnkarnsujarit, N. (2022). Blown film extrusion of PBAT/TPS/ZnO nanocomposites for shelf-life extension of meat packaging. Colloids and Surfaces B: Biointerfaces, 112472. DOI: 10.1016/j.colsurfb.2022.112472

[22] Phothisarattana, D., Wongphan, P., Promhuad, K., Promsorn, J., & Harnkarnsujarit, N. (2021). Biodegradable Poly (Butylene Adipate-Co-Terephthalate) and Thermoplastic Starch-Blended TiO2 Nanocomposite Blown Films as Functional Active Packaging of Fresh Fruit. Polymers, 13(23), 4192. DOI: 10.3390/polym13234192

[23] Rindlav-Westling A, Stading M, Hermansson A. M. and Gatenholm P. (1998). Structure, mechanical and barrier properties of amylose and amylopectin films. Carbohydrate Polymers, 36, 217–224. DOI: 10.1016/S0144-8617(98)00025-3

[24] Yue Y, Liu J, Yao M, Yao X, Fan J. and Ji H. (2012). The investigation of the binding behavior between ethyl maltol and human serum albumin by multi-spectroscopic methods and molecular docking. Spectrochimica Acta Part A: Molecular and Biomolecular Spectroscopy, 96, 316-323. DOI: 10.1016/j.saa.2012.05.041

[25] Tako M, Tamaki Y, Teruya T. and Takeda Y. (2014). The principles of starch gelatinization and retrogradation. Food and Nutrition Sciences, 2014. DOI: 10.4236/fns.2014.53035

[26] I. G. Donhowe, O. Fennema: The effects of plasticizers on crystallinity, permeability, and mechanical properties of methylcellulose films. J. Food Process. Preserv. 1993, 17, 247–257. DOI: 10.1111/j.1745-4549.1993.tb00729.x

[27] T. H. McHugh, J. M. Krochta: Sorbitol-vs glycerol-plasticized whey protein edible films: Integrated oxygen permeability and tensile property evaluation. J. Agric. Food Chem. 1994, 42(4), 841–845.

AG-O09

Industrial By-Products as Raw Materials for Active Films: The Approach of the ValICET Project

Raquel Sendón[1]*, Sandra Mariño-Cortegoso[1], Ana Sanches Silva[2,3,4], Ana Rodríguez Bernaldo de Quirós[1], * Letricia Barbosa-Pereira[1]

[1]*Analytical Chemistry, Nutrition and Food Science Department, Pharmacy Faculty, University of Santiago de Compostela, 15782 Santiago de Compostela, Spain;*
[2]*National Institute for Agrarian and Veterinary Research (INIAV), I.P., 4485-655 Vairão, Vila do Conde, Portugal;*
[3]*Center for Study in Animal Science (CECA), ICETA, University of Porto, 4051-051 Porto, Portugal;*
[4]*University of Coimbra, Faculty of Pharmacy, 3000-548 Coimbra, Portugal;*

***Correspondence to:** [1]Analytical Chemistry, Nutrition and Food Science Department, Pharmacy Faculty, University of Santiago de Compostela, 15782 Santiago de Compostela, Spain.*
E-mail: raquel.sendon@usc.es

ABSTRACT: Worldwide the agri-food industry offers a large number of food products, but the main drawback is the large amounts of waste produced, whose disposal represents a critical hurdle to sustainability, the economy. and the environment. Particularly, the Mediterranean Basin is an agroecological diversity hotspot with an extraordinary variety of raw materials that allow the agri-food industry to offer a very large number of traditional food products. The ValICET project (Valorise foods and Improve Competitiveness through Emerging Technologies applied to food by-products within the circular economy framework) has been funded by PRIMA (Section 2), framed in the Thematic Area 3-Agrofood chain, according to Goals 2 and 12 of Agenda 2030 of the European Union to ensure "Zero hunger" and "Sustainable consumption and production patterns". ValICET aims at developing innovative and sustainable processing solutions capable of converting agri-food residues into sustainable natural high-added value compounds to be reincorporated into the food supply chain to extend the shelf-life of traditional Mediterranean food products within the framework of the circular economy. Among different strategies, it is included the development of active coatings and other food contact materials using bioactives/extracts obtained from different by-products. During the development of these new materials, and as a part of the evaluation of their effectiveness as active materials, migration assays are an important tool to study the release of active compounds intended to act as antioxidant or antimicrobial compounds. In this communication, the project ValICET will be presented, focusing on the different materials developed as a potential strategy to contribute to the circular economy and decrease the negative environmental impact of the traditional food contact materials. Several examples will highlight the value of the migration assays and the analytical methods developed to determine the key compounds when the activity of these materials is evaluated.

Keyword: active materials, by-products, phenolic compounds, migration assays

1. INTRODUCTION

Worldwide the agri-food industry offers a large number of food products, but the main drawback is the large amounts of wastes produced, whose disposal represents a critical hurdle to sustainability, the economy and the environment.

The Mediterranean Basin is an agri-ecological diversity hotspot with an extraordinary variety of raw materials that allow the agri-food industry to offer many traditional food products. This agri-food value chain represents a complex system which involves several local actors (long supply chain). It is included farmers, small-scale food manufacturers, and local distributors, among others. All of them are challenged by the dual objective of (i) enhancing competitiveness, which is currently challenged by non-optimized organizational model, long supply chain and imported low price products of global markets and (ii) improving sustainability of the productions and increase safety and quality of food products.

Thus, in agreement with the Goals 2 and 12 of Agenda 2030 [2] to ensure "Zero hunger" and "Sustainable consumption and production patterns", respectively, it is highly recommended that all the stakeholders of the agro-food supply chain, including the Agro-industrial and Scientific communities work together to set up innovative technological and organizational strategies to (i) optimize agro-food chain models to reduce post-farming losses and the environmental impact of the production processes; (ii) implement the circular economy concept along with responsible production and consumption concepts, and (iii) increase productivity of high quality food products with health beneficial properties with fair price for consumers while enabling reasonable profit for producers.

In this line arises the ValICET project (Valorise foods and Improve Competitiveness through Emerging Technologies applied to food by-products within the circular economy framework), funded by PRIMA [1] (Section 2), framed in the Thematic Area 3-Agrofood chain; the project focuses on the development of novel strategies for the valorisation of wastes and by-products from post-farming processes of Mediterranean agro-food systems and the design of eco-friendly procedures to recover high-added value bioactive compounds to be reintroduced in the agro-food chain as food ingredients/additives within the circular economy framework. The functional ingredients/additives and coatings with high antioxidant capacity and antimicrobial activity can contribute to reducing post-farming food losses during transportation and storage, improve the nutritional and sensory quality, and extend the shelf-life of Mediterranean food products, such as fruits, bakery, wines and dairy products. ValICET uses an innovative approach (inter-linking expertise from academia and the food industry) to hypothesize innovative solutions, leading to the maximal exploitation of food wastes and food by-products potentialities, through their conversion into value-added foods for human consumption. To achieve this goal, a team including research institutions and SMEs from five countries of the Mediterranean basin (Spain, Italy, France, Tunisia and Portugal) are applying an integrated interdisciplinary approach.

The valorisation of the residues produced in the Mediterranean area, is very promising, because they still contain molecules of interest such as of proteins, fibers, carotenoids, and polyphenols, such as anthocyanins. For example, grape by-products are particularly rich in catechins, gallic acid, and procyanidins [3]. Date fruit seeds have been demonstrated to possess high antioxidant activities due to their high content in phenolic acids such as hydroxycinnamic acids, and flavonoids like flavonols, flavones, and flavan-3-ols [4]. Cherries' by-products contain high amounts of anthocyanins, catechins, quercetin and caffeic acid [5-6]. Citrus peels are rich in hesperidin, a potent flavonoid with high anti-inflammatory and potential antiviral capacity against SARS-CoV-2 [7] and also with interest for coatings production. Whey contains proteins and peptides with several applications: (i) preparations of food emulsions, (ii) encapsulation of bioactive compounds (e.g. polyphenols), and (iii) formulation of active and edible coatings. Pomegranate by products extracts have shown high antioxidant capacity due to the presence of phenolic compounds, mainly anthocyanins, hydrolysable tannins, phytosterols and phenolic acids, among others. Moreover pomegranate peel has shown antimicrobial capacity and also an interesting therapeutic potential in carcinoma prevention and in cardiovascular protection [8].

Several new ingredients can be obtained from the food chain by-products that could be reincorporated into the food chain. Thus, among the objectives of ValICET, one is the application of natural ingredients and bio-compounds in Mediterranean food products (e.g. pastries, dairy products, wines, fruits) to produce healthier foods with improved functionality and longer shelf-life. A strategy to extend the shelf-life of different food is the development of active coatings and other food contact materials using bioactives/extracts obtained from numerous by-products.

In this communication, the project ValICET will be presented, focusing on the different materials developed as a potential strategy to contribute to the circular economy and decrease the negative environmental impact of the traditional food contact materials.

In line with the aforementioned objective, a by-product obtained from the juice industry, namely pomegranate by-product, was used to prepare different pectin based active materials. The analytical methods developed to determine the by-product key components and the different films prepared, jointly with the results of the migration assays carried out are summarized in this document.

2. MATERIALS AND METHODS

2.1 Active compounds identification and quantification by liquid chromatography-diode array detector (LC-DAD).

The used methodology was described by Andrade et al. [9] and Barbosa et al. [10]. The analytical determinations were performed using an Agilent HPLC 1200 (Hewlett-Packard, Waldbronn, Germany) equipped with an autosampler, a pump, a degassing system, a thermostatted column system, and a DAD that were all controlled by HP ChemStation software. Compounds were separated using a reverse-phase Kinetex EVO C18 100 Å column (150 × 3 mm internal diameter, 5 mm of particle size) (Phenomenex, Torrance, CA, USA) at 30 °C.
The mobile phase was composed of two solvents: water (solvent A) and methanol (solvent B) both with 0.1% acetic acid. The working gradient was: 0 min, 95% of A; 3 min, 90% of A; 10 min, 80%

of A; 18 min, 70% of A; 25 min, 30% of A; 33 min, 0% of A; 33–40 min, 0% of A. The mobile phase flow was 0.6 mL/min, and the injection volume was 20 □ L. Scanning was performed continuously at wavelengths between 200 nm and 400 nm. The phenolic compounds in the samples were identified comparing their retention times and the UV spectra with those obtained by the injected standards under the same conditions. Quantification of compounds was performed using the external standard method at the maximum absorbance of the characteristic wavelengths of the different chemical families of the phenolic compounds (278, 325, and 360 nm).

2.2 Active compounds identity confirmation by liquid chromatography-mass spectrometry (LC-MS/MS).

An LC-ESI-MS/MS (Thermo Fisher Scientific, San José, CA, USA) was used to confirm the presence of phenolic compounds. This instrument was equipped with an Accela quaternary pump, a degassing system, an autosampler, a column, and a TSQ Quantum Access max triple quadrupole mass spectrometer with an electrospray ionization (ESI) source that operated in negative mode. The chromatographic conditions and the column used were the same as those described above for LC-DAD determination.

MS data were acquired in selected reaction monitoring (SRM) mode. Nitrogen gas with a purity of 99.98% was used as the ion sweep gas, envelope, and auxiliary gas, and argon was the collision gas (1.5 mTorr). The MS/MS conditions selected were vaporization temperature set at 340 °C, 2500 V for the electrospray voltage, capillary temperature set at 350 °C, 25 psi of envelope gas, and five arbitrary units of the pressure for the auxiliary gas. The identity of the phenolic compounds was confirmed by comparing the retention time, precursor ion, and fragmentation ions to those obtained from the injection of pure standards.

2.3 Pomegranate extracts

Pomegranate byproduct was generously supplied by the Portuguese juice company, Frubaça—Cooperativa de Hortofruticultores.

The sample (approx. 5 g) was extracted with 50 mL of absolute ethanol on an IKA shaker (VXR basic Vibrax®, IKA®-Werke GmbH & Co. KG, Staufen, Germany) at 2000 rpm for 30 min at room temperature (25 ± 1°C), protected from the light. Then, the mixture was centrifuged (Hettich® Universal Zentrifugen 320R, Tuttlingen, Germany) at 3500 rpm at 15 °C for 15 min. The supernatant was removed and was concentrated on a rotary evaporator (Büchi model R-210 Labortechnik, Switzerland) at 40 °C to a final volume of 10 mL. The extract was held at □20 °C overnight and then evaporated to 5 mL under a nitrogen flow. This extract (E1) was held at -20 °C protected from the light until further analysis. To the solid residue that remained after the extraction with ethanol, 40 mL of acetone were added and a second extraction step applied (E2).

2.4 Films preparation

For the preparation of films, different 2 % pectin solutions with 0.5 % of glycerol were used. For the incorporation of the extracts to the casting solution, different nanoemulsions (NEM) were prepared with the following composition: NEM1: 2% of E1 extract, NEM2: 2% of E2 extract, NEM3: 1% of E1 extract and 1% of E2 extract. The water phase was composed of water (81%) and Tween 20 (15%). The oil phase was elaborated with food-grade oil sunflower at 2% and the corresponding percentage of extracts. The final solution was mixed at 9,000 rpm using Ultraturrax (IKA T25 digital, Spain) for 5 min. Then, the nanoemulsion was obtained by ultrasonic homogenization (Branson Digital Sonifier, SSE-1, Branson, USA). To the casting solution 2 and 4 % of the different nanoemulsions prepared were added resulting in six different films: film A1: 2% of NEM1; A2: 4% of NEM1; B1: 2% of NEM2; B2: 4% of NEM2; C1: 2% of NEM3; C2: 4% of NEM3.

2.5 Total extraction and migration assay of films

Active compounds extraction from the polymeric matrix was performed by immersing 0.1 dm^2 of film in 10 mL of methanol, kept for two days at 70 °C. Then, for further analysis an aliquot was concentrated under nitrogen flow at 40 °C.

Migration assay was performed based on the 10/2011 (UE) Regulation and later modifications using 95% of ethanol (v/v) as food simulant.

3. RESULTS AND DISCUSSION

Pomegranate by-product is mainly composed of seeds, peels, and pulp. The main compounds identified and quantified by LC-DAD in both extracts (E1 and E2) were ellagic acid (77.74 and 151.05 mg/L respectively), punicalagin A (87.01 and 718.20 mg/L) and punicalagin B (90.11 and 992.32 mg/L) that were quantified by LC-DAD. Moreover, a large number of related polyphenolic compounds were identified in the extracts by LC-MS/MS. The methodology followed that the same used by Barbosa et al. [10]. These are summarized in Table 1. The identity of some of them was confirmed by the corresponding commercial standard (Std in Table 1), while in other cases was based on the elemental composition data determined from mass measurements in negative ionization mode found in the literature.

Table 1 Compounds identified in the pomegranate extracts.

Peak	R_t (min)	λ_{max}	$[M-H]^{-1}$ (m/z)	MS/MS (m/z)	Tentative identification	Confirmation
1	2.51	290	331	169	Galloyl-hexoside	[11,12]
2	2.66	270	169	125	Gallic acid	[11], Std
3	3.74	260	541	367	Punicalagin isomer	[11]
4	4.31	260, 293	153	109, 108	Protocatechuic acid	[11,13], Std
5	4.89	325	1082	600, 781, 300	**Punicalagin A**	[11-13], Std
6	6.43	280	577	407, 289	PCB	[11,13]
7	6.5	325	197	123, 182	Hydroxycaffeic acid	[11]
8	7.04	325	1082	600, 781, 300	**Punicalagin B**	[11], Std
9	7.15	280	577	407, 289	PCB1	[11,13] Std
10	7.94	280	289	245, 203	Catechin	[11,13], Std
11	8.65	240, 330	799	781	Ellagic acid der (Granatanin A)	[11,13]
12	8.82	250, 325	561/525	363	Vanillic acid der	[12]

Peak	R_t (min)	λ_{max}	[M-H]$^{-1}$ (m/z)	MS/MS (m/z)	Tentative identification	Confirmation
13	9.34	250, 325	561/525	363	Vanillic acid der	[12]
14	9.59	325	179	134, 135	Caffeic acid	[13], Std
15	9.99	250, 325	353	191, 173	Chlorogenic acid	[11], Std
16	10.13	280	577	407, 289	PCB	[13]
17	10.32	260, 370	783	301	Pedunculagin I	[11]
18	10.9	260, 370	785	301	Pedunculagin II	[11]
19	11.42	240, 330	355	192, 95	Ferulic acid-*O*-hexoside	[11,12]
20	11.64	280	289	245, 203	Epicatechin	[11,13], Std
21	12.04	280	457	169, 125	Epigallocatechin Gallate (EGCG)	[13], Std
22	12.88	280	577	289, 407	PCB	[11,13]
23	13.39	310	163	119, 93	*p*-Coumaric acid	[11,13], Std
24	15.26	325	193	134	Ferulic acid	[13], Std
25	15.51	325	475	163	Coumaric acid der	[12]
26	17.62	360	463	301	Ellagic acid hexoside	[11, 12]
27	18.61	280	305	179	Epigallocatechin	[11], Std
28	18.61	280	577	289, 245	PCB	[13]
29	21.21	360	463	301, 300	Ellagic acid hexoside	[11, 12]
30	21.25	360	463	300, 271	Isoquercetin	[13], Std
31	21.37	278	435	273, 167, 123	Phlorizin	[11], Std
32	21.39	510	595	273	Cyanidin-3-O-rutinoside	[11]
33	21.5	255, 360	301	283, 145	**Ellagic acid**	[11-13], Std
34	21.66	360	447	301, 300	Ellagic acid deoxyhexoside	[11,13]
35	22.1	360	433	301, 300	Ellagic acid pentoside	[11,13]
36	22.34	360	447	301, 300	Ellagic acid deoxyhexoside	[11,13]
37	22.69	280, 500	625	301	Delphinidin-3,5-O-diglucoside	[11,12]
38	23.84	360	274	70, 88	Quercetin	Std
39	24.2	360	425	301	Ellagic acid der	[11]
40	24.63	365	579	271	Naringenin-7-O-rhamnoglucoside	[13]
41	24.9	280, 360	269	117	Apigenin	Std

E1 and E2 extracts yielded from the pomegranate by-product are very rich in phenolic compounds, including punicalagin A and B and ellagic acid that showed high radical scavenging capacity in previous works [14].

Regarding the composition of the prepared nanoemulsions, only ellagic acid was detected, being NEM2 the one with higher amounts of this active compound (NEM1: 1.89 mg/L, NEM2: 4.69 mg/L, NEM3: 2.66 mg/L).

Films prepared by casting were subjected to two assays: the extraction allowed to know the compounds available to migrate and with the migration assay was possible to quantify the compounds that would migrate to the food under packaging conditions. Observing the data (Table 2) obtained from the extraction of films prepared with the higher concentrations of ellagic acid were found in the films prepared with 4 % of each nanoemulsion. Considering the data obtained in the migration, C2 film showed the higher migration rate, being the more suitable formulation to be used as active packaging.

Table 2. Quantification of ellagic acid in the extraccion and migration assays of prepared films.

Films	Ellagic acid (μg/dm²)	
	Extraction	**Migration**
A1	6.40 ± 1.48	<LOQ
A2	52.78 ± 12.44	0.59 ± 0.12
B1	7.56 ± 1.46	<LOQ
B2	23.44 ± 3.41	1.12 ± 0.14
C1	7.56 ± 1.24	<LOQ
C2	21.63 ± 2.71	3.06 ± 0.70

4. CONCLUSIONS

Extracts from pomegranate by-products are rich in phenolic compounds, including punicalagin A and B and ellagic acid. Nanoemulsions could be an effective system to be used as a carrier of ellagic acid.

Pectin films developed with pomegranate extracts allowed the release of ellagic acid for the estimated time contact between the active film and fatty food.

5. ACKNOWLEDGEMENT

ValICET project is supported by the PRIMA program (Partnership for research and innovation in the Mediterranean area, Section 2 Funded Projects 2020, Agri-Food Value Chain) and funded in Spain by the Spanish Ministry of Science and Innovation and the State Research Agency project PCI2021-121965 (MCIN/AEI/10.13039/501100011033 and by the European Union "NextGenerationEU"/PRTR). A. Sanches Silva is grateful to the Foundation for Science and Technology (FCT) for funding the ValICET project in Portugal (PRIMA/0001/2020).

6. REFERENCES

[1] PRIMA Foundation, "Partnership for Research and Innovation in the Mediterranean Area". https://prima-med.org (accessed on 19 April 2022).

[2] European Commission, Sustainable Development Goals. https://ec.europa.eu/international-partnerships/sustainable-development-goals_en (accessed on 19 April 2022).

[3] Adámez, J. D., Samino, E. G., Sánchez, E. V., & González-Gómez, D. (2012). In vitro estimation of the antibacterial activity and antioxidant capacity of aqueous extracts from grape-seeds (Vitis vinifera L.). Food Control, 24(1-2), 136-141.

[4] Habib, H. M., Platat, C., Meudec, E., Cheynier, V., & Ibrahim, W. H. (2014). Polyphenolic compounds in date fruit seed (Phoenix dactylifera): characterisation and quantification by using UPLC□DAD□ESI□MS. Journal of the Science of Food and Agriculture, 94(6), 1084-1089.

[5] Yılmaz, F. M., Görgüç, A., Karaaslan, M., Vardin, H., Ersus Bilek, S., Uygun, Ö., & Bircan, C. (2019). Sour cherry by-products: Compositions, functional properties and recovery potentials-a review. Critical reviews in food science and nutrition, 59(22), 3549-3563.

[6] Pataro, G., Carullo, D., Bobinaite, R., Donsì, G., & Ferrari, G. (2017). Improving the extraction yield of juice and bioactive compounds from sweet cherries and their by-products by pulsed electric fields. Chemical Engineering Transactions, 57, 1717-1722.

[7] Bellavite, P., & Donzelli, A. (2020). Hesperidin and SARS-CoV-2: new light on the healthy function of citrus fruits. Antioxidants, 9(8), 742.

[8] Andrade, M. A., Lima, V., Silva, A. S., Vilarinho, F., Castilho, M. C., Khwaldia, K., & Ramos, F. (2019). Pomegranate and grape by-products and their active compounds: Are they a valuable source for food applications?. Trends in Food Science & Technology, 86, 68-84.

[9] Andrade, M. A., Ribeiro-Santos, R., Costa Bonito, M. C., Saraiva, M., & Sanches-Silva, A. (2018). Characterization of rosemary and thyme extracts for incorporation into a whey protein-based film. LWT - Food Science and Technology, 92(January), 497-508. DOI: 10.1016/j.lwt.2018.02.041

[10] Barbosa, C. H., Andrade, M. A., Séndon, R., Silva, A. S., Ramos, F., Vilarinho, F., Khwaldia, K., & Barbosa-Pereira, L. (2021). Industrial Fruits By-Products and Their Antioxidant Profile: Can They Be Exploited for Industrial Food Applications?. Foods, 10(2), 272. DOI: 10.3390/foods10020272.

[11] Ambigaipalan, P., de Camargo, A. C., & Shahidi, F. (2016). Phenolic compounds of pomegranate byproducts (outer skin, mesocarp, divider membrane) and their antioxidant activities. Journal of agricultural and food chemistry, 64(34), 6584-6604.

[12] Brighenti, V., Groothuis, S. F., Prencipe, F. P., Amir, R., Benvenuti, S., & Pellati, F. (2017). Metabolite fingerprinting of Punica granatum L.(pomegranate) polyphenols by means of high-performance liquid chromatography with diode array and electrospray ionization-mass spectrometry detection. Journal of Chromatography a, 1480, 20-31.

[13] Wu, S., & Tian, L. (2017). Diverse phytochemicals and bioactivities in the ancient fruit and modern functional food pomegranate (Punica granatum). Molecules, 22(10), 1606.

[14] Mariño-Cortegoso, S., Yáñez-González, R., Rodríguez-Bernaldo de Quirós, A., Raquel Sendón, R., Barbosa-Pereira L., Development of edible and active films using pomegranate byproducts, in Proceedings of 14th World Congress on Polyphenols Applications - September 22 - 24, 2021.

AG-O10

Pyrogallol Incorporated Thermoplastic Starch Based Films as Oxygen Scavenging Active Bioplastic Packaging

Juthathip Promsorn[1], Nathdanai Harnkarnsujarit[1, 2*]

[1]Department of Packaging and Materials Technology, Kasetsart Univresity, Bangkok, 10900, Thailand
[2]Center for Advanced Studies for Agriculture and Food, Kasetsart Univresity, Bangkok, 10900, Thailand

***Correspondence to:** Department of Packaging and Materials Technology,*
Kasetsart Univresity, Bangkok, 10900, Thailand.
E-mail; nathdanai.h@ku.ac.th

ABSTRACT: Food industries have begun to develop a sustainable packaging to stabilize the nutrients and quality of packaged food. Oxygen is one of the major factors accelerating deterioration which attributes to residual gas in the package headspace and permeation. Preventing food deterioration supports sustainable food production as delaying or reducing degree of food loss. Normally, oxygen scavenger sachets are used to protect oxygen sensitive food products but it's difficult to manage and recycle after their use. Alternatively, natural substances can be used as oxygen scavengers including plant extracts, tocopherol, ascorbic acid, and especially polyphenols, which are known for their ability to react with oxygen. Pyrogallol is a polyphenol which is potentially suitable to be oxygen scavenger for packaging applications. This research aimed to develop thermoplastic starch (TPS) based films containing pyrogallol for oxygen scavenging packaging. The active films were prepared via extrusion and investigated for morphology, mechanical properties, light transmission and oxygen scavenging activity. Scanning electron microscope (SEM) indicated that pyrogallol gave the non-homogeneous and rougher surface of the film. Fourier transform infrared spectroscopy suggested hydrogen bonding between pyrogallol and TPS phase. In addition, the tensile strength also increased with higher pyrogallol contents in the films due to rigidity of pyrogallol particles. Light transmission depended on the amount of pyrogallol in the films. Oxygen scavenging capacity of the materials were determined at different temperature. Results indicated that the increased storage time and temperature gave higher oxygen scavenging capacity. The rate of oxygen absorption and final residual oxygen contents were highly dependent on the amount of pyrogallol and storage temperature which controlled the reaction rates between pyrogallol and oxygen. Accordingly, incorporation of pyrogallol into TPS effectively produced sustainable active packaging which can be used for stabilization of oxygen sensitive product.

Keyword: Thermoplastic starch, Pyrogallol, Oxygen scavenger, Active packaging

1. INTRODUCTION

Sustainable packaging is one of major concern in the food industries nowadays. Numerous food businesses have begun to develop a sustainable packaging which can stabilize quality of packaged food. Oxygen is the main reason for quality deterioration in packaged products. Oxygen contents in the packaging accelerate quality loss including color, smell, taste and quality of food change. Therefore, it is necessary to remove oxygen in the packaging. Normally, oxygen scavenger sachets which contain iron based powder are used to protect oxygen sensitive food products [1]. However, they are difficult to manage after use and had limitation when use with metal detector.

Alternatively, Non-metallic oxygen scavengers have also used as oxygen scavengers including plant extracts, tocopherol, carnosol, sesamol, ascorbic acid, and especially polyphenols [2]. The extracts have ability to react with oxygen. Pyrogallol is a polyphenol that are commonly found in fruits and plants, which is potentially suitable as oxygen scavenger for packaging applications [3]. Polyphenols are known for their antioxidant properties due to the ability to scavenge free radicals [4]. It has been confirmed that some portion of pyrogallol is an active flavonoid constituent and presents a high oxygen absorbing activity [5]. Therefore, this research aimed to produce and characterize the films containing pyrogallol which is organic based oxygen scavengers and measured the oxygen scavenging activity.

2. MATERIALS AND METHODS

2.1 Cast film extrusion

Acetylate cassava starch (SMS Corporation Co., Ltd, Thailand) was mixed with glycerol (Asian Scientific Co., Ltd., Thailand) using a dough mixer (HM-275; Otto King Glass Co., Ltd., China) for 10 min and subjected to twin-screw extrusion (Labtech Engineering, Samut Prakan, Thailand) at temperatures ranging from 90 to 145°C. The ratio of TPS and LLDPE is 50:50. The concentration of pyrogallol (Xi'an Ceres Biotech Co., Ltd., China) is 0, 3 and 5 % of materials. The ratio of amount of potassium carbonate (Laboratory reagent, Kemaus, Australia) to pyrogallol is 2:1. The mixture of thermoplastic starch, LLDPE, pyrogallol and potassium carbonate were subjected to a twin screw extruder with a sheet die to convert into films.

2.2 Morphology

Scanning electron microscopy (JSM-6610LV, Jeol Ltd., USA) was used to determine morphology of the films. The sample was prepared by immersed in liquid nitrogen and cut with a sharp blade. The specimens were placed on the stub and coated with gold using a sputter coater (SPI-Module, Structure Probe, Inc., USA). The surface morphology of the films was characterized at magnification of 200x and 10 Kv

2.3 Fourier transform infrared spectroscopy (FTIR)

Infrared absorption spectra of films were determined using a Bruker Tensor 27 FTIR Spectrometer (Bruker OPTIK GmbH, Germany) in attenuated total reflectance (ATR) mode. IR spectra were collected in five replicates and averaged at a resolution of 4 cm^{-1} with 64 scanning times. IR absorbance spectra were recorded between 500 and 4000 cm^{-1} by software OPUS 6.5.

2.4 Mechanical properties

The tensile strength (TS) and elongation at break (EB) values represented mechanical properties of the film as ASTM D882-88 using a 5965 universal testing machine (Instron, USA). The films were cut into 2.5×10 cm^2 and determined thickness using a digital micrometer. The film samples were kept at 50%RH for 2 days before measurement. Distance between the handle was 5 cm and the speed was 500 mm/min. TS and EB were collected and averaged in nine replicates

2.5 Light transmission

The samples were cut into rectangle shape and kept at 50 % RH for 2 days. Light transmission was measured in triplicate using a UV-vis spectrophotometer (Evolution 300 UV-vis, Thermo Scientific, Germany). Films were clamped between metal plates with an oval-shaped hole and scanned within the wavelength of 200-800 nm.

2.2 Oxygen scavenging activity

Films were cut into 5×5 cm^2 and placed in a glass bottle containing ambient air (20.6% of oxygen). The bottles contained 10 mL of water to generate 100%RH. The concentration of oxygen in the package headspace was measured in four replication using gas headspace analyzer (Dansensor®, CheckPoint3, Dansensor, Ringsted, Denmark). Samples were stored at different temperature namely 4 and 25°C. Septa were adhered to each package. Gas components were collected through the septum using a needle connecting with the headspace analyzer. The oxygen-scavenging capacity was calculated from Equation.

$$\text{Oxygen-scavenging capacity (\%)} = (C_0 - C_t/C_0) \times 100$$

Where C_0 and C_t are the concentrations of oxygen at initial ($t = 0$) and storage time, respectively.

2.7 Statistical analysis

One-way analysis of variance (ANOVA) was performed using SPSS version 22 (SPSS Inc., USA). Significant difference among samples were determined using Duncan's multiple range tests at a confidence interval of 0.05 ($P \leq 0.05$).

3. RESULTS AND DISCUSSION

3.1 Morphology

Surface and cross section morphology of the films were determined by scanning electron microscopy (SEM) (Figure 1). TPS/LLDPE film showed non-homogeneous microstructures between TPS phase and LLDPE phase due to their different polarity. Cross section image shows finely dispersed pores in films containing pyrogallol, indicating better compatibility between hydrophobicity. In addition, surface morphology showed larger fiber-like layer when increased pyrogallol with the direction as extrusion process. These results suggested that pyrogallol enhanced dispersion of TPS in LLDPE matrices.

TPS/LLDPE TPS/LLDPE/PG 3% TPS/LLDPE/PG 5%

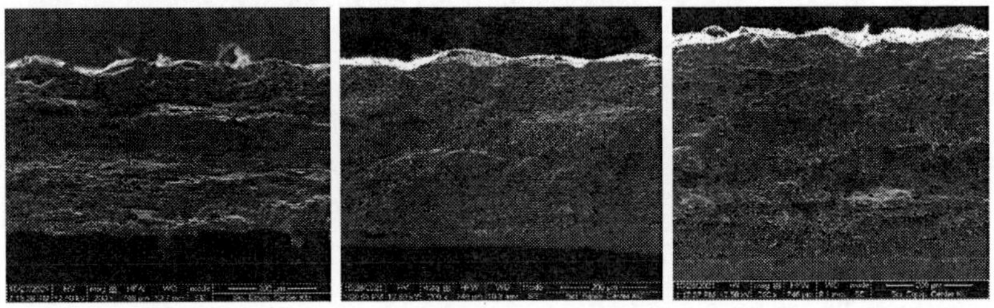

TPS/LLDPE TPS/LLDPE/PG 3% TPS/LLDPE/PG 5%

Figure 1 SEM images showing surface and cross section of TPS and LLDPE film containing pyrogallol (PG) at 0, 3 and 5%.

3.2 Fourier transform infrared spectroscopy (FTIR)

FTIR spectra of TPS/LLDPE containing PG are shown in Fig. 2. The absorption intensity ranged between 3000 - 3600 cm^{-1} attributed to O-H stretching vibration increased when adding pyrogallol. The results suggested interaction between pyrogallol and starch occurred via hydrogen. LLDPE absorption peaks at 2916 and 2948 cm^{-1} attributed to C-H stretching vibration. Intensity of the C-H band decreased with adding pyrogallol indicating interaction between PG and LLDPE. The intensity at 1650 cm^{-1} increased with increasing pyrogallol due to water absorption. Increased hydroxyl groups of materials, because of pyrogallol structure containing 3 hydroxyl groups, formed H-bonding with water.

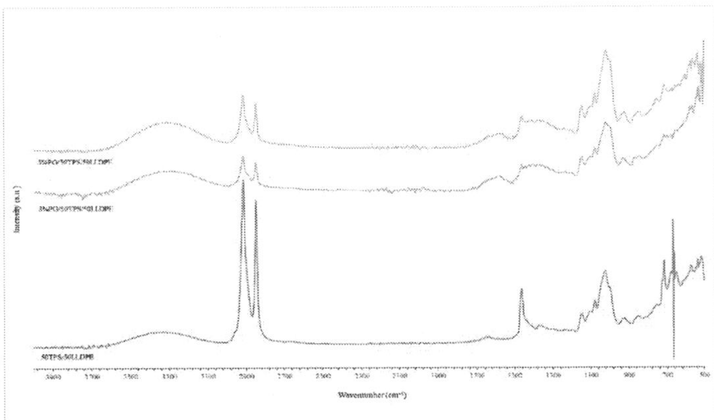

Figure 2 FTIR spectra of TPS and LLDPE film containing pyrogallol (PG) at 0, 3 and 5%.

3.3 Mechanical properties

Tensile strength (TS) and elongation at break (EB) of the films are shown in Table 1. The TPS and LLDPE film incorporated with pyrogallol had increase tensile strength (TS) due to hydrogen bonding between pyrogallol and TPS phase. Elongation at break (EB) decreased with added pyrogallol 3% because adding pyrogallol with small content. However, increased pyrogallol 5% was increasing the compatibility of the pyrogallol with TPS and LLDPE.

Table 1 Mechanical properties of TPS and LLDPE film containing pyrogallol (PG) at 0, 3 and 5%.

Films	Tensile Strength (MPa)	Elongation at break (%)
TPS/LLDPE	13.68 ± 2.28^{a}	680.84 ± 55.22^{a}
TPS/LLDPE/PG 3%	16.48 ± 0.80^{b}	441.88 ± 21.55^{b}
TPS/LLDPE/PG 5%	17.02 ± 1.28^{b}	657.60 ± 16.44^{a}

3.4 Light transmission

LLDPE showed the highest light transmission and transparency at 200 and 800 nm (Figure 1). Light transmission at 600 nm decreased to 0.3, 0.005% with addition of 3, 5% PG, respectively. Increased pyrogallol contents decreased film transparency. The reaction of pyrogallol in alkaline condition with potassium carbonate formed dark pigments.

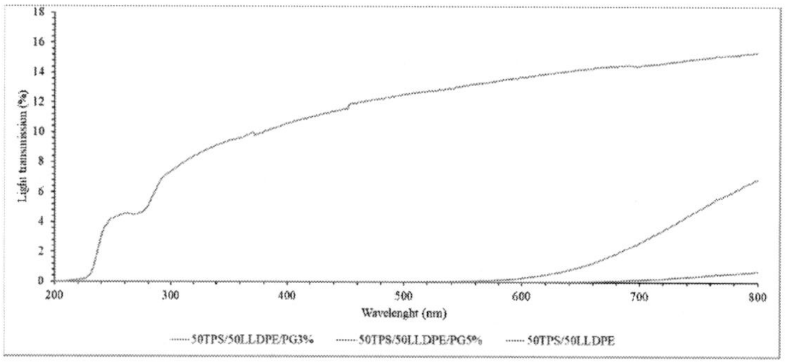

Figure 3 Light transmission of TPS and LLDPE film containing pyrogallol (PG) at 0, 3 and 5%.

3.5 Oxygen scavenging activity

The oxygen scavenging activity of the TPS and LLDPE blend film containing pyrogallol at 0, 3 and 5% was determined at different temperature (4 and 25 °C) and shown in Figure 4. The oxygen scavenging activity was dependent on the pyrogallol content, storage temperature and storage time. Higher pyrogallol gave higher levels of oxygen removal. The Pure TPS and LLDPE blend film show no reduction of oxygen for the whole storage time. The film containing pyrogallol at 5% showed the highest oxygen scavenging capability at all storage temperature (4 and 25 °C). The TPS and LLDPE blend film containing pyrogallol 3 and 5% effectively absorbed oxygen for 2.30 and 3.03% at 4°C. In addition, TPS and LLDPE blend film containing pyrogallol 3 and 5% which stored at 25°C are effectively absorbed oxygen up to 9.91 and 10.65%.

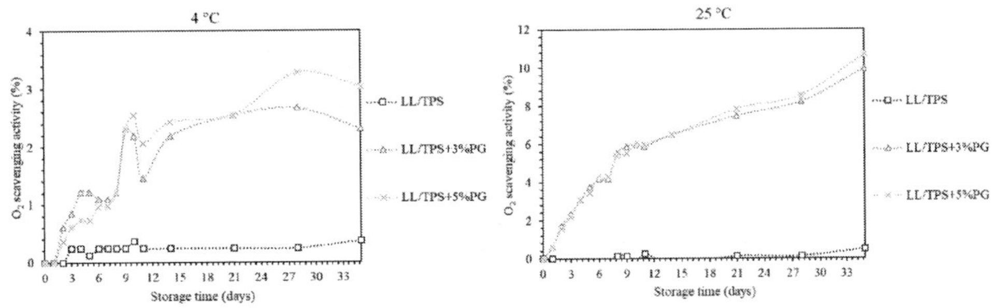

Figure 4 Oxygen scavenging activity of TPS and LLDPE blended films containing different contents of pyrogallol (PG) 0, 5 and 10% as a function of storage time.

4. CONCLUSIONS

TPS and LLDPE blend film containing pyrogallol produced by extrusion process had dark purple color which gave low light transmission. Films with pyrogallol showed oxygen absorption efficacy depending on temperature. Oxygen absorption capacity increased with increasing temperature. However, pyrogallol gave the non-homogeneous and rougher surface of the film due to their different polarity with the blend matrices. In addition, the strength of the films increased due to hydrogen bonding between pyrogallol and TPS phase. These results demonstrate that the TPS/LLDPE containing pyrogallol films can be used as an oxygen scavenging polymers for food products with high moisture content including juice, energy drinks, and raw meat.

5. ACKNOWLEDGEMENT

This research was funded by Kasetsart University Research and Development Institute, KURDI grant no. FF(KU)1.65.

6. REFERENCES

[1] Day, B. P. (2008). Active packaging of food. *Smart packaging technologies for fast moving consumer goods, 1.*

[2] Pokorný, J. (1991). Natural antioxidants for food use. *Trends in Food Science & Technology, 2*, 223-227.

[3] Gaikwad, K. K., & Lee, Y. S. (2016). Novel natural phenolic compound-based oxygen scavenging system for active packaging applications. *Journal of Food Measurement and Characterization, 10*(3), 533-538.

[4] Gaikwad, K. K., & Lee, Y. S. (2017). Effect of storage conditions on the absorption kinetics of non-metallic oxygen scavenger suitable for moist food packaging. *Journal of Food Measurement and Characterization, 11*(3), 965-971.

[5] Jing, G., Huang, H., Yang, B., Li, J., Zheng, X., & Jiang, Y. (2013). Effect of pyrogallol on the physiology and biochemistry of litchi fruit during storage. *Chemistry Central Journal, 7*(1), 1-11.

AG-O11

Pilot-scale Production of Bioactive EVOH-Based Films Containing Methyl Anthranilate: Processing, Functional Properties and Antifungal Activity Evaluation

Alejandro Aragón-Gutiérrez[1]*, Raquel Heras-Mozos[2], Antonio Montesinos[1], Miriam Gallur[1], Carmen Sánchez Reig[1], Daniel López[3], Rafael Gavara[2], Pilar Hernández-Muñoz[2]

[1]*Grupo de Tecnología de Envases y Embalajes, Instituto Tecnológico del Embalaje, Transportey Logística, ITENE, Unidad Asociada al CSIC, calle de Albert Einstein 1, 46980 Paterna, Valencia, Spain;*
[2]*Instituto de Agroquímica y Tecnología de Alimentos, IATA-CSIC, calle del Catedrático Agustín Escardino Benlloch 7, 46980 Paterna, Valencia, Spain.*
[3]*Instituto de Ciencia y Tecnología de Polímeros, ICTP-CSIC, calle Juan de la Cierva 3, 28006 Madrid, Spain.*

***Correspondence to:** Grupo de Tecnología de Envases y Embalajes, Instituto Tecnológico del Embalaje, Transporte y Logística, ITENE, Unidad Asociada al CSIC, calle de Albert Einstein 1, 46980 Paterna, Valencia, Spain. alejandro.aragon@itene.com*

ABSTRACT: In the present study, new formulations based on ethylene vinyl alcohol (EVOH) copolymer were developed by incorporating methyl anthranilate at different concentrations (3, 5 and 8 wt.%) as antifungal compound with the aim of obtaining active films for food packaging applications. To this end, a twin screw extruder with a specifically designed screw configuration for this process, was employed to produce films at pilot scale. The presence of the active ingredient was confirmed by FTIR spectroscopy and the quantification analyses of methyl anthranilate in the films after the extrusion process showed a high retention capacity of the substance in the polymer matrix. Then, the optical, thermal, mechanical and water vapour barrier performance, as well as the antifungal activity in vitro of the active films, were evaluated. The incorporation of methyl anthranilate did not affect the high transparency of EVOH copolymer. Differential scanning calorimetry (DSC) analysis showed a decrease in the glass transition temperature of the active films promoted by the incorporation of methyl anthranilate, indicating a plasticizing effect over the polymer matrix. This behaviour was confirmed by an increase in the elongation values of the cast-extruded films, while Young Modulus was not greatly modified. The water vapour permeability, tested at 33% and 75% RH, slightly increased when methyl anthranilate was incorporated above 5 wt.%, possibly due to the interactions between water and this substance. Finally, the evaluation of the antifungal activity of the active films containing methyl anthranilate showed a great effectiveness against *P. expansum* and *B. cinerea*, demonstrating the potential applicability of the developed films for active food packaging.

Keyword: Methyl anthranilate; ethylene vinyl alcohol copolymer; active packaging; melt-extrusion; antifungal properties.

1. INTRODUCTION

Packaging is positioned as one of the main tools to reduce food spoilage and, therefore, many efforts from both the academia and the food industry are being directed to improve and optimize food packaging systems [1]. Among others, the development of active antimicrobial packaging has emerged as an innovative technique to extend the shelf life by inhibiting microbial growth in foods while maintaining their quality and safety [2,3]. Thanks to their low costs, wide availability and versatile properties, the research and developments on active packaging systems have been mainly focused on the use of plastic materials [4]. In this line, ethylene vinyl alcohol (EVOH) copolymers are among the best candidates used in the design and development of active materials intended for food packaging applications. The high susceptibility of EVOH to water, caused by its hydrophilic nature, results in severe plasticization of the copolymer chain when it is in contact with food products with high water activity. This phenomenon can be utilized as an activation mechanism to promote the release or migration of active agents previously added to the EVOH copolymer to the food media [5]. On the other hand, current trends in active packaging systems restrict the use of synthetic additives in favor of active ingredients extracted from natural bioresources such as essential oils or polyphenol-rich extracts [6]. In this context, we considered to extend the use of methyl anthranilate (MA), a natural ingredient already employed in flavouring foods and drugs, to prepare active packaging films based on EVOH copolymer. Methyl anthranilate is a metabolite responsible of the characteristic odour of Concord grapes and is also present in several essential oils such as neroli, ylang-ylang and jasmine [7]. The direct incorporation of active components into polymeric matrixes through conventional melt-processing technologies such as extrusion, compounding and injection moulding is positioned as an explorable and encouraging strategy to prepare active materials scalable in industrial environments. The main limitation of these routes is that the high temperatures and the shear forces generated during the process usually result in significant losses of the bioactive agents due to their degradation or volatilization [8,9].

Therefore, the objective of this work was to prepare cast-extruded EVOH-based films containing methyl anthranilate as active compound for their use in packaging applications. To this aim, a copolymer composed of 44% ethylene molar content ($EVOH_{44}$) was selected within the EVOH family, since it presents a lower melting temperature due to its crystallinity structure and, hence, the extrusion process can be carried out at softer conditions. In addition, a specific screw configuration was designed in order to optimize the melt-processing and minimize the potential losses of MA. The influence of different concentrations of methyl anthranilate over the main properties of interest of EVOH films in food packaging applications was studied. Finally, the ability of methyl anthranilate incorporated in the active films to inhibit the growth of *P. expansum* and *B. cinerea* was investigated.

2. MATERIALS AND METHODS

Materials: Ethylene vinyl alcohol copolymer composed of 44% ethylene molar content was kindly provided by The Nippon Synthetic Chemical Company (Osaka, Japan). Methyl anthranilate was purchased from Sigma-Aldrich (Sigma-Aldrich Corp., St. Louis, Mo., U.S.A.). For the microbiological assays, potato dextrose agar (PDA) was purchased from Scharlau (Scharlab S.L., Barcelona, Spain). The fungal strains *Penicillium expansum* and *B. cinerea* were supplied by the

Spanish Type Culture Collection (CECT) and were maintained in sterile glycerol at -80 °C until its use.

Samples preparation: Film formulations were prepared in a pilot-scale co-rotating twin screw extruder Brabender TSE 20/40 with a 20 mm diameter screw and a length-diameter ratio (L/D) of 40. EVOH pellets, previously dried at 90 °C for 4 hours, were fed through the main hopper employing a gravimetric feeder. On the other hand, a loss weight injection pump was used to feed methyl anthranilate into the polymer melt, taking advantage of its low viscosity. The pumping rate (mL/min) was adjusted for each formulation in order to obtain a final content of MA of 3, 5 and 8 wt.% with respect to the polymer.

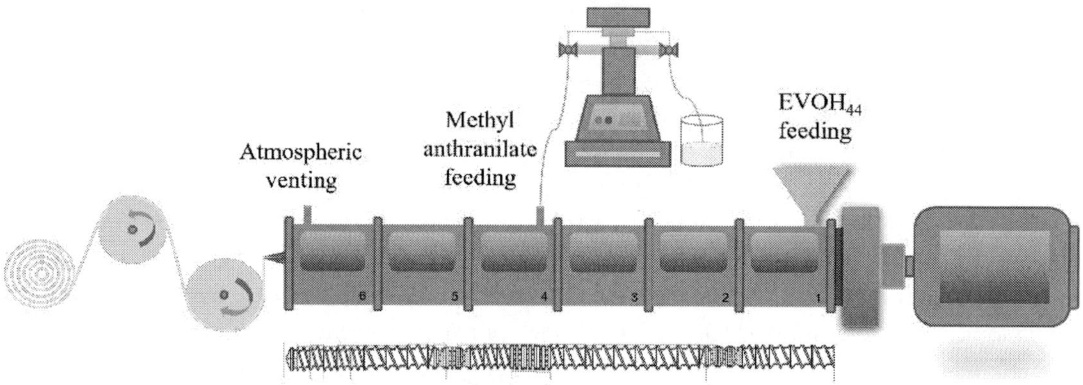

Figure 1. Schematic representation of the extrusion process

The active ingredient was fed in the fourth barrel of the extruder with the aim of reducing its residence time and avoid potential losses of the compound due to its volatility (see Figure 1). The temperature profile along the 6-barrel zones from hopper (zone 1) to die (zone 6) was 175-180-180-180-180-180°C, the applied screw rotation speed was set at 40 rpm, and the throughput was 1.5 kg/h. The compounded material was transformed into films of around 50 μm thickness and10 cm width using a cast film line.

Identification and quantification of methyl anthranilate in EVOH-based films: The presence of methyl anthranilate in EVOH-based films after the extrusion process was qualitatively confirmed by Fourier transform infrared spectroscopy in attenuated total reflectance mode (ATR-FTIR), using a TENSOR 27 Spectrophotometer (Bruker, Massachusetts, USA). Then, the final content of MA remaining in the bioactive films was quantitatively determined by solid–liquid extraction followed by liquid chromatography mass spectrometry (LC/MS) analysis, as described elsewhere [9]. A calibration curve of methyl anthranilate was created by injecting known concentrations of the agent into the LC/MS equipment.

Thermal analysis: A DSC Q-2000 analyzer from TA Instruments (New Castle, DE, USA) was used to evaluate the thermal transition, i.e., glass transition temperature (T_g) of the developed films. The measurements were performed following the procedure described elsewhere [10].

Mechanical properties: The mechanical characteristics of the films were evaluated in terms of elastic modulus (E); and strain at break (ε_B%) by submitting the specimens to stress–strain tests in a universal tensile test machine Testometric M350–20CT (Rochdale, UK), using the conditions described earlier [9].

Water vapor permeability (WVP): The water vapour permeability tests (WVP) were carried out at room temperature and at two different RH, 33% and 75% RH, following the gravimetrically method based on the standard ASTM E96M, as described earlier [11].

Film thickness, and color properties: The film thickness of the films was determined employing a micrometre MiniTest 7200FH (ElektroPhysik Dr. Steingroever GmbH & Co, Germany) with a low range resolution of 0.1 μm. The color properties of the films were measured in the CIELab space using a KONICA CM-2500d (Konica Minolta Sensing Americas, Inc., Ramsey, NJ, USA), following the method described in our previous work [9].

Efficacy of EVOH films containing methyl anthranilate against fungal growth *in vitro*: The antimicrobial activity of cast-extruded EVOH films with methyl anthranilate against both fungal strains, *Penicillium expansum* and *Botrytis cinerea,* was carried out by micro-atmosphere assay, which is based on the effectiveness of the volatile vapour phase generated inside the plate. To conduct this assay, 0.3 g of each film of 8 cm of diameter, with different amount of MA (0, 3, 5, 8 *wt.* %), was placed on the lid of the petri dish, which contained 15 mL of PDA and was inoculated at 3 equidistant points with conidial suspension. The plates with active film and fungal inoculation were incubated for 10 days at 26 °C. The growth of the fungal colony was monitored by measuring colony diameter at 3, 5, and 7 days, after which the film was removed and incubated for 3 days more to observe the recovery of the fungal colony. A control without film was also carried out. The radial growth of colony was measured in cm and inhibition of fungal growth with respect to control was calculated and expressed as percentage. All assays were performed in triplicate.

3. RESULTS AND DISCUSSION

3.1 Identification and quantification of methyl anthranilate in active EVOH-based films.

The FTIR spectra of the EVOH films with various methyl anthranilate contents and without agent confirmed the presence of a significant amount of the active ingredient after the extrusion process by the observation of several absorbance peaks related to MA such as the C=O stretching vibration band at 1690 cm^{-1}, the bending of the N-H at 1620 cm^{-1}, and by the C-O stretching vibration at 1248 cm^{-1}. Moreover, the intensity of these bands augmented as the concentration of methyl anthranilate increased in the active extruded films.

The LC/MS quantification analysis results shown in Table 1 revealed that the loss of the active compound during the extrusion process was lower than 25% for all formulations, indicating a high retention capacity of the substance in the polymer matrix. The employed screw configuration, and the optimization of the processing parameters (i.e., screw speed, residence time of the bioactive compound, temperature profile) allowed methyl anthranilate to remain inside the barrel of the extrusion equipment just the necessary time to get a good dispersion and diffusion of active ingredient through the EVOH polymer matrix and avoiding unnecessary losses.

Table 1. EVOH-based formulations, final concentration of MA determined by LC-MS, and glass transition temperature obtained from DSC analysis.

Code	Methyl anthranilate (wt.%)	Retention capacity (%)	Glass transition temperature (°C)
EVOH	n.d.	-	52
EVOH-MA3	2.52 ± 0.36	84	43
EVOH-MA5	3.84 ± 0.20	76	40
EVOH-MA8	6.42 ± 0.17	80	37

3.2. Thermal analysis

In the control EVOH film, it was possible to observe during the first heating scan of DSC a low-intensity endothermic peak at 52 °C, associated with the glass transition region of the copolymer. With respect to the bioactive EVOH films in which methyl anthranilate was incorporated, the glass transition temperature values decreased in temperature, ranging between 37 and 43 °C, indicating an additive-induced plasticization over the EVOH copolymer. This phenomenon was more evident with the highest incorporated amount of MA in the films, that is, 8 wt.%, resulting in a decrease in the T_g of 15 °C (see Table 1).

3.3. Mechanical properties and water vapor permeability.

The mechanical properties of the films are reported in Table 2. The incorporation of methyl anthranilate slightly decreased the Young Modulus (*E*) from 2.5 ± 0.2 GPa of EVOH control film to a value of 2.2 ± 0.1 GPa of EVOH film incorporating up to an 8 wt.% of methyl anthranilate, indicating that the bioactive film formulations had a less rigid behaviour. On the other hand, the plasticizing effect of methyl anthranilate over the EVOH matrix observed by DSC, resulted in a noticeable increase in the elongation at break values, shifting from 174% in EVOH film to 253% in EVOH-MA8 formulation, thus, demonstrating their suitability for flexible packaging applications.

Table 2. Mechanical properties and water vapor permeability of developed bioactive films.

Reference	E (GPa)	Elongation @ break (%)	WVP·10^{15} (kg·m/m^2 s Pa)	
			33 % RH	75% RH
EVOH	2.5 ± 0.2	174 ± 16	0.12 ± 0.01	1.08 ± 0.02
EVOH-MA3	2.4 ± 0.1	224 ± 18	0.11 ± 0.01	0.99 ± 0.04
EVOH-MA5	2.3 ± 0.2	260 ± 21	0.15 ± 0.02	1.14 ± 0.02
EVOH-MA8	2.2 ± 0.1	253 ± 12	0.18 ± 0.01	1.38 ± 0.06

The WVP was tested at 33% and 75% relative humidity at room temperature and the values are gathered in Table 2. The WVP of pure $EVOH_{44}$ shifted from a value of 0.12 $x10^{-15}$ $kg{\cdot}m\ m^{-2}\ s^{-1}\ Pa^{-1}$ at 33% RH to 1.08 x10-15 $kg{\cdot}m\ m^{-2}\ s^{-1}\ Pa^{-1}$ at 75%, which are in good concordance with those reported in previous works [12]. The incorporation of 3 wt.% of methyl anthranilate slightly reduced de WPV with respect to the control EVOH film at both tested humidity, possibly due to the low affinity and less sensitivity of the active ingredient to water. However, films containing 5 and 8 wt.% of MA seemed to be more water sensitive than the control EVOH film and EVOH-MA3 sample at 33 and 75% RH. This result can be explained as a consequence of the significant plasticizing effect observed in the thermal analysis for the active films.

3.4. Optical properties

Table 3 reports the main optical and colour parameters of the cast-extruded active films based on EVOH copolymer incorporated with methyl anthranilate, having all of them an average thickness in the range of 50-55 µm. EVOH control film was characterized by a high transparency and with the addition of MA, the film formulations maintained their original appearance. The developed films exhibited very similar *L* values, 92-93, indicating a high lightness of the colour of the active films. Furthermore, the colour parameter *a** of EVOH copolymer film, which indicates the green (negative) or red colour (positive), was not significantly affected by the presence of methyl anthranilate in the active formulations, shifting towards negative values as the content of the active ingredient augmented in the films. On the contrary, a clear tendency to positive values of *b** parameter provoked by the incorporation of methyl anthranilate was observed in the active EVOH-MA films, indicating a deviation toward yellow chromaticity

Table 3. Thickness, color parameters and transparency of EVOH-based films

Reference	Thickness (µm)	*L*	*a**	*b**
EVOH	50 ± 2	93.1	0.04	-0.49
EVOH-MA3	53 ± 3	92.6 ± 0.1	-0.59 ± 0.02	3.6 ± 0.04
EVOH-MA5	55 ± 3	92.5 ± 0.1	-0.81 ± 0.01	4.12 ± 0.02
EVOH-MA8	54 ± 4	92.6 ± 0.1	-0.78 ± 0.05	4.31 ± 0.04

3.5. Antifungal activity of cast-extruded active EVOH-based films

Finally, the effectiveness of the active films containing methyl anthranilate with 3, 5 and 8 *wt.%* was evaluated against *Penicillium expansum* and *Botrytis cinerea* by generation micro-atmosphere inside petri dish. The inhibition of fungal colony was calculated in presence of different active films. As discussed above, EVOH is a highly hydrophilic material due to its great hydroxyl group content, increasing aroma permeation in presence of high relative humidity. The EVOH-based films in the petri dish were subjected to high humidity environment, which allowed MA to be released from the polymer matrix to the plate headspace, allowing to exert its activity. Fungicidal activity was observed for films with 5 and 8 wt.% of MA. Moreover, films with 3 wt.% of MA showed an inhibition around 80% during all days of incubation. On the other hand, EVOH film incorporated with the highest amount of MA (8 wt.%) also resulted in fungicidal activity on *P. expansum*, while this fungi incubated in the presence of EVOH-MA3 and EVOH-MA5 exhibited a great inhibition, and colonies resulted whiteness and small, which could be associated with an inhibition of sporulation [13]. The visual aspect of fungal colonies after seven days of incubation is represented in Figure 2, which shows the great effectiveness of the developed active films against both fungi. The antimicrobial activity of thermo-processed films may be compromised by losses occurred during film processing [14]. In our study, cast-extruded EVOH films incorporated with MA showed a great antifungal activity due not only to the effectiveness of MA, but also to the high retention capacity of the active ingredient in the films after the extrusion. Thanks to the hydrophilic nature of EVOH copolymer, the compound release from the polymer matrix is mainly due to exposition of films to high relative humidity since fungal growth medium produce high humidity atmosphere into plates, which lead to a rapid release of the volatile substance, allowing a strong inhibition of microorganism [3].

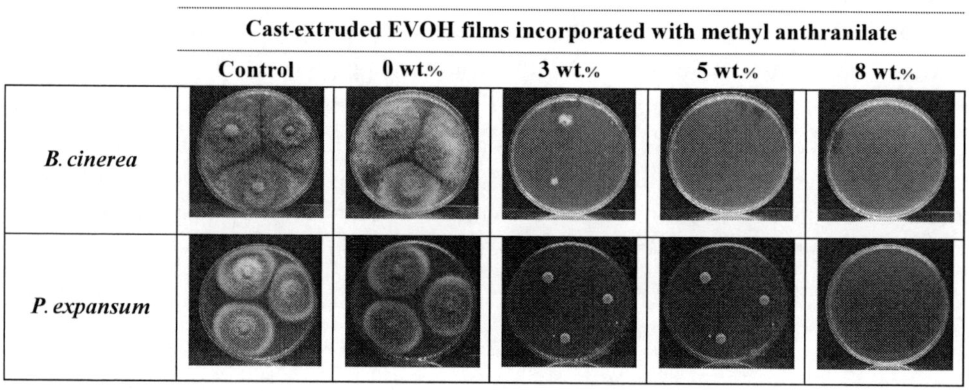

Figure 2. Visual aspect of fungal colony of P. expansum and B. cinerea co-incubated with cast-extruded EVOH films incorporated with methyl anthranilate after 7 days at 26 °C.

4. CONCLUSIONS

In the present work, novel active EVOH-based films containing 3 wt.%, 5 wt.%, and 8 wt.% of bioactive methyl anthranilate compound were prepared at pilot-scale by melt extrusion processing in a twin-screw extruder and fully characterized. The results described in this work show the feasibility of the employed method to obtain bioactive films by an industrial processing extrusion technique with enhanced functionalities thanks to their antifungal properties, which are of great interest in food packaging applications.

5. ACKNOWLEDGEMENT

Gavara and Hernández-Muñoz want to thank the Spanish Ministry of Science and Innovation for financial support (project RTI2018-093452-B-I00) and Heras' predoctoral contract.

6. REFERENCES

1. Han, J.-W.; Ruiz-Garcia, L.; Qian, J.-P.; Yang, X.-T. Food Packaging: A Comprehensive Review and Future Trends. *Compr. Rev. Food Sci. Food Saf.* **2018**, *17*, 860–877, doi:https://doi.org/10.1111/1541-4337.12343.
2. Gavara, R.; López-Carballo, G.; Hernández-Muñoz, P.; Catalá, R.; Muriel-Galet, V.; Cerisuelo, J.P.; Dominguez, I. *Practical Guide to Antimicrobial Active Packaging*; iSmithers Rapra Publishing: Surrey, UK, 2015; ISBN 9781910242735.
3. Muriel-Galet, V.; Cerisuelo, J.P.; Lopez-Carballo, G.; Lara, M.; Gavara, R.; Hernandez-Munoz, P. Development of antimicrobial films for microbiological control of packaged salad. *Int. J. Food Microbiol.* **2012**, *157*, 195–201, doi:10.1016/j.ijfoodmicro.2012.05.002.
4. David J. Ettinger Active and Intelligent Packaging: A U.S. and EU Perspective Available online: https://www.packaginglaw.com/special-focus/active-and-intelligent-packaging-us-and-eu-perspective.
5. Muriel-Galet, V.; Cran, M.J.; Bigger, S.W.; Hernández-Muñoz, P.; Gavara, R. Antioxidant and antimicrobial properties of ethylene vinyl alcohol copolymer films based on the release of oregano essential oil and green tea extract components. *J. Food Eng.* **2015**, *149*, 9–16, doi:https://doi.org/10.1016/j.jfoodeng.2014.10.007.
6. Carpena, M.; Nuñez-Estevez, B.; Soria-Lopez, A.; Garcia-Oliveira, P.; Prieto, M.A. Essential Oils and Their Application on Active Packaging Systems: A Review. *Resour.* 2021, *10*.
7. Wang, J.; De Luca, V. The biosynthesis and regulation of biosynthesis of Concord grape fruit esters, including "foxy" methylanthranilate. *Plant J.* **2005**, *44*, 606–619, doi:10.1111/j.1365-313X.2005.02552.x.
8. Nobile, M.A.; Conte, A.; Buonocore, G.G.; Anna Lucia, I.; Massaro, A.; Panza, O. Active packaging by extrusión processing of reciclable and biodegradable polymer. *J. Food Eng.* **2009**, *93*, 1–6, doi:10.1016/j.jfoodeng.2008.12.022.
9. Aragón-Gutiérrez, A.; Heras-Mozos, R.; Gallur., M.; López, D.; Hernández-Muñoz., P. Hot-Melt-Extruded Active Films Prepared from Packaging Applications. *Foods* **2021**, *10*, 1–18.
10. Aragón-Gutiérrez, A.; Rosa, E.; Gallur, M.; López, D.; Hernández-Muñoz, P.; Gavara, R. Melt-Processed Bioactive EVOH Films Incorporated with Ferulic Acid. *Polymers (Basel).* **2021**, *13*, 68., doi:https://doi.org/10.3390/polym13010068.
11. Ramos, M.; Fortunati, E.; Peltzer, M.; Jimenez, A.; Kenny, J.; Garrigós, M. Characterization and disintegrability under composting conditions of PLA-based nanocomposite films with thymol and silver nanoparticles. *Polym. Degrad. Stab.* **2016**, *132*, doi:10.1016/j.polymdegradstab.2016.05.015.
12. López-de-Dicastillo, C.; Gómez-Estaca, J.; Catalá, R.; Gavara, R.; Hernández-Muñoz, P. Active antioxidant packaging films: Development and effect on lipid stability of brined sardines. *Food Chem.* **2012**, *131*, 1376–1384, doi:https://doi.org/10.1016/j.foodchem.2011.10.002.
13. Ma, W.; Zhao, L.; Zhao, W.; Xie, Y. (E)-2-Hexenal, as a Potential Natural Antifungal Compound, Inhibits Aspergillus flavus Spore Germination by Disrupting Mitochondrial Energy Metabolism. *J. Agric. Food Chem.* **2019**, *67*, 1138–1145, doi:10.1021/acs.jafc.8b06367.
14. González-Estrada, R.R.; Calderón-Santoyo, M.; Ragazzo-Sánchez, J.A.; Peyron, S.; Chalier, P. Antimicrobial soy protein isolate-based films: physical characterisation, active agent retention and antifungal properties against Penicillium italicum. *Int. J. Food Sci. Technol.* **2018**, *53*, 921–929, doi:https://doi.org/10.1111/ijfs.13664.

AG-P01

Development of Electrospun Chitosan Nanofibers for Antimicrobial Food Packaging

Edwin Ochieng[1], Lamin Kassama[1]*, Armitra-Jackson Davis[1], Joong min Shin[2]

[1]Food and Animal Sciences, Alabama A&M University, Normal Al, 35762, USA
[2]Industrial technology and Packaging, California Polytechnic State University, San Luis Obispo CA, 93407, USA

***Correspondence to:** [1]Food and Animal Sciences, Alabama A&M University, Normal Al, 35762, USA*
Lamin.kassama@aamu.edu

ABSTRACT: Electrospinning (EN) process utilizes an electrically charged jet of a polymer solution that is atomized into a substrate of nanofiber. The polymer (chitosan) solution is influenced by its cationic structure, which inversely affects the hydrodynamic and viscosity of the solution, which are key parameters for the EN process. Hence, the objective of this study was to investigate the effect of chemical hydrolysis of Chitosan (CS) polymer on the viscosity of chitosan solution's spinnability into functional antimicrobial packaging film. Medium molecular weight CS was hydrolyzed with NaOH at 90°C for 12, 24, 36, and 48 h. Fourier transform infrared (FTIR) was used to analyze the hydrolyzed samples. Brookfield rheometer was used to test the viscosity of hydrolyzed CS samples at a controlled temperature of 25°C. The electrospun fiber morphology was characterized using a scanning electron microscope (SEM). The antimicrobial properties of the film were tested against *E.coli* and *Salmonella* spp., using the disc diffusion method. The FTIR results showed that hydrolysis does not affect ($p > 0.05$) the molecular structure of chitosan. However, the consistency index of the unhydrolyzed CS solution decreased significantly with the hydrolysis. The initial value (55857±37 Pa.s) was decreased to 35972 ± 68, 28657 ± 91, 2232 ± 84, and 2002 ± 61 Pa.s by 12, 24, 36, and 48 h of hydrolysis, respectively. Chitosan solution generally exhibited pseudoplastic ($n<1$) behavior and the Power law equation was found to be the best fit ($R^2 = 0.96$). An inverse correlation between the electrical conductivity with viscosity was also observed. The morphology of electrospun nanofiber was observed using SEM. The data showed that over 36 h of hydrolysis produced the most consistent and fine nanofiber, while no significant effect ($p > 0.05$) on the antibacterial properties were observed.

Keyword: Antimicrobial, Chitosan, Electrospinning, Hydrolysis, Consistency

1. INTRODUCTION

Electrospinning is a facile technique of fabricating nanofibers from synthetic and natural polymers. The method involves a polymer solution that is stretched through a high electric field to fabricate fibers with diameters between tens of nanometers to submicron [1]. A large surface area is characteristic of a high diameter electrospun fibers to volume ratio due to size and high porosity [2]. According to Ataei et al. [3], the electrospun fiber membranes exhibit fibrous morphology with a large surface area per unit mass (10 - 500 m^2/g), tailored surface texture, a range of diameters between nano to a micron, high porosity (80-90%), and high loading capacity [2,4]. Additionally, electrospinning is a non-thermal process thus suitable for encapsulation of heat-sensitive bioactive compounds, preserving their structure while maintaining stability and functionality during storage [5]. Zhang and and others further outlined the rapid evaporation of the organic solvents used in the process reduces the amount of residue in the packaging film. These attributes make this a potential technology for the sustained release of bioactive compounds for the development of active and functional films desirable for food packaging.

Renewable materials such as biopolymers are preferred for Electrospinning over the traditional petroleum-based packaging polymers due to their biocompatibility and environmentally friendly nature. Chitosan is one of the biopolymers that has been widely explored in the encapsulation of essential oils due to its biocompatible capacity and cost-effective application in food packaging film [5]. Studies have also shown that chitosan has some antibacterial properties [5-7]. Fabrication of chitosan fiber through electrospinning requires the polymer to be dissolved in a solvent, which, by extension, influences the nature of the fiber textures formed [6]. Concentrated acetic acid solutions, acrylic acid, hydrochloric acid, lactic acid, formic, trifluoroacetic acid, and diluted acetic copolymerized with polyethylene oxide are some of the solvents widely used to dissolve chitosan and successfully used to spun encapsulated nanofiber [7].

The viscosity of the polymer solution plays an important role in the ectrospinnability of polymers and hence defines the morphology of the nanofiber. Like other polymers, the spinnability of chitosan depends on the concentration and consistency of the solution, its molecular weight, and the type of solvent used, which in turn affects the viscosity [8]. Salas et al. [6] reported that chitosan of different molecular weights, 30, 106, and 398 kDa on 90% acetic acid, exhibited an increase in viscosity in different concentrations, respectively. Deacetylation and deprotonation of the chitosan molecular chain length using alkali treatment at a specific temperature is one of the methods that influence the viscosity of the polymer to achieve the best electrospinning results. Thus, the objective of this study was to investigate the effect of chemical hydrolysis of chitosan (CS) polymer on the viscosity of the solution's spinnability into functional antimicrobial packaging film.

2. MATERIALS AND METHODS

2.1 Sample Preparations

Medium molecular weight chitosan and sodium hydroxide pellets were purchased from Sigma Aldrich Co. (St. Louis MO. USA). Glacial acetic acid was obtained from Gardena (New Brunswick, NJ. USA).

2.2 Chitosan hydrolysis

The hydrolysis process started with adding NaOH pellets to deionized water in a beaker to make 50% (wt/wt) under constant stirring and heating while gradually increasing the temperature to 95°C. Chitosan was then added to the solution in the ratio of 1:25 (w/v) and the temperature was maintained at 95°C for 12, 24, 36, and 48 h with constant stirring [9]. The samples were then vacuum filtered with a vacuum pump (DVP 2C-TYRO 12, Wertheim Germany) and repeatedly washed with deionized water. The sodium hydroxide was completely removed from the samples by rinsing with hot water several times. The filtered chitosan powdered samples were then oven-dried at 60°C for 12 h.

2.3 Fourier Transform-Infrared (FT-IR) Spectroscopy Analyses

The dried chitosan sample treatments (0, 12, 24, 36, and 48 h) were analyzed using the FTIR (Nicolet Magna-IR 560, Thermo-Electron Corp, Verona WI. USA) with a resolution of 2 cm^{-1} and 32 scans in micro-ATR mode. Time 0 h is the control and represented unhydrolyzed, pure chitosan. The spectrometer was equipped with a Specac Quest single reflection ATR accessory containing a diamond crystal sample plate. The sample was pulverized into a fine powder and blended with 2% KBr (wt/wt). The blended mixtures were compressed into pellets and the FTIR analysis was conducted to determine the composition of the sample.

2.4 Viscosity determination

The viscosities of the samples were measured using a temperature-controlled and programmable Brookfield Viscometer (HBDV-III U, Brookfield Engineering Laboratories, Middleboro, MA, USA). The various hydrolyzed samples and the pure chitosan powders were dissolved in 90% acetic acid to make up 3% (w/v) chitosan solution and placed in the sample cup to measure the viscosity at different shear rates. The samples were characterized at 25°C using the spindle 21, and the rheological properties of chitosan solutions were determined at a shear rate ranging between 50 to 230 s^{-1}.

2.5 Electrical Conductivity determination

The electrical conductivity of the chitosan solutions was determined using the Hanna conductivity meter (HI 4321, Hanna Instruments, RI. USA).

2.6 Electrospinning process

The sample treatments (Hydrolyzed and unhydrolyzed) were dissolved in 90% acetic acid 3% (w/v) in a beaker under constant magnetic stirring at room temperature until all the samples were dissolved. The chitosan solution was placed in a 10 mL volume syringe with a 14.43 mm internal diameter. The syringe was fitted with a steel needle with an internal diameter of 0.91 mm. The electrospinning system (Labmate P301, Spruce Science, CA. USA) consisted of the high voltage generator which was set at 24.5 kV, and at a flow rate of 2.5 mL/h. The distance between the syringe pump needle tip and the collection drum was adjusted to 80 mm, while the collection drum speed was at a constant speed of 100 rpm. The electrospun film samples were collected on the rotating drum collector.

2.7 Scanning Electron Microscopy Analyses (SEM)

The morphology of the electrospun films was characterized using scanning electron microscopy (SEM). The films were coated with a gold layer using an automatic sputter coater under a vacuum. The SEM (Zeiss, Germany), operating at an accelerated voltage of 20 kV was used to analyze the morphology of the nanofibers.

2.8 Antimicrobial analysis

The antimicrobial properties of nanofibers films electrospun from different treatments were evaluated. Circular discs of 6 mm diameter were cut from the various treatment electrospun films and were used to examine the antimicrobial properties against *Salmonella ssp.* The bacterial colonies were suspended in Mueller-Hinton broth to reach turbidity equal to the 0.5 McFarland standard. The bacterial suspension was inoculated on the surface of Mueller-Hinton agar plates using a sterile swap [10]. All sample discs of nanofibers and film were placed on the surface of the plates incubated at 37°C for 24 h. After incubation the zone of inhibition for each sample was determined by measuring the diameter in millimeters. The analysis was conducted in triplicate.

3. RESULTS AND DISCUSSION

3.1 Effect of hydrolysis

Electrospinning is one of the most efficient and effective methods of encapsulating bioactive compounds continuously into functional nanofiber packaging films with polymers. However, this method is affected by many variables which directly influence the morphologies of the nanofibers. Producing a defined morphology requires an understanding of the process of electrospinning and the associated parameters that affect it, such as polymer solution viscosity, electrical conductivity, electric field strength, feeding rate, and the distance between the needle tip to the drum collector. There is a direct correlation between the polymer molecular weight, solution viscosity, and spinnability. Hydrolysis of CS is a significant step that helps in the reduction of the molecular weight of the polymer to enhance the spinnability under the optimum electrospinning process setup. Figure 1 represents an FTIR analysis illustrating the effect of hydrolysis time on the molecular chain length of the medium molecular weight CS powder. The control (unhydrolyzed CS) exhibited a different spectrum at 500-4000 cm^{-1} compared to the hydrolyzed samples, and similar results were also reported by [11]. The absorption band in the range of 2250-2400 cm^{-1} is prominent in pure chitosan, while its intensity is weaker with an increase in the hydrolysis time of the samples, thus confirming the removal of the acetyl group. The characteristic peak for CS, 1599 cm^{-1} of the (-NH_2 tension) is present in both the pure CS and the hydrolyzed samples as was also shown by [9,11] in their analysis of CS. Thus, FTIR results indicated that chemical hydrolysis does not affect the structure of chitosan.

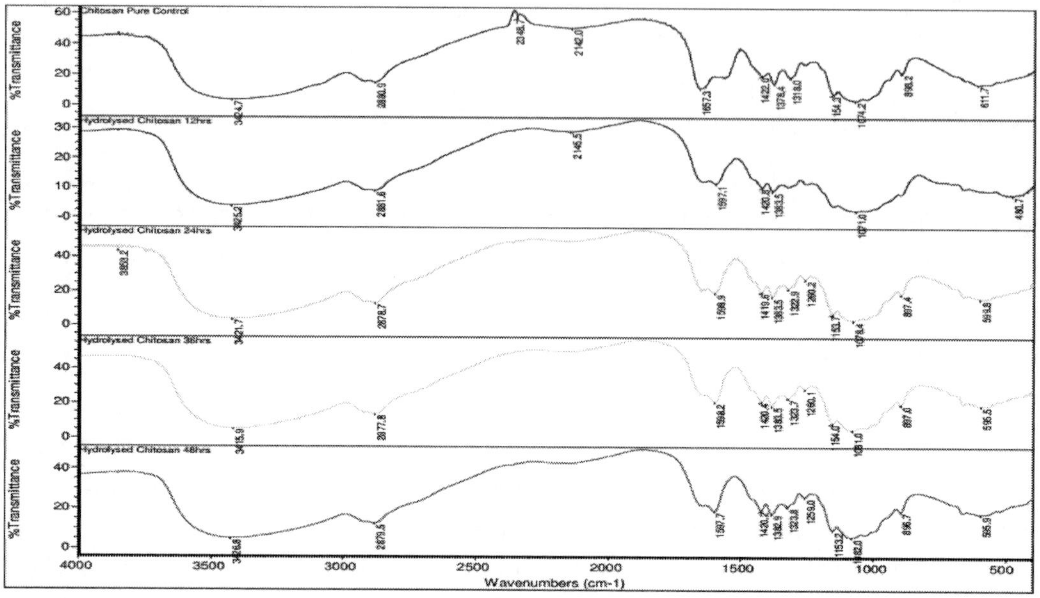

Figure 1: FTIR spectroscopy for different CS solutions hydrolyzed at 0, 12, 24, 36, and 48 h.

3.2 Effect of hydrolysis time on the viscosity

The polymer solution viscosity is an important determinant in the electrospinning process and fiber morphology. Deacetylation of CS chain length through hydrolysis enhances the solubility in 90% acetic acid for a short time. The rheological properties of all the treatments exhibited shear thinning fluid behavior ($n < 1$), hence they can all be characterized as pseudoplastic (Figure 2). The hydrolysis treatment of the samples may have contributed to the ease in disassembly of the macromolecular aggression when external forces are applied as shown by the rheological behavior. The power-law equation was found to be the best fit ($R^2 = 0.96$) for all the samples. The consistency index of the unhydrolyzed CS solution was 55857 ± 37 Pa.s compared to 35972 ± 68, 28657 ± 91, 2232 ± 84, and 2002 ± 61 Pa.s for solutions hydrolyzed for 12, 24, 36, and 48 h, respectively. An increase in hydrolysis time significantly ($p < 0.05$) decreased the apparent viscosity of the CS solution when the shear rate increased from 50 to 140/s. The unhydrolyzed CS had higher apparent viscosity due to its longer molecular chains length and the larger relative movement resistance between the molecules as compared to hydrolyzed samples as also reported by Fatima [11]. Hydrolysis time between 36 and 48 h showed no significant difference ($p > 0.05$) in the apparent viscosity of the solutions. An inverse correlation between the electrical conductivity with apparent viscosity was also observed.

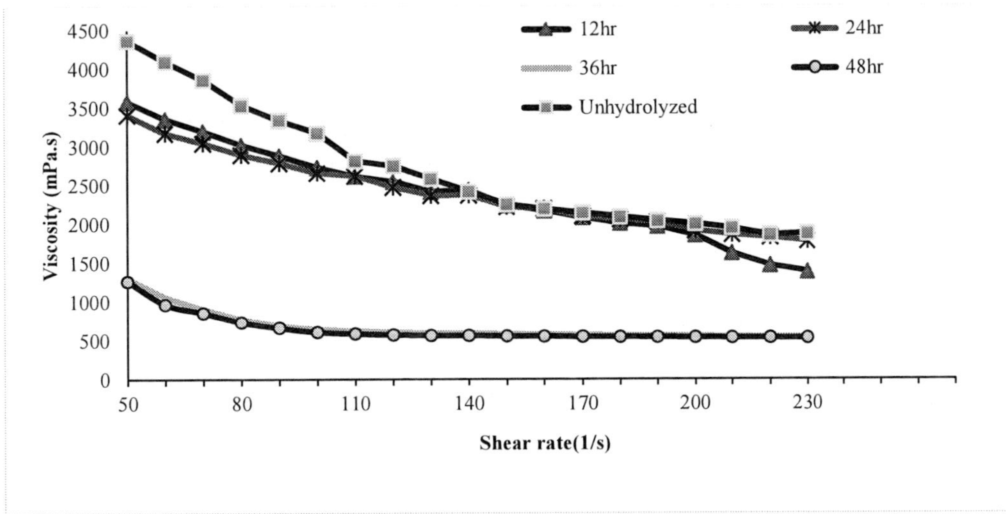

Figure 2: Effect of different hydrolysis time treatments on the apparent viscosity of CS solution.

3.3 Scanning Electronic microscopy results

The morphology of the electrospun films was examined under the SEM, Figure 3 shows the micrograph of 12, 24, and 36 h, respectively. The electrospinning of unhydrolyzed (pure chitosan) was not possible due to its inherent high viscosity. Hydrolyzed chitosan (12 h) resulted in more droppings and electrospraying instead of electrospinning, while the 24 h of treatment shows bead formation during electrospinning as shown in Figure 3. The electrospun mat from the 36 h treatment exhibited a more randomly oriented nanofiber with distinct morphology and less bead formation.

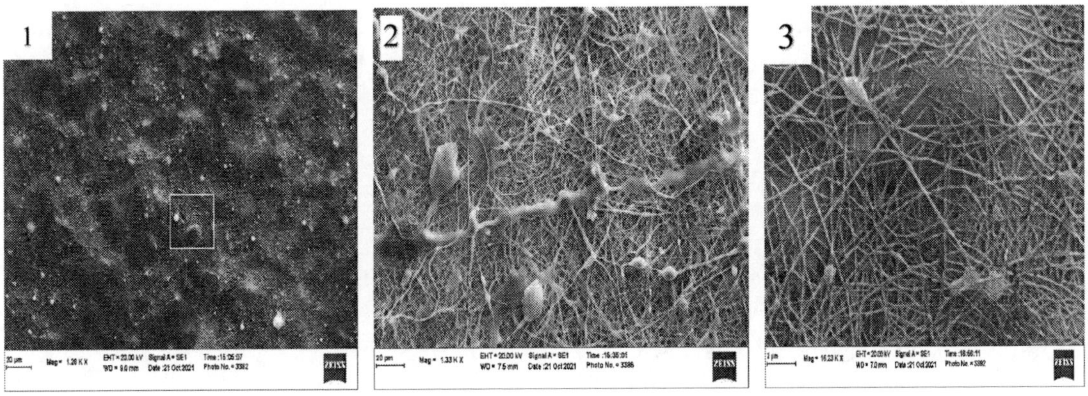

Figure 3. Morphology of chitosan films for samples prepared from 3% polymer solution in 90% AcOH 1) 12 h hydrolysis, 2) 24 h hydrolysis, 3) 36 h hydrolysis.

3.4 Antimicrobial susceptibility analysis

Escherichia coli and *Salmonella spp.* are the most common causes of foodborne illness in the United States [12]. Developing a packaging material with the potential to inhibit the growth of these microorganisms would contribute to the reduction of incidences of foodborne illnesses. The antimicrobial activity of chitosan nanofiber disc was tested against *Salmonella spp.* and the results show some evidence of microbial sensitivity. However, there was no significant difference ($p > 0.05$) in the zone of inhibition diameters among the different treatment hydrolysis times of the chitosan samples.

4. CONCLUSION

Chitosan nanofibers were successfully prepared through electrospinning of hydrolyzed samples. This study demonstrated that the apparent viscosity of the chitosan solution is the main determinant factor for a successful nanofiber fabrication when keeping the other electrospinning parameters constant. The FTIR results showed alkali treatment through hydrolysis helps with the deacetylation of the chitosan structure, which directly influences the consistency of the polymer solution, hence facilitating easy spinnability. The longer the hydrolysis time, the easier it is to spin chitosan solution. However, there was no significant difference ($p > 0.05$) in the viscosities of the polymer solution found after 36 h of hydrolysis. There was also no significant difference ($p > 0.05$) in the zone of inhibition against *Salmonella spp.* across the hydrolyzed samples, hence hydrolysis of chitosan was also found to have no significant effect ($P > 0.05$) on the antimicrobial properties of chitosan.

5. ACKNOWLEDGEMENTS

The authors wish to acknowledge the USDA National Institute of Food and Agriculture, [USDA-NIFA CBG- Project Title: Develop Capacity in the Application of Nanotechnology in Active and Intelligent Food Packaging to Enhance Experiential Learning in Food Safety, Accession number: 1018224], and the Agriculture Experimental Research Station, Alabama A and M University for providing financial assistance to support this research project.

6. REFERENCES

[1] Altan, A., Aytac, Z., & Uyar, T. (2018). Carvacrol loaded electrospun fibrous films from zein and poly(lactic acid) for active food packaging. *Food Hydrocolloids*, *81*, 48-59. https://doi.org/10.1016/j.foodhyd.2018.02.028

[2] Topuz, F., & Uyar, T. (2020). Antioxidant, antibacterial, and antifungal electrospun nanofibers for food packaging applications. *Food Research International*, *130*, 108927. https://doi.org/10.1016/j.foodres.2019.108927

[3] Ataei, S., Azari, P., Hassan, A., Pingguan-Murphy, B., Yahya, R., & Muhamad, F. (2020). Essential oils-loaded Electrospun Biopolymers: A future perspective for active food packaging. *Advances in Polymer Technology*, *2020*, 1-21. https://doi.org/10.1155/2020/9040535

[4] Zhang, C., Li, Y., Wang, P., & Zhang, H. (2020). Electrospinning of nanofibers: Potentials and perspectives for active food packaging. *Comprehensive Reviews in Food Science and Food Safety*, *19*(2), 479-502. https://doi.org/10.1111/1541-4337.12536

[5] Cui, H., Bai, M., Rashed, M. M., & Lin, L. (2018). The antibacterial activity of clove oil/chitosan nanoparticles embedded gelatin nanofibers against *escherichia coli* O157:H7 biofilms on cucumber. *International Journal of Food Microbiology*, *266*, 69-78. https://doi.org/10.1016/j.ijfoodmicro.2017.11.019

[6] Salas, C., Thompson, Z., & Bhattarai, N. (2017). Electrospun chitosan fibers. *Electrospun Nanofibers*, 371-398. https://doi.org/10.1016/b978-0-08-100907-9.00015-5

[7] Qasim, S., Zafar, M., Najeeb, S., Khurshid, Z., Shah, A., Husain, S., & Rehman, I. (2018). Electrospinning of chitosan-based solutions for tissue engineering and regenerative medicine. *International Journal of Molecular Sciences*, *19*(2), 407. https://doi.org/10.3390/ijms19020407

[8] Antaby, E., Klinkhammer, K., & Sabantina, L. (2021). Electrospinning of Chitosan for antibacterial applications—Current trends. *Applied Sciences*, *11*(24), 11937. https://doi.org/10.3390/app112411937

[9] Talebian, S., Afifi, A. M., & Khanlou, H. M. (2014). Fabrication and characterization of chitosan/polyvinyl alcohol nanofibres via electrospinning. *Materials Research Innovations*, *18*(sup6), S6-331-S6-335. https://doi.org/10.1179/1432891714z.000000000979

[10] Abbaspour, M., Sharif Makhmalzadeh, B., Rezaee, B., Shoja, S., & Ahangari, Z. (2015). Evaluation of the antimicrobial effect of Chitosan/Polyvinyl alcohol Electrospun nanofibers containing Mafenide acetate. *Jundishapur Journal of Microbiology*, *8*(10). https://doi.org/10.5812/jjm.24239

[11] Fatima, B. (2020). Quantitative analysis by IR: Determination of chitin/Chitosan DD. *Modern Spectroscopic Techniques and Applications*. https://doi.org/10.5772/intechopen.89708

[12] Bacteria and Viruses. (2022). Retrieved 4 April 2022, from https://www.foodsafety.gov/food-poisoning/bacteria-and-viruses

CONFREENCE PROCEEDINGS :

DISTRIBUTION AND TRANSPORT PACKAGING

• ORAL PRESENTATION

• GENERAL STREAM

DB-O01 Impact of the horizontal acceleration phenomenon duration on the tilt or toppling of the load

DB-O02 Structural damage analysis of the contaminating insects subjected to drop shock

DB-O03 Measurement of the Vertical Vibration Experienced by Pallets During Material Handling

DB-O05 Development of a Friction-Driven Finite Element Model to Simulate the Load Bridging Effect of Unit Loads Stored in Warehouse Racks

• PEER REVIEWED STREAM

DP-O01 Response of a packaging system to a transient event: real-time replication VS SRS simulation

DP-O02 Effect of the sustained forces during transportation on the creep behavior of stretch film affecting unit load stability

DP-O03 A Review and Update of the Corrugated Fiberboard Cushioning

DP-O05 On Friction Measurements and Certification of Slip-Preventing Materials for Road Freight Usages

DP-O06 Measurement and Analysis of low-acceleration and long-duration longitudinal events using delivery van

• POSTER PRESENTATION

DB-P01 Record and analysis of Europe / China Rail transportation

DB-P02 The new silk road – Study of transloading events

DB-P03 Evaluating Consumer Package Failure from Review Images for Using Machine Learning

DB-O01

Impact of the horizontal acceleration phenomenon duration on the tilt or toppling of the load

Baptiste Boutrige[1,2,3]*, Elora Leguebe[1,2,3], Victor Huart[1], Jean-Baptiste Nolot[3], Nicolas Krajka[1], Jérôme Pellot[1], Marie Le Baillif[2,3], Serge Odof[2,3], Damien Erre[2,3]

[1]*METROPACK 30-32 Rue du Capitaine Georges Madon ZAC Croix Blandin, 51100 REIMS, FRANCE*
[2] *ITheMM/ Université de Reims Champagne-Ardenne, BP 1039, 51687 REIMS CEDEX 2, FRANCE*
[3]*ESIREIMS-ESIEC École nationale Supérieure d'Ingénieurs de Reims, 3 esplanade Roland Garros, 51100 REIMS, FRANCE*

***Correspondence to:** ESIREIMS-ESIEC École nationale Supérieure d'Ingénieurs de Reims, 3 esplanade Roland Garros, 51100 REIMS, FRANCE.*
E-mail: baptiste.boutrige@metropack.fr

ABSTRACT: During transport, carried products can be subject to long lateral acceleration due to roundabouts, accelerations, brakes, etc. The tilt of the load is the result of a multitude of parameters. Below sufficient application time and intensity, the stress would cause no effect on the load. However, sufficient stress would cause the tilting and eventually the toppling of the load. These stress levels depend on many parameters including the load and the applied acceleration profile. A theoretical approach is proposed to estimate the acceleration duration required to cause the toppling. While an acceleration of 0.6g applied for 50ms would cause no tilt of the load, the same acceleration applied for 400ms would inevitably result in the toppling of the same load. In this study, tilting conditions were defined for a given load and tilting predictions related to a given acceleration profile were investigated. Accelerations profiles were the combination of constant acceleration and acceleration ramps. Acceleration tests were performed with different ramp and constant acceleration durations. Reducing the ramp or the constant acceleration duration was found to significantly reduce the tilting of the load.

Keyword: Acceleration, Tilting, Eumos, Load stability

1. INTRODUCTION

The palletized and transported products need to be stabilized to avoid any load instability transportation such as tilting. Transportation constraints are more and more investigated [1-7]. The study and the comprehension of the acceleration phenomena and their impact on the studied load are to be conducted. Accelerations profiles are the combination of constant acceleration and acceleration ramps. It is this combination which is responsible of potential load instability. The present study was focused on the tilting angle of a load [8] when it was subjected to an acceleration pressure and on the impact of different acceleration profile applied on a load.

2 METERIALS AND METHODS

2.1 Material

The present investigation was performed using cardboard filled with different loads. Two heterogeneous loads (HE1 and HE2) were made out of partially loaded boxes or stakes of the previously described boxes. Details of the different loads are presented in Table 1.

Table 1. Description of the load

		HE1	*HE2*
Height h[cm]		140	141
Length L [cm]		120	120
Width l [cm]		80	80
Weight [kg]		234	274.1
G [cm]	X_G	37	36
	Y_G	67	73
	Z_G	60	60

G (X_G; Y_G; Z_G): gravity center of the load

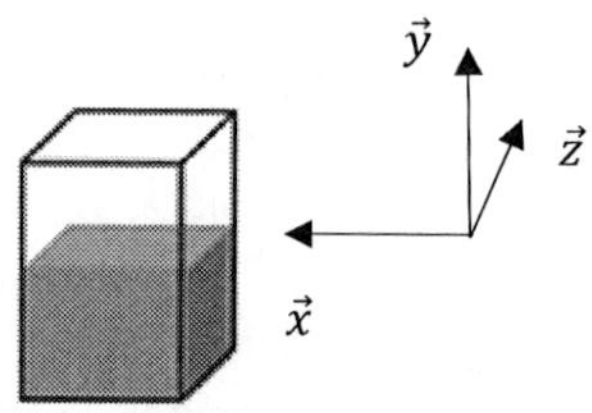

Figure 1 Heterogeneous box load

Acceleration tests were performed using a Lansmont TruMotion acceleration bench (Monterey, United-States) according the EUMOS 40509:2020 standard, without any blocking device. The acceleration (γ) was set to different acceleration profiles. The movement resulting from the chosen acceleration will cause a tilting angle (θ) recorded.

During these tests, two Lansmont saver 3D15 (Monterey, United-States) were used to record data:

Saver 3D15 are used for measurement applications where constant acceleration recording capabilities are required. The record of the 3D15 was a continuous recording with a sampling frequency of 50 Hz.

2.2 *Methods*

a. Theoretical approach

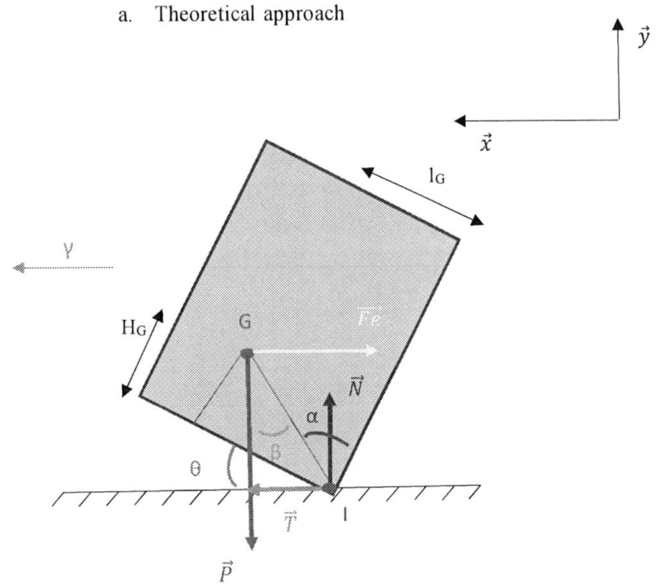

H_G: height of the gravity center
l_G: width of the gravity center
G: gravity centre
I: contact point between the floor and the box
α: angle between the IG segment and the height of the box
β: angle between the base of the box and the GI segment
θ: tilting angle
γ: acceleration value
$\vec{P}$: force of gravity
$\vec{Fe}$: inertia force
$\vec{N}$: normal force
$\vec{T}$: tangential force

Figure 2 : Model used to study the impact of an acceleration on the load tilting

According to physical study, the impact of an acceleration on the box tilting can be described by figure 2 : where H_G is the height and Y_G the width of G the gravity centre of the box, I is the contact point between the floor and the box, α is the angle between the IG segment and the height of the box, β is the angle between the base of the box and the GI segment, θ is the tilting angle, γ is the acceleration value, $\vec{P}$ is the force of gravity, $\vec{Fe}$ is the inertia force, $\vec{N}$ is the normal force, $\vec{T}$ is the tangential force.

By applying the "angular momentum theorem" on the state described in Figure 7, we can easily obtain Equation 4.

$$\ddot{\theta} = -\frac{3.g}{4.GI}\sin(\alpha - \theta) + \frac{3.\gamma}{4GI}\cos(\alpha - \theta) \quad (1)$$

Two situations could be induced from this equation:

<u>Case 1:</u> the tilting angle θ and the angular acceleration $\ddot{\theta}$ are nulls. It corresponds of the balance box situation. γ_{min} can then be defined as the minimum acceleration required for the box to tilt. With an acceleration value less than γ_{min}, the box stays at its initial position. (See equation 2).

$$\gamma_{min} = g.\tan(\alpha) \quad (2)$$

<u>Case 2:</u> the tilting angle θ and the angular acceleration $\ddot{\theta}$ are not nulls. It means that the box tip over because of a constant acceleration value. By using Taylor series to the first order, equation 4 can be simplified to the equation 3.

$$\ddot{\theta} - \left(\frac{3.g}{4.GI}.\cos(\alpha) + \frac{3.\gamma}{4GI}.\sin(\alpha)\right).\theta = -\frac{3.g}{4.GI}.\sin(\alpha) + \frac{3.\gamma}{4GI}.\cos(\alpha) \quad (3)$$

Once the differential equation 3 is solved, the tilting angle can be expressed as a function of the acceleration value and its application time (see equation 4, where H(t) is the Heaviside function).

$$\theta_{(t)} = H_{(t)}.\frac{-g.\sin(\alpha)+\gamma.\cos(\alpha)}{g.\cos(\alpha) + \gamma.\sin(\alpha)}.\left(-1 + \cosh\left(\frac{\sqrt{3.g.\cos(\alpha)+3.\gamma.\sin(\alpha)}}{2.\sqrt{GI}}.t\right)\right) \quad (4)$$

b. Experimental approach

For the acceleration test, one of the 3D15 sensors was put inside a box on the top load, the second was put on the acceleration bench as presented on figure 3.

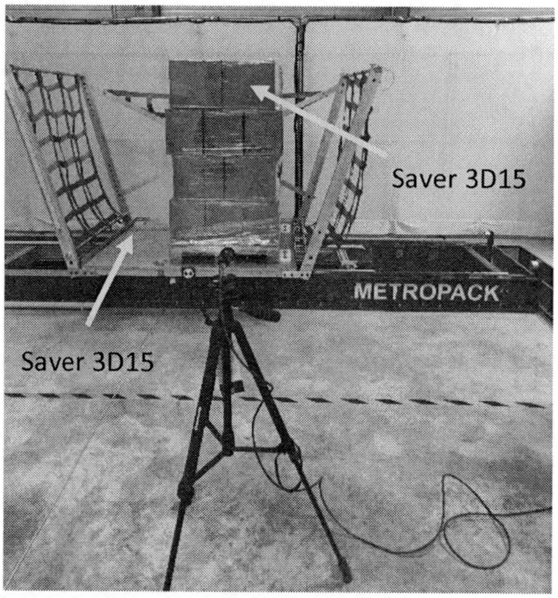

Figure 3 : Representation of the test configuration

Once the configuration is done, acceleration profile was sent to the acceleration bench to study the impact of different profile on the load. During all test, a camera is used to record the movement of the load. Using a tracking video system, it's possible to measure the tilting angle of the load in function of the time.

3. Results and discussion

3.1 Impact of acceleration profile on the load tilting

Different acceleration profiles were sent to HE1 and HE2. The figures 4 and 5 represent two acceleration profiles used, and the tilting angle of the load in function of the time.

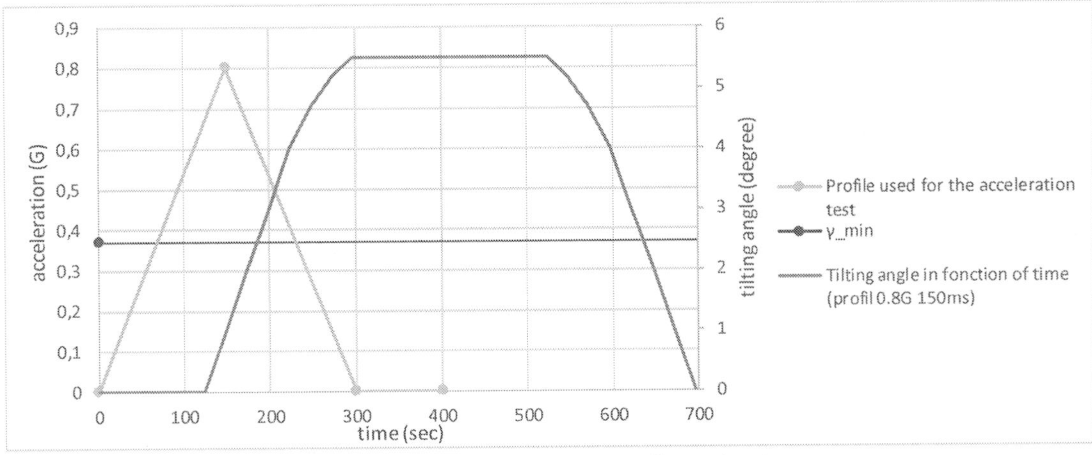

Figure 4 : Used acceleration profile number 1

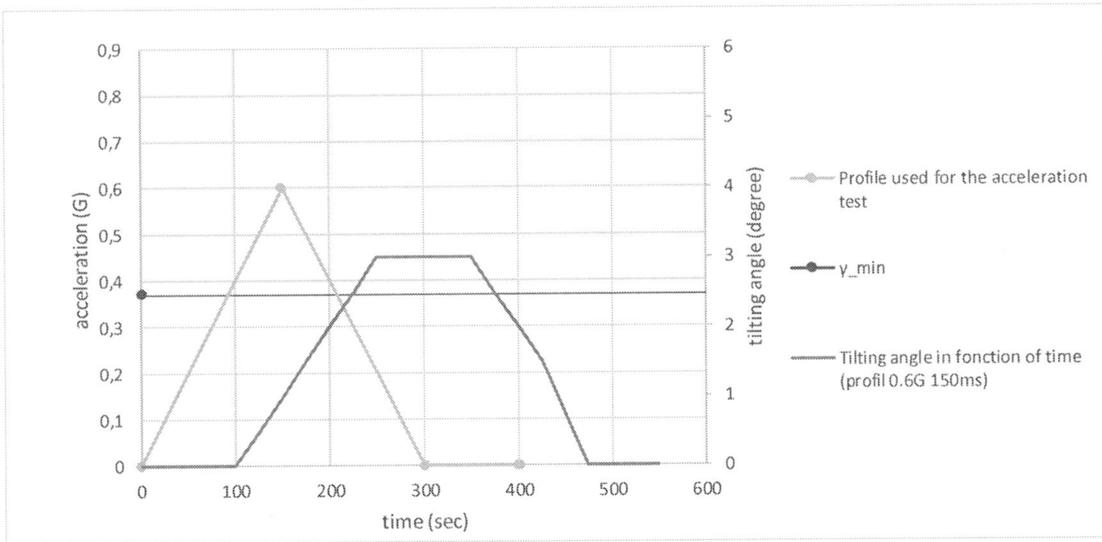

Figure 5 : Used acceleration profile number 2

We can see, on the figure 4 and 5, that the tilting of the load begins just once the acceleration is higher than the minimal acceleration required for the box to tilt. So, no matter what the application time of the acceleration is long, if the acceleration isn't higher than the minimal required acceleration, no tilting of the load is possible. Moreover, the representation of the tilting angle evolution in function of the time is the same: an augmentation of the tilting angle follows by a constant phase of the tilting angle and finish by a decrease of the tilting angle. Just the duration of the phase and the value of the tilting angle change between the different tests.

How the angle of inclination changes depending on the profile used? The following profile, on figure 6, was the profile used for the test.

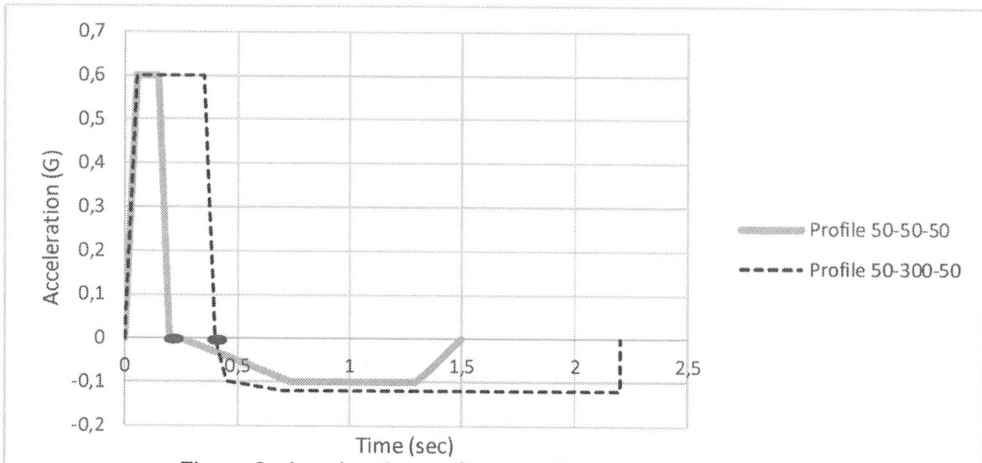

Figure 6 : Acceleration profile used for the test

The two acceleration ramps have a duration of 50ms. In this case, we only change the constant acceleration phase duration. The duration of the phase at constant acceleration varies from 50ms to 300ms.

For each profile tested, the tilting angle is determined when the acceleration pass 0G again (at the red point on the figure 6). We evaluated the angle at this moment because the braking of the acceleration bench is not constant depending on the acceleration test and if we just taken the maximal tilting angle, maybe we had the influence of the different braking. So, we evaluate the tilting angle at this point to test just the influence of the duration of the phase at constant acceleration. With these different tilting angle value, the figure 7 can be made.

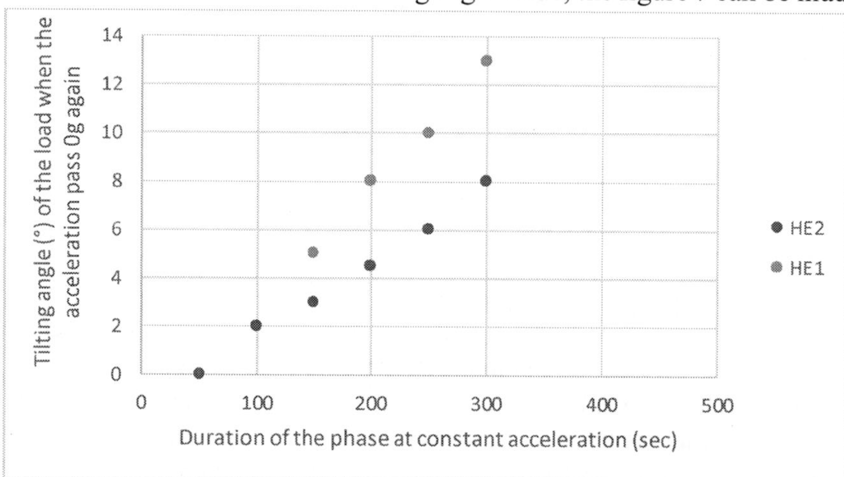

Figure 7 : Tilting angle in function of the duration of the constant acceleration phase

3.2 Recording of acceleration received by the load on the top of the load

To understand what the difference between the signal sent by the acceleration bench and the signal received two saver 3D15 were used. On the following figures 10 and 11, we can see the data recorded after two different tests: An Eumos 0.4G test and a ramp of 200ms to reach 0.4G follow by a ramp of 200ms to reach 0G are performed.

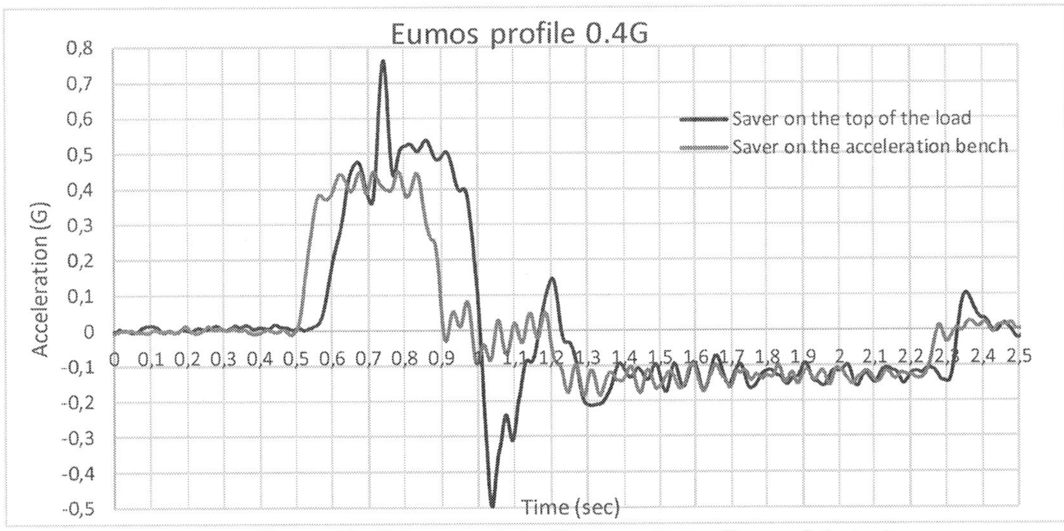

Figure 10 : Data recorded with saver during Eumos 0.4g test

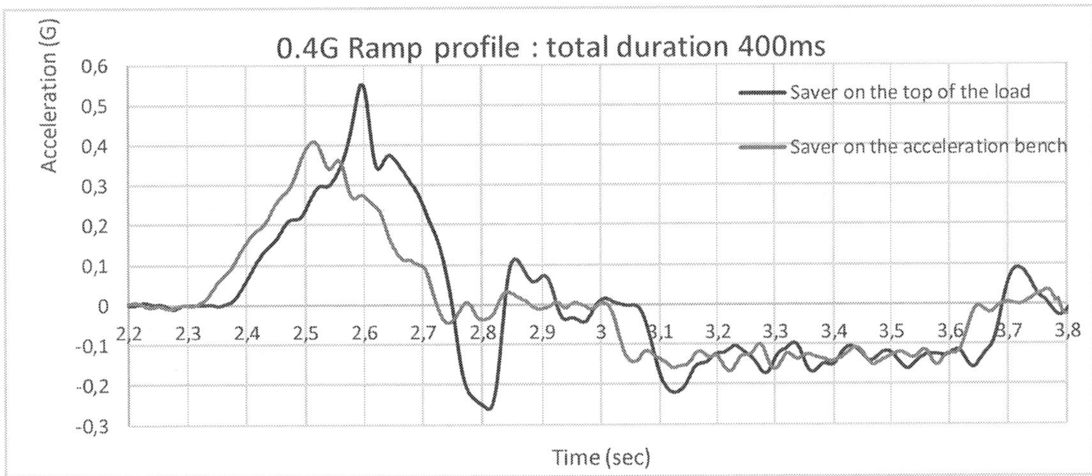

Figure 11 : Data recorded with saver during 0.4g Ramp profile of 200ms per ramp

The saver on the acceleration bench represents the signal sent by the acceleration bench. We can see an interval between the two signals. In fact, there is 50ms between the beginning of the two signals for the two different tests. We can observe there is a phase difference if we observe the signal at the bottom of the load or on the acceleration bench or if we observe it on the top of the load. The second thing that we can see, is a little amplification of the signal recorded on the top of the load. The acceleration maximal should be 0.4G but for Eumos we are at 0.5G (If we exclude the acceleration peak at 0.75G due to interference). So, the signal received by the load is 50ms out of phase and is amplified. It confirms that the profile sent is not the same than the profile received by the load.

4. Conclusion

The tilting angle of a load when is subjected to an acceleration pressure and the impact of different acceleration profiles applied on a load were investigated. We saw than each load has a minimum required acceleration to tilt. If the sent acceleration is lower than it, no tilting is possible. An Eumos profile is composed of an acceleration ramp and a constant acceleration phase. If we change the duration of one of this two parameters, the tilting angle of the load is changed too: the more the duration of the profile will be, the higher the tilting angle will be. With the different Data Logger, we saw that we have a phase difference between the profile sent by the acceleration bench and the profile received at the top of the load. It represents the time response of the load, and it depends on the place we put the sensor. If the sensor is set on the top of the load, the response time will be a little bit longer. These data should be considerate to simulate real event in laboratory.

5. Reference

[1] Huart, V., et al. "Check analysis of horizontal stresses in road transport", 27th IAPRI Symposium Conference on Packaging–Valencia 2015 SPAIN

[2] Rouillard V. Generating Road vibration test schedules from pavement profiles for packaging optimization. Packag Technol Sci. 2008.

[3] Kipp B. Environmental data recording, analysis and simulation of transport vibrations. Packag Technol Sci. 2008.

[4] Juwet M, Esprit E, Vanden Berghe G. On horizontal dynamic effects on palletized goods during road transport: measurement of horizontal dynamic effects. Packag Technol Sci. 2018.

[5] Kudarauskas N. Analysis of emergency braking of a vehicle. Transport. 2007.

[6] Nei D, Nakamura N, Roy P, et al. Wavelet analysis of shock and vibration on the truck bed. Packag Technol Sci. 2008.

[7] Kyle Dunno – Characterization of low-acceleration, long-duration horizontal events using a spring suspension straight delivery box truck, 2021.

[8] Charlie Mathey – Contribution à l'étude du comportement de structures libres, rigides, élancées, glissantes et basculantes sous séisme, 2017

DB-O02

Structural damage analysis of the contaminating insects subjected to drop shock

Saki Matsumoto[1]*, Hiroaki Kitazawa[1], Masayasu Nagata[1], Akihiro Miyanoshita[1]

[1]*Institute of Food Research, National Agriculture and Food Research Organization, Tsukuba, 305-8642, Japan*

***Correspondence to:** Institute of Food Research, National Agriculture and Food Research Organization, 2-1-12 Kannondai, Tsukuba, Ibaraki, 305-8642, Japan.*
matsumotos489@affrc.go.jp

ABSTRACT: Pest contaminants, such as insects or pathogenic microorganisms, are a major threat to the food industry. Invasion by pest insects can occur at any stage during transportation of agricultural products, and identifying the timing and route of invasion is essential in order to prevent further contamination. Contaminated pests can be fragmented by the impact of cargo handling during transportation. However, there have been few methods to access the invasion history of foreign substances in the transported contents using the breakage as a clue. If the shock history of a contaminated insect can be analyzed, the timing and location of contamination may be identified. In the present study, we conducted drop tests and evaluated the damage to insect bodies by examining the total length of cracks on the pronotum and the number of fragments across the whole body. The damage sustained in response to drop shock was found to depend on the state (alive or dead) of the insect at the time of impact. Drop height and total crack length were found to be positively correlated, as was drop height and fragment number. Multiple regression analysis was performed to estimate the drop height from crack length and fragment number. We found that analysis of damage to the insect body may enable estimation of the drop height and could be used to elucidate infestation history during food distribution and in relation to the food distribution route.

Keyword: Drop shock, Insect fragments, Insect pest contamination

1. INTRODUCTION

Contamination of food by insect pests is an unacceptable problem in the food industry. When insect pests are detected in a food supply chain, it is essential to take reasonable precautions to prevent further infestation through migration and reproduction [1–3]. Identifying the specific route and timing of invasion is necessary in order to prevent further invasion into other areas. Investigation for the invasion route begins with gathering information about the contaminated insect such as the location where the pest was found, species, and feeding habits. However, there is a limitation to identifying the invasion route from the above information alone.

Pests are damaged by shocks caused by truck vibrations or cargo handling during transit of contaminated food. It is well known that such shocks cause fresh fruits and vegetables and industrial products to deteriorate [4–7]; however, there has been little research into the damage caused to contaminating pests. If the shocks received by pests can be evaluated from analysis of their damage, it may be possible to identify the location and timing of contamination by relating it to the distribution history of the food products. The present study aimed to evaluate the damage caused to of insect bodies in response to shocks that occurred during transportation. Specifically, we analyzed the number of fragments and total crack lengths of *Tribolium castaneum*, which are important pests of food grains, to clarify the relationship between these factors when the insects are subjected to drop shock. We also evaluated whether drop height can be estimated from the number of fragments and total crack length.

2. MATERIALS AND METHODS

2.1 Insects

The insects, *T. castaneum* (body size: 3–4 mm) were reared in the Institute of Food Research of National Agriculture and Food Research Organization in Tsukuba, Japan. The effects of drop height were tested on alive insects and insects that had been dead for 1 day. For the latter condition, adult insects were killed by placing in boiling water and then stored at 25°C and 70% relative humidity for 1 day.

2.2 Packaging and drop tests

Insects were packaged and drop tests were performed as per previous reports [8]. To prevent shock attenuation, a 2-mm thick aluminum plate (138 × 138 mm, weight: 98.4 g) was attached to the bottom surface of a cardboard box (inside dimensions: 140 × 140 × 47 mm, lid: 148 × 148 × 35 mm, weight: 90.2 g) using double-sided tape. A 1-mm-thick aluminum plate (138 × 138 mm, weight: 49.1 g) was placed over the insect to simulate compaction stress caused by raw grains, and a lid was placed over the cardboard box (Figure 1). Single shocks were induced by dropping the box containing the insect using a drop tester (DTS-50; Shinyei Technology, Tsukuba, Japan). The drop height was set as 0.2, 0.4, 0.6, or 0.8 m. The peak acceleration (ms^{-2}) for each drop height was measured using a published method [5]. Four or five replicates were performed for each drop height and condition (living or dead).

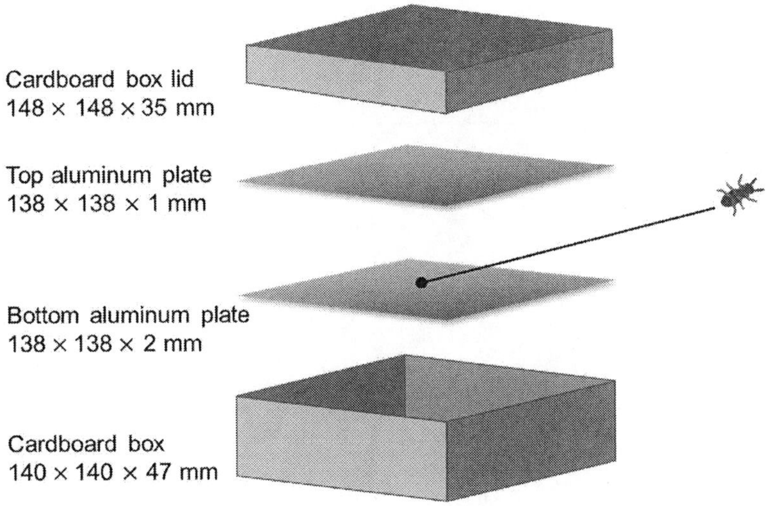

Figure 1 Illustration of the packaging used for the drop test [8]

2.3 Damage analysis

The degree of damage to the insect body was evaluated by analyzing the number of fragments and the total crack length. The number of fragments was counted all separate pieces from which the body had been completely detached. The total crack length was measured the sum of the lengths of all cracks.

Dropped insects were observed under a stereoscopic microscope (VHX-900; Keyence, Osaka, Japan). The total crack length was analyzed using a Fiji software package provided by the U.S. National Institutes of Health [9].

2.4 Statistical analysis

Associations between the number of fragments/total crack length and drop height were determined using Pearson's simple linear correlation. Correlation values were deemed significantly different from zero when $P < 0.05$.

3. RESULTS AND DISCUSSION

3.1 Damage analysis

The average peak acceleration values were 2359.5 ms^{-2} (corresponding to 240.6 G) for 0.2 m, 2650.7 ms^{-2} (270.3 G) for 0.4 m, 3156.8 ms^{-2} (321.9 G) for 0.6 m, and 5194.6 ms^{-2} (529.7 G) for 0.8 m [8]. Images of *T. castaneum* after dropping from 0.2–0.8 m are shown in Figure 2. Living insects exhibited few fragments following exposure to drop shock, whereas many fragments were identified on dead insects following dropping. Hemolymph was observed on the body surface of insects when they were dropped while alive, which made it difficult to identify cracks on the body surface due to coagulation of hemolymph. Although it was possible to confirm the presence of cracks (Figure 2), further analysis of these cracks was difficult. Thus, if the body surface of a dead pest found in food products is found to be coagulated with hemolymph, it can be concluded that the insect was alive when it was impacted, although it may not be possible to estimate the intensity of the impact.

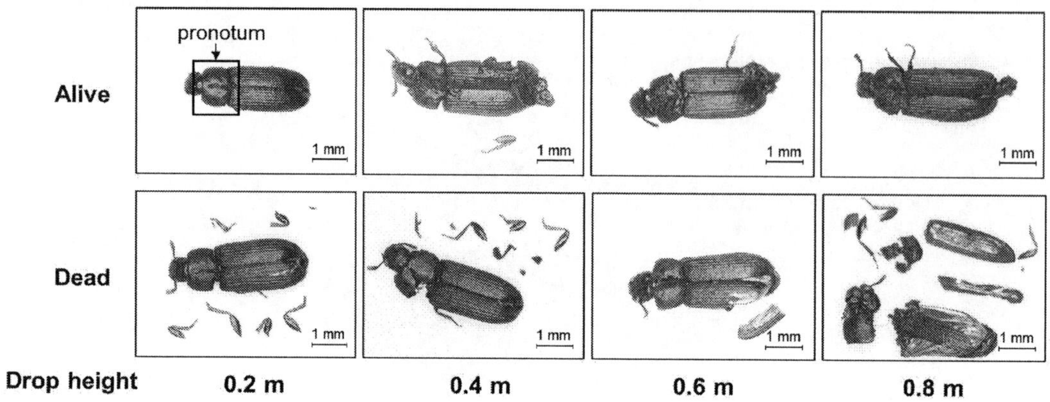

Figure 2 Microscopic images of insects following drop shock sustained when alive (upper images) and dead (lower images)

The total crack length on the pronotum increased with increased drop height when drop shock was induced in dead insects (Figure 3), and drop height and total crack length were found to be significantly correlated. Thus, a long total crack length indicates that insects received a strong drop shock.

The number of fragments was found to be higher when dead insects were dropped compared with living insects, regardless of drop height (Figure 4).

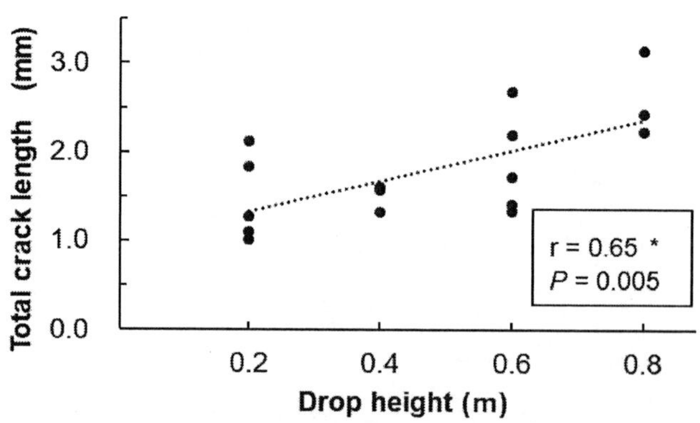

Figure 3 Total crack length on the pronotum sustained after drop tests.
Notes: *significant correlation ($P < 0.05$)

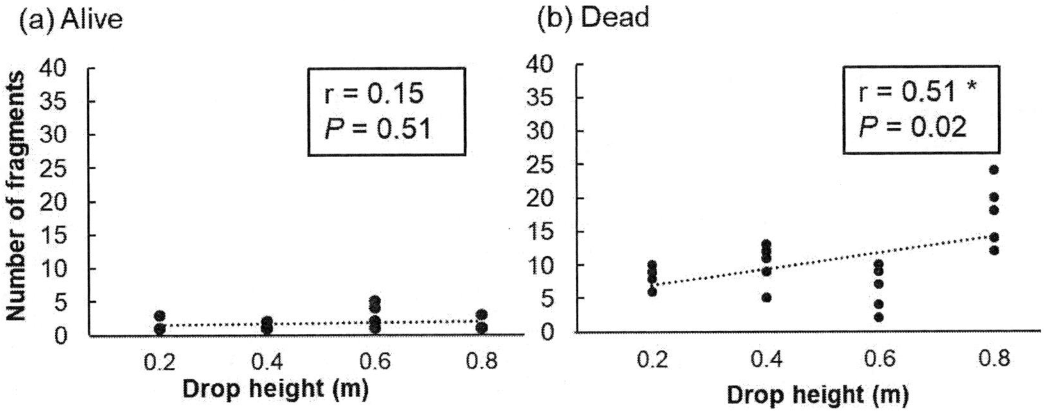

Figure 4 Graphs showing the number of fragments following drop tests with (a) alive or (b) dead insects. Notes: *significant correlation ($P < 0.05$) [8]

3.2 Prediction of drop height from fragment number and total crack length

The result of calculating drop height from the two parameters (number of fragments and total crack length) by multiple linear regression analysis are presented in Figure 5. We identified a moderate correlation between experimental results and the estimated drop height from both parameters. Only total crack length was found to have a statistically significant effect on the dependent factor ($P = 0.02$).

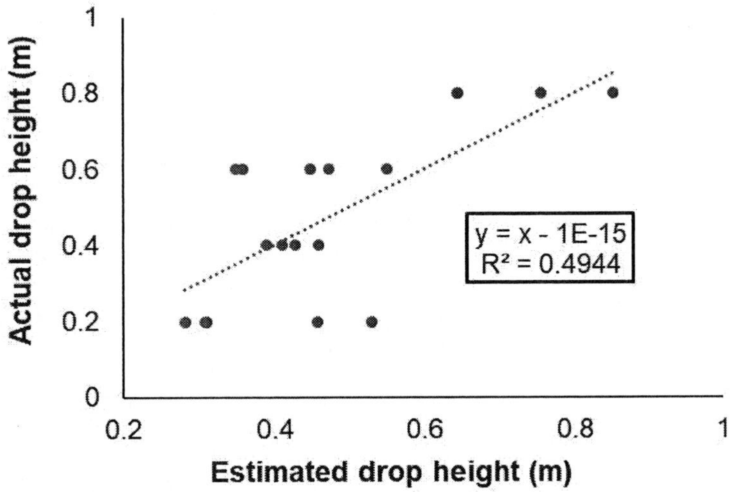

Figure 5 Graph showing the positive correlation between actual and calculated drop height.

4. CONCLUSIONS

If the dead body of a contaminating pest is coagulated with hemolymph, it can be assumed that the insect was alive when it received the shock, although it may be difficult to estimate the intensity of the impact. This may enable identification of the causative transportation events, such as freight handling at the distribution station or at a wholesale market. Analysis of the damage caused by shock events could make it possible to infer when and where insect invasion occurred in the food supply chain. Furthermore, high numbers of fragments or large total crack length indicates the occurrence of high-impact shock after the insect died. Aspects of physical appearance, such as the presence of hemolymph, can help to determine whether the insect was dead or alive when it was subjected to impact. If the distribution route of a food product is known, the timing of contamination can be estimated by comparing damage information with the points along the distribution route where drop impacts may occur, such as unloading and reloading.

5. ACKNOWLEDGEMENT

This work was supported by JSPS KAKENHI Grant (Grant Number: 19K23698) and the Tojuro Iijima foundation for food science and technology.

6. REFERENCES

[1] Mattos EC, Marciano MAM, Daros VDSMG, Faccini CC, Lourenço AM. and Pereira-Chioccola VL. (2020). Fragment detection of Coleopteran and Triatomine insects in experimentally contaminated acai pulp and sugarcane juice. Revista Da Sociedade Brasileira De Medicina Tropical, 53, p. e20190119. DOI: 10.1590/0037-8682-0119-2019.

[2] Johnson, J. B. (2020). An overview of near-infrared spectroscopy (NIRS) for the detection of insect pests in stored grains. Journal of Stored Products Research, 86, p. 101558. DOI: 10.1016/j.jspr.2019.101558.

[3] Perez-Mendoza J, Throne JE, Dowell FE. and Baker JE. (2003). Detection of insect fragments in wheat flour by near-infrared spectroscopy. Journal of Stored Products Research, 39(3), pp. 305-312. DOI: 10.1016/S0022-474X(02)00021-8.

[4] de Q, De F, De C. and Vaccari M. (2018). A brief review on palletized products transportation technology. Fme Transactions, 46(1), pp. 101-107. DOI: 10.5937/fmet1801101T.

[5] Kitazawa H, Li L, Hasegawa N, Rattanakaran J. and Saengrayap R. (2018). Evaluation of shock-proof performance of new cushioning system for portable packaging of apples. Environmental Control in Biology, 56(4), pp. 167-172. DOI: 10.2525/ecb.56.167.

[6] La Scalia G, Aiello G, Miceli A, Nasca A, Alfonzo A. and Settanni L. (2016). Effect of vibration on the quality of strawberry fruits caused by simulated transport. Journal of Food Process Engineering, 39(2), pp. 140-156. DOI: 10.1111/jfpe.12207.

[7] Van Zeebroeck M, Lombaert G, Dintwa E, Ramon H, Degrande G. and Tijskens E. (2008). The simulation of the impact damage to fruit during the passage of a truck over a speed bump by means of the discrete element method. Biosystems Engineering, 101(1), pp. 58-68. DOI: 10.1016/j.biosystemseng.2008.06.003.

[8] Matsumoto S, Kitazawa H, Nagata M. and Miyanoshita A. (2022). Effects of drop shock on fragmentation of two stored grain pest species, *Tribolium castaneum* and *Sitophilus zeamais*. Journal of Stored Products Research, 95, p. 101916. DOI: 10.1016/j.jspr.2021.101916

[9] Schindelin J, Arganda-Carreras I, Frise E, Kaynig V, Longair M, Pietzsch T, Preibisch S, Rueden C, Saalfeld, S, Schmid B, Tinevez JY, White DJ, Hartenstein V, Eliceiri K, Tomancak P. and Cardona A. (2012). Fiji: an open-source platform for biological-image analysis. Nature Methods, 9(7), pp. 676-682. DOI: 10.1038/nmeth.2019.

DB-O03

Measurement of the Vertical Vibration Experienced by Pallets During Material Handling

Clark Sabattus[1], Laszlo Horvath[1]*, Marshall S. White[1], John Bouldin[1]

[1]*Department of Sustainable Biomaterials, Virginia Tech, Blacksburg, 24060, USA*

***Correspondence to:** Department of Sustainable Biomaterials, Virginia Tech, 1650 Research Center Dr, Blacksburg, 24060, USA. lhorvat@vt.edu*

ABSTRACT: Pallets are ubiquitous in the supply chain with over 2.6 billion in circulation in the United States alone. Although often overlooked, pallets can become costly for a company if not designed correctly for their specific supply chain. Durability is an essential characteristic of pallets; it defines the expected life of the pallet in the supply chain. Forklifts are the dominant mode of material handling for palletized products, and they are responsible for the majority of damages experienced by pallets. Despite the prominence of forklifts in the supply chain and their importance in pallet design, there is a lack of research focusing on this interaction between forklifts and pallets. The objective of this research paper was to investigate the intensity of the vibrations that forklifts exert on pallets during material handling. Forklifts in multiple warehousing and manufacturing facilities were instrumented with Lansmont Saver 3x90 data loggers to characterize and record acceleration, duration, and GRMS values during forklift handling in the field. The highest vibration levels were observed for distribution facilities with an average acceleration (peak, g) of 0.353 g. The results of this study will be used to revise the durability testing procedures used in pallet testing standards in order to better represent the current material handling processes found in modern supply chains.

Keyword: Shock, durability, pallet, impact, forklift

1 Introduction

1.1 Pallet Industry

Pallets are ubiquitous in the supply chain with over 2 .6 billion pallets in circulation in the United States. Although pallets can be made of different materials, pallets made of wood dominate the market with a 93% market share [1]. In 2016, 839 million wooden pallets were produced; they consisted of 513 million new pallets and 326 million repaired pallets [2]. This output represents an overall 14% increase compared to 2011 results (22% increase for new pallets and 5% for recycled) [2]. Wood pallets consume 21.8% of new lumber used in the United States, which translated to 9.16 billion board feet of lumber in 2016 [3]. However, millions of pallets end up in landfills each year; 25.39 million alone in 2016. Each year nearly 41 million each year are recycled, and this has gone up recently, indicating there are increasing efforts put forward by landfills to recover wooden pallets [4]. Since the pallet industry uses such a high volume of wood, it is important to look at the historical data in regards to wood usage as this will give us a sense of the state of the pallet industry [2]. This increase in the use of pallets worldwide demonstrates the effectiveness and acceptance of pallets in different industry applications.

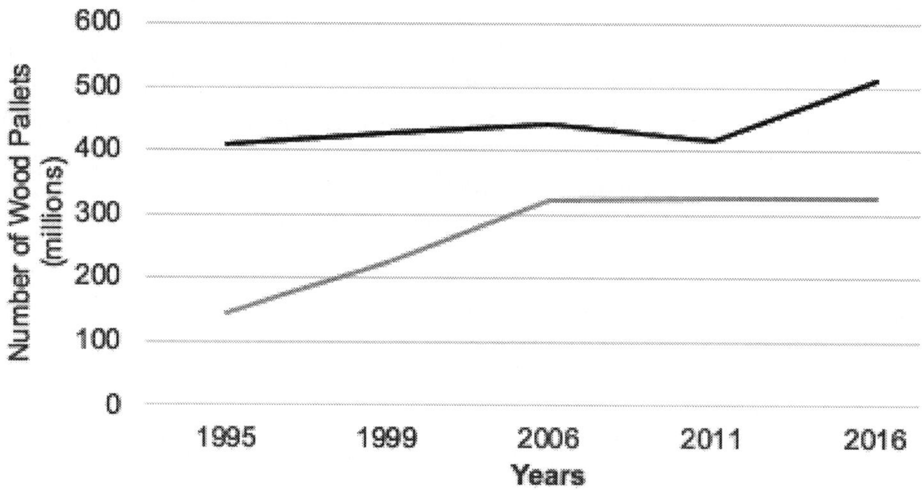

Figure 1: *Estimation of Wooden Pallet Production* [2].

1.2 Warehousing

Warehouses may be defined as a space where goods are received, stored, and distributed [5]. Warehousing is an essential part of most supply chains. There is large variability between warehouses' operations, and this plays a role in the environment that a pallet experiences. The three most common warehousing systems are manufacturing, distribution centers, and food warehouses, respectively, as shown in Figure 2 [6].

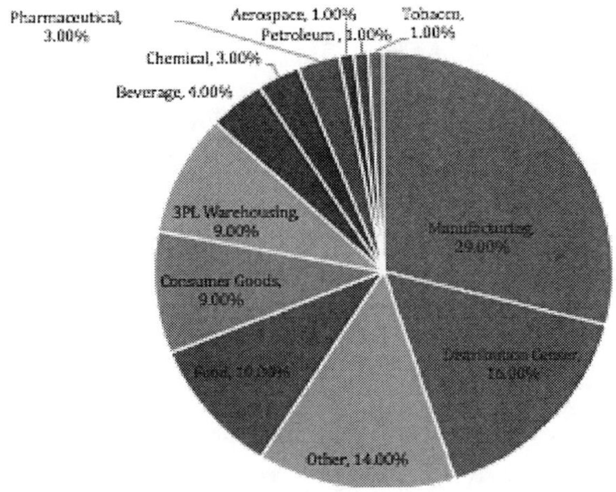

Figure 2: Survey responses received from the different industry sectors. [6].

1.3 *Forklifts*

The movement of goods within a warehouse is conducted using different forms of material handling equipment. The most typical form seen across all warehouses is the forklift; around 91% of respondents indicated forklift use [6]. Forklifts are motorized vehicles with fork tines on the front which allow for the lifting of heavy unit loads [6]. Forklifts come in a variety of unique designs based on its intended use within the warehouse [7]. There are different types of forklifts, such as counterbalance, reach truck, order picker, multidirectional forklifts, and other types that may be designed specifically for their intended applications. The most popular forklift, however, is the counterbalance forklift [8].

1.4 Hazards

A pallet is exposed to a variety of hazards during its life in the supply chain. Hazards leading to pallet damages may create problems within warehousing facilities, and this can result in additional expenditures during operation. In analyzing the useful life of a pallet, it is important to identify the main types of warehousing hazards. One of the hazards that may result in additional vulnerability to a pallets' durability is vibration.

1.5 Vibration

There is a wide variety of knowledge that can be found already which investigates the vibration levels during truck shipment routes, but little information regarding vibration levels within warehouse operations during material handling. However, recently the characteristics of the vibrations that occur from a forklift under different conditions were investigated. The purpose of this study was to investigate the vibrations that industrial forklifts experience. This study looked at gas, reach, and electric forklifts. Each forklift went through a series of trials to determine the different variables such as unit load weight, speed, road conditions, and sensor location [9]. It was found that the highest vibration intensities were in the 3-4 Hz range, and the gas forklift saw the highest intensity vibrations. The study also found that, with an increase in payload, the vibration levels seems to decrease; conversely, an increase in speed increased vibration levels [9]. This investigation helps to build a basic understanding of the vibration levels seen during forklift handling and how different variables effect that vibration; however, there is still a large gap of knowledge relating to what these forklift vibration levels look like in the field.

Objective

The objective of this research was to characterize the vibration acceleration and G_{rms} levels in the field between different warehouse facilities in order to improve unit load testing protocols.

2 Materials and Methods

To measure the impact vibration levels during forklift handling, Saver 3D15 (Lansmont Corp, Monterey CA, USA) and Saver 3X90 (Lansmont Corp, Monterey CA, USA) data loggers were used for data collection. The data loggers were attached to different forklifts found in the facilities that participated in this study. The sensors were left in the facilities for 1-2 weeks to record the acceleration, change in velocity, duration, and grms values of vibrational behavior. Upon retrieval of the data loggers, a statistical analysis was performed to identify the behavior of vibration intensities.

2.1 Forklifts

Forklift type and model varied between the facilities: there were a Yale ERP040 electric forklift, a Hyster CT2 electric forklift, and a Toyota 8FDU gas forklift. The Yale ERP040 is a solid, pneumatic, 3-wheel, counterbalanced, sit-down, electric forklift with a capacity of 3,000 - 4,000lbs. The Hyster CT2 is a solid, pneumatic, 4-wheel, sit-down, electric-controlled powershift, counterbalanced forklift with a capacity of 4,000 - 5,000lbs. The Toyota 8FDU is a 4-wheel, air-pneumatic, sit-down, gas forklift with a load capacity of 3,000 - 6,500lbs. Figure 3 displays each forklift model used during field data collection.

1. Yale ERP040 **2.** Hyster CT2 **3.** Toyota 8FDU

Figure 3: Representative forklifts.

2.2 Floor Characteristics

Floor characteristics varied between facilities. Floor materials such as concrete, asphalt, and dirt were identified. In addition, the condition of the floors was identified as a variable between facilities.

2.3 Load Characteristics

Due to the high payload variability each forklift experienced during duration of each data collection period, payload ranges were not able to be properly identified.

Characteristics:	**Facility 1**	**Facility 2**	**Facility 3**
Facility type	Distribution	Manufacturing	Pallet Manufacturer
Forklift	Yale ERP040	Hyster CT2	Toyota 8FDU
Floor condition	Concrete/Asphalt	Concrete	Concrete/Dirt

Table 1: Summary table of the characteristics of the investigated warehouses.

2.4 Sensor Setup & Settings

Lansmont 3X90 and 3D15 data loggers were used to record vibration measurements. The significant difference between the 3D15 and 3X90 data loggers is the internal accelerometer type. The Lansmont 3D15 utilizes a triaxial MEMS accelerometer, and the Lansmont 3X90 utilizes a triaxial piezoelectric sensor. Piezoelectric accelerometers have the capability of capturing data more efficiently at higher frequencies in comparison to MEMS accelerometers. The sensor settings used in the study are shown in Table 2.

The sensors were mounted using two strips of double-sided mounting tape (Scotch Corp, Saint Paul MN, USA) on the back of the fork tine carriage as seen in Figure 4.

Figure 4: Location of sensor of fork tine carriage.

	3D15		**3X90**	
Settings	**Signal Triggered**	**Timer Triggered**	**Signal Triggered**	**Timer Triggered**
Memory Allocation	30%	70%	30%	70%
Record Time	2s	2s	2s	2s
Sampling Rate	500 samples/sec	500 samples/sec	500 samples/sec	500 samples/sec
Sample Size	1,000 samples	1,000 samples	1,000 samples	1,000 samples
Trigger Level	1.2 g	1.2 g	0.2 g	0.2 g
Filter	250 Hz	250 Hz	250 Hz	250 Hz
Pre-Trigger	50%	50%	50%	50%

Table 2: Sensor Settings.

2.5 Statistical Methods

Data was analyzed using SaverX (Lansmont Corporation, Monterey, CA, USA), JMP (SAS, Cary, NC, USA) and Matlab (MathWorks, Natick, MA, USA) statistical softwares. Vibration events with less than 0.02 G_{rms} and acceleration events with more than 2 g were filtered out using a SaverX summary selector feature. A Tukey's analysis was then run on the mean values to determine if there were any significant differences.

3 Results and Discussion

3.1 Peak Acceleration

The field data was collected from three different types of warehouses. Warehouses 1 (WH1), 2 (WH2), and 3 (WH3) were a distribution center, a manufacturing facility, and a pallet manufacturer, respectively. The mean peak acceleration values observed in this study ranged from 0.264-0.353 g which is comparable to the values measured by Huang et al. (2021) and Borocz et al. (2022) [9] [10]. However, the 98th percentile of the acceleration values observed during field measurement was 70% greater than what was observed during laboratory handling.

As seen in Table 3 below, the mean peak acceleration values in the field were relatively similar. However, the mean acceleration for the gas forklift type, when compared to the other forklifts used in field data collection, were different from Huang et al.'s findings [9]. The gas-powered forklift (WH3) saw a lower average of 0.264 g's when compared to Huang's study. WH1, which utilizes an electric forklift, saw an average peak acceleration value similar to what Huang et al. found for gas forklifts in his study [9]. In an uncontrolled environment, this difference may be attributed to differing forklift operators; they may cause a difference in values from warehouse to warehouse.

The mean peak acceleration values were the lowest for Warehouse 3 (0.264g) and the highest for Warehouse 1 (0.353g). The peak acceleration values at 98% also revealed that peak accelerations experienced by the forklift can be as high as 1.65 g. The distribution plots of the peak acceleration values for the three warehouses are presented in Figure 5. The distribution of the values is skewed to the right showing that most of the peak acceleration values occurred in a low acceleration range below 0.5g. The observed distribution is the same across all three warehouses.

When the CDF plots for the three warehouses are compared to each other (Figure 6), it can be observed that the measured peak acceleration values were similar between Warehouses 1 and 3 while the peak acceleration values were consistently lower for Warehouse 2.

Using a 95% confidence interval, Tukey's HSD shows no significant difference in the mean between Warehouse 2 and 3; however, Warehouse 1 is significantly different from both Warehouse 2 and 3. Warehouse 2 had the highest standard error of all the warehouses at 0.016, compared to WH1 and WH3 which had a standard error of 0.008 and 0.007 respectively.

Warehouse:	Mean (g)	98% (g)	95% (g)	90% (g)
#1	0.353 g's	1.51 g's	1.01 g's	0.67 g's
#2	0.274 g's	1.65 g's	0.84 g's	0.46 g's
#3	0.264 g's	1.13 g's	0.81 g's	0.57 g's

Table 3: Summary of acceleration data measured during forklift vibration for the three investigated warehouses.

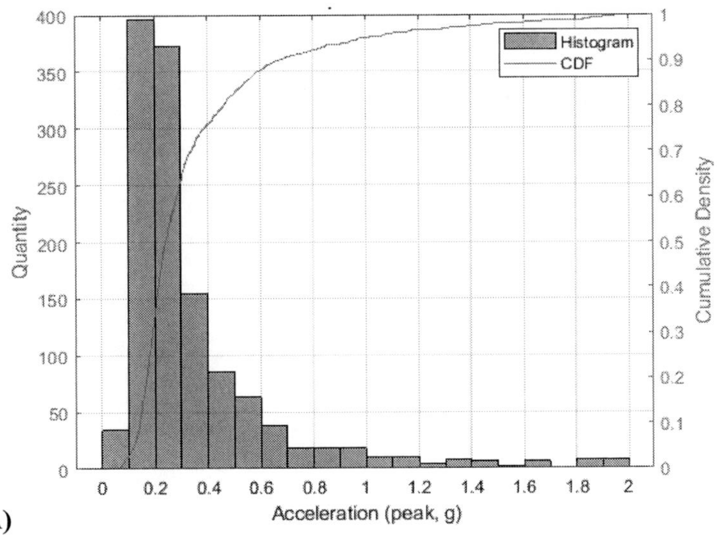

Histogram
CDF
Quantity
Cumulative Density
Acceleration (peak, g)

(A)

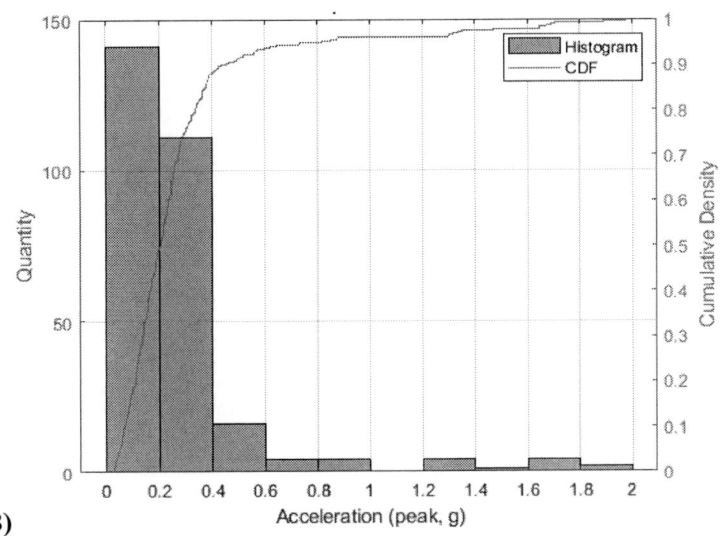

Histogram
CDF
Quantity
Cumulative Density
Acceleration (peak, g)

(B)

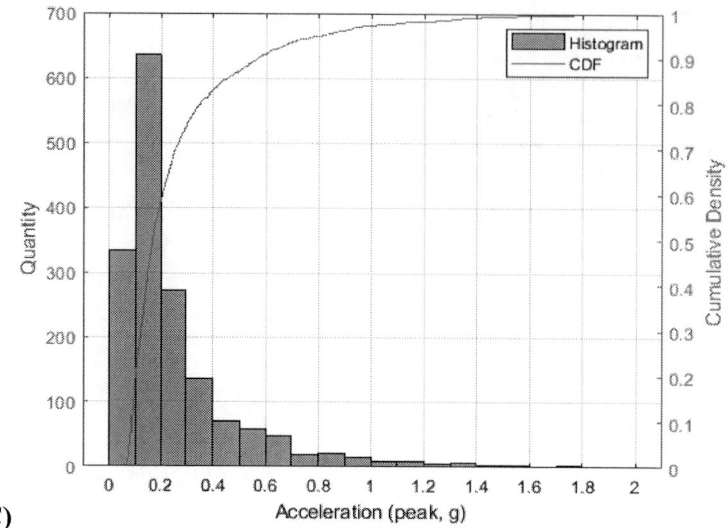

(C)

Figure 5: Histograms and cumulative distribution function plots for the peak acceleration values measured during forklift vibration for (a) Warehouse 1, (b) Warehouse 2, (c) Warehouse 3.

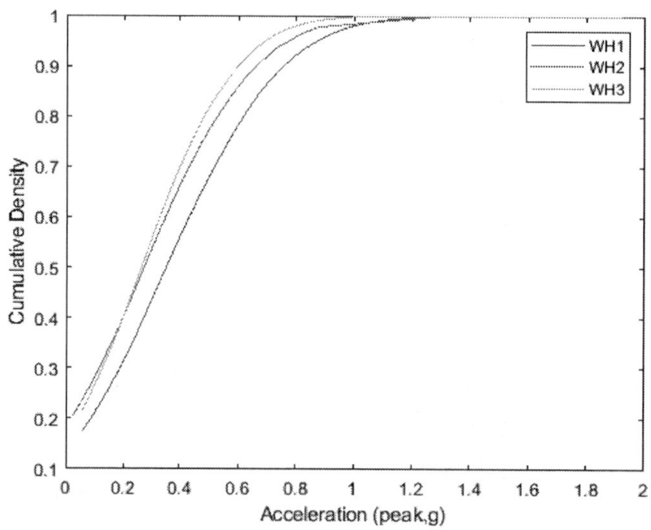

Figure 6: Combined Warehouse Acceleration CDF plots.

3.2 RMS (g)

The mean RMS (g) ranged between 0.060-0.062g while the maximum values were between 0.421-0.497g. The mean RMS (g) values were 24% lower than was reported by Borocz et al. [10]. Borocz et al. reported values close to 0.09 g. Meanwhile, the maximum and the 98th percentile of the RMS values measured during this study were greater than what was observed under laboratory conditions. The vibration data collected in the field shows a much lower intensity than previously studied. An explanation for this trend could be that, during the field forklift handling operations, the forklift travels over uneven floors and thresholds which results in low-probability but high-intensity events that were recorded by the sensor.

The distribution plots for the RMS (g) values showed a similar trend as was observed in the peak acceleration values for Warehouses 1 and 3 (Figure 7). However, it showed much more pronounced concentrations of events for Warehouse 2, in the 0.1-0.05g range. The trend observed for WH2 had a lower number of low-intensity (<0.05g) vibration events and more events in the 0.05-0.1g range.

Using a 95% confidence level with the Tukey's analysis, there were no significant differences in RMS (g) mean levels measured for the three investigated warehouses. The combined CDF for RMS (g) values can be found in Figure 8.

Huang et al. (2021) observed GRMS (G^2/Hz) values ranging between 0.024-0.090 GRMS (G^2/Hz) for data collected when the sensor was attached to the carriage of the forklift. The lowest GRMS value occurred from a combination of electric forklifts, concrete road conditions, a speed of 3.22 km/h, and a payload of 2,500 lbs. The highest GRMS value occurred from a combination of the gas forklifts, asphalt road conditions, a speed of 4.82 km/h, and a payload of 1,500 lbs. The GRMS range for data collected during this study was 0.084-0.098 GRMS when attached to the carriage of the forklift (Table 5). This falls relatively close to the overall range Huang et al. found during his study for data collected from the forklift carriage [9]. The lowest GRMS value came from WH3 which utilized a gas forklift. These findings were different than what was found by Huang et al. as he measured the highest vibration results for the gas forklift [9]. All warehouses had a positive skewness value, indicating that more low-intensity vibration events were observed in the field (Table 5).

Warehouse:	Maximum RMS (g)	Mean RMS (g)	98% RMS (g)	95% RMS (g)	90% RMS (g)
#1	0.497 RMS (g)	0.060 RMS (g)	0.222 RMS (g)	0.164 RMS (g)	0.113 RMS (g)
#2	0.421 RMS (g)	0.060 RMS (g)	0.150 RMS (g)	0.100 RMS (g)	0.093 RMS (g)
#3	0.436 RMS (g)	0.062 RMS (g)	0.226 RMS (g)	0.168 RMS (g)	0.120 RMS (g)

Table 4: RMS (g) values measured during forklift vibration.

Warehouse:	Mean GRMS	Skewness	Kurtosis
#1	0.091 GRMS	5.59	35.57
#2	0.098 GRMS	4.88	26.38
#3	0.084 GRMS	6.55	49.11

Table 5: Mean GRMS, skewness, and kurtosis values.

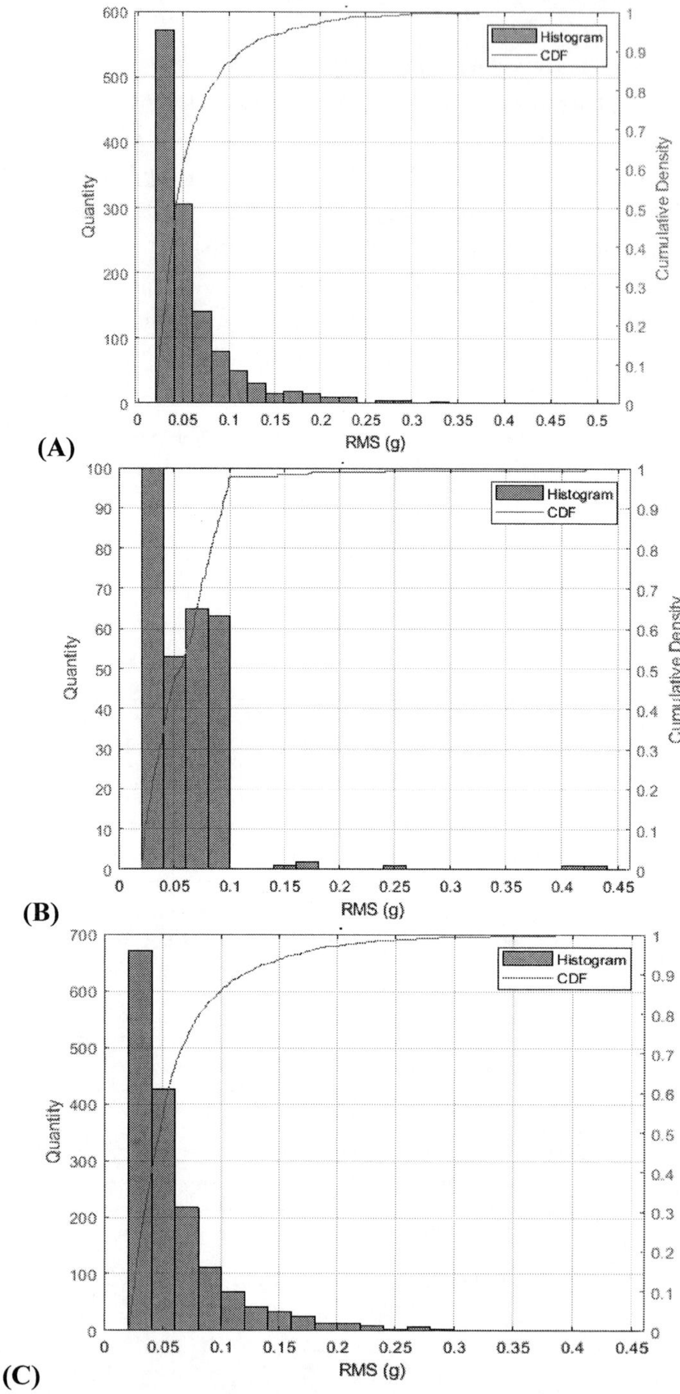

Figure 7: Histograms and cumulative distribution function plots for the RMS (g) values measured during forklift vibration for (a) Warehouse 1, (b) Warehouse 2, (c) Warehouse 3.

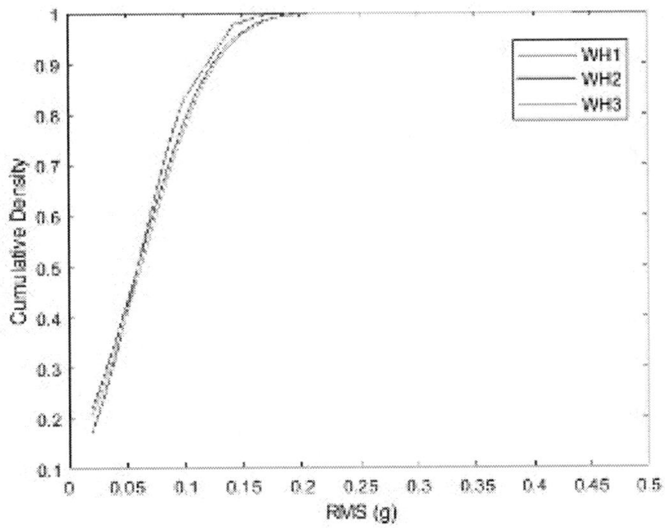

Figure 8: Combined Warehouse RMS (g) CDF plots.

3.3 PSD

The Power Spectral Density (PSD) plot for the field measurement is presented in Figure 9. The highest peak value occurred between 4-8 Hz which is slightly higher than what was reported by Huang et al. [9]. Although the GRMS values calculated for the field measurements were mostly in-line with what was reported by Huang et al., the intensity of the highest peaks show vibrational intensity values nearly 10 times lower than Huang et al.'s former study [9]. Forklifts in the field drive to and from different locations within the facility, and they are not always handling unit loads. This could be responsible for the lower peak intensities and higher frequency vibrations seen between the warehouses.

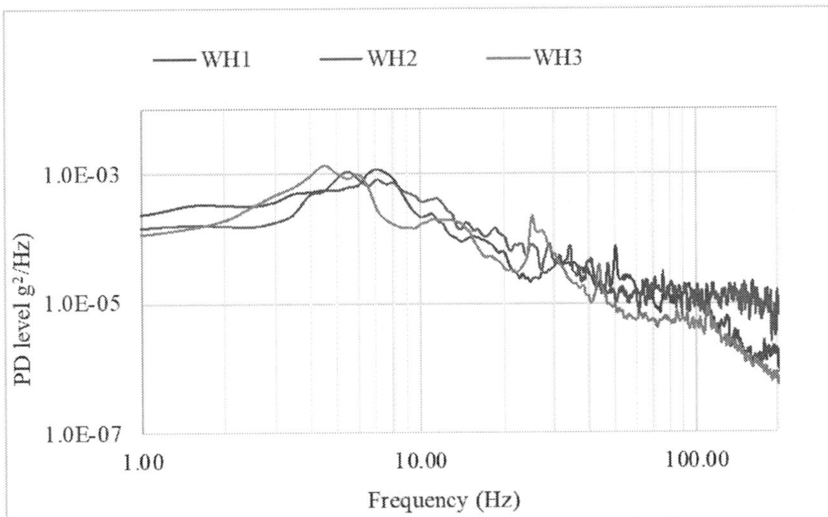

Figure 9: PSD plots for the three investigated warehouses.

4. Conclusion

The highest intensity vibrations were found to occur between 4-8 Hz which is slightly higher than was observed in previous laboratory studies. This may be a result of inconsistent warehouse conditions as compared to the previous laboratory studies which were all conducted in the same facility, only differentiating between forklift variables.

The mean GRMS values for the measured forklift vibration were in the same range of 0.084-0.091grms that was found for the laboratory conditions in previous studies.

The peak acceleration values observed in the field showed high-level skewness towards the lower values but still exhibited much greater maximum values than what was measured for laboratory conditions. The increased number of lower-intensity, peak-acceleration values could be caused by more frequent empty travel by the forklift.

More forklift traveling when compared to previous studies may have an effect on the much smaller values that were seen in the field. Contrary to previous studies, the highest vibration levels occurred in WH1 which utilized a sit-down, electric forklift.

The highest peak acceleration values occurred in the distribution center, the manufacturing facility, and the pallet manufacturer, respectively.

The RMS (g) values found during the study were also within close proximity to the ranges in previously mentioned studies; however, they were consistently toward the upper limit of those ranges. The pallet manufacturing facility saw the highest RMS (g) mean value of all facilities studied which utilized an outdoor, gas-powered forklift.

Random vibration events found during field data collection followed a non-gaussian distribution for all warehouses.

5. Acknowledgements

The research was supported financially by The Center for Packaging and Unit Load Designs' Industrial Affiliate Program

6. References

1. McCrea, B. (2020). *Annual pallet report: 2020's market evaluation. Modern Materials Handling* Retrieved from https://www.mmh.com/article/annual_pallet_report_2020s_market_evaluation
2. Gerber, N., Horvath, L., Araman, P., & Gething, B. (2020). Investigation of new and recovered wood shipping platforms in the United States. *BioResources*, *15*(2), 2818–2838. https://doi.org/10.15376/biores.15.2.2818-2838
3. Madison's Lumber Reporter (2018). "North America full-year 2016 lumber production: WWPA," *Lumber Track*, (https://madisonsreport.com/2017/03/16/north-america-full-year-16-lumber-production-wwpa/)
4. Shiner, Z., (2018). An Investigation of Wood and Wood Packaging Waste in the United States. Virginia Polytechnic Institute and State University, Virginia, USA.
5. Bartholdi III, J. J., & Hackman, S. T. (2017). *Warehouse & Distribution Science* Retrieved from https://www.warehouse-science.com/book/index.html
6. Rojas, A., (2019). Case Study: The effect of pallet design on the performance of semi-automatic & fully-automatic warehouses. Virginia Polytechnic Institute and State University, Virginia, USA.
7. Howie, A. (2008). Fundamentals of warehousing & distribution: an introductory course in material handling. United States: Material Handling Industry of America.
8. Tareq, K., Amr, M., & Abdallah, S. (1970, January 1). *Forklift design*. An-Najah National University. Retrieved from https://repository.najah.edu/handle/20.500.11888/13958
9. Huang, Y. Y., Horvath, L., & Böröcz, P. (2021). Measurement and analysis of industrial forklifts vibration levels for unit load testing purposes. *Applied Sciences*, *11*(7), 2901. https://doi.org/10.3390/app11072901
10. Böröcz, P. Y., Horvath, L., & Huang, Y. Y. (2022). Measurement and Analysis of Industrial Forklifts Vibration Levels. *ASTM International*. 50(2): 1236-1246. DOI: 10.1520/JTE20210293

DB-O05

Development of a Friction-Driven Finite Element Model to Simulate the Load Bridging Effect of Unit Loads Stored in Warehouse Racks

Eduardo Molina[1]*, Laszlo Horvath[1], Robert L. West[2]

[1]*Department of Sustainable Biomaterials, Virginia Tech, Blacksburg, 24061, USA*
[2]*Department of Mechanical Engineering, Virginia Tech, Blacksburg, 24061, USA*

***Correspondence to:** Department of Sustainable Biomaterials, Virginia Tech, Blacksburg, 24061, USA.*
E-mail: molina@vt.edu

ABSTRACT: Current pallet design methodology frequently underestimates the load capacity of the pallet by assuming the payload is uniformly distributed and flexible. By considering the effect of payload characteristics and their interactions during pallet design, the structure of the pallets can be optimized, and raw material consumption reduced. The objective of this study was to develop and validate a finite element model capable of simulating the bending of a generic pallet while supporting a payload made of corrugated boxes and stored on a warehouse load beam rack. The model was generalized in order to maximize its applicability in unit load design. Using a two-dimensional, nonlinear, implicit dynamic model, it allowed for the evaluation of the effect of different payload configurations on the pallet bending response. The model predicted the deflection of the pallet segment and the movement of the packages for a unit load segment with three or four columns of boxes supported in a warehouse rack support with a 7.8% accuracy. Further refinement of the model would be required to predict the behavior of unit loads carrying larger boxes. The model presented provides an efficient solution to the study of the affecting factors to ultimately optimize pallet design. Such a model has not been previously developed. The model successfully acts as a tool to study and predict the load bridging performance of unit loads requiring only widely available input data, therefore providing a general solution.

Keyword: pallets; unit load interactions; package; simulation; packaging optimization

1. INTRODUCTION

Most consumer products are handled and transported predominantly in unit load form, commonly using wooden pallets and corrugated fiberboard boxes [1], [2]. The design of pallets for material handling has evolved from a trial and error method to the development of advanced modeling systems [4]–[8]. The need for better models is supported by the material consumption present in pallet manufacturing and by the overall resources utilized in transportation operations. Estimates place the number of pallets in circulation at over 2 billion units in the United States and 6.8 billion units globally [3], [9]. In 2016 in the United States, more than 500 million new pallets were manufactured, and 341 million were repaired for continued use. The volume consumed for pallet manufacturing represented 21.8% of the total lumber production of the United States. Pallets for material handling represent one of the largest segments in consumption of sawn hardwoods in the United States, with 45% of the total production being dedicated to them [10]. The need for an efficient use of the natural resources consumed to support supply chains is critical and requires the proper design of each component in the system.

The load carrying capacity of most pallets is determined using a uniformly distributed load for each support condition of use, such as for warehouse racking or floor stacking [11], [12]. Research conducted as early as 1982 shows that when pallets are loaded with specific payloads, instead of uniform loads, pallet bending is reduced due to the redistribution of the stresses on the pallet towards the supports [4], [13]. Further exploration of this redistribution of stress across the pallet's top deck, known as the load bridging effect, has been conducted mostly through physical tests and by measuring the change in pallet deflection. The international standard, ISO 8611-Part 3 [14], recognizes the effect of load bridging by allowing for the determination of a maximum working load of a pallet for a specific payload.

Research efforts have been made to further understand which aspect of the pallet, or its load, significantly influences load bridging for specific payload configurations. One of the first studies was conducted in 1997, by first determining five differing levels of rigidity for a payload and then measuring how much the deflection changed under each level [15]. Although a clear effect in the results was observed (deflection decreased in direct relation to the payload rigidity), the results can only be considered preliminary since the effect of individual factors were not isolated from each other. The isolated effect of multiple palletizing characteristics such as containment force [16], box size [17]–[19], flute size [17], box contents [20], and interlock stacking [21] were investigated by many researchers. All of these studies used physical testing, which involved the construction of unit loads or segments of unit loads and subsequent laboratory measurements.

The application of the finite element method to study the load bridging effect can be a cost and time efficient alternative. FEA allows researchers to analyze the effect of multiple variables and their interactions with a scope not possible in physical testing. The finite element method has not been previously utilized for the modeling of unit loads and their effects on pallet deformation.

The main objective of this investigation was to develop and validate a two-dimensional finite element model that can accurately predict the deflection of a pallet analog carrying stacked boxes in a simulated warehouse storage rack. Subsequently, an analog model was developed using the Gaussian Process method to characterize the effects, significant factors, and interactions influencing load bridging on unit loads.

2. MATERIALS AND METHODS

2.1. Finite Element Model Development

This research project limits the support conditions studied to warehouse racking across the pallet width. Previous research studies have identified this support condition as the one that most often limits the pallet load carrying capacity, given that it commonly presents the greatest pallet deformation [18], [19], [21]. The warehouse racking condition where the pallet is racked across its width will be referred to as simply "racking support." Additionally, it is assumed that under this condition, as shown in Figure 1, the displacement due to bending of the supporting pallet and box movement is primarily along the *xy*-plane, with negligible displacement along the z-axis. This assumption is supported by the results in the study by Molina et. al [21]. A single row of boxes, shown shaded in Figure 1, was modeled to represent the unit load and the payload's interactions. The pallet geometry was simplified to be just one board deep enough to support a single row of boxes. A similarly simplified unit load segment has been used for physical testing in previous studies [16], [17], [20]. All measurements of pallet bending were considered initial, ignoring the effects of long-term creep on the bending response.

Corrugated boxes were modeled as nondeformable solid objects, subject to gravitational forces and frictional interactions. The boxes used for the experimental validations contained rigid OSB containers closely fitted inside, preventing any buckling deformation from handling and loading during the tests. The pallet was represented in the model using a board as a pallet analog; thus, the effect of the pallet structure was not considered. Pallet dimensions were based on the Grocery Manufacturers Association specification of 1,219 mm. x 1,016 mm.

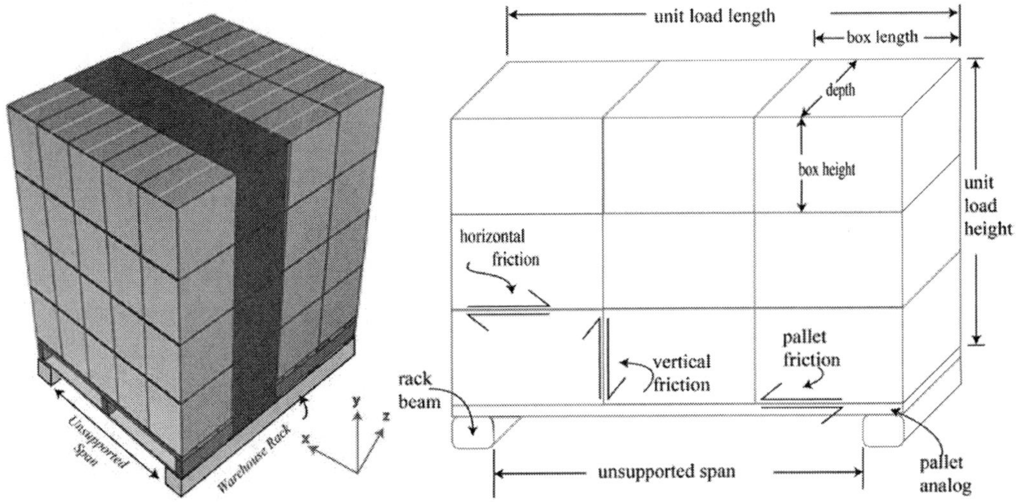

Figure 1. Unit load supported on a warehouse rack across the width (left). Shaded segment represents the section modeled. Diagram of the unit load segment modeled (right).

A two-dimensional finite element model was developed, to replicate the segment of a unit load of stacked corrugated boxes on a board, as shown in Figure 1. The initial model was developed for a unit load segment of 3 columns with 3 layers of boxes and then modified to present different combinations.

FEA Material and Section Assignments

The pallet analog, or board, was modeled as a two-dimensional object with dimensions of 1,016 mm along the length and a thickness of 19 mm, made out of Poly-methyl methacrylate (PMMA). Boxes were also 2D objects with dimensions 338 mm x 318 mm. A homogeneous solid section thickness of 254 mm was used for all instances as the material depth along the z-axis. A summary of the model's material properties is presented in Table 1. The aluminum rack beams of a warehouse rack were modeled as 2D discrete rigid objects measuring 50.8 mm x 50.8 mm with their corners rounded to a 5 mm radius in order to replicate the same contact properties experienced by the board, representing the pallet analogue, and the beams in the racking condition simulated.

Table 1. Material properties for the unit load analogue components used for the model development.

Component	Property	Value	Source
PMMA board	Dimensions	1,016 x 254 x 19 mm	Experimental measurement
	Young's modulus	3.34 GPa	Experimental measurement using ASTM D198 [22]
	Poisson's ratio	0.35	[23]
	Mass density	1180 kg m^{-3}	Experimental measurement
Corrugated fiberboard boxes	Dimensions	338 x 254 x 318 mm	Experimental measurement
	Young's modulus	600 GPa	Model input for non-deformable object
	Poisson's ratio	0.44	[24]
	Mass density	438 kg m^{-3}	Experimental measurement

Considering that the model being replicated was affected mostly by weight and not by the application of specific externally applied loads, the loading was simulated through the addition of gravitational forces along the negative *y*-axis, using the standard acceleration on the surface of the Earth of 9.81 m s^{-2}. The specification of mass density for each instance in the model was required. The finite element model was solved with a nonlinear incremental solver to account for the nonlinear mechanics associated with the frictional contact mechanics as well as the large deflections developed in the model. Racking supports, or in this case the aluminum rack beams, were fixed in place using the Abaqus encastre condition, not allowing for displacement or rotation on any axis. A summary is shown in Table 2.

Table 2. Loading and boundary conditions used for the unit load segment model.

Property	Location	Type/Value
Loading	Entire model	Gravitational Acceleration - 9.81 m s^{-2} along *y*-axis
Boundary conditions	Rack beams (2)	Fixed (encastre)
	Pallet analog	Frictional contacts
	Boxes (9)	Frictional contacts

In order to model the contact properties of the unit load segment, the static friction coefficient between the different components were measured. The coefficient of friction was measured for the corrugated fiberboard before it was converted into boxes following the slide angle method from the industry standard TAPPI T-815 (2018). Figure 1 shows the location of the contacts for the horizontal, vertical and pallet frictions.

In order to establish boundary conditions for the model and assure proper loading of the parts, surfaces were generated so every contact region could be explicitly declared. Every contact interaction was defined with friction properties, with tangential behavior using the penalty friction formulation. Small sliding was permitted in the formulation of the model to represent the small relative motions expected between boxes and between the boxes and the pallet analog. Each contact interaction was defined as a surface-to surface contact with a node-to-surface discretization method.

Meshing was assigned individually to each object in the model. The pallet analog was discretized using the Abaqus CPE4I element type, which includes incompatible deformation modes, to include additional degrees of freedom in order to capture the bending kinematics more accurately. The quadrilateral elements for the pallet segment were sized at 2.54 mm, with a total of 3,200 elements.

In order to arrive at the optimal number of elements for the pallet analog, a mesh convergence study was conducted for a unit load consisting of 3 columns of 3 layers of boxes. Boxes were generated as 2D solid elements, formulated as a plane/strain linear elastic material with reduced integration, specifically the

Abaqus CPE4R element. The elements were globally sized at 12.7 mm for every box, regardless of its size. To select this element size, a mesh convergence study was conducted. The midspan pallet analog deflection results were considered converged after the number of elements reached 6,075 for the 9 boxes, or when the seed size was 12.7 mm. Decreasing the box element size further, to a 5.08 mm seed, changed the results by 1.09% but increased the number of elements 6.3 times, making it unnecessarily demanding. The racking beam supports were discretized as 2-node, linear, discrete, rigid elements with a length of 2.54 mm and rigid link formulation, represented in Abaqus as rigid 2-noded 2D (R2D2) element.

The overall model was formulated as nonlinear, implicit dynamic. The dynamic formulation includes the representation of inertial forces into the system equations of dynamic equilibrium in order to account for the transient characteristics of the model, including the quick movements of all boxes, the large deflections experienced by the pallet segment, and all of the frictional contacts. An extended time step of 5 seconds was used to aid the convergence of the nonlinear contacts and deformations across the unit load segment and they were considered initial board deflections, not accounting for creep effects on the structure.

2.2. Model Validation

In order to validate the accuracy of the model, a series of experimental tests were conducted and then simulated with the same input values in the finite element model.

Pallet analogs were simulated using a board of Poly-methyl methacrylate (PMMA), with dimensions of 1,016 mm length, 254 mm width and a thickness of 19 mm. Regular-slotted container-style (RSC) corrugated boxes made of single-wall C-flute corrugated board with 7.2 kN m^{-1} Edge Crush Test (ECT) value were used. The box sizes and weights used for the study are shown in Table 3. Each corrugated box contained a box insert made of 12.7 mm thick oriented-strand-board (OSB) built to fit tightly the inside of the corrugated boxes and avoid any box deformation from repetitive loading. Weight was added using wooden pellets inside the OSB box in order to have boxes with a consistent density of 438 kg/m^3. The boxes were manufactured on a computerized cutting table (Esko Kongsberg XL44, Manufacturer: Esko, Miamisburg, Ohio, US). Top and bottom flaps of the boxes were sealed using a 50.8 mm packaging tape (3M, Saint Paul, MN, United States).

Table 3. Weight and dimensions of boxes used for model validation.

Box	Dimensions (mm)	Weight per box (kg)
Small	254 x 254 x 318	9.07
Medium	338 x 254 x 318	12.11
Large	508 x 254 x 318	18.14

The model validation for this research was conducted in two sequential phases. First, the experimental analog of a pallet bending while in a warehouse rack support across its width was validated. This initial step allowed the researchers to confirm the accuracy of the model in replicating the pallet analog bending, before including additional complexities from the payload characteristics. Second, the complete unit load segment was tested, validating the model's ability to replicate the pallet analog's deflection response while under the different payloads being studied.

In order to measure pallet deflection while under each unit load configuration, a simulated pallet rack was built. The objective of this was to validate the support model without including boxes. Two square, 50.8 mm x 50.8 mm, hollow aluminum beams acted as direct supports and the pallet simulator rested on them. A free span of 914 mm between the inside ends of the aluminum beams simulated the common warehouse racking condition where the pallet is racked across its width. The aluminum beams were placed on top of two I-beams (101 mm wide flange, 101 mm high), which were secured to a rigid, flat platen on the floor.

Unit loads with different configurations were developed to measure the bending moment response of the pallet. Different configurations were achieved by changing the unit load's main variables such as the number of columns, number of layers, and the total weight supported by the pallet. Representing a wider range of unit loads during the experimental studies allows for more robust model validation afterwards. The number of layers and total weight of each unit load evaluated are shown in Table 4. Each test was replicated five times. Figure 2 shows an example of a loaded unit load segment of two columns and two layers, supported to prevent bending while under deflection.

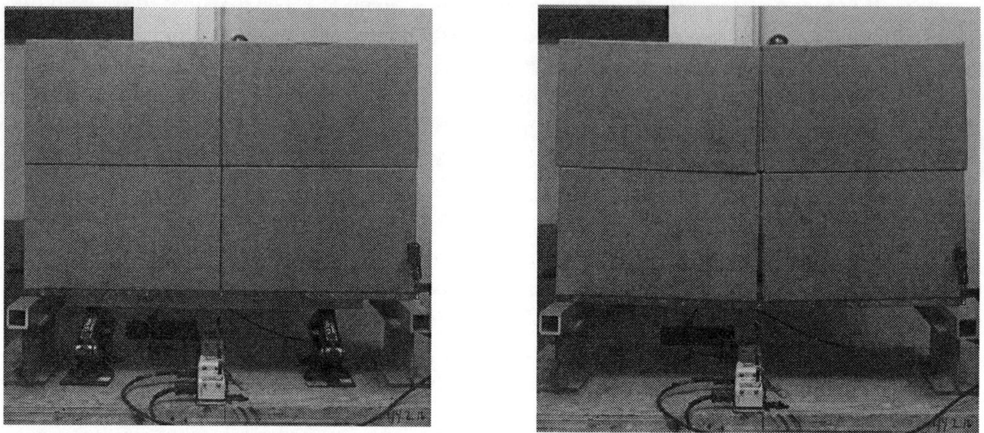

(a) Supported Unit Load (b) Unit load under load

Figure 2. Picture of unit load segment with 2 layers of large boxes, fully supported (a) and showing deflection under load (b).

To conduct the validation of the finite element model of the unit load segment, each of the seven analogues studied was simulated and compared directly to the results of the experimental tests.

2.3. Gaussian Process Model

From the results of the finite element analysis of the unit load bending model, a simplified, surrogate model was developed. The goal of this model was to allow for a faster calculation and generation of data to evaluate the effect of the different factors affecting load bridging.

To fit the deterministic responses from the finite element simulations, a Gaussian process model was used. The Gaussian Process model is $y = \mu + z(\mathbf{X})$, where $z(\mathbf{X})$ is Gaussian stochastic process with covariance matrix $\sigma^2\mathbf{R}(\boldsymbol{\theta})$. $\mathbf{R}(\boldsymbol{\theta})$ is a correlation matrix with the elements shown in (8). Values for r_{ij} are the correlations between responses at two design points. The parameters μ and θ_s, s = 1, 2,…,k were estimated using the method of maximum likelihood.

$$r_{ij} = e^{-\sum_{s=1}^{k} \theta_s (x_{is}-x_{js})^2} \qquad (8)$$

In order to generate the most accurate model and to validate the prediction results from the Gaussian Process model, a k-fold cross validation was conducted. The model with the data set that provided the highest accuracy was selected as the resulting statistical model. Goodness of fit and predictive performance was evaluated by comparing the Root Mean Square Error (RMSE, **Error! Reference source not found.**) of the predictions against the finite element model simulation results, as well as the Mean Absolute Percent Error (MAPE). Both error measurements together provided enough confidence to determine model accuracy.

3. RESULTS AND DISCUSSION

3.1. Model Validation

For this initial validation, the main response of interest after running the simulation was the deformation of the pallet segment along the *y*-axis. The largest deflection was identified at the midspan of the pallet analogue board. The predicted deflection was 9.25 mm (0.364 in) and the measured deflection on the experiments was 9.40 mm with a standard deviation of 0.56 mm. Figure 3 shows the finite element model results of the deformed structure. When compared with the experimental test measurement, the result had a deviation of -1.62%, which was deemed acceptable for the validation of the test setup and confirmed the model's ability to accurately replicate the racking across the width pallet support test analogue.

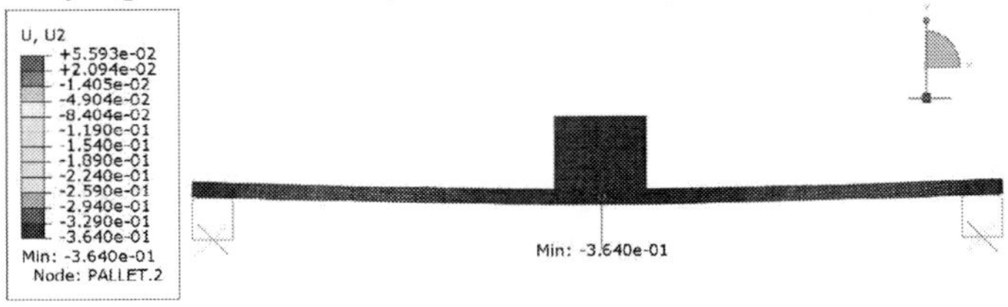

Figure 3. Two-dimensional FE model result for the simulated pallet segment and weight box, deformed. Figure values shown in inches.

Using the support test setup that has already been validated and the methods previously described, experimental trials were conducted by loading the board with different box sizes. Small (254 x 254 x 318mm), medium (338 x 254 x 318mm), and large (508 x 254 x 318mm) boxes were tested. The initial deflections recorded at the end of the simulation at the board midspan are shown in Table 4 for the seven different unit load combinations of columns and layers of boxes.

Parallel to this, the finite element model was adjusted to the same box sizes and number of layers combinations. Table 4 shows the results for each of the analyses. The mean absolute percent error (MAPE) of the model is 22%. The greatest error with 59% MAPE was observed for the unit loads with the largest boxes (two columns). By limiting the prediction scope of the model to those unit loads with three or four columns, model accuracy improves to a MAPE of 7.8%, which is considered an acceptable accuracy for such a general unit load bending model. Unit loads with three columns presented a MAPE of 12% and those with four columns a MAPE of 1.3%, suggesting an increased accuracy with smaller boxes.

The scenario of unit loads of two columns present a special case that needs to be formulated as a different model in order to accurately predict pallet deflection. The assumptions and simplifications in this project do not properly represent the actual unit load behavior. The boxes in the designed experiment present a large 'rotation' movement when they are under load. This movement is accurately predicted by the model, as seen in Figure 4 (a). The absolute value predicted, on the other hand, diverges by 29% from the measured experimental value for the 2x2 unit. A similar case occurs for the unit load of 2x3. For this scenario, the model underpredicted the overall pallet deflection, with an error of 88%. This specific two-column scenario presented significant differences from the rest of the unit loads studied. When using the friction values obtained by evaluating the linerboard contact properties, as conducted for this study, the actual properties of the contact might not have been properly represented. Corrugated boxes have imperfections. One example could be that there was compression of the flutes or of the linerboard between the fluting structure, albeit small, but that can affect the characteristics of the horizontal or pallet frictions. Additional variations could come from the use of packaging tape for the closures and any other environmental impurities when testing, such as dust, that could affect more directly the horizontal surfaces. When measuring the friction with the slide angle method, only the actual contact properties between the linerboards are evaluated, with low weight and perfect samples.

For all the five scenarios where the unit loads had three or four columns, the model was capable of accurately predicting the deflection and the movement of the boxes. This movement is closely linked to their load bridging behavior. Figure 4(b) and 6(c) show the behavior of the boxes in the experimental tests and in the FE simulation for the 3x3 and 4x3 units. Visually, this model reasonably replicates the box movement. The model was able to simulate a unit load's bending accurately for three and four-column scenarios without requiring the further complexity that would be added by having to simulate corrugated panels and their potential buckling.

Overall, the model can simulate, with acceptable margins of error, the deflection of unit load segments and the box movements for unit loads with three and four columns of boxes. This can translate into better understanding of the load bridging effect by conducting further simulations with a wider range of variables, such as different numbers of columns, layers of boxes, payload heights, pallet stiffnesses, and all of the friction characteristics of each contact.

Table 4. Physical properties of each experimental unit with its corresponding deflection (mm) result for the experimental measurement and the finite element model simulation deflection result (mm) for each source of friction and its corresponding error.

Box Size	Unit Load (Columns x Layers)	Payload Weight (kg)	Height (mm)	Experimental Test Results			FEA Simulation Results	
				Average Deflection (mm)	Std. Dev. (mm)	COV (%)	Deflection (mm)	Percent Error
Large	2x2	72.6	636	4.38	0.48	11.11	3.12	29%
	2x3	108.9	954	2.65	0.41	5.45	0.31	88%
Medium	3x2	72.6	636	7.43	0.41	5.45	5.65	24%
	3x3	108.9	954	6.85	0.30	4.56	6.18	10%
	3x4	145.1	1272	8.38	0.41	4.93	8.09	3.4%
Small	4x2	72.6	636	6.89	0.2	3.11	6.93	-0.6%
	4x3	108.9	954	9.92	0.43	4.35	9.79	1.3%

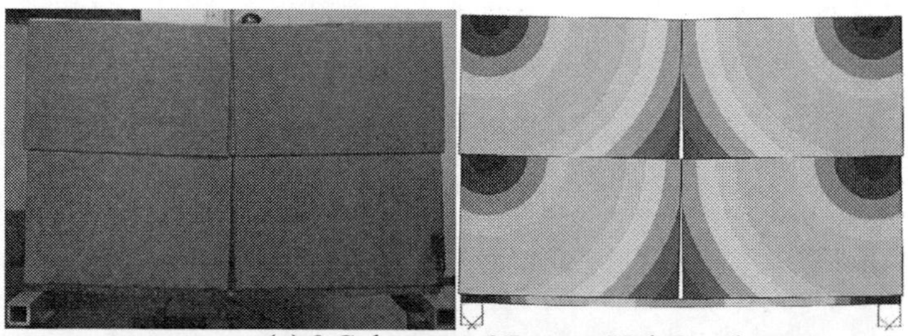

(a) 2 Columns – 2 Layers Unit Load

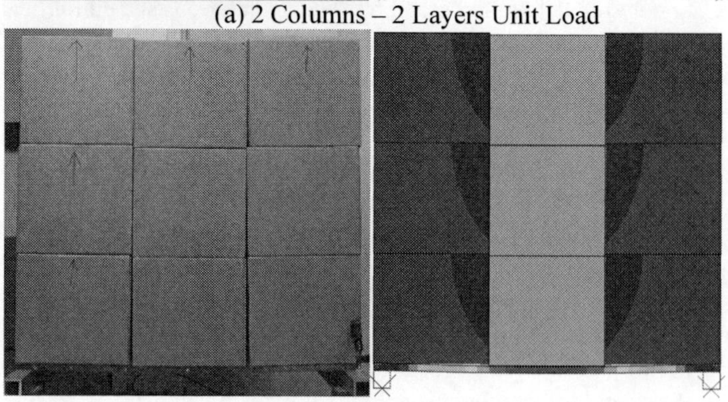

(b) 3 Columns – 3 Layers Unit Load

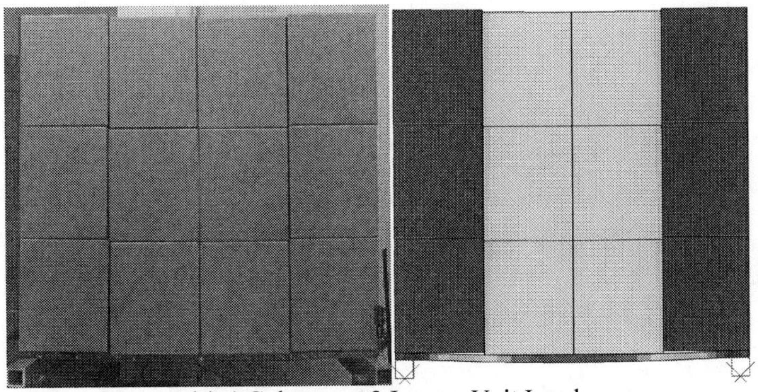

(c) 4 Columns – 3 Layers Unit Load

Images shown for the model (left) are colored based on deflection. Red equals large displacements, blue represents no movement.

Figure 4. Comparison of movement of unit load components under deflection for experimental (left) and finite element analysis (right) for two layers and (a) two columns, (b) three columns, and (c) four columns of packages.

3.2. Analysis significant factors influencing load bridging

With the goal of better understanding of each of the significant factors influencing load bridging, a detailed analysis was conducted. The results inform and support the decision-making process of unit load design. Given the many interactions acting upon the bending response of the board, the full model should be applied to specific unit load evaluations. The analysis was limited to the dimensions studied in the Latin Hypercube Design in the previous section, since that was the range of the input data of the model.

3.2.1. Effect of number of columns on the bending of a board

Previous researchers have identified that the size of the box has a very significant influence on the load bridging levels that are generated [17], [18]. In general, it is commonly assumed that smaller boxes generate higher bending. This hypothesis is based on the application of the discrete load on the pallet deck. Infinitely small boxes will resemble a uniformly distributed load, and therefore, generate the same bending as uniform load applicators. Figure 5 shows that same overall trend for most scenarios studied. It is of high relevance to note that when boxes are longer along the length (three columns), but the payload is relatively short (700 mm) and the box-to-pallet CoF is low (0.30), this observed trend is no longer valid. In this scenario, the unit load with three columns of boxes presented a deflection ratio of 0.65, while that for four columns was 0.43; five columns was 0.59; and six columns was 0.64. Unit loads with these characteristics present a higher deflection than those with smaller boxes, almost matching the same deflection level as a unit load with six columns. The interaction of the box length dimension with low friction properties generated higher deflection than one driven mostly by box size. As observable in Figure 6, the rotational movement of the boxes when the board was under load was different for the unit load of three columns (Figure 6 (a)) than for the ones with a higher number of columns. The same change in the movement of the boxes was generated by maintaining the number of columns but either increasing the friction coefficient of the pallet (Figure 6 (b)) or the overall height of the payload (Figure 6 (c)). This confirmed that no generalization can be made regarding the effect of box size on load bridging without considering the remaining interacting factors as well. A key aspect in reducing the board deflection was to prevent the rotating movement of the packages. This was possible through many different combinations of factors. Because of this, no specific attempts to quantify a threshold point were made, avoiding the misinterpretation of the model results.

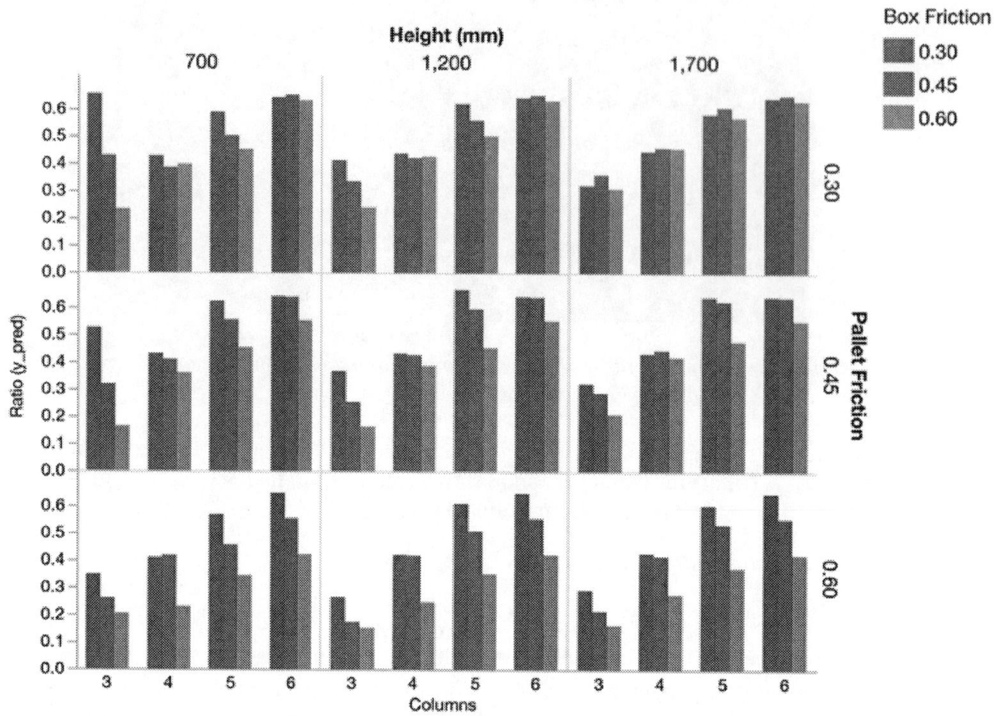

Figure 5. Effect of number of columns on board bending ratio by height (mm), box friction and pallet friction coefficients.

3.2.2. Effect of the payload height on the bending ratio of a board

The height of the payload was identified as a significant factor influencing the bending of the pallet deck. When analyzing the height, on a range between 500 mm and 2 m and for a range of friction coefficients and number of columns, it can be observed that the effect of height is highly dependent on the other variables, as shown in Figure 7. Park [16] conducted preliminary experimentations on the effect of payload height. Using large, corrugated boxes (arranged in two columns), Park identified the fact that by increasing the height of the unit load, there was a significant decrease in the overall board bending response. A similar trend can be observed in the model's response for unit loads consisting of three columns. But, as the boxes become smaller, the effect of height was no longer relevant. The movement of the boxes, such as in Figure 6 for six columns, was mostly downwards. There was no rotational movement on unit loads with low payload heights having the same total weight per column, so the load perceived by the board was not changed, therefore it generated the same board deflection ratios. For the majority of boxes used in the field, changing the height of the boxes or adding additional layers of the same package will not significantly change the resulting deflection.

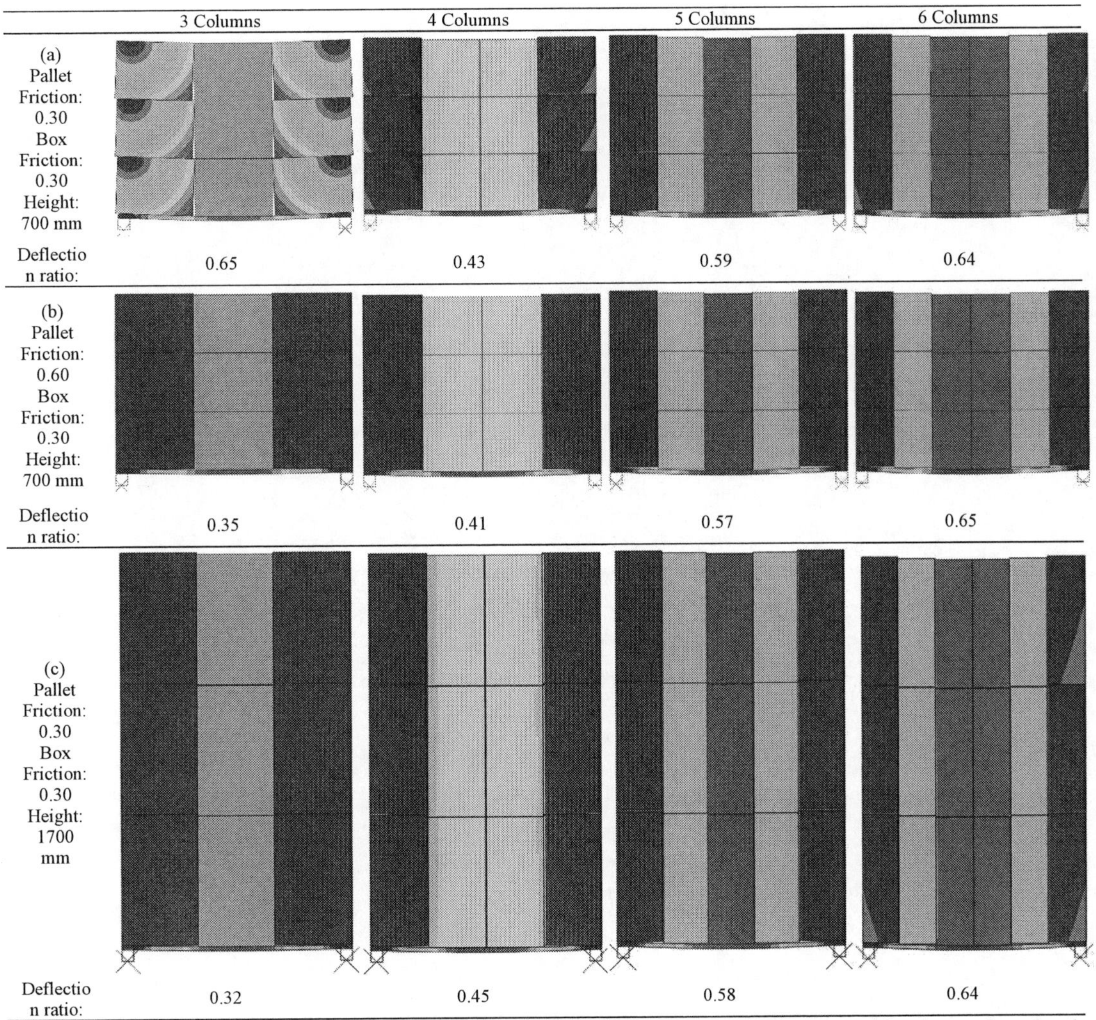

Figure 6. Finite element model simulations for 3, 4, 5, and 6 columns and three combinations of payload height and pallet and box friction coefficients. Colors shown represent displacement (U). Blue equals no movement, red equals large displacements.

3.2.3. Effect of the pallet friction coefficient on the bending ratio of a board

As can be observed in the results (Figure 8), there was an inverse relationship between the deflection ratio of the board and the pallet CoF for most scenarios studied. When the friction of the pallet deck increased, the bending of the board decreased. This trend has the potential to influence the design of plastic pallets. Pallet safe load carrying capacity can be determined by finding the maximum deflection for safe handling or determining the load at which certain pallet components begin to fail [11], [12]. Plastic pallets, given the commonly low stiffness of their structures, tend to have a maximum carrying capacity determined by the bending level under load. Affecting the resultant bending of the boards by simply increasing the friction can potentially increase the carrying capacity of a pallet. As with other factors, the interactions between number of columns, payload height, and box friction determine the resulting effects, and therefore, it all must be studied as a system. However, unit loads carrying packages with low friction levels, such as the equivalent to returnable plastic containers (RPC), did not present such changes in board deflection, except in the case of larger boxes.

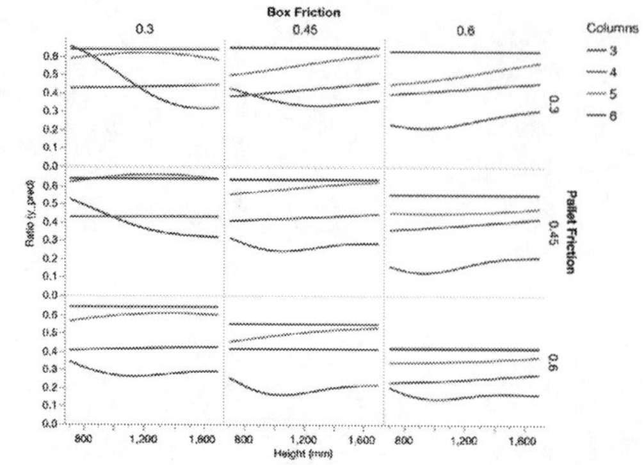

Figure 7. Effect of payload height (mm) on board bending ratio by pallet friction, box friction and number of columns.

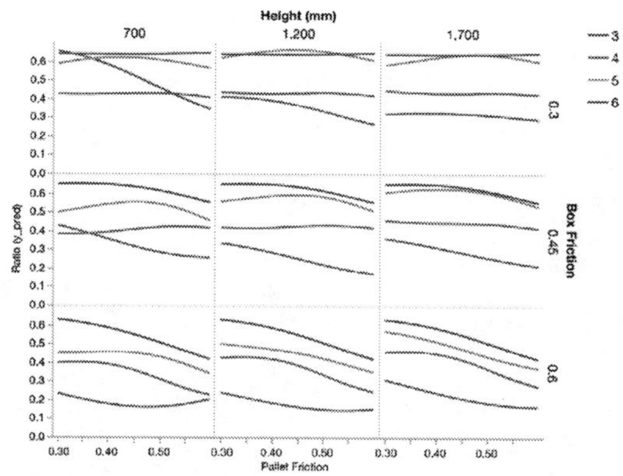

Figure 8. Effect of pallet friction coefficient on board bending ratio by payload height (mm), box friction and number of columns.

3.2.4. Effect of the package friction coefficients on the bending ratio of a board

The friction coefficient of the packages was observed to be a factor that when increased, tended to reduce the bending response of the boards. The rate of change was more pronounced for larger boxes, when there was larger contact area between the boxes. This can be clearly observed in Figure 9. Furthermore, the friction coefficient for the packages were modeled differently for the horizontal and the vertical box surfaces. These components do not affect the bending response equally. Increasing the friction properties of the vertical package surfaces generated a reduction in the bending of the boards for most of the studied scenarios. Alternatively, changes on the frictional properties of the horizontal box surfaces only generated a change in board deflection for unit loads with large boxes. As seen in Figure 6 (a), the unit load with three columns presented a rotational motion of the boxes. Increasing the horizontal friction between boxes reduced this motion and affected the resulting deflection by decreasing it. Unit loads where the aspect ratio or other factors already prevented rotational motion, were not affected by changes in the horizontal CoFs. A feature that could impact the overall effect of package friction on the bending of a unit loads is the application of containment mechanisms. Methods such as stretch wrapping can restrain the boxes and prevent any rotation or displacement. The increased unitizing forces would increase the overall load bridging effect and potentially reduce the deflection experienced by the unit load. By not considering stretch wrapping or other containment methods commonly utilized, this model provides a conservative estimation of the load bridging effect. Further refinements can be developed in order to take into account all of the many possible modifications available for unit loads.

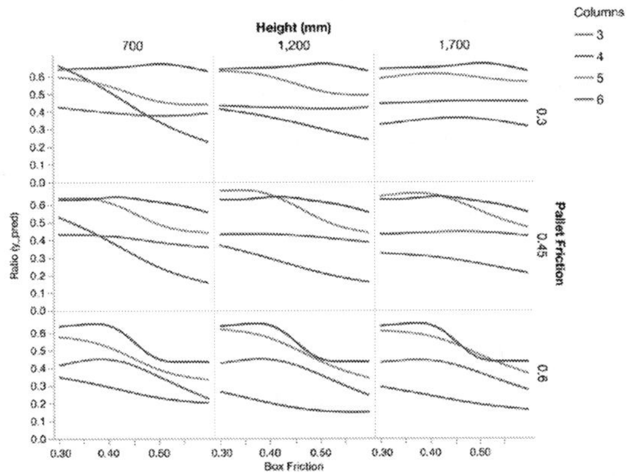

Figure 9. Effect of the friction coefficient of packages on board bending ratio by pallet friction, payload height (mm) and number of columns.

3.3. Discussion on the factor interactions

After conducting the analysis for the individual factors that influence the load bridging effect, it was evident that the existent interactions between the factors increased the complexity of the trends of the response. Traditional approaches to packaging design have focused on simple adjustment factors for specific characteristics, such as reducing box strength by a certain percent due to environmental conditions. This simplified approach, although useful for dissemination purposes, cannot be applied to the unit load development process. Some factors presented certain trends for common conditions, but when the values of other factors were modified, the trend changed. Due to the complexity of these interactions, it is not reliable to propose individual adjustment factors for each variable, that can estimate the effects of load bridging. To predict the load bridging effect, a simplified model is required to calculate a resulting adjustment. The Gaussian Process model developed fulfils the expectations of an approachable model that can be utilized in everyday scenarios as part of the unit load design process.

4. CONCLUSIONS

Following the main objective of the research project, this study focused on the understanding of the load bridging effect in unit loads. Taking into consideration the scope of the model, the following can be concluded.

A two-dimensional finite element model of a simplified unit load was successfully developed and validated for the most common loads. The developed model is capable of accurately predicting the bending response change of a pallet segment when the payload characteristics are modified.

Given the acting complex interactions in the studied phenomena, load bridging on unit loads must be studied as a system. Analyzing individual variables, while ignoring the effects of others, can prevent unit load designers from predicting the actual changes in pallet deflection.

In order to simplify the implicit dynamic finite element model, a Gaussian Process Model was developed. The resulting formulation for the changes in the ratio of pallet deflection can be more easily adopted.

Dimensions of the boxes, height of the unit load and the coefficient of frictions among the different components were identified as the factors influencing load bridging and the resulting pallet deflection.

The model developed presents an opportunity to improve the pallet and unit load design methodology. Accounting for the factors identified can allow designers to select potentially problematic unit loads and further investigate.

The results presented provide a guideline towards the change in trends of pallet deflection as payload configuration is modified. It is not the intent of the author to directly modify the safe carrying capacity of pallets. Pallet designs must be properly verified by the currently accepted methodologies. This model allows for the identification of payloads that can potentially generate larger deformation. Pallet designs must be tested with such payloads.

5. REFERENCES

[1] Fibre Box Association, *Fibre Box Handbook*, 22nd ed. Elk Grove Village, IL, 2015.

[2] D. Twede, S. E. M. Selke, D.-P. Kamdem, and D. Shires, *Cartons, crates and corrugated board: handbook of paper and wood packaging technology*, 2nd. DEStech Publications, Inc., 2014.

[3] D. McGinley, "Wood Pallets & Skids Production in the US," Melbourne, Australia, 2019.

[4] B. Fagan, "Load-support conditions and computerized test apparatus for wood pallets," Master's Thesis, Virginia Polytechnic Institute and State University, Blacksburg, VA, 1982.

[5] J. R. Loferski, T. E. Mclain, and S. T. Collie, "Analysis of racked wood pallets," *Wood Fiber Sci.*, vol. 20, no. 3, pp. 304–319, 1988.

[6] S. Samarasinghe, "Predicting rotation modulus for block pallet joints," Master's Thesis, Virginia Tech, Blacksburg, VA, 1987.

[7] R. LeBlanc and S. Richardson, *Pallets: A North American perspective*, 1ed. Ontario, Canada, Canada: PACTS Management Inc., 2003.

[8] J. R. Loferski, "A Reliability Based Design Procedure for Wood Pallets," PhD Dissertation, Virginia Tech, 1985.

[9] Research and Markets, "Pallet Market: Global Industry Trends, Share, Size, Growth, Opportunity and Forecast 2019-2024," *Research and Markets*, Dublin, Ireland, Jun. 2019.

[10] N. Gerber, L. Horvath, P. Araman, and B. Gething, "Investigation of New and Recovered Wood Shipping Platforms in the United States," *BioResources*, vol. 15, no. 2, pp. 2818–2838, 2020, doi: 10.15376/biores.15.2.2818-2838.

[11] ASTM International, "ASTM D1185-98a(2009) Standard test methods for pallets and related structures employed in materials handling and shipping." West Conshohocken, PA, 2009. doi: 10.1520/D1185-98AR09.

[12] ISO, "ISO 8611-1:2011(E) Pallets for materials handling — Flat pallets." Geneva, Switzerland, 2011.

[13] S. T. Collie, "Laboratory verification of pallet design procedures," Master's Thesis, Virginia Polytechnic Institute and State University, Blacksburg, VA, 1984.

[14] ISO, "ISO 8611-3:2011(E) Pallets for materials handling - Maximum Working Loads," *ISO*. Geneva, Switzerland, 2011.

[15] Center for Unit-Load Design, "The effect of load bridging on unit-load deflection," Blacksburg, VA, 1997.

[16] J. Park, "Investigation of fundamental relationships to improve the sustainability of unit loads," Ph.D. Dissertation, Virginia Polytechnic Institute and State University, 2015.

[17] J. Park, L. Horvath, M. S. White, S. Phanthanousy, P. Araman, and R. J. Bush, "The influence of package size and flute type of corrugated boxes on load bridging in unit loads," *Packag. Technol. Sci.*, vol. 30, no. 1–2, pp. 33–43, Jan. 2017, doi: 10.1002/pts.2279.

[18] A. P. Clayton, L. Horvath, J. Bouldin, and B. Gething, "Investigation of the effect of column stacked corrugated boxes on load bridging using partial four-way stringer class wooden pallets," *Packag. Technol. Sci.*, vol. 32, no. 9, pp. 423–439, 2019, doi: 10.1002/pts.2438.

[19] S. M. Morrissette, L. Horvath, and K. DeLack, "Investigation into the load bridging effect for block class pallets as a function of package size and pallet stiffness," *Packag. Technol. Sci.*, pp. 1–19, Oct. 2020, doi: 10.1002/pts.2539.
[20] S. Phanthanousy, "The Effect of the Stiffness of Unit Load Components on Pallet Deflection and Box Compression Strength," 2017.
[21] E. Molina, L. Horvath, and M. S. White, "Investigation of pallet stacking pattern on unit load bridging," *Packag. Technol. Sci.*, vol. 31, no. 10, pp. 653–663, Oct. 2018, doi: 10.1002/pts.2406.
[22] ASTM, "D198-15 Standard Test Methods of Static Tests of Lumber in Structural Sizes." West Conshohocken, PA, 2015. doi: 10.1520/D0198-15.
[23] Altuglas International, "Plexiglass General Information and Physical Properties," 2006. http://www.plexiglas.com/export/sites/plexiglas/.content/medias/downloads/sheet-docs/plexiglas-general-information-and-physical-properties.pdf (accessed May 28, 2019).
[24] A. C. Gilchrist, J. C. Suhling, and T. J. Urbanik, "Nonlinear finite element modeling of corrugated board," *Am. Soc. Mech. Eng. Appl. Mech. Div. AMD*, vol. 231, pp. 101–106, 1999.

DB-P01

Record and analysis of Europe / China Rail transportation

Nolot Jean-Baptiste[1], Delloque Philippe[1], Huart Victor[2], Odof Serge[1,3], Krajka Nicolas[2], Pellot Jérôme[2], Erre Damien[1,3]

[1]ESIREIMS Ecole nationale Supérieure d'Ingénieurs de Reims / University of Reims Champagne-Ardenne, esplanade Roland Garros, 51100 REIMS, France
[2]METROPACK, 30-32 Rue du Capitaine Georges Madon ZAC Croix Blandin, 51100 REIMS, France
[3]ITHEMM / University of Reims Champagne-Ardenne, 51100 REIMS, France

ABSTRACT: World's largest economy, the European Union accounts for more than 20% of the wealth produced worldwide. China became in 2020 the leading partner of the European Union in terms of goods trade. The majority of this trade in goods today is carried out by sea. However, a new supply chain is now operational by land. This new Silk Road crosses twenty countries and is operated by rail and truck. This 10,000 to 15,000 km journey uses different types of trains (track gauges). European manufacturers are becoming users of this route for the export of their product to Asia. Many examples of exports are reported in the press (bottled water for infants, polymer granules, information material, industrial equipment, pharmaceutical products, etc.). In this work, Europe-China transport studies were carried out. These routes have been analyzed. The first observation from the analysis shows that even if the type of transport equipment is known, the duration and sequence of modes of transport are much more numerous than on "usual" trip. In this study, we will focus on the railway part because it shows a more-than-7-cumulative-days duration. Several phases are detected and represent vertical but also horizontal vibratory stresses. The PSDs of the vertical and horizontal vibrations identified will be presented and commented. Some phases of the studied transport present horizontal transient phenomena. All these events begin with a shock followed by oscillations. From the recorded signals, we built a lab-simulation test protocol. It is based on the replication of an average time signal modulated at different levels to fit with the actual measured distribution. This protocol was applied to a pallet of palletized water bottles to see the impact on the load stability. The deformation measurements will be presented

Keyword: Rail environment Recording, Vertical and Horizontal vibration analysis, Transient events, Silk Road, Transport simulation

1. INTRODUCTION

Drawing abundantly on the historical aspects of the 2000-year-old silk road, connecting China to Europe through Central Asia, the Chinese president suggested that a terrestrial economic belt was built in order to "open a logistic passageway going all the way from the Pacific Ocean to the Baltic sea". This project from year 2014 is called OBOR, standing for "One Belt One Road". In the facts, it is about building an economic and commercial road allowing a connection between China, Europe, Africa and Central Asia. This project enlists 68 countries, which represents 4,4 billion people, 40% of world's GDP, and requires over tens of thousands of dollars. World's largest economy, the European Union accounts for more than 20% of the wealth produced worldwide.
China became in 2020 the leading partner of the European Union in terms of goods trade. The majority of this trade in goods today is carried out by sea. However, a new supply chain is now operational by land.

This new Silk Road (Figure 1) crosses twenty countries and is operated by rail and truck. This 10,000 to 15,000 km journey uses different types of trains (track gauges).

Figure 1 : Silk Road United Kingdom – China [1]

The road is not easy. The trains must be changed 2 to 3 times because of the rails gauges in Russia, Bielorussia, Mongolia and Kazakhstan, which are different from China's and Europe's (Table 1).

Country	China	Russia	France	Germany	Spain	Finland
Standard gauge (mm)	1435	1520	1435	1435	1668	1524

Table 1 : Standard Gauge for Rail Way VS Country [2]

European manufacturers are becoming users of this route for the export of their product to Asia. Many examples of exports are reported in the press (bottled water for infants, polymer granules, information material, industrial equipment, pharmaceutical products, etc.).

2. MATERIALS AND METHODS

Many LANSMONT Datalogger SAVER 3X90 & SAVER 3D15 were placed in containers and various trips were recorded. (X axis = Horizontal Forward / Y axis = Horizontal Lateral / Z axis = Vertical). The containers left from France and Belgium and were sent to various cities in China. The road used is not precisely defined. Two towns on the way are known to be unavoidable logistic rail hubs: DUISBURG – Germany & XI'AN – China.

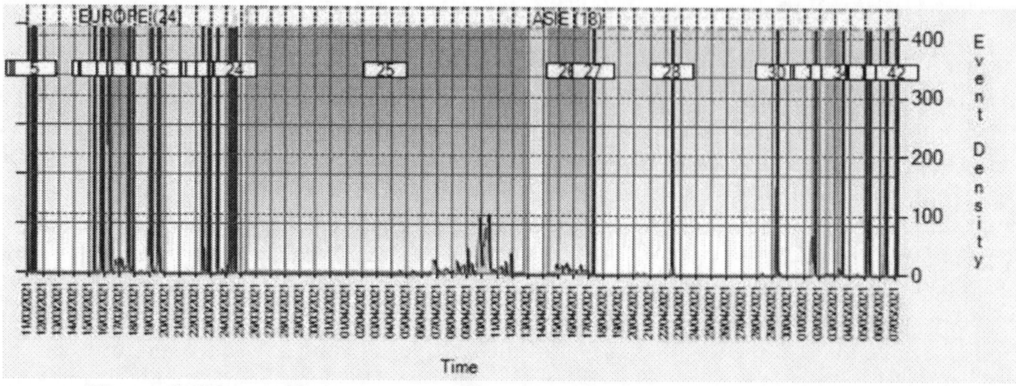

Figure 2: Phases Chronogram – Blue=Handling, Pink=Truck, Orange=Rail

The chart (Figure 2) compiles the accelerometer data recorded during the journey. Our expertise allows us to define its phases. We calculate, based on the Signal events (here the trigger is set over 0,81g), the apparition rate per hour. This density shows that during phase 25 (rail phase 4), the events are particularly current. In almost 24 hours, 1300 events are recorded. It is confirmed by every recording. This vibratory phase (rail phase 4) is the roughest of this trip.

The recorded journeys last between 21 and 65 days (Manufacturer to Customer). In the case of pharmaceutical products, the custom phases are highly reduced.
The first observation from the analysis shows that even if the means of transportation is known (Rail), the duration and sequence of modes of transport are much more numerous than on a "usual" trip. We were able to measure and identify 5 rail phases.
We have measured and identified five Rail phases.

Phase	x Grms	y Grms	z Grms	X Grms	Y Grms	Z Grms	Duration	Freq. 1	Freq. 2
	TIMER event (Dt=15 mn)			SIGNAL event (>0.8g)			(min)	(Hz)	(Hz)
Rail 1	0,0309	0,041	0,0653	0,0458	0,0677	0,117	60	5,3	10,8
Rail 2	0,0272	0,0245	0,0508	0,0529	0,0417	0,095	1200	2,4	10,5
Rail 3	0,0227	0,0301	0,0564	0,0459	0,0822	0,141	120	2	13,6
Rail 4	0,0326	0,0401	0,0613	0,0639	0,0933	0,188	7000	2,9	10
Rail 5	0,0287	0,033	0,0549	0,144	0,0426	0,0766	2440	7,6 (X)	-

Table 2 : PSD's Level & duration Rails phases
(Green color = small vibration level – Red = High vibration level)

In this study, we will focus on the railway part because it shows a duration of more than 7 cumulative days (Table 2). Several phases are detected and represent vertical but also horizontal vibratory stresses.
The PSDs of the vertical and horizontal vibrations identified will be presented and commented.

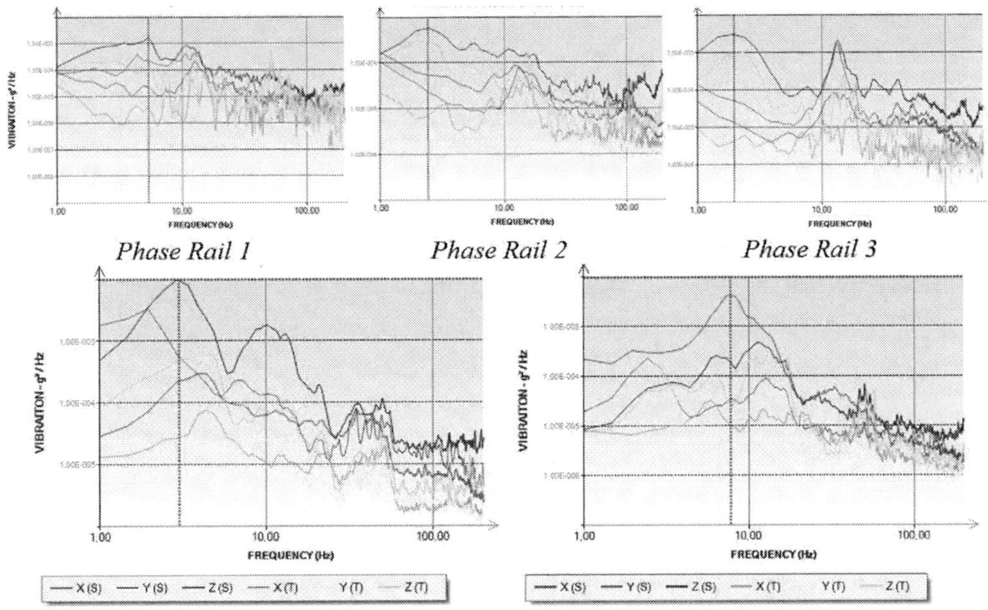

Phase Rail 1 *Phase Rail 2* *Phase Rail 3*

Phase Rail 4 *Phase Rail 5*

Figure 3 : PSD's Signal & Timer from Rail phases

Usually the fundamental vibrations for a rail road (figure 3) are vertical and located in the lower frequency's levels [2-5Hz] [3]. It checks out for the trains on this Silk Road. A [10-15Hz]-secondary peek of vertical vibrations is also noticeable.

The prevailing vibration on rail phase N°5 (the last rail phase before arriving to Xi'an) is horizontal. During this 2440-minute phase (40,7 hours), the X-axis records many 7,6Hz horizontal vibrations. This phenomenon is not rare. Various recorded journeys show between 200 and 600 similar events.

All these events begin with a shock followed by oscillations (Figure 4).

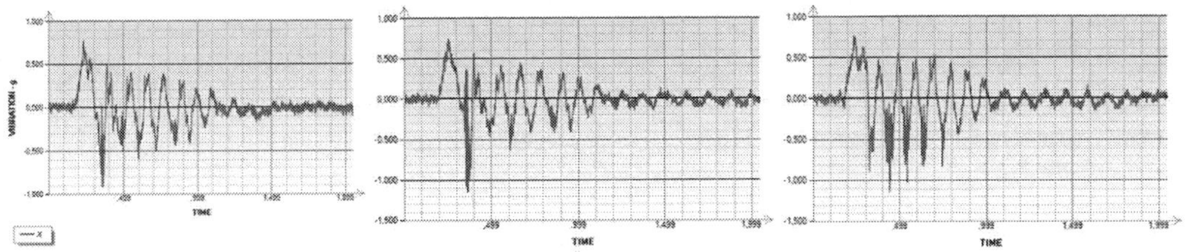

Figure 4 : Examples of Horizontal X Accelerogram of Rail Phase 5 events

Hypothesis: The link between the wagons probably present a mechanical play; during traction or braking phases, the wagons collide. The convoy then enters a damped-oscillation phase.

We isolated the excitatory shocks. A few shocks (125) are presented in the Figure 5 chart (Acceleration VS Duration).

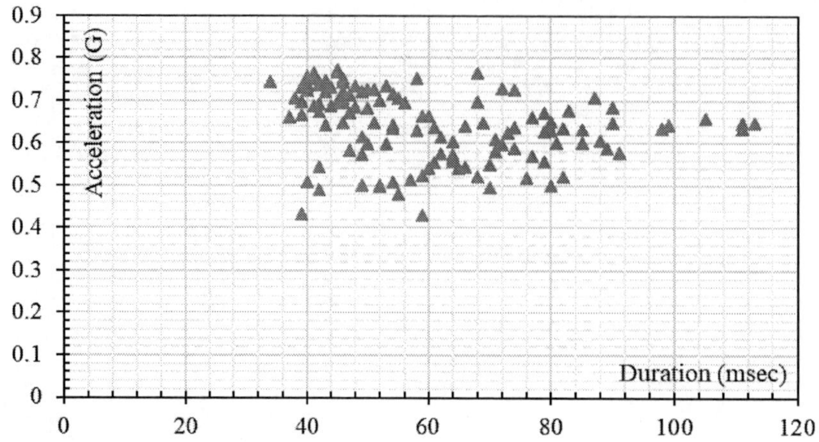

Figure 5 : Diagram of X Horizontal Shock events measured on Rail Phase 5

These shocks have various durations (from 35 to 115ms). The acceleration rates are, however, more homogeneous (from 0,5 to 0,75g). The excitatory shocks are due to the cushioning pads on the wagons. They are always oriented in the same direction (X is positive in this recording). It is therefore a repetition of the wagons' movements.
The oscillations following an excitatory shock are extremely regular. They are clearly identified in the rail 5 PSD – (X axis @ 7,6 Hz).

3. EFECT OF RAIL HORIZONTAL EVENTS ON PALLET

To simplify this particular horizontal stress, we measured the effect this oscillatory shock has on a pallet

We chose a pallet of bottled water. This packaging is particularly unstable (because not very rigid). The 1.5L bottles are made of PET. They are grouped in packs of 6 bottles maintained by a shrinkable film. The pallet is constituted of 4 layers of 21 packs. This pallet setting (800mm x 1200mm - mass = 784 kg) is maintained by 2 layers of stretch film. The pallet is not very rigid and has a low natural frequency [1,5Hz]. The first harmonic is in the 4-8Hz range. The acceleration amplification does not exceed x2. It is crucial to know the dynamic behavior of a specimen to properly evaluate its horizontal deformability [4] during dynamic tests.

From the recorded signals, we built a lab simulation test protocol SRHP (Special Rail Horizontal Protocol). We chose 5 different time events based on the excitatory shock's distribution presented in Figure 5. This set is presented in Figure 6 and the ignition shocks data are compiled in Table 3.

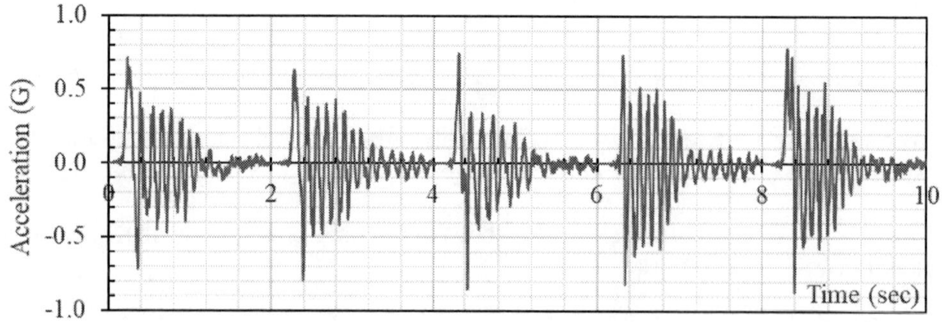

Figure 6 : Accelerogram SRHP

This accelerogram (Figure 6) was injected in the SEREME hydraulic vibration table following the horizontal axis. The M+P "Road Load" control module makes it possible to control our test equipment with a file in real time, as is practiced, for example, on earthquake resistance tests. This protocol was applied to the pallet of palletized water bottles to see the impact on the load stability (Figure 7).

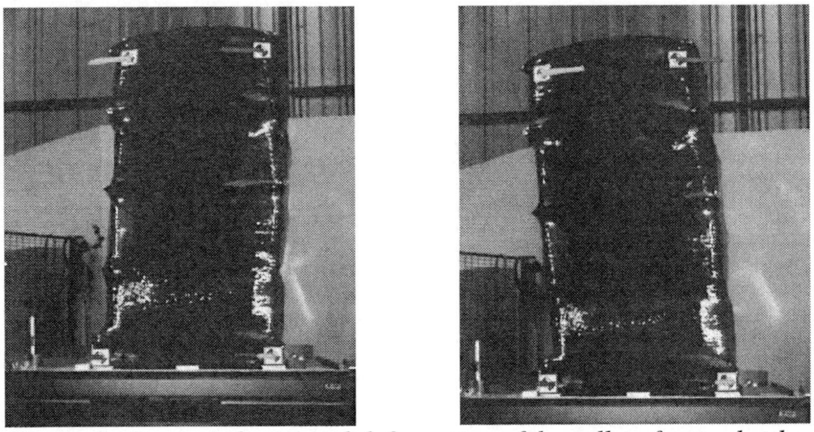

Figure 6 : Dynamic horizontal deformation of the pallet of water bottles

With an OLYMPUS high-speed camera (i-speed TR - 400 frames/sec), we were able to measure the horizontal deformation of the top of the pallet when the test protocol is launched. The characteristics of the five SRHP events are given in Table 3. The deformation measurements will be presented on Table 3.

Ignition Shock	Event 1	Event 2	Event 3	Event 4	Event 5
Acceleration (G)	0.65	0.63	0.75	0.73	0.72
Duration (ms)	115	95	57	44	54+40
Shock Freq (Hz)	4.3	5.3	8.8	11.4	5.3
Horizontal Deflection (mm)	between 50 to 112 mm				

Table 3: Ignition horizontal shock characteristics & Pallet deflection

Analyzing carefully the videos of the tests, we can see that the horizontal deformation of the pallet due to the ignition shock. The following oscillations do not induce further deformations.

4. CONCLUSION

The major part of the Silk Road is a rail road (7 days of continuous vibrations). The PSDs are very common but show frequential differencies (first peak between 2Hz and 5,3Hz) due to the different kinds of trains. During the last rail phase in China, significative, non-negligible horizontal events are recorded. They are due to shocks followed by damped oscillations. These shocks must be taken into account to ensure the stability of the palletized goods.

5. REFERENCES

[1] Herve Rebillon - Journalist and founder from TRM24 – LE MONDE Journal France

[2] CIA.gov Railway Factbook

[3] P. Böröcz, S. P. Singh - Measurement and Analysis of Vibration Levels in Rail Transport in Central Europe – PTS 2017; 30; 361-371

[4] JB. Nolot, A. Copinet, V. Huart, S. Odof, JC. Candore, N. Krajka, P. Delloque, D.Erre - Dynamical response of pallets subjected to horizontal stresses - Proceedings of 21th IAPRI World Packaging Conference - Zhuhai, China, 2018 – Oral presentation

DB-P02

The new silk road – Study of transloading events

Victor HUART*, Jean-Baptiste NOLOT[2], Baptiste BOUTRIGE[1,3], Elora LEGUEBE[1,3]
Nicolas KRAJKA[1], Jérome PELLOT[1], Damien ERRE[2,3]

[1]METROPACK, 30-32 Rue du Capitaine Georges Madon, 51100, Reims, FRANCE
[2]ESIReims, Ecole nationale Supérieure d'Ingénieurs de Reims / University of Reims Champagne-Ardenne, esplanade Roland Garros, 51100 REIMS, France
[3]ITHEMM / University of Reims Champagne-Ardenne, 51100 REIMS, France

***Correspondence to:** METROPACK, 30-32 Rue du Capitaine Georges Madon, 51100, Reims, FRANCE, victor.huart@metropack.fr*

ABSTRACT: The economic recovery following the covid crisis is causing major disruptions in the transport of goods. We observe in maritime transport, a global disorganization. This disruption is causing longer lead times and transport prices increasing. At the same time, in recent years we have observed a strong increase in land transport by rail between China and Europe. This is caused by the construction of a new Silk Road. This route has different interests, particularly in terms of duration (around 3 weeks), price (cheaper than the plane, and more and more competitive due to the maritime prices increasement), environmental (less polluting than the plane), in addition, the transport unit remains usual by the use of containers. In this study we will focus on the many load breaks generated by this transport on 10,000 - 15,000 km of network. Data recorders have been placed in containers to record these journeys. Analysis of the data collected allows us to identify between 20 and 30 breaks in loads. These events have been categorized (transloading pallet, transloading container, maneuver for securing the wagons to the train). This analysis focuses on the various stresses generated by container handling equipment. We will present the machines potentially used on these routes. From field data, we will present the different actual PSD to observe the single-frequency vibratory nature of handling gears (mainly composed of low frequencies between 1.9 and 12 Hz), the accelerations generated and their duration of application. A test protocol was developed in laboratory to encompass these different breaks of load. For the vibrations, a customized sweep frequency is applied to cover these different mono-frequential stresses. Horizontal long acceleration phenomena are also detected and reproduced in laboratory to observe the effect on a pallet.

Keyword: Silk Road, train, vibration, handling, transloading

1. INTRODUCTION

The new silk road is a Chinese strategic project to economically connect China to Europe through Central Asia via a wide network of land and rail roads. On the seaside, this network of commercial routes includes the African shoreline on the Indian Ocean side.
Called the "Project of the Century" by Xi Jinping, the OBOR program (OBOR standing for "One Belt, One Road") aims to create a new generation of transnational trade routes.
It includes 68 countries, representing 4.4 billion of inhabitants and almost 40% of the world's GDP. [1]. The economic recovery following the covid crisis is causing major disruptions in the transport of goods. We observe in maritime transport, a global disorganization. This disruption is causing longer lead times and increasing transport prices.
At the same time, in the recent years we have observed a strong increase in land transport by rail between China and Europe. This is thanks to the construction of a new Silk Road. This route has different interests, particularly in terms of duration (around 3 weeks), price (cheaper than the plane, and more and more competitive due to the increased overseas shipping prices) and environment (less polluting than the plane). In addition, the transport unit remains usual by the use of containers. All the European countries are affected by this route, both for imports and exports. This mode of intercontinental land transport is a novelty for European manufacturers. European manufacturers are wondering about the qualification plans linked to this distribution cycle. In this study we will focus on the many load breaks generated by this means of transportation on the 10,000 - 15,000 km of network. Data recorders have been placed in containers to record these journeys. Analysis of the data collected allows us to identify between 20 and 30 breaks in loads. These events have been categorized (transloading pallet, transloading container, maneuver for securing the wagons to the train).

2. MATERIALS AND METHODS

Many Lansmont data loggers, Saver 3X90 et 3D15, were placed in a freight container in order to record a Europe (France) – Asia (Xi'an) journey on the new Silk Road.
The figure below (Figure 1) shows the different phases we were able to observe on this trip.

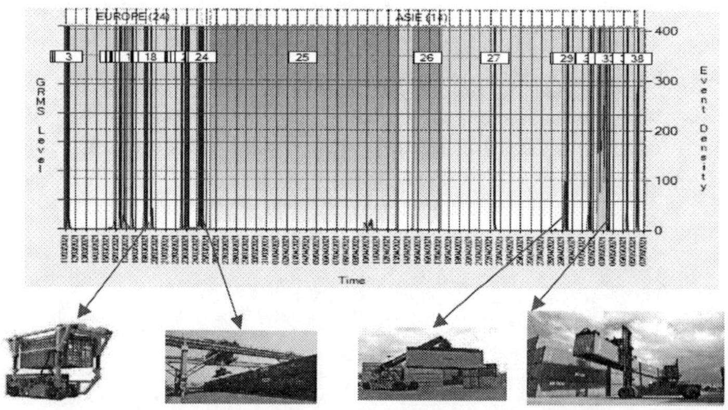

Figure 1 : Transports pahses – Blue=Handling, Pink=Truck, Orange=Rail

The phasing is based on the accelerometer data recorded on the Silk Road. The total duration of the logistic journey is 57 days. This duration is very high compared to normal transport's.
This study is here based on the blue phases due to handling and load breaks. We could detect 27 handling phases. These handling operations are carried out by "heavy" handling machines quite

similar to what can be observed in sea transportation. This similarity comes from the fact that the transportation unit is the same : a container. Different kinds of machinery are used to carry out the handling operations (Figure 2) : Forklift truck with vertical grip and with or without rotation movement, forklift truck with lateral grip, straddle carrier (particularly used for transloading due to the rail gauges).

Figure 2 : Handling gear examples.

3. RESULTS AND DISCUSSION

3.1 Vibrations

In this part of the study, the PSD of each vibrations phase was calculated. Most of the events in these phases are mono-frequential. For example, the Figure 3 shows raw data recorded on european and asian portions of the Silk Road; the vibrations are mainly mono-frequential and vertical (following the Z-axis – blue curve) centered around 2.9, 5.4 and 9.0Hz.
These phases also include a wide range of high frequencies [50-130Hz], of lower intensities, corresponding to the metallic noise of the container. This study won't focus on this noise rather than on the low frequencies generated by the handling machinery.

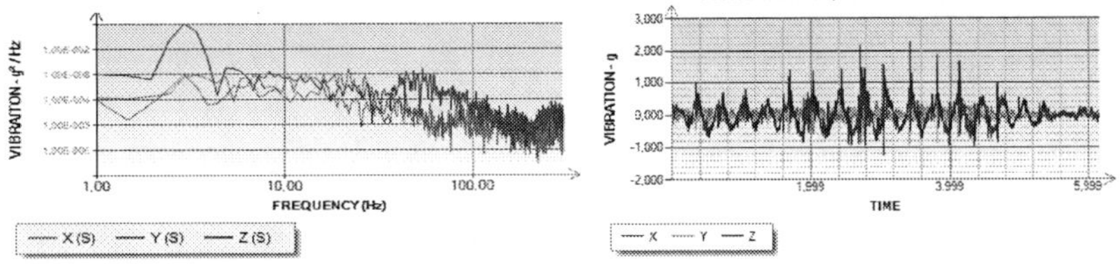

Main frequency : 2.9 Hz

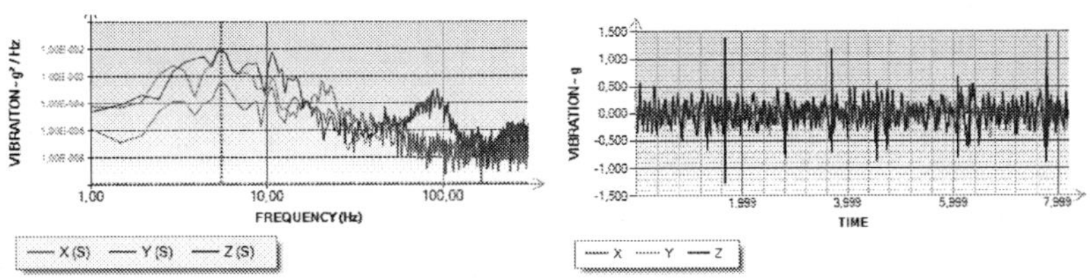

Main frequency : 5.4 Hz

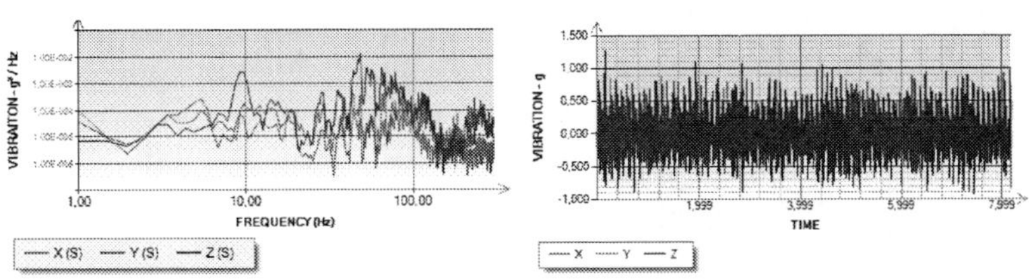

Main frequency : 9.0 Hz

Figure 3: Example of mono-frequential vibrations generated by handling gear

The following table (Table 1) shows the main frequencies we could observe during each detected handling phase and the stress duration.

For Example, during the phase "Handling 1" , the main frequency is 5.60Hz. This stress is applied during 3.55min.

Table 1 : Frequential analysis of each detected handling phases

Phase	Main frequency (Hz)	Constraint duration (min)
Handling 1	5.60	3.55
Handling 2	11.70	1.00
Handling 3	9.14	1.00
Handling 4	10.50	0.05
Handling 5	2.18	0.45
Handling 6	1.96	3.00
Handling 7	10.00	1.00
Handling 8	4.02	0.50
Handling 9	3.80	3.58
Handling 10	5.30	0.65
Handling 11	6.28	6.00
Handling 12	-	-
Handling 13	4.90	1.00
Handling 14	5.00	4.00
Handling 15	6.20	5.00
Handling 16	11.50	3.00
Handling 17	4.40	1.00
Handling 18	4.00	4.00
Handling 19	-	-
Handling 20	2.90	6.00
Handling 21	3.00	1.00
Handling 22	-	-
Handling 23	9.00	1.00
Handling 24	2.60	0.50
Handling 25	3.20	2.85
Handling 26	3.50	0.50
Handling 27	3.30	2.50

We can see thanks to this table that the phases mainly present a low frequency level going from 1.96Hz to 11.70Hz. The handling phases 12, 19 and 22 are not vibration phases. Figure 4 compiles the whole set of average PSDs associated to each handling phase. We can note that the fundamental oscillations are focused on a frequency range going from 2Hz to 12Hz.

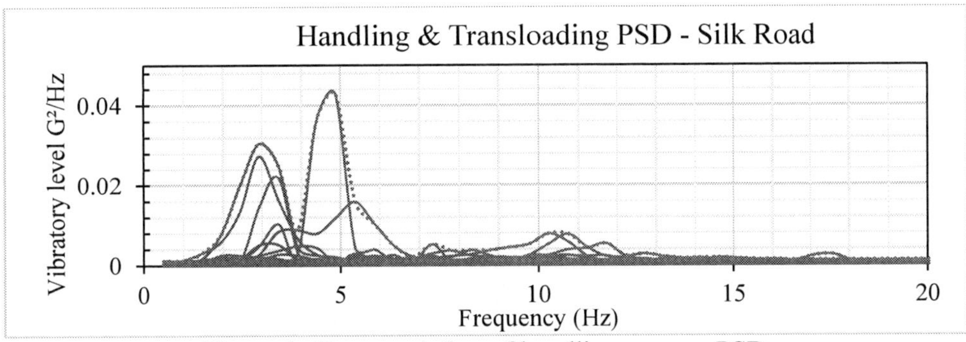

Figure 4 : Representation of handling average PSDs

The representation is a linear scale; it allows us to better visualize each and every mono-frequential peak associated with each handling phase.

3.2 Long accelerations

On the non-mono-frequential handling phases, we also identified long acceleration phenomena due to the handling machinery. The main directions are now horizontal. These accelerations are low but continuous (Table 2). The movements associated with these accelerations can be measured in meters.

Table 2 : Main long acceleration phenomenon

Main event	Mangnitude (G)	Duration (ms)
1	0.5	150
2	0.2	700
3	0.2	1000

On figure 5 below, we showed the horizontal accelerogram following the X and Y axes, typical of these « non-vibratory » handling phases. Here the event starts with a long acceleration phenomenon (Y axis in green) which lasts about 150ms; it is followed by a damped oscillation following the Z axis (vertical) and the Y axis (horizontal) with a 4Hz frequency. The movement is therefore a damped rotation initiated by a horizontal movement (braking, container handling...)

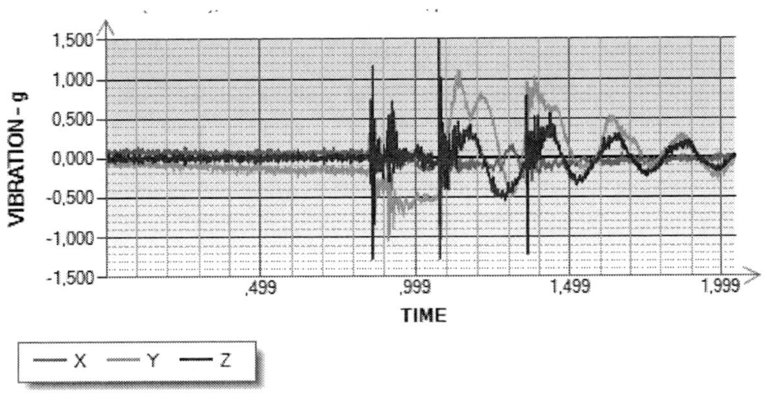

Figure 5 : Long acceleration (event 1)

These events mainly impact the load stability [2].

4 PROTOCOL

4.1 Vibrations

We saw in the previous part that the handling phases mainly consist of mono-frequential vibrations going from 2 to 12Hz. In order to reproduce these phases in the laboratory, we chose to apply a frequential sweep instead of a PSD in order to keep the sequential aspect.
To define our frequential scan protocol, we first determined the distribution of the frequencies in ascending order and associated the stress duration. It allows us to establish a frequency test perimeter. We broke this bandwidth down in 2.5-octaves ranges (Table 3):

Octave 1 : [2-4 Hz[
Octave 2 : [4-8Hz [
½ Octave 3 : [8-12 Hz]

Table 3: frequencies groups and duration by octave range

Phase	Main Frequency	Constraint duration	Duration per octave
1	1.96	3.00	24.38 min
	2.18	0.45	
	2.60	0.50	
	2.90	6.00	
	3.00	1.00	
	3.20	2.85	
	3.30	2.50	
	3.50	0.50	
	3.80	3.58	
	4.01	4.00	
2	4.02	0.50	21.70 min
	4.40	1.00	
	4.90	1.00	
	5.00	4.00	
	5.30	0.65	
	5.60	3.55	
	6.20	5.00	
	6.28	6.00	
3	9.00	1.00	7.05 min
	9.14	1.00	
	10.00	1.00	
	10.50	0.05	
	11.50	3.00	
	11.70	1.00	

In order to define the amplitude of the oscillations we must apply, the effective power (RMS) of each octave range is calculated based on the PSD envelope of the whole handling phases of the transport (Figure 6).

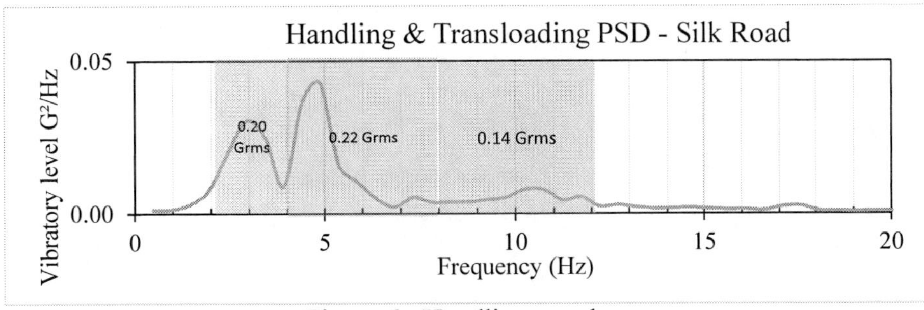

Figure 6 : Handling envelope

Table 4 shows the sweep speed and amplitude of the oscillation for each studied frequency range.

Table 4 : Sweep frequency characteristics

Step	Duration per octave	Sweep speed and magnitude
Octave 1 (2-4 Hz)	24.38 min	2-4 Hz : Sweep 0.041 octave/min -0.28G Peak
Octave 2(4-8 Hz)	21.70 min	4-8 Hz : Sweep 0.048 octave/min – 0.31G Peak
½ Octave 3(8-12 Hz)	7.05 min	8-12 Hz : Sweep 0.06 octave/min – 0.20G Peak

We intend to reproduce this representative handling event.

4.2 Acceleration

The most critical long acceleration profile for this journey (Figure 7) was injected in the pallet (pallet weight : 274kg, pallet dimensions : height :1413mm, length 1200mm, width:800mm)

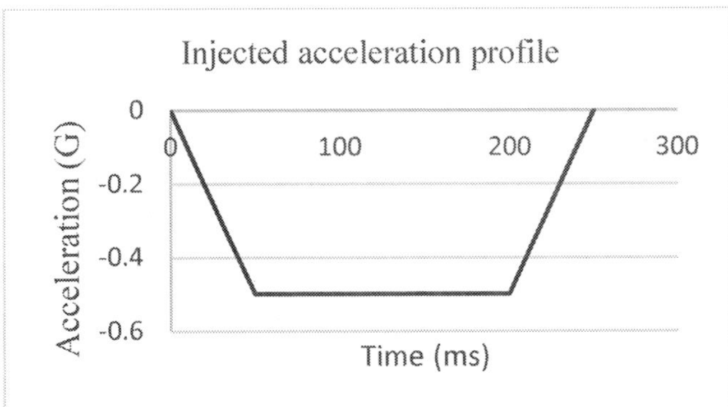

Figure 7 : Long acceleration profile.

This profile created a 1-degree-movement of the load, an insignificant value during the test (Figure 8).

Figure 8 : positioning and measurement of the pallet deformation

This does not mean that these events should not be taken into account on the Silk Road, but that complementary measurements should be considered in order to confirm or deny this observation.

CONCLUSION

This analysis focuses on the various stresses generated by container handling equipment present on the new Silk Road. We have seen that the different handling phases (transloading) are composed of single-frequency vibrations generated by handling gears (mainly composed of low frequencies between 1.9 and 12 Hz). We have defined a test which permits us to reproduce the stress typology of the Silk Road transloading using sweep frequencies. The sweep frequency speed was determined using the stress duration for the actual transloading duration for each observed frequency bandwidth. The sweep magnitude was determined from the Grms value of a particular bandwidth (2-4 Hz, 4-8 Hz, 8-12 Hz) of the global PSD; In this trip we also have found long acceleration phenomena. This type of events induces potential problems of load stability. More data must be provided in order to refine this analysis and complete the different results.

Bibliography

[1] James Griffiths, 2017, “Just what is this One Belt, One Road thing anyway?”, https://edition.cnn.com, May 12, 2017.

[2] B.Boutrige and al, Determination of the impact of the acceleration phenomenon on a load stability in packaging transport, 30th IAPRI Member Conference.

DB-P03

Evaluating Consumer Package Failure from Review Images for Using Machine Learning

Sarah Holland[1], Mahsa Tavasoli[1], Euihark Lee[1]*

[1]*School of Packaging, Michigan State University, East Lansing, 48823, United States of America*

***Correspondence to:** School of Packaging, Michigan State University, 448 Wilson Rd, East Lansing, 48823, United States of America, leeeuiha@msu.edu*

ABSTRACT: Products that endure the stresses of E-commerce should undergo various tests to determine their protection ability. These tests include, but are not limited to, vibration, drop, compression, and shipping tests. Due to the large number of products that are shipped directly from a warehouse to the consumer, it becomes very expensive and time-consuming to retest each product to the full extent if any damage occurs during distribution. When working with E-commerce, there are more touchpoints to consider, resulting in a potentially rougher distribution environment. This paper proposes a novel packaging evaluation method using an Artificial Neural Network . An Artificial Neural Network works in a way similar to that of neurons in a human brain: by making connections between a trained dataset and the new images provided after training. By evaluating consumer review images with an Artificial Neural Network, damaged products can be identified allowing a decision to be made as to what type of package failure occurred. One of the benefits to using an Artificial Neural Network to evaluate consumer review images is that it allows for the evaluation of package protection in a real distribution environment as compared to a controlled lab setting. In turn, this reduces the cost of testing, as the package failure will have been identified with the assistance of the Artificial Neural Network, rather than full retesting to identify where damage occurred. This process would also reduce costs associated with the usage of materials for testing, due to the lower number of test samples required.

Keyword: Packaging Evaluation, Artificial Neural Network works, E-commerce, Image Training

1. INTRODUCTION

E-commerce has recently become a heavily utilized resource in the supply chain for a variety of products. This has been even further utilized since the start of the COVID-19 global pandemic [1]. A study conducted by Shi et. al [2] discusses the importance of consumer's perception on packaging and utilizes deep learning in doing so. This paper shows that package appearance is just as important as the product protection quality in the eyes of the consumer. Due to this, and the increase in products that consumers are willing to purchase online, products and their packaging must undergo various tests to determine their protective abilities. These tests include, but may not be limited to, vibration, drop, compression, and shipping tests. Retesting that must occur if large amounts of damages are reported can become very expensive and time consuming. Esfahanian & Lee introduced a packaging evaluation method from consumers' text review data by implementing machine learning tools [3]. This research expands to evaluate the image data from the reviews. In this study, an Artificial Neural Network (ANN) is adapted as an image training tool to predict the cause of the packaging damage. An ANN works similarly to neurons in a human brain: Connections are made between a trained dataset and new images provided after training. Similar use of an ANN can be seen from Davatalab et. al. [4], where concrete layers are evaluated for layer deformation. One benefit of the proposed system is that packages can be tested in a real distribution environment after the controlled lab test has been completed and shows the reality of what a package will go through in e-commerce. This, in turn, will reduce costs related to investigations and retesting.

2. MATERIALS AND METHODS

The packaging evaluation process using review image data is composed of 4 main processes as shown in Figure 1. The first step is a data collection process. In this process, the damaged images are collected and labeled with the cause of the damage from pretested results. After the data is collected, the data preparation process is required. The image data has a lot of noise including image backgrounds, different colors, orientations, and other factors. To minimize the effect of the noise in each image, the proposed method utilizes multiple different tools including product detection, cropping, and removing background. Once data preparation is completed, the training process is performed with ANN. In the training process, images were embedded then pushed through a data sampler set up to split the images randomly into 80% for training and 20% for testing. The ANN process is using 200 maximum iterations and two layers containing 50 neurons each. The ANN and remaining 20 images were used for verification, which provided accuracy of the prediction.

To completely remove the backgrounds of both the original and cropped images, python code was utilized. The images were put through this code to yield three versions of the image. One original image, one with the background removed, and one showing the background area that had been removed. The second of these was then manually downloaded and put through the same process outlined above.

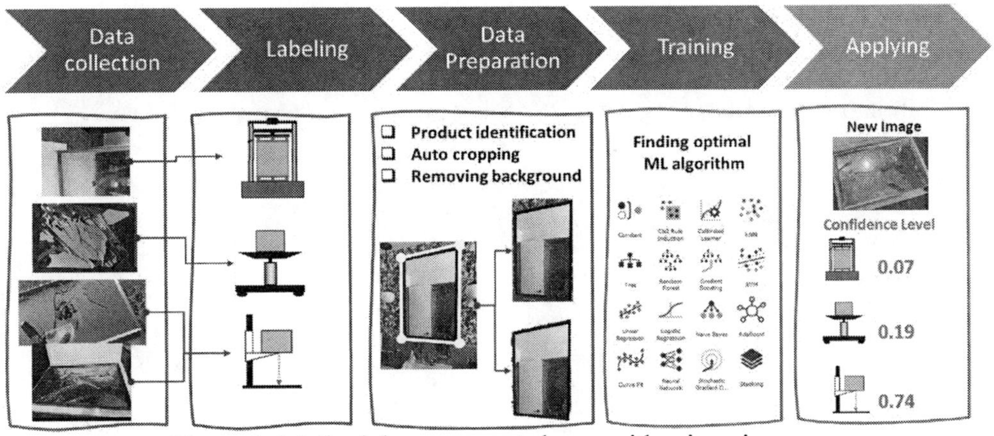

Figure 1. Methodology concept shown with mirror images

While the method was developed, it was also important to find the optimal number of neurons in two hidden layers to increase the accuracy. An exhausted search method was implemented to find the best neuron combinations for 1st and 2nd hidden layers. The neurons per layer were processed in increments of 10, starting with 30 and ending with 100 based on different image process methods, as shown in Figure 2. From here, the best option was chosen based on the highest accuracy rate. This step was performed for each photo category, but the same activation and solver were utilized for all four categories.

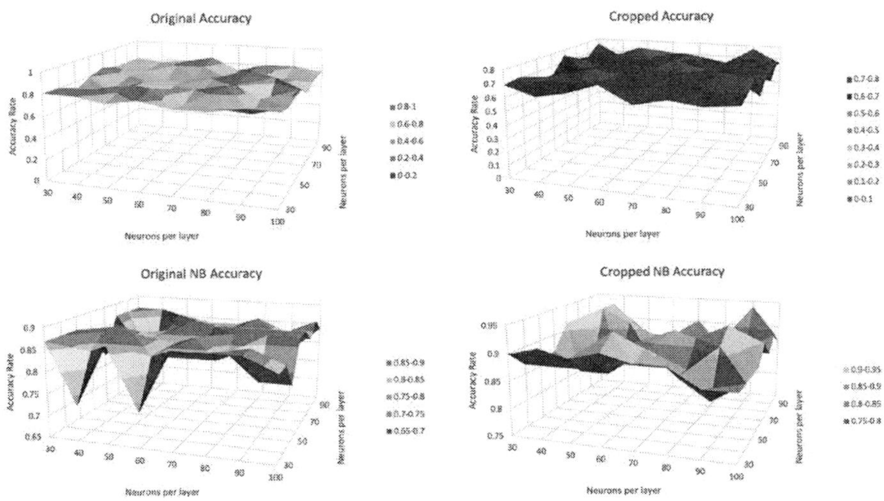

Figure 2. Neurons per layer accuracy with mirrors

3. RESULTS AND DISCUSSION

Methodology utilizing television review images is shown in Figure 3. In total, there were 129 images, split into 103 training images and 26 testing images, for the same 80/20 split.

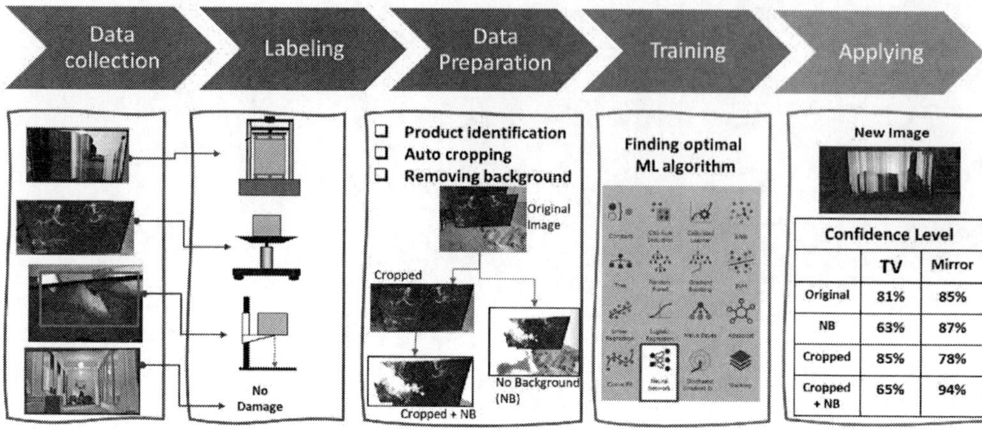

Figure 3. Methodology shown with television images

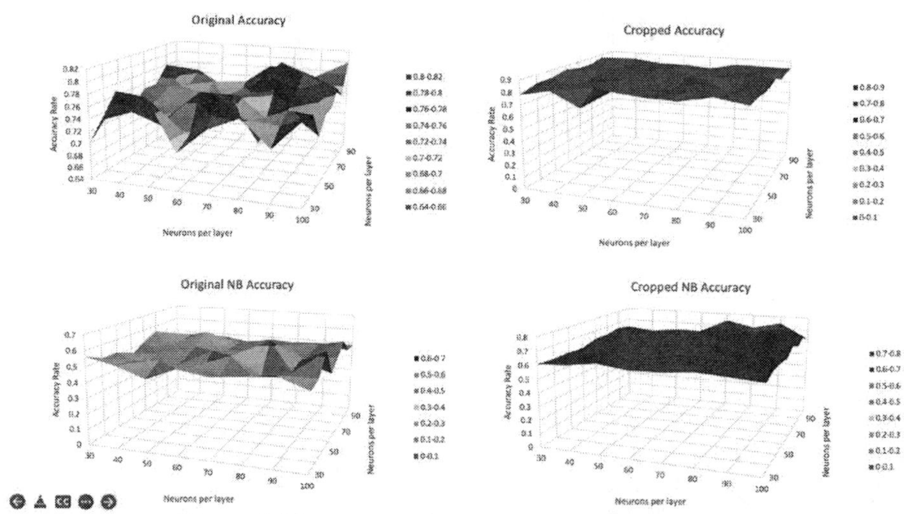

Figure 4. Neurons per layer accuracy with televisions

Figure 4 shows the confidence levels, or accuracy rates, of each workflow for different combinations of neurons per layer. These results show that while the presence of a background can affect the predictive ability of the ANN greatly, the television was best predicted with a cropped photo that still contained its background. This differs from the mirror, which showed the best results in cropped images without their backgrounds. This demonstrates the importance of evaluating images through all four categories, as the most accurate one is not always the same. The proposed method can provide increased insight on real life package performance, detect packaging problems in early stages of distribution, and increases the efficiency of the packaging evaluation process.

4. CONCLUSIONS

Consumer review images can be extremely beneficial to the long-term quality of a product and package. By evaluating these for damage, both time and money can be saved and put towards improving packaging instead of testing to determine why failure is occurring. Further study is required to improve both data preparation process and accuracy. However, the process discussed in this paper is showing a new approach to become a resource that many industries will be able to utilize in conjunction with e-commerce.

5. REFERENCES

[1] Jílková, P., & Králová, P. (2021). Digital consumer behaviour and ecommerce trends during the COVID-19 crisis. *International Advances in Economic Research*, *27*(1), 83-85.

[2] Shi, L. (2022). Design of Packaging Design Evaluation Architecture Based on Deep Learning. *Scientific Programming, 2022.*

[3] Esfahanian, Shiva., Lee, Euihark. (2021). "Packaging Evaluation Using NLP for Consumer Reviews" *IAPRI 2021 Member Conference, ISBN: 978-1-63821-160-0*

[4] Davtalab, O., Kazemian, A., Yuan, X. & Khoshnevis, B. (2022). Automated inspection in robotic additive manufacturing using deep learning for layer deformation detection. *Journal of Intelligent Manufacturing*, *33*(3), 771–784.

DP-O01

Response of a packaging system to a transient event: real-time replication VS SRS simulation

Prof. ERRE Damien

ESIREIMS / ESIEC, UNIVERSITE DE REIMS

ABSTRACT: Vibrations and shocks are found among mechanical phenomena undergone by packaging systems. In order to verify the behavior of packaging systems under shocks, test protocols advocate the execution of drop tests, inclined plane tests, forklift handling courses, etc. in laboratories. Nevertheless even if these protocols are efficient, they do not replicate the wide variety of transient events occurring in a transport. That is why there is a need in developing a method capable of replicating actual transient events in a reproductive way in laboratories. This work concerns the comparison of the responses of a packaging system subjected to two different simulations of the same transient event with the aim to forward develop specifications. The definition and elaboration of Shock Response Spectra (SRS) are reviewed in the first part of this work. Afterwards, a data acquisition enables the recording of classical transient events generated by a forklift. Then the calculation of the SRS of one of those transient events enables the generation of a time signal different from the transient event recorded. Finally, a comparison between the response of a packaging system subjected to the time transient event as it is recorded and the response to the time transient event calculated from the SRS is carried out. The first results based on the study of autocorrelation functions, intercorrelation functions, Energy Spectral Densities (ESD) and damage calculation conclude towards an equivalence between a real-time replication and an SRS simulation in the case of one transient event and one packaging system.

Keyword: packaging system, transient events, shock response spectrum

DP-O02

Effect of the sustained forces during transportation on the creep behavior of stretch film affecting unit load stability

Manuel García-Romeu Martínez

ABSTRACT: This paper shows the effect of horizontal sustained stress on the stretch film used for palletizing loads when those forces are produced by vehicles driving in curves and roundabouts or making manoeuvres like braking and turning. Stretch film used for wrapping goods in a palletized load should hold the system from the factory to the retailer's shelf. Holding load integrity is critical not only from a business point of view, but it is also a safety question. If the stretch film does not hold the verticality of the load, it could become a deadly issue. If the load cannot withstand the strain, it could fall down inside the vehicle and onto the road, resulting in traffic accidents. Typically, the unit load design considers a perfectly square, trapezoidal shaped load. The sides of the load on pallets can be considered flat and vertical. When the stretch film is applied, it constrains the load by applying a force on the load. But what happens when the load applies a force on the film? Sustained acceleration over a prolonged period of time, such as during braking, turns or circulating on roundabouts, can damage the stretch film causing a sustained film elongation, typically known as creep, and this seriously affects the stability of the unit load. This study evaluates the creep behaviour of the stretch film due to sustained forces applied for a long time as it occurs on the road, and the outcome obtained has been used to understand its application in the design of the right wrapping scheme for a simple case of stacked boxes.

Keyword: creep, stretch film, truck braking, load stability, steady force

DP-O03

A Review and Update of the Corrugated Fiberboard Cushioning

Changfeng Ge*

***Correspondence to:** Department of Packaging Science, Rochester Institute of Technology, 78 Lomb Memorial Drive, Rochester, NY 14623, USA*

ABSTRACT: The corrugated fiberboard was the dominated cushioning material before the advent of expanded plastic foams in the 1960s. With recent advances in packaging sustainability, the corrugated pads are again the research focus. This paper reviewed studies of corrugated fiberboard cushioning in areas of design and experimental evaluation. Since most of these research works were performed more than half a century ago, the historical data were reexamined and compared with the current practices. Previous and recent structure designs, represented by layered corrugated pads, tube, spring and zig-zag components, were displayed and discussed. This study also reviewed studies investigating the crush resistance and cushion curves in hope to provide some practical guidance on prediction of cushioning performance. As an updated development by the author, this study presented cushion curves developed under three loading patterns, namely column, underhang and overhang. Finally, the study pointed out that double dipping points and shift-to-right tendency on the cushion curves, as well as the long plateau on the stress- strain curves are unique properties of corrugated fiberboard cushioning. Packaging engineers are suggested to factor in these properties in design for different applications.

Keyword: Corrugated fiberboard, Cushion curves, Corrugated pads, Cushion test, Crush resistance.

DP-O05

On Friction Measurements and Certification of Slip-Preventing Materials for Road Freight Usages

Julien Lépine[1]* and André Bégin-Drolet[2]

[1]*Department of Operations and Decision Systems, Université Laval, Québec city, Québec, G1V 0A6, Canada*
[2]*Department of mechanical engineering, Université Laval, Québec city, Québec, G1V 0A6, Canada*

***Correspondence to:** Department of Operations and Decision Systems, Université Laval, Québec city, Québec, G1V 0A6, Canada, julien.lepine@fsa.ulaval.ca*

ABSTRACT: Slip-preventing materials are often used as tertiary packaging to improve cargo securement and simplify truck loading. There are many national and international standards on their safe usage based on the extra friction they can provide. However, it is challenging to quantify the coefficient of friction (CoF) of slip-preventing materials. This paper presents CoFs of three slip-preventing materials and one baseline condition measured on different surfaces and contact pressures. These measurements show interesting phenomena that affect the result variability. For instance, 60% of the pairs of surfaces tested were affected by the contact pressure. Hence, the load of the freight seems to affect performance of the slip-preventing materials. This interaction happened in different ways such as changes in the friction mode (constant slip, and stick slip at difference frequencies and magnitudes). The typical CoF values available in standards as well as standardized measurement methods do not consider very well these effects nor how to deal with high variability of CoF measurements. This can lead to either over or under CoF estimations. To overcome these limitations, this paper presents a statistical method to quantify slip-preventing material CoFs based on variability and confidence intervals of the measurements.

Keyword: Slip-preventing materials, Anti-slip mats, Coefficient of friction measurements, Cargo securement, Freight stability, Tertiary packaging

DP-O06

Measurement and Analysis of low-acceleration and long-duration longitudinal events using delivery van

Péter Böröcz[1]*, Csaba Pánczél[1], Kyle Dunno[2], S. Paul Singh[3]

[1]*Széchenyi István University, Győr, Hungary*
[2]*Rochester Institute of Technology, Rochester, USA*
[3]*Michigan State University, East Lansing 48824, USA*

***Correspondence to:** boroczp@sze.hu*

ABSTRACT: This study investigates longitudinal acceleration events during freight transportation characterized as low-acceleration and long-duration using delivery van type vehicles. In the past several decades, there has been an increase in shipments requiring only single or small pallet load quantities and mixed palletized unit loads comprised of different goods. These loads are often transported in delivery vans without load securing devices, increasing the risk of product loss and damage due to load failures resulting from unit loads shifting. A field data acquisition system was used to observe and record the random acceleration events from five vehicles for five days, explicitly targeting the vehicles' braking and acceleration maneuvers. The study aimed to understand the physical phenomenon and provide new information that can be used during pre-shipment tests to prevent damage to goods and ensure unit load integrity is maintained throughout the supply chain. The events were statistically analyzed to understand their probability of occurrence, severity level, and quantify critical parameters such as event rise and hold times. For the braking maneuver, the average deceleration peak was 0.26 g with a corresponding rise and hold duration of 0.75 and 1.12 s, respectively. During the vehicle's acceleration maneuver, the average acceleration peak was 0.31 g with a rise and hold time of 1.26 and 1.28 s, respectively. Utilizing the field data, composite profiles were developed, and these profiles were compared to the currently available test procedures and previous results of other studies.

Keyword: braking, acceleration, jerk, unit load stability

CONFREENCE PROCEEDINGS :
LOGISTIC & SUPPLY CHAIN

• ORAL PRESENTATION

• GENERAL STREAM

LG-O01

The Role of Digital Specification Data Management in Streamlining Packaged-Product Supply Chains

Adam Armstrong[1]*, Daniel Klemm[2], Jay Singh[3]

[1]*Specright, Tustin, California, 92782, USA*
[2]*EVS Sports, Rancho Dominguez, California, 90220, USA*
[3]*Cal Poly State University, San Luis Obispo, California, 93407, USA*

***Correspondence to:** Specright, Tustin, California, 92782, USA. aarmstrong@specright.com*

ABSTRACT: Specifications represent the DNA-level information of packaged products and their associated components. Items such as the bill of materials, technical drawings, and supplier requirements are stored together to create a traceable trail of information for stakeholders along the supply chain to refer to in the face of events such as recalls, sustainability mandates, and supply chain disruptions. The goal of this paper is to determine the value of migrating specification data management to a digital, internet-enabled environment. To evaluate this, a combination of primary research (first-person interviews) and secondary research (evaluating software systems, prior academic papers, and general market research) was conducted to reach the following results. Companies at all points of the supply chain – from packaging manufacturers to consumer goods producers to retailers - are currently undergoing a digital transformation to change the way they manage data. Specifically, purpose-built cloud-based software is an emerging technological innovation entirely focused on harnessing the power of packaging/product specifications to create efficiencies and generate value. These applications coordinate the data that millions of stock keeping units (SKUs) worldwide are generating into a harmonized system. Data-driven insights from these systems facilitate increased speed to market for new product development, reduced costs through easily identified economies of scale for procurement, and decreased risk via complete, accurate specifications for quality assurance teams. Innovative cloud-based solutions enable processing of data and workflow automation, ultimately reducing the time skilled individuals are spending on administrative tasks. This also allows for fast querying, advanced analytics, and visualizations that drive improved decision making – for sustainability, material optimization, and more. Using several real-world examples, this paper details the framework and advantages of a digitized specification system for global supply chain demands.

Keyword: Packaged-Product, Specifications, Supply Chain, Specification Data Management, Cloud-based Software

1. INTRODUCTION

The packaging and product supply chain is currently undergoing a digital transformation that changes the way organizations manage data. Cloud-based software is an emerging technological innovation entirely focused on harnessing the power of packaging/product specifications to create efficiencies. These applications coordinate the data that millions of SKUs worldwide are generating into a harmonized system that will organize the SKUs and produce valuable information that will allow stakeholders to make decisions based upon data-driven insights.

Specifications are the DNA-level information of packaging and products. Items such as the bill of materials, technical drawings, and inventory are stored together to create a traceable trail of information for stakeholders along the supply chain to refer to in the face of recalls, sustainability reports, and root cause analysis of procurement delays. Organizations have gone from storing this data on paper to creating a digital trail with manual processes and legacy systems on the computer. However, these systems cannot contend with the sheer amount of data companies now possess, wasting time and money trying to organize it all. In addition, packaging and product specification data lacks a common language that creates consistency and reduces errors. Tracing an item to its source is a laborious endeavor, and resource investment is trying to solve the problem with existing manual processes and legacy systems when funding towards an innovative, cloud-based solution will most often generate more significant, long-term value.

Such a system would be able to process data and create a standardized template for specifications. This organization would allow for fast querying and advanced analytics that turn into insight-driving visualizations. This framework would create a single point of truth for specifications, enhancing how companies along the supply chain collaborate and share information and streamline packaging and product creation workflow. Software solutions for specification data management exist in varying levels of involvement and installation that allow stakeholders to find a model that fits their needs in an ever-changing supply chain management landscape.

2. DIGITIZATION AND ITS EFFECTS

A digital transformation refers to technological innovations in the form of software solutions that fundamentally change how businesses operate and create value for consumers [1]. Often, this means switching paths completely when it comes to business practices and legacy systems that have been in place for decades and starting fresh with operations that still have yet to flesh themselves out entirely. Technological innovations such as "cloud computing, Subscription-as-a-service (SaaS), augmented data management, augmented analytics, artificial intelligence (AI), APIs and mobile devices" are creating a multi-faceted environment for businesses to manage their data and fast-track decision-making processes [2]. The software solutions market expanded to $3 billion in 2020 and, based on projections, will grow by at least 10% year over year throughout the next five years [3]. Organizations are in the early stages of applying these innovations to packaging and product specifications; Gartner's 2021 "Market Guide for Packaging and Product Specification Content Management" polled companies across five industries to determine how they manage their packaging specifications and found that only 23% of the participants across all industries use software from a "specialized software provider [3]." This poll puts the current market at the beginning of the early adopter phase of Everett's innovation adoption curve, further emphasizing the potential for the market's growth. Therefore, companies will continue to rapidly incorporate more digital solutions to their specifications as a direct response to the increased demand for businesses to manage corporate social responsibility, SKU growth and complexity, and consumer trends-- all while attempting to increase speed-to-market and minimize costs [3].

The expansion of SKUs has made managing various packaging and product specifications an administrative nightmare. There are roughly 200+ million packaging SKUs worldwide and 60 million in the United States [4]. As this number continues to grow, the demand to manage all of the packaging and product specifications also increases. Currently, many supply chains still operate based on paper, manual processes, and legacy systems. However, a transition towards the digital transformation of SDM systems is inevitable to respond and adapt to sudden changes in consumer behavior, health codes, and government mandates [1]. This change was magnified during the Covid-19 pandemic, starting in 2020, as supply chains faced more frequent disruptions. Additionally, working remotely in the product and manufacturing world is challenging without digitization. Packaging, quality assurance, and operations employees have vital roles in the supply chain, and SDM networks can make these jobs possible in a suddenly more flexible way that encourages innovation.

Digitization has allowed for more rapid dissemination of information through the cloud. By transitioning packaging and product specification data into a digitized SDM, employees have greater flexibility to access data and manage projects [5]. With that in mind, the primary focus of IT departments has shifted from finding ways to generate cost savings within their company's current data management systems to now transferring data and workflows via digital transformation to a more specialized cloud-based software solution, enabling competitive advantages [1].

3. DEFINING SPECIFICATIONS

A specification is "a detailed instruction of work to be done and/or the materials to be used to complete the job" and could otherwise be known as a set of instructions, guidelines, parameters, requirements, or conditions [5]. Packaging and product specifications are composed of everything that goes into creation-- from ideation, raw material extraction, processing and manufacturing, distribution, all the way to the consumers' hands [6]. Whether it be a bag, box, or bottle, specifications have numerous unique characteristics; the same goes for an SKU's primary, secondary, and tertiary packaging [2]. For example, a carrot purchased in a grocery store has numerous levels of specifications, including the components of the dirt it is grown in, the tractor used to farm the produce, the box and pallet used to ship the carrot, as well as the characteristics of the carrot itself.

Any product that is "developed, grown, or manufactured" has specifications that make up its DNA, and each point along the supply chain adds to the specification in the creation of the product. Vital to a specification is also the bill of materials (BOM), or a detailed list of the raw materials, sub-assemblies, intermediate assemblies, sub-components, parts, and the quantities of each needed to manufacture an end product. Different types of BOMs break down a product differently based upon the process it represents. For example, an engineering BOM, or EBOM, breaks down a product based upon the Computer-Aided Design (CAD) or Electric Design Automation (EDA) created during the design phase. Therefore, the EBOM would look different from a manufacturing BOM (MBOM), whose components would be the pieces necessary to create a finished product, along with the quantity of each needed component.

The BOM and other specifications illustrate product expectations to stakeholders along the supply chain and give them a model for comparison. Specifications provide standards for function and aesthetics and minimize defects by conforming to a detailed uniform specification. In addition, consumers are growing in their desire to do business with transparent companies in how they produce goods, how those goods are sustainable, and their general social responsibility standards [7]. Ultimately, specifications are the DNA of supply chains, and specification data management allows industries to operate efficiently and safely.

4. THE CURRENT STATE OF SDM

4.1 History of Specification Data Management

The management of specification data has, for the most part, transferred from manual recording on paper to the computer with the invention of radio frequency identification (RFID) technologies and various software systems, making tracking and data storage a lot more reasonable [8]. Currently, Excel spreadsheets, homegrown systems, ERPs, PDFs, and shared drives comprise how most organizations manage their packaging and product specifications. However, as SKUs have increased and become more complex, IT leaders have employed the concept of "lift and shift," where they add on to existing systems such as ERPs whose original purpose was not to manage specification data [9]. Many companies have specific software for their relationships within the individual business, the suppliers they work with, and the production of goods that houses different pieces of a complete specification for a package and product in respective software systems [10].

4.2 Costs of Archaic Specification Data Management Systems

Manual processes and legacy systems cannot create transparency to scale; frequently, data will flow from "the developer to procurement to the supplier," and it is not shareable across the entire supply chain at a moment's notice for every single specification in question [2]. Manual, legacy systems rely upon user input which inevitably leaves more room for errors and miscommunications. John Blake's 2019 survey of IT professionals found that 97% agreed that delays in strategic decision-making, missed revenue and cost savings, and slow product rollouts all occurred within their organizations due to poor data management. Blake's study concluded that, among the same group of IT professionals, almost half of them were using applications such as Excel and Word or leaving it to their suppliers to manage their specifications.

Consumers desire a combination of personalization, quality, and urgency in the products they purchase [11]. Organizations are also growing more conscious of the products they buy, and supply chain transparency must be prevalent so that companies may display whom they source from and where they produce goods [6]. In addition, globalization creates an incredibly complex digital trail along the supply chain from a product's inception to the consumer, and it necessitates a high level of traceability and transparency.

There is a great deal of opportunity for improvement in the management of specifications along the supply chain as the data in question has historically seen little investment in its storage and analysis [2]. Instead of incurring the costs of these inefficiencies, companies can employ a digital SDM system to streamline their process. A digitized SDM system acts as the foundation for supply chain data that harmonizes and ensures the accuracy and efficiency of PLM, ERP, and BI systems.

Transitioning specification data to a cloud-based software service allows organizations to maintain their legacy systems by integrating and extracting only the necessary packaging and product specification data. The addition of a digital SDM creates a digital trail of specification information and automated workflows. As more companies begin to transition to this data digitization, the deployment of blockchain will further embed a community of traceability and transparency across industries that rely upon specifications [11].

5 A DIGITIZED SDM

5.1 Framework

As the amount of available data increases, the sorting and sharing of information becomes even more challenging. By introducing a cloud-based digitized SDM network architecture that focuses on accumulating data across the entire supply chain, the application removes the data from the individual software applications within a legacy system. It allows for data to move between producers, suppliers, and any other stakeholders in the supply chain [10]. The ultimate goal is a data platform that allows users to pivot from data searches to analysis - and a greater understanding of the data in question [12]. Such a cloud-based software has a framework made up of the following categories: identification and classification, transformation and modeling, processing, and presentation of traceability data [8]. This digitized SDM must be easy for users to operate, give timely feedback to ensure more effective communication along the supply chain, and create a visual story by utilizing data to create queries that become models, charts, and graphs [12].

5.2 Single source of truth

A digitized SDM system creates the common language necessary to ensure that specifications shared across critical stakeholders in the supply chain are uniform. Because the specifications are the same across the supply chain, there are fewer purchasing and supplier errors; suppliers know what to deliver, and the product arrives as specified with less chance of delay. Moreover, the entire supply chain's ability to operate with the same specification information mitigates the chance of miscommunications and results in cost savings from reduced errors and product recalls [13]. A cloud-based SDM system ensures that all stakeholders look at the same specifications, including critical documents such as drawings and certifications, creating less cross-department confusion. This improvement contributes to further down the supply chain where customers happily receive quality products on time, maintaining or even improving brand reputation.

Utilizing a specification data management system lets employees waste less time dealing with inaccuracies and out-of-date specifications and focus more on value-added activities, resulting in streamlined workflows. Forrester's "Economic Impact of Specright" study showed that having specifications in one online location freed up employees along the supply chain so that they could focus on value-added activities, creating three extra hours per week for employees to focus on such activities, as well as increased employee satisfaction with their job [13].

5.3 Collaboration

A great example of the power of collaboration along the supply chain is the transformation in the relationship between the retail superstore, Walmart, and the consumer goods manufacturer, Procter & Gamble (P&G). According to a case study written by industry experts Grean and Shaw in 2002, the two companies sold $375 million of product in 1988, equivalent to $870 million today, and the companies' relationship was struggling because they were both competitors and collaborators. As a result, they shared minimal business data and instead focused solely on transactions from P&G to Walmart and then to the consumer [14].

However, this soon changed when P&G and Walmart decided to form a strategic relationship. They created a standardized system for sales tracking and a system whereby data was shared freely on all business functions rather than just sales [14]. The formation of an "extranet" created a private network between P&G and Walmart to share data about all business functions. This collaboration between the two companies ensured a single source of truth through a standardized template. As a result, Walmart carried less of P&G's inventory and focused on what consumers were buying (based on data analytics), resulting in the businesses doing $4 billion of sales in 2002. In the context of a digitized SDM system, all parties gain from employing a standardized way of disseminating specification data to reduce inventories, cut costs, and better understand consumers.

5.4 Accessibility

SDM allows companies to be in charge of their specifications rather than having a vendor along the supply chain manage them. This ownership creates a level of independence that allows companies to act with more agility in seeking out the most competitive bids for a particular specification on a more frequent basis. Owning specifications is also more cost-effective than having suppliers hold specifications [2]. Moving specification data from hard copies to the cloud enables greater accessibility from more locations and more devices, ensuring quicker decision-making improving the time-to-market of a product in the future. As consumer trends go towards product customization, agility in changing product and packaging specifications becomes crucial, and owning specifications makes the information searchable and easy to find [13].

5.5 Automating Workflows

Digitized SDM systems are the missing link between specifications and SKUs. They give stakeholders the ability to follow a product as it is revised and receive instant notifications, giving organizations a greater chance of reducing errors by ensuring that every different stakeholder has one single reference point [15]. Additionally, the development of checkpoints along the approval process creates a counterbalance that stifles neither creativity nor efficiency. Project timelines can also be made more accurately by automating workflows that otherwise cause unforeseen time delays and reduce development cycle time for future designs. As a result, there is a reduction in shipment delays and the administrative costs that go along with them.

Accessibility to specifications and the automation of their approval processes creates a digital trail through which higher levels of traceability are made possible. Many product recalls occur due to simple mistakes such as ingredients lists not being updated [13]. However, the SDM system will ensure that any change in ingredients or components will result in quick updates to specifications. Moreover, if there is a recall, a digital trail allows for faster root cause analysis by illustrating a clear history of the specification. SDM systems also automatically notify businesses when it is time to renew regulatory documents; organizations can be more agile in responding to regulatory changes and decrease their risk of not adhering to those changes [13]. A practical and accessible SDM system will identify stakeholders throughout the supply chain from the product source to the packaging to the consumer, promoting transparency and retaining brand equity [2].

5.6 Organizing specification data

Digitized SDM systems can automatically clean specification data and organize it with an analytical categorization system, removing errors and duplicates. Key specification attributes will also be required to make database querying easier and save time. As a result, stakeholders will no longer struggle with incomplete or missing data.

SDM applications improve speed to market, cost optimization, and collaboration through features built to give otherwise disorganized and unused data a purpose [2]. Disconnected data will transform into a transparent and accurate web of specifications that all necessary stakeholders can access. In addition, harnessing this newfound data collection will help focus on productive SKUs

– more easily recognizing instances where packaging needs can be reduced while maintaining a high quality, and create optimal palletization setups.

6 SOFTWARE SOLUTIONS FOR SDM

Digitized SDMs go by many different names and acronyms. Whether it be a supply chain collaboration network (SCCN), a multi-enterprise supply chain business network (MESCBN), or a packaging and product specification content management system (PPSCM), these digital packages manage product and packaging data for its entire lifecycle, from the gathering of initial resources through to the creation of a final product and the bill of materials to go along with it [2, 3]. SDM software creates an elevated level of insight through the use of data visualization tools such as models and dashboards, creating new ways to manipulate previously disorganized data into metrics or even crucial performance indicators. SDM applications also illustrate a more transparent picture of an organization's supply chain, allowing users to manage the development of a package and product specification through the different geographical locations, SKUs, and organizations involved [2]. There are currently three categories of applications for the management of specification-related data in the market:

"Specification content management enablers" are individual applications with unique attributes that focus on specific scenarios organizations may deal with in the management of their packaging and product specifications [2]. Some examples of the different functionalities of specification content management enabler applications include:

- Document control: DocXellent's ENSUR application creates a cloud-based environment through which specification documents like BOMs and diagrams can be transferred to from legacy systems to create a more streamlined workflow.
- Artwork and specification synchronization: In the ManageArtworks application, all graphics that pertain to a packaging or product will be attached to their corresponding specifications to maintain harmonization of the full scope of a product through its lifecycle.
- Audit vendor compliance with regulations: With the QADEX SDM application, organizations can determine whether vendor specifications align with corresponding regulations.

Product Lifecycle Management (PLM) applications operate within a pre-existing software service and manage a range of information through the lifespan of the product and its packaging. This includes managing the conception and progression of packaging and product ideas, storing and updating specifications, and the automated auditing of specifications for regulatory compliance. For PLM applications to achieve a DNA level of granularity for specification management, further customization is necessary [2].

Finally, “purpose-built” SDM Applications are focused solely on managing product and packaging specifications, a key difference between other applications that offer packaging and product specification management as an add-on [2]. This focus has resulted in packaging and product specification-focused applications. GlobalPKG, IntellaPac, and Interspec are all specification management systems built for this purpose. Specright is a cloud-based SDM platform with automated workflow management, packaging optimization analytics, and sustainabilty metric reporting. These and other functions are specifically designed to leverage the advantages of cloud-based specification collaboration - from suppliers to manufacturers to brands to retailers.

According to Gartner's Market Guide for Packaging and Product Specification Content Management, “tracking sustainability metrics for products and packaging, documenting production line equipment and spare parts, quality assurance, supplier quality audits and project management workflows" are a few other advanced functionalities that these differentiate SDM applications from their historical PLM and Document Control predecessors.

7 CONCLUSION

As technology progresses, frequent digital transformation is necessary to modernize information management and collaboration within a business. At a time when supply chain and manufacturing efficiency is as critical as ever – and the regulatory environment is changing quickly - specification data management ranks high on the priority list of digital transformation initiatives. Specification data must be utilized to create efficiencies in sourcing, processing, production, and transportation workflows, and to better understand consumer trends. Greater precision and shorter delays in production will increase profits and result in more sustainabile outcomes. With the ubiquitous nature of the internet, cloud-based solutions are essential for SDM, creating a real-time connection between stakeholders along the supply chain.

Digital transformation is changing the way companies operate. As a result, employees have more flexibility, and businesses are more agile than ever before. The evolution of specification data management is emblematic of this transformation. Ultimately, cloud-based SDM platforms are an increasingly essential part of any packaged product company's technology stack – driving supply chain efficiency, increasing product quality, and creating a sustainable future.

8 REFERENCES

[1] The Enterprisers Project. (2020). What Is Digital Transformation? enterprisersproject.com/what-is-digital-transformation. Accessed April 18, 2022

[2] Blake, J. (2020). Market Guide for Packaging and Product Specification Content Management. Gartner Research. Report ID: G00464835.

[3] Banker, S. (2021). A Multibillion-Dollar Software Market Growing at a Double Digit Rate That Nobody Knows About. Forbes. August 6, 2021. www.forbes.com/sites/stevebanker/2021/08/06/a-multibillion-dollar-software-market-growing-at-a-double-digit-rate-that-nobody-knows-about/?sh=43a87e395561. Accessed April 4, 2022

[4] Specright. (2016). Best Practices for Managing Packaging Specifications. White Paper. https://www.manufacturing.net/home/whitepaper/13120024/specright-best-practices-for-managing-packaging-specifications. Accessed April 2, 2022

[5] Shoukry, A. (2020). The 'New' Normal: The Future of Work for Product and Manufacturing Companies. Forbes. July 2, 2020. www.forbes.com/sites/forbestechcouncil/2020/07/03/the-new-normal-the-future-of-work-for-product-and-manufacturing-companies/?sh=172c521e1cdd. Accessed March 20, 2022

[6] Wright, M. & Armstrong, A. (2021). Specification Management 101: How to Reach Your Goals Using Packaging Data. IoPP Standard Webinar. March 18, 2021

[7] Forrester. (2020). The Total Economic Impact™ Of Specright. A Forrester Total Economic Impact™ Study. January 2020. https://specright.com/wp-content/uploads/2020/02/Forrester-Report-Specright-Summary.pdf. Accessed March 8, 2022

[8] Folinas, D., Manikas, I., & Manos, B. (2006). Traceability data management for food chains. British Food Journal.

[9] Shoukry, A. (2020). Specifications are the Littlest Form of Big Data. Forbes. April 6, 2020. www.forbes.com/sites/forbestechcouncil/2020/04/06/specifications-are-the-littlest-form-of-big-data/?sh=518626cd79c7. Accessed March 8, 2022

[10] Reichenbach, C & Roese, M. (2021). From Automated to Autonomous – Manufacturing Industry Digital Transformation. CIO. July 19, 2021. www.cio.com/article/3625523/from-automated-to-autonomous-manufacturing-industry-digital-transformation.html. Accessed March 14, 2022
[11] Shoukry, A. (2020). The Intersection of Blockchain and Supply Chain. Forbes. July 31, 2020. www.forbes.com/sites/forbestechcouncil/2020/07/31/the-intersection-of-blockchain-and-supply-chain/?sh=7b7c6cb412ee. Accessed March 8, 2022
[12] Zhang, G., Deng, X., & Zhang, Q. (2020, October). Construction and Design of Resource Management System of Packaging Design Information Big Data Platform. In Journal of Physics: Conference Series (Vol. 1648, No. 2, p. 022064). IOP Publishing.
[13] Finnerty, K. & Gaffney, M. (2020). The Total Economic Impact of Specright Specification Data Management. Forrester Total Economic Impact Study Webinar. May 21, 2020
[14] Grean, M., & Shaw, M. J. (2002). Supply-chain partnership between P&G and Wal-Mart. In E-business management (pp. 155-171). Springer, Boston, MA.
[15] Esko-Graphics BV. (2022). WebCenter For end-to-end digitization of business processes. https://www.esko.com/en/Products/Webcenter. Accessed March 14, 2022

LG-O02

Impacts of the local E-commerce

Ana Gabriela Flores[1], Nadia Garcia[1], Ana Zorrilla[1,3], Diana Chapa[2]*, Cristina Guzman[1,2,3*]

[1]*Department of Industrial Design, School of Architecture and habitat sciences, Universidad de Monterrey, San Pedro N.L., 66238, Mexico.*
[2]*Graduate Department of Art and Design, Design and Development Packaging Master Universidad de Monterrey, San Pedro N.L., 66238, Mexico.*
[3]*Innovation Center ABRE-UDEM, Universidad de Monterrey, Apodaca N.L, 66600, Mexico.*

***Correspondence to:** Department of Industrial Design, School of Architecture and habitat sciences, Universidad de Monterrey, San Pedro N.L., 66238, Mexico.*

ABSTRACT: Nowadays, everyone talks about e-commerce and how distribution and supply chain has changed—distinguishing the acceleration of the deliveries is one of the strategies to achieve the market goals. But what has happened in the micro scenarios? What is happening in the cities? How has the pandemic changed the commerce of food and groceries? How do the restaurants of fast and slow cooking change paradigms? The new restaurant trend called "dark kitchens" has revolutionized food deliveries and interaction between food and consumers' experiences. User experience and the unboxing trend in social media had been promoters of the specialized packages, converting the packaging as part of the brand into an eye-catching product. Even though not all restaurants have invested in packaging, disposable packaging rouses exponentially, and the real problem translates into collecting and recycling. This paper aims to discuss this social problem, for example, the growth of food-to-go in Europe increased by 85% from 2019 to 2021. Reviewing how packaging companies are developing biodegradable plates and containers which require special conditions to dispose of them, even though the local garbage dump doesn't know or have a correct treatment. And finally, how social media had impulse the packaging design to gain markets and consumers. This study is a literature review of case studies during the last three years, focused on the changes in food delivery, user experience in the marketing field, and the new sustainable commercial materials and their way of disposing of them, on the circular economy trend.

Keyword: e-commerce, packaging design, biodegradable materials, disposable packaging, circular economy

1. INTRODUCTION

Pandemic effects, change in consumption habits

The last two years have been marked globally by the pandemic SARs-CoV-2. When people talk about globalization, they talk about prosperity, exchanges and evolution, but was discovered that globalization involves problems such as global warming, climate change, ocean contamination, economic impacts, and global pandemic.

Consumer habits changed in many forms, from the way they thought, the way they feel, and the way they buy. It was not only on a personal level but the changes also affected the industry in the way of promoting food and services, the way of purchasing from home, delivery of all types of products, and selling and collecting through electronic methods.

The way stores, supermarkets, and restaurants were required to make changes was unexpected. In three months electronic commerce became the way to do business and delivery systems as the way to receive food at home. "Electronic commerce, best known as E-commerce, is defined as the transaction of goods and services through electronic communication services" [1]. Consumers were forced to use electronic resources, therefore the electronic adoption of the consumers was accelerated at least 10 years in only a few months.

Not just as the business changed, the consumers changed their habits and their preferences. For example, consumers make fewer visits with higher purchasing, the frozen food got valued once again, the preference for local food products rose and others' habits had changed [2]. Therefore, the marketing departments need to get to know the new consumer's preferences, through the advance in electronic apps, which are a big source of information that can accelerate the results. "As a result of the single-person household, new products have been developed for the single-family household, consumer market to satisfy their needs representing smaller, individually-portioned sized, of fully cooked items, offered at local convenience stores as prepare food ready to go, as an alternative of daily boxed lunches" [3].

"The pandemic is gutting the restaurant industry's traditional dine-in business, but the crisis has also presented an opportunity for digital income channels to accelerate its growth - especially for ghost kitchens" [4, 5]. After governments mandated the restrictions in restaurants and public places, combined with the fear of the population getting sick, electronic platforms presented a raise. As well as delivery services, such as Uber Eats which grew more than 100% [6]. During the current year, restrictions have been reduced, now people began to go out to social meetings, but the commodity of food delivery did not and will not disappear.

The accelerated growth of electronic sales has presented good and bad impacts on the environment. Buying online does not represent more impact on the environment, but electronic marketing strategies have turned consumerism into a way of art. The last-mile delivery could make the difference. Therefore to be a sustainable consumer in shopping online there are three recommendations: Buy local, the closer the better to reduce footprint. Select ecological products and packages, as much as possible. Try to support small businesses and the neighborhood, before the big chains.

The study was focused on local e-commerce because the last-mile delivery is mostly related to transportation impacts, air pollution, emission of greenhouse gasses, noise pollution, and traffic congestion [7]. Even though e-commerce has risen exponentially in the last few years, for example, in 2018 online retail in the US represented 14.3 percent of the total retail sales in the US [8] and it is expected to continue growing in the global market.

The local e-commerce gave an option to survive the pandemic restrictions, to avoid contact with people. The population changed crowded places to stay at home. The world is constantly changing and people are looking for options that make life easier due to the daily bustle. The COVID-19 pandemic accelerated some changes in the use of electronic gadgets and applications [9- 11]. Food Delivery Applications (FDAs) were of great help, preventing people from going out and enjoying a delicious meal while staying home. With this, online purchases became much more common and e-commerce grew exponentially throughout the world, including food delivery, as well as the acquisition of pharmacy products, supermarket provisions, and other products. Covaliu, A. [12] mentioned that apps are notifying if it is preferred that the food be left at the door, lobby, or where the client prefers to avoid contact. Delivery companies were involved in a daily competency, known as the Delivery war [13]. All those transformations had an impact on consumption, production of new materials and new packaging, the development of new technologies, and impact on energy consumption and people's health.

2. METHODOLOGY

For this study, a qualitative approach was achieved to evaluate how e-commerce at the local level had been adapted by the consumers, and how the business had changed to satisfy the needs of their consumers. As a result of the increasing food consumption, the consumption of packaging rises, and the waste of packaging and food needs to be considered because of the increase of garbage in landfills.
This research is a comprehensive literature review mainly of the past 3 years (2019 -2021), and some from 2000 to 2018 to evaluate the changes in the business models, and how the industry is dealing with the rising problems resulting from those changes. The review consists of **92 papers,** which are represented in Table 1 references reviewed. The research reviews academic papers, commercial informatión, electronic pages, institutional e-pages, and other resources. In previous studies, a conflict between the packaging and the way it was disposed of arose the question: How the food industry and the consumers are dealing with package disposal? How sustainability and the circular economy had been implemented at the local level?

Table 1. A summary of the references reviewed in this research: academic documents, white papers, electronic pages, and newspaper articles (not all are referenced in this paper).

	ACADEMIC	WHITE PAPER	ELECTRONIC PAGES	TOTAL
2011 - 2018	3	4	0	7
2019	11	6	0	17
2020	6	15	0	21
2021	12	22	0	34
2022	0	1	0	1
UNKNOWN	0	0	12	12
	32	48	12	92

PRE-COVID 19	29
DURING COVID 19	51
UNKNOWN	12
TOTAL	**92**

This paper was developed focused on the interest of the local e-commerce on daily bases using electronic transactions, therefore as the biggest environmental impact normally happens on the last mile delivery. The research focuses on e-commerce food delivery because the food represents the major daily transactions. The major topics reviewed were food delivery, food business, packaging trends, food distribution, technology, sustainability, and circular economy.

2. LITERATURE REVIEW

3.1 The market

E-commerce has been in the market for more than a decade, and it has been growing each year, but what happened with the pandemia broke up all the prognostics.

Torregrosa in 2016 [14] pointed out the rapid increase of immediacy in everyday life consumers need things to be easy, rising the sales of mobile devices and PC. Between 2016 to 2018 e-commerce has had a disproportionate growth in the total retail sales in the US, from 11.6% in 2016 to 14.3% in 2018 (2.7%) at the total online retail sales in US; making US the third biggest e-commerce market in the world, bested by the UK and German [8]. In 2019, the food packaging global market was estimated at over USD 300 billion, with a prediction of an increasing rate of 5% (this was before the pandemic) [15]. Just in Los Angeles, California has increased orders by 3-35% in 2019 [16].

Even though the pandemic was distinguished for seeking sustainability and the circular economy as the actual concern of the industries and the consumers. Three aspects are pointed out: First, sustainable packaging, where the plastic reduction was stopped to assure hygiene and security to comply with security measures. At the same time, the industry had circular goals to achieve for 2025. Sustainable packaging is a crucial value for brands and their image. Trends in sustainable packaging are designed for and focused on recyclability, the replacement of plastic, biodegradable and compostable materials, and higher quality [17].

Second, reusable packaging, where the consumers will be the force to do the change, the preference of the consumers for reusable packaging and reusable materials will push the industries to accelerate the process. Consumers have expressed their concerns about sustainability through changes in their lifestyles and consuming habits, for example, 61% of UK consumers have limited the use of single-use plastic and 34% prefer brands that follow environmentally friendly practices and values [17]. Despite the pressures of the Covid-19 pandemic to fulfill hygienic measures, sustainable packaging is also becoming a growing priority for consumers. Trivium and Boston Consulting Group mentioned that 67% of consumers considered seeking a reduction in environmental footprint through the recycling of packaging of products they buy [18].

Third, grab-and-go packaging was a trend before pandemia [19] at supermarkets, c-stores, hotels, colleges, etc., and will still be a trend after the pandemia. Consumers will spend more time outside places and will buy it at kiosks, carts, window delivery or drive-thru [20]. The importance of knowing the consumer is a vital task to develop the right packaging. "Without the tactile and personal experience of purchasing a product, storytelling on the packaging is a good opportunity for a brand to stand out and create a valuable customer relationship." [21]

Now, after the pandemic, the custom of maintaining sanitation as a high level of care and health of products, people demand quality services that take care of and keep their orders in the best conditions. "Because fewer customers intend to use public services, the traditional catering industry has suffered dramatically during the COVID-19 pandemic" [22]. Therefore the contactless feature and convenience of FDAs have significantly contributed to users' perceived it as technological and mental benefits under COVID-19 pandemic conditions. The way in which the market growth has been translated to the food industry in the development of new business models, like dark kitchens or ghost kitchens, and business strategies focused on food delivery. Dark kitchens demand less investment compared to a restaurant. The cost-benefit attracts more people and operators to this business model because they can enjoy the cost cuts in terms of supplies, furniture, interior design, rent, etc. [23, 24].

Consequently, due to the minimum time invested by the clients in stores, the packaging is now in charge of sending and representing the brand by itself, without human interaction. "21st-century packaging trends will focus on caring and empathy, displaying color and brand messages through accessible materials, designs and print possibilities. Packaging that is more environmentally friendly and uses fewer resources and less waste will become more popular."[25]

3.2 Functionality /ergonomic aspects

Packaging functionality and ergonomics aspects are really important, as Weiss mentioned: The key benefits are defined as protection, transportation, influencing consumer purchasing habits, and making your brand stand out. But the two key points to choosing the right food packaging: First, the packaging should keep the food product safe and secondly it should catch the customer's eye. Nevertheless, the packaging design is responsible for the function that it provides, including the visual communications, portability, and easy to use. "Food packaging not only makes a product stand out on the shelf but it also protects them from chemical, physical and environmental factors that can contaminate them" [26].

Also, for functionality, the packaging seeks to create a barrier that keeps the product fresh and quality for a longer period of time. "The sustainable extended shelf-life packaging products ensure consumers experience the same quality and freshness whether food is produced in-store or upstream" [27]. As a result of the prevalence of the product integrity once it is received by the customer it will be a decisive factor in the repurchasing of the same product or the same brand in the future, and due to the confidence that this packaging will provide to the customer's perception. The graphics in packaging design are intended to gain the customer's trust by the use of soft colors and showing the product by using transparent windows; the minimalism is present to attract the attention of the client. The use of technological elements such as QR codes and social networks allows producers to have direct contact with the customer even after their purchase, through reviews, unboxing experiences and surveys [17, 28].

3.3 Logistic / Distribution

In 2015, according to US research, for the first time in human history, global restaurant spending surpassed grocery store food purchases. In economic terms, it has been estimated that the global restaurant industry accounts for 5% of the total global economy [29], during 2019 and 2020 the market changed, increasing the restaurant orders by delivering services and in 2022 the trend continues. During the covid 19 pandemic, the food market was greatly affected due to movement restrictions; according to IGD.com [30] operators of takeaways, food was the biggest driver and is expected to continue to be for the next years, as a result of the changing customer habits post-pandemic. Wells, L. [31] mentioned that takeaway food in grocery stores and convenience was the largest takeaway channel. Customers are enjoying the online food business because of the convenience, the time savings, and the ordering and delivery of ready-to-eat food through mobile and digital platforms [21, 32].

The use of FDA´s had become necessary to find out the intentions and Gratifications behind their use, its popularity was also fueled by customers [33]; as a result of the convenience of receiving food at home through delivery systems, it has become a trend and a preference for 62% of the consumers; the convenience, has become more important than the price. Of all the rainbow of products online, the most demanded is the food online. Considering the strong binomial between packaging and food, it facilitated a new model of business: the dark kitchen [34]. Dark kitchens, Ghost kitchens, or Digital kitchens had stolen the food market base on four advantages [35, 36]: Low cost of inversion and low operational cost. No waiter needed, therefore it reduces personal, more flexibility in the process, and adaptability to small spaces. Automatization through technology. And the last depended on future purchasing, the food quality, its delivery presentation, package, and the user experience.

According to Li, Mirosa, & Bremer [37], food delivery through digital media brings economic, social, and environmental impacts: The economic aspect has been positively impacted: for example, new jobs opportunities have been opened for chefs, restaurant administrative staff, delivery men and programmers, also to who sell and repair motorcycles or who manufacture food packaging [38]. On the market side e-commerce has hastily a trend for hunting for local businesses which imply the strategic search of locations since proximity to consumers is an aggregated value at the moment of distributing food products. Fast delivery or same-day delivery is the goal, Amazon and Zappos are now offering this kind of delivery in big cities. At the same time, Omnichannel marketing techniques have been increasing to create an impact too [14]. Steever & Murray [39] propose a new framework of multiple restaurants included in a single order, complying with customer service and providing effective handling of the demand. According to Makwana [40] delivery firms turning to dark kitchens, typically require 6,000 sq ft (55 m^2) of space to serve five to six branded eateries.

In the end, the only contact with the consumer is presented at the last mile, all the e-commerce business, more in the food delivery system where the time and quality become factors of efficiency on the sale. To achieve better results the technology is the clue, to planning the strategic routes, communicating with 3P logistics, tracking in real-time the delivery, and the opportunity of having feedback from the consumers [32, 34].

3.4 Technology and Apps.

The market reacted to the pandemia restrictions, it took less than 30 days for the population to change from supermarket shopping to electronic shopping through all the delivery platforms. But for some of the big companies, it took longer, an example is Woolworths Australia´s biggest supermarket, who teamed up with Uber Eats by August 2021 offering an unprecedented feature "same-hour delivery" as a strategy around speed and convenience for their customer's demands [41]. The popularity of FDAs has grown in recent years, as mentioned by Singh [42] in the year 2020 the applications had revenues of 26.5 billion dollars, showing a growth of more than triple compared to the year 2015.

As Lee et al. pointed out, the growth of e-commerce has resulted in new businesses online to offline (O2O), which is a marketing method based on information and communication technologies. Small-scale restaurants that have lower advertising and marketing abilities can use delivery apps as a convenient and highly efficient sales and marketing tool [16, 43- 45]. Technology in the food business is the most relevant sector for growth. Update in technologies has proven to boost and improve the profits and the engagement of users because of satisfactory experiences [21]. “Consumers will have more positive attitudes when their expectations toward these food delivery apps are matched or exceeded by their perceived value” [3].

The research through mobile devices, m-commerce consists in accessing mobile services to buy products, and this is what big companies and businesses will bet on [14]. Internet and smart devices have connected consumers to food services, “food entertainment” on social media, and third-party food delivery [45]. Nevertheless, some countries were not prepared for the Internet of Things (IoT). Mexican internet users are still only 67% of the total population of 131 million [46].

In the research of technology trends, we found several tools such as barcodes and RFID labels, and the increasing use of Near Field Communication (NFC) and Quick Response (QR). Emerging technologies will have significant relevance in packaging, much greater use of artificial intelligence (AI), internet in packaging, augmented reality (AR), and virtual reality (VR) [47].

Software developers companies with designed algorithmic control are used to monitor the delivery workers as to the consumer's satisfaction with the service [48]. Proposing Genetic Algorithms (GA) is a method through computational tests which validate the feasibility and rationality of efficient combinations of parameters for efficient cold chain logistics. Due to the high demand for healthy and convenient diets, there is a higher need for refrigerated and frozen food [49].

“Restaurant-borrowed innovations are speeding up efficiency and sophistication in convenience food service” [50]. Food apps are designed for promotions, types of restaurants, food experience, cooking, and restaurant reviews [51- 55]. Digital kitchens are those which have all types of technology appliances to accelerate cooking, control, and manage remotely. They also cook dishes in parallel, saving time and money and reducing food waste [56]. Some chains have developed a new store concept consisting of consumers ordering by app, but they will have to pick the order in a lobby with the sound, and the smell of their traditional spaces to create an experience. This new Digital Kitchen concept will allow companies such as *“Chipotle”* to enter more urban areas that would not support full-size restaurants and adds flexibility to future locations [57, 58].

The biggest shift has happened in the last year with exponential growth in the use of technology in all fields, but food service has become imperative. Therefore restaurants will need to continue investing in technology and digital media. Today in India it is possible to find restaurant management by robots, recipes under Alexa services, or video marketing from marketeers and restaurants [59].

Since the drones for delivery were tested, they had created higher expectations, including fear or negativeness, also called “perceived risk”. Perceived risks are defined as financial, time, privacy, performance, and psychological risks [60]. Hwang et al. found that in addition to the ecological benefit, it generates a better attitude of the customer and makes them feel that efforts are being made to improve the experience, resulting in an intention to consume again [61]. Industries such as UPS, Google, and Amazon were the first companies interested in using drones for product delivery because the benefits include a reduction in time delivery, reduction in production cost, and fast delivery with less fuel and labor [60].

3.5 Sustainability and Environment features

Environmental problems have been growing in recent years and also the population's awareness of using products that do less damage to the planet is growing, even if this entails additional costs. Companies are looking for new alternative materials for the packaging industry. Some alternatives look to reuse the containers and the return packaging from the final consumer. Another way is by reducing packaging waste and developing plastics based on biodegradable resources [62], as well as proper handling of packaging waste [63].

Ellen MacArthur Foundation (EMF) in 2022 [29] reports that only 14% of the plastic package is recycled, and 40% is finished in landfills. Biodegradable paper-based materials represent approximately 32% of the total value [15]. Materials recycling has been proved as a way to reduce cost expenses of disposition and reduce environmental impact [64]. As mentioned in United Nations for Sustainable Development Goal 12: Sustainable consumption and production encourage more sustainable patterns through various measures. This study is interested in 12.2.2 Domestic Material Consumption (DMC) to evaluate how the consumers are disposing of the packaging materials/ garbage [65].

Song et. al. found that plastic containers made of polypropylene & polystyrene (EPS) foam account for around 75% of the total food delivery packaging waste in weight, followed by wood chopsticks and plastic bags. The rest of the materials plastic spoons and paper only account for 5% of the weight [66]. In the end, after the consumption of the food delivery, the food packaging results in a massive amount of plastic, paper, and cardboard, and the management turns into a global problem, specifically of plastic. The problem is present in developing and in undeveloped countries [67]. Today, Mexico generates more than 44 million tons of waste per year and expects this number to reach 65 million by the year 2030 [68] and the recycling facilities are not enough. In this context, it is necessary to develop an action plan where organizations, government, and society are committed to transit to the CE [69].

According to the European Parliament, Circular Economy (CE) is a model of production and consumption, which implies reducing the environmental impact. Their goal is to reduce waste to a minimum. The CE model involves recycling, reusing, repairing, and remanufacturing making the most of the means [69]. The study carried out by Chunguang, B., Joseph, S., Fengfu, Y., & Yijie, D. [70] showed that statistical data is relevant for those seeking a circular economy within their companies. The strategic points are: having a wide range of green suppliers, having green products and services, having ranges of delivery times, and several renewable resources available.

Under the trends of packaging, the research found three main areas: sustainability, technology, and user experience [71]. By breaking down each area, we found at the sustainable level three main topics, such as zero waste, less plastic, and eco-friendly actions. All of them are looking to reduce food and material waste through diverse actions. The concern of this study is the growth of the food delivery industry, which has risen to disturbing levels and the cities are not prepared to manage all the waste produced by the food industry at home, with domestic collectors. DS Smith foresees the need for new and sustainable packaging solutions will increase and will continue applying the principles of circular design to create packaging circular solutions for the market challenges [72]. Han, Ruiz-García, Qian, & Yang [73] there are possibilities and new approaches to improving passive features of food packaging.

Gheorghita Puscaselu, Besliu, & Gutt [74] made test studies for the development of new materials for food packaging, however, there are still problems to be solved and the high cost that they currently represent. On the other hand, the increasing demand for online goods has created difficulties in how to meet online shopping preferences without adding to the waste generation [75]. Causing local and international waste management problems.

Implementation of CE practices, emphasizes the green and sustainable procurement of materials and components and also provides economic benefits to the organizations by facilitating proper waste management, conservation of resources, and efficient utilization of finance [76]. Some companies such as Winnow, DyeCoo, Close the Loop, Enerkem, Schneider Electric, Cambrian Innovation, Lehigh Technologies, HYLA Mobile, TriCiclos, Miniwiz, and AB InBev were part of the winners of The Circular Awards 2019, where 450 applicants from 45 countries participate [77].

4. RESULTS

4.1 -The Market

Dark kitchens and Digital kitchens had exponential growth and will be growing around the world, bringing opportunities to entrepreneurs, nevertheless, exclusive creations with creative packaging are the key tools to ensure the reception of a fresh product ready to eat. The development of food packaging for delivery to ensure that the product arrives in excellent condition is a challenge, much more from a sustainable perspective. The work is allocated to the pack performance, representing new opportunities to develop better packages.

4.2 - Functionality / Ergonomics aspects

In the last few years, technological advances and developments have driven the growth of e-commerce, the industry of food delivery, and local commerce. The sector of food delivery and local commerce has increased exponentially and the technologies and apps for this industry allow consumers to have better and faster access to products and services.

4.3 - Logistics / Distribution

The distribution and delivery system changes the consumer's habits of online food products, supermarket consumption, or buying prepared food from restaurants. Dark kitchens, Ghost kitchens, and Digital kitchens had revolutionized the food market, all of them using delivery systems that require packaging delivery to improve time, quality, and experience and achieve efficiency in sales.

4.4- Technology & Applications

The mobile-commerce with all the technological tools such as applications had increased communications, monitoring workers and users, improve strategies and logistics, and transformed the user experience. The new delivery systems with drones will enhance the performance of the services.

4.5 - Sustainability and Environmental aspects.

The study found two main topics in the circular economy: food, looking to reduce food waste, and reducing plastic packaging waste. The rest of the packaging materials need to be recycled but are not the global target.

As Lerner hub of the Ellen MacArthur foundation points out "design is key to the first principle of the circular economy, "design out waste and pollution." The reality is that most things today are still designed for the linear model. Which means that almost everything needs to be redesigned by the principles of the circular economy" [29].

5. DISCUSSION AND CONCLUSIONS

The study found that most papers are focused on the product, in this case on the food and system delivery but scarcely talk about the packaging. On recycling, most of the documents talk about plastic and aluminum bottles, but there is not much information about food packaging, paper cardboard packs, and plastic packaging for food delivery.

One consideration is to increase the recycling materials, where the polymers classification is needed to be reviewed to include the new materials, develop the recycling facilities that are needed to support the amount of trash in every city, and develop policies to recycle bio-polymers and other new materials.

Encourage the industry to reduce materials waste, classify trash, recycle it in a proper manner, and present reports of impact. On the other hand, household trash separation needs to be mandatory to facilitate recycling. But each local government needs to open a recycling facility to reuse, recycle, or incinerator to produce energy.

6. REFERENCES

[1] Chapa-Chapa, D., Zorrilla-Rodriguez, A. I., & Guzmán-Siller, C. (2021). Conference IAPRI online. *Universidad de Monterrey*. Monterrey, Nuevo León, Mexico.

[2] Severson, K. (2020, September). *7 Ways the Pandemic Has Changed How We Shop for Food*. Retrieved from New York Times [January 30, 2022]

[3] Cho, M., Bonn, M. A., & Li, J. (2019, January). *Differences in perceptions about food delivery apps between single-person and multi-person households*. Retrieved from Science Direct: S0278431918301762 [April 26, 2022]

[4] Kelso, A. (2020, October 5). *How the pandemic accelerated the US ghost kitchen market '5 years in 3 months'*. Retrieved from Restaurant Dive [February 12, 2022]

[5] Camargo, B. (2020, December 28). *Dark kitchens: parte de la nueva normalidad*. Retrieved from UDEM:https://www.udem.edu.mx/es/institucional/noticia/[January 25, 2022]

[6] Greenpeace. (2020, November). *Impactos ambientales y alternativas al*. Retrieved from Greenpeace: https://es.greenpeace.org/es/wp-content/uploads/sites/3/2020/11/ [February 28, 2022]

[7] Demir, Emrah, Huang, Y., Scholts, S., & Woensel, T. V. (2015). *A Selective Review on the Negative Externalities of the Freight Transportation: Modeling and Pricing*. Transportation Research Part E: Logistics and Transportation 77:95-114.

[8] Ecommerce Guide. (2020, February 18). *Top 10 Ecommerce Sites in the USA*. Retrieved from Ecomerce Guide: https://ecommerceguide.com/top/top-10-ecommerce-sites-in-the-usa/ [February 15, 2022]

[9] Elandroide. (2021, January 8). *Las 10 mejores apps para pedir comida a domicilio*. Retrieved from Elandroide: https://elandroidefeliz.com/mejores-apps-de-comida-a-domicilio/ [April 11, 2022]

[10] Corpuz, J., & Woo, K. (2022, May 6). *Best food delivery services in 2022: Grubhub vs Uber Eats vs Doordash*. Retrieved from Tom's Guide: https://www.tomsguide.com/best-picks/best-food-delivery-services [January 29, 2022]

[11] Fakfare, P. (2021, October 25). *Influence of service attributes of food delivery application on customers' satisfaction and their behavioural responses: The IPMA approach*. Retrieved from Science Direct [March 29, 2022]

[12] Covaliu, A. (2022, April 6). *Las mejores apps para pedir comida*. Retrieved from Digital Trends: https://es.digitaltrends.com/celular/apps-para-pedir-comida/ [March 30, 2022]

[13] La Hostelera. (2021, March 17). *Quién es quién en Food Delivery*. Retrieved from La Hostelera: https://www.lahostelera.com/blog/listado-de-plataformas-de-entrega-a-domicilio-food-delivery/ [March 23, 2022]

[14] Torregrosa, J. (2016, January 19). *eCommerce local, el último tren para el pequeño negocio*. Retrieved from IMF Blog de Marketing: https://blogs.imf-formacion.com/blog/marketing/ecommerce-local-negocios-pequenos/ [March 5, 2022]

[15] Motelica, L., Ficai, D., Ficai, A., Oprea, O., Alpaslan Kaya, D., & Andronescu, E. (2020, October 9). *Biodegradable Antimicrobial Food Packaging: Trends and Perspectives*. Retrieved from National Library of Medicine [Feb. 18, 2022]

[16] Lee, S. W., Sung, H. J., & Jeon, H. M. (2019, June 4). *Determinants of Continuous Intention on Food Delivery Apps: Extending UTAUT2 with Information Quality*. Retrieved from MDPI [March 30, 2022]

[17] Stanley, J. (2021, August 23). *What are the new trends in packaging?* Retrieved from The Drum: https://www.thedrum.com/opinion/2021/08/23/what-are-the-new-trends-packaging [April 23, 2022]

[18] SyndiGate Media Inc. (2021, June 19). *Btab Group Revitalizes Global Ecommerce Operation with Biodegradable and Recyclable Packaging Materials*. Retrieved from Legal Monitor Worldwide [April 6, 2022]

[19] Nestlé Professional. (2018, March 1). *Get in on Grab-and-Go*. Retrieved from Nestlé Professional: https://www.nestleprofessional.us/trends/get-grab-and-go [March 23, 2022]

[20] García, G. (2021, January 28). *Tendencias en packaging 2021: Empaques y envases en la industria alimentaria*. Retrieved from The Food Tech [March 15, 2022]

[21] Bourassé, K. (2021, June 18). *TOP 5 PACKAGING TRENDS IN 2021*. Retrieved from Geneva Business News: https://www.gbnews.ch/top-5-packaging-trends-in-2021/ [March 1, 2022]

[22] Zhao, Y., & Bacao, F. (2020, October). *What factors determining customer continuingly using food delivery apps during 2019 novel coronavirus pandemic period?* Retrieved from Science Direct: S0278431920302358 [April 3, 2022]

[23] Thamaraiselvan, N., Jayadevan, G., & Chandrasekar, K. S. (2019, July). *Digital Food Delivery Apps Revolutionizing*. Retrieved from International Journal of Recent Technology and Engineering [April 4, 2022]

[24] Sánchez, S. (2020, July 10). *¿Qué son las 'dark kitchens' y cómo revolucionarán la industria del consumo?* Retrieved from Forbes [March 1, 2022]

[25] Design Hugger. (2022, February 2). *Tendencias en diseño de packaging 2022*. Retrieved from Tendencias Diseño Hugger: https://trendsdesignhugger.com/packaging-design-trends-2022/ [April 26, 2022]

[26] Weiss, D. (2019, August 16). *Your guide to food packaging.* Retrieved from New Food Magazine: https://www.newfoodmagazine.com/article/90810/food-packaging-types-importance-trends/ [January 31, 2022]
[27] Kazarian, K. (2021, July 21). *Recyclable fresh food to-go packaging extends shelf life.* Retrieved from Food Engineering [April 1, 2022]
[28] In the bag. (2021, February 1). *Latest packaging trends 2021-2022.* Retrieved from In the bag: https://inthebag.com/2021/02/01/latest-packaging-trends-2021-2022/ [March 19, 2022]
[29] Ellen Macarthur Foundation . (2022, Enero 26). *El papel de los restaurantes en un sistema alimentario urbano circular.* Retrieved from Ellen Macarthur Foundation [February 6, 2022]
[30] IGD.com. (2021, July 27). *Food-to-go market recovering faster than expected, according to IGD.* Retrieved from IGD.com [March 29, 2022]
[31] Wells, L. (2021, April 7). *Food-to-go market to grow by 32% in 2021, research finds.* Retrieved from Talking Retail [February 9, 2022]
[32] Ramesh, R., Venkatesa Prabhu, S., Sasikumar, B., Kiruthika Devi, B., Prasath, P., & Praveena Rachel Kamala, S. (2021, June 12). *An empirical study of online food delivery services from applications perspective.* Retrieved Science Direct
[33] Ray, A., Dhir, A., Kumar Bala, P., & Kaur, P. (2019, November). *Why do people use food delivery apps (FDA)? A uses and gratification theory perspective.* Retrieved from Science Direct [February 26, 2022]
[34] Bee Track. (2020, August 6). *Dark kitchen: ¿qué es y por qué son tan eficientes estos e-commerce?* Retrieved from BeeTrack: https://www.beetrack.com/es/blog/dark-kitchen [January 19, 2022]
[35] Dephna. (2020, August 25). *The Kitchen Takeover: Dark Kitchen Business Model & Benefits.* Retrieved from Dephna: https://www.dephna.com/insights/what-is-a-dark-kitchen-benefits-business-models [February 9, 2022]
[36] Pymnts. (2021, June 1). *Virtual Kitchens Go Big As Digital Demand Remains Elevated.* Retrieved from Pymnts [April 5, 2022]
[37] Li, C., Mirosa, M., & Bremer, P. (2020, July 8). *Review of Online Food Delivery Platforms and their Impacts on Sustainability.* Retrieved from MDPI: https://www.mdpi.com/2071-1050/12/14/5528 [April 3, 2022]
[38] Stephens, J., Miller, H., & Militello, L. (2020, February 20). *Food Delivery Apps and the Negative Health Impacts for Americans.* Retrieved from Frontiers in Nutrition [February 24, 2022]
[39] Steever, Z., Karwan, M., & Murray, C. (2019, July). *Dynamic courier routing for a food delivery service.* Retrieved from Science Direct: https://www.sciencedirect.com/science/article/abs/pii/S0305054819300681 [March 7, 2022]
[40] Makwana, R. (2019, November 14). *Can dark kitchens really deliver?* Retrieved from Property Week: https://www.propertyweek.com/insight/can-dark-kitchens-really-deliver/5105181.article [April 25, 2022]
[41] Sharma, R. (2021, August 22). *Woolworths, Uber Eats join hands to meet same-hour delivery demand.* Retrieved from Nasdaq [February 16, 2022]
[42] Singh, A. (2021, October 11). *Top 10 Successful Online Food Delivery Apps in the World.* Retrieved from Net Solutions: https://www.netsolutions.com/insights/top-10-successful-food-delivery-apps-in-the-world/ [March 22, 2022]
[43] Forbes Advertorial. (2020, August 24). *Wopen: la solución e-commerce que reactiva el comercio local en México.* Retrieved from Forbes [April 7, 2022]
[44] Correa, J. C., Garzón, W., Brooker, P., Sakarkar, G., Carranza, S., Yunado, L., & Rincón, A. (2019, January). *Evaluation of collaborative consumption of food delivery services through web mining techniques.* Retrieved from Science Direct
[45] Jagtiani, R., & Fazira, E. (2019, June 7). *Connected Consumers in Southeast Asia: Digital Lives Extend to Cooking and Dining.* Retrieved from Euromonitor Internacional [February 5, 2022]
[46] Disfold. (2021, July 9). *Top 10 e-commerce sites in Mexico 2019.* Retrieved from Disfold: https://disfold.com/top-e-commerce-sites-mexico/ [April 16, 2022]
[47] Mohan, A. M. (2021, January 21). *Diez tendencias del mercado de envases para 2021.* Retrieved from Mundo PMMI [January 27, 2022]
[48] Griesbach, K., Reich, A., Elliott-Negri, L., & Milkman, R. (2019, August 30). *Algorithmic Control in Platform Food Delivery Work.* Retrieved from Sage Journals [January 30, 2022]
[49] Zhang, Y., & Chen, X. (2014, April). *An Optimization Model for the Vehicle Routing Problem in Multi-product Frozen Food Delivery.* Retrieved from Scielo [March 28, 2022]
[50] Covino, R. M. (2021, August). *New technology adoption in foodservice equipment is reshaping the food-to-go landscape.* Retrieved from file:///Users/mariajose/Downloads/ContentServer%20(1).pdf
[51] Bryan, A. (2022, July 19). *10 Best Food Delivery Service Apps That You Must Try in 2022.* Retrieved from Urban Taste Bud: https://urbantastebud.com/best-food-delivery-service-apps/ [February 14, 2022]
[52] Fried, L. (2020, April 10). *9 of the Best Food Apps and Sites for Travelers.* Retrieved from Too Many Adapters: https://toomanyadapters.com/best-food-apps-sites-travel/ [February 22, 2022]
[53] *Ultimate eats: the world's top 10 food experiences.* (2020, October 7). Retrieved from Lonely Planet: https://www.lonelyplanet.com/articles/best-food-experience [January 31, 2022]
[54] Bryan, A. (2021, November 13). *The 11 Best Apps Every Foodie Must Have On Their Phone.* Retrieved from Urban Taste Bud: https://urbantastebud.com/11-best-apps-every-foodie-must-have-on-their-phone/ [April 28, 2022]
[55] Novoseltseva, E. (2020, January 29). *28 Food Apps Every Foodie Should Try.* Retrieved from Apiumhub: https://apiumhub.com/tech-blog-barcelona/food-apps-foodie/ [March 31, 2022]
[56] Business Opportunities. (2020, June 29). *Digital Kitchen: The Cloud as Chef de Cuisine.* Retrieved from Business Opportunities [March 5, 2022]
[57] Danley, S. (2020, November 13). *Chipotle introduces 'Digital Kitchen'.* Retrieved from Meat + Poultry: https://www.meatpoultry.com/articles/24099-chipotle-introduces-digital-kitchen [April 5, 2022]
[58] Teck Chai, L., & Ching Yat, D. N. (2019). *Online Food Delivery Services: Making Food Delivery the New Normal.* Retrieved from https://www.researchgate.net/profile/Teck-Chai-Lau- [April 9, 2022]

[59] Kapoor, S. (2021, May 12). *Future of Food—Hygiene, naked kitchens, digital shift new trends of F&B: Sanjeev Kapoor.* Retrieved from Forbes India [March 13, 2022]
[60] Hwang, J., & Young (Jacey) Choe, J. (2019, July 11). *Exploring perceived risk in building successful drone food delivery services.* Retrieved from Emerald [April 15, 2022]
[61] Hwang, J., Lee, J.-S., & Kim, H. (2019, August). *Perceived innovativeness of drone food delivery services and its impacts on attitude and behavioral intentions: The moderating role of gender and age.* Retrieved from Science Direct
[62] Jerelyn, A. (2021, July 13). *7 Sustainable Packaging Trends for 2021.* Retrieved from Packaging Digest: https://www.packagingdigest.com/sustainability/7-sustainable-packaging-trends-2021 [April 18, 2022]
[63] Guarneros Olmos, F. (2021, March). *El auge del e-commerce generó más basura.* Retrieved from Expansion: https://expansion.mx/tecnologia/2021/03/12/el-auge-del-e-commerce-genero-mas- [March 11, 2022]
[64] Dixon, J. (2011, March). *Compactación y reciclaje de bidones plásticos y cartón.* Retrieved from Red Global de Hospitales Verdes y Saludables [January 31, 2022]
[65] UN Enviroment Programme. (2021, July 7). *GOAL 12: Sustainable consumption and production.* Retrieved from UN Enviroment Programme [March 9, 22]
[66] Song, G., Zhang, H., Duan, H., & Xu, M. (2018, March). *Packaging waste from food delivery in China's mega cities.* [April 17, 2022]
[67] Modi, B., Timilsina, H., Bhandari, S., Achhami, A., Pakka, S., Shrestha, P., . . . Parajuli, N. (2021, August 25). *Current Trends of Food Analysis, Safety, and Packaging.* Retrieved from Hindawi [April 26, 2022]
[68] Secretaria de Medio Ambiente y Recursos Naturales. (2019, January). *Visión Nacional hacía una gestión sustentable: cero residuos.* Retrieved from SEMARNAT [April 27, 2022]
[69] Córdova Preciado, M., Salgado Beltrán, L., & Bravo Díaz, B. (2021, June). *Economía circular y su situación en México.* Retrieved from Universidad de Sonora [April 19, 2022]
[70] Chunguang, B., Joseph, S., Fengfu, Y., & Yijie, D. (2020, October). *Sustainable supply chain flexibility and its relationship to circular economy-target performance.* Retrieved from International Journal of Production Research
[71] Karl Knauer. (2021, March 29). *The packaging mega trends for 2021.* Retrieved from Karl Knauer: https://www.karlknauer.com/en/innovation-and-trends/packaging/packaging-trends [March 12, 2022]
[72] García, G. (2021, January 14). *Packaging sostenible: Crece la demanda en consumidores por empaques que generen menos residuos.* Retrieved from The Food Tech [March 14, 2022]
[73] Han, J.-W., Ruiz-García, L., Qian, J.-P., & Yang, X.-T. (2018, May 15). *Food Packaging: A Comprehensive Review and Future Trends.* Retrieved from IFT: https://onlinelibrary.wiley.com/doi/10.1111/1541-4337.12343 [April 1, 2022]
[74] Gheorghita Puscaselu, R., Besliu, I., & Gutt, G. (2021, November). *Edible Biopolymers-Based Materials for Food Applications—The Eco Alternative to Conventional Synthetic Packaging.* Retrieved from Polymers [April 13, 2022]
[75] BNP Media. (2019, September). *DRIVING E-COMMERCE GROWTH THROUGH SUSTAINABLE PACKAGING: Brands seek sustainable packaging that also shows their organizational values.* Retrieved from Packaging Strategies
[76] Syed A. Rehman, K., Hafiz Muhammad, Z.-u.-h., Muhammad, U., & Zhang, Y. (2021, December). *Digital technology and circular economy practices: An strategy to improve organizational performance.* Business Strategy & Development
[77] Teichert, T., Rezaei, S., & Correa, J. C. (2020, March 9). *Customers' experiences of fast.* Retrieved from Emerald [February 4, 2022]

CONFREENCE PROCEEDINGS :

MEDICAL, COSMETIC & PHARMACEUTICAL PACKAGING

• ORAL PRESENTATION

• GENERAL STREAM

MD-O01

Usability Study of aluminum pouch of PTP (Press Through Package) among patients and pharmacists

Kiyomi sadamoto1*, Hiroyuki Ura1, Kiyoshi Kubota2

[1]*Department of Clinical Pharmacy, Shonan University of Medical Science, Yokohama, 244-0806, Japan*
[2]*General incorporated association MECSION, Guzo, Hiratsuka, Kanagawa, Japan242-3*

***Correspondence to:** Department of Clinical Pharmacy, Shonan University of Medical Science, 16-48 Kamishinano Totsuka Yokohama Kanagawa, 244-0806, Japan.*
Kiyomi.sadamoto@gmail.com

ABSTRACT: PTP (Press Through Package) is the most popular type of package in prescription drugs in Japan. To keep several PTP stable, aluminum pouches (pillow packaging) are given as the secondary packaging. There are many opportunities for pharmacists to open pillow packing by the composition work, so the easy opening and handling are necessary for their efficient works. On the other hand, if the patients are given drugs with pillow packages, easy opening is also essential, particularly for the elderly and patients with some handy capped in hands. So, easy opening is demanded from the viewpoint of work efficiency in the pharmacist, and stressless drug taking for the patient. We conducted the study which include ergonomic study on opening status, and objective and subjective studies. There were three samples: sample ① general packaging which is used commonly and two new types of packaging that are designed for improving usability (sample ② packaging with improved openability with vertical lines and sample ③ packaging that can be opened easily with a pull-up method). There were four groups in the sensory test: the ordinal patient, the older patient, patients with Rheumatoid Arthritis and Patients with sarcopenia (weak hand power). All group members participated the study. In the result, some patients who had difficulty to open with sample ①, succussed to open sample ③. In addition, sample ③ was highly evaluated by all the target groups with low grip and pinch strength. In addition, understandable design and storage stability of sample ③ were also evaluated by patients. In the examination among pharmacists, we tried plural votes [Quadratic Voting] for an index of the general likableness numerically, in the results, superiority of the packaging development product (②, ③) was confirmed. All the results suggest that sample ③ could become a universal package for any user.

Keyword: PTP (Press Through Package), Pillow Package, pharmacist, patients, universal package

1. INTRODUCTION

Drug packages are commonly handling by both drug users and the pharmacist. So, easy handling pharmaceutical products are essential for any user patients including the elderly and patients with handy capped with their hands[3]. In daily works, pharmacist makes an effort to quick set up medicine for the patients, they need to open drug packages many times. If the package of pharmaceutical products is complex in handling, it takes time to set up prescription drugs for patients[1,2]. Therefore, both patients and pharmacists need packages which are reasonable to use. The usability and preference of these packages has not been studied. To study usability and users' preference for these packages are interesting and useful for developing universal packages for users.

2. MATERIALS AND METHODS

For the material of the study, we had prepared three types of pillow packages ① general packaging which used commonly in general and two other new types of packaging that are designed for improving usability (sample ② packaging with improved openability with vertical lines and sample ③ packaging that can be opened easily with a pull-up method).

After an explanation of three type pillow packages, subjects were asked to answer evaluate score of NRS after opening three type pillow packages. And they also answered their preference. If they have some comments, it was also noted.

Subjects include patients and pharmacists. Patients include older patients, patients with Rheumatoid Arthritis, and patients with sarcopenia (patients who have weakness of power: Grip strength: <26 ♂, <18♀).

Pharmacists are mainly working in the hospital. We measured the pinch power and grip strength among patients. Pharmacists did NRS and QV.

※QV is a method of collective decision-making in which a participant votes not just for or against an issue, but also expresses how strongly they feel about it.

Equipment: 1. Isoforce GT-300 (100N) OG Giken Co., Ltd

2. Smedley-type Hand Dynamometer TKK5401(100kg) Takeikikikogyo Co., Ltd

2.1 Table

NRS		Instruction of how to open pouch	Openness of pillow package	Taking out PTP from pillow package
0	impossible		to open pouch	to take out PTP sheets
1	very difficult (higher)	to understand the method of cutting	to open pillow package	To take out PTP sheets from pillow package
2	very difficult			
3	difficult (higher)			
4	difficult			
5	not so easy			
6	not so difficult			
7	quite easy			
8	quite easy (higher)			
9	very easy			
10	very easy (higher)			

NRS of Sensual Test for Patients and pharmacists

2.1 Table shows the NRS scale
Patients and pharmacist evaluated packages followed this scale.

2.2 Table

Number of Votes	1	2	3	4	5
"Vote credit" cost	1	4	9	16	25

Scale of QV

2.2 Table shows vote pricing in this research. The simplified formula on how quadratic voting functions is credit cost to the voter = (number of votes)^2.

3. RESULTS AND DISCUSSION

Subject of patients:

Ordinal patients 26 (male: female = 5: 21). Age 68.0±11.1 Older
Patients: 10 (male: female = 4: 6) Age65.5±12.5. RA
Patients: 8 (male: female = 0: 8) Age71.8±9.8 Sarcopenia
Patients: 8 (male: female = 0: 8) Age67.5±11.0

Subjects of pharmacists:

20 (male: female = 7:13). Years of Experience of Pharmacist 15.6±13.6 (R: 1～37).

3.1 Figure

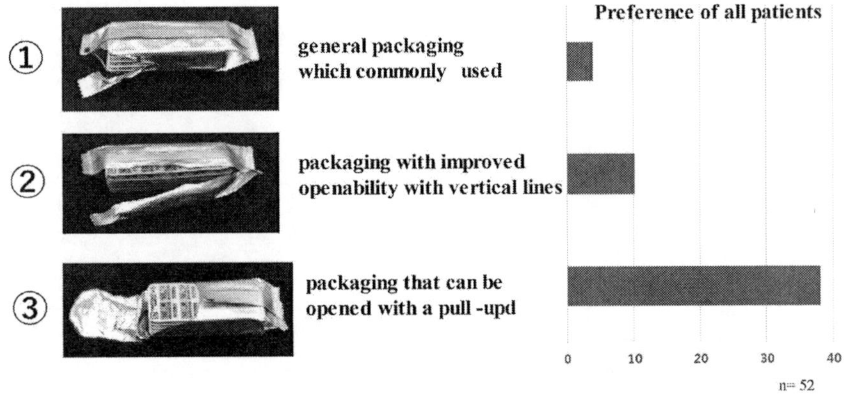

Different pillow packages and patients' preference

Fig3.1 shows the difference in opening status of three packages and preference of all patients. About 73% of patients prefer ③ Package. More than 80% of patients were able to open all type of packages, however they prefered ③ most. In addition to the hearing comments among 11 patients, ③ is most understandable package and it is also able to keep PTPs after they use several sheets of PTPs. So patients tended to evaluate overall usability of packages. Even patients with weak pinch power including salcopenia, could open ② and ③, so both packages seems to be universal packages for all patients.

3.2 Figure

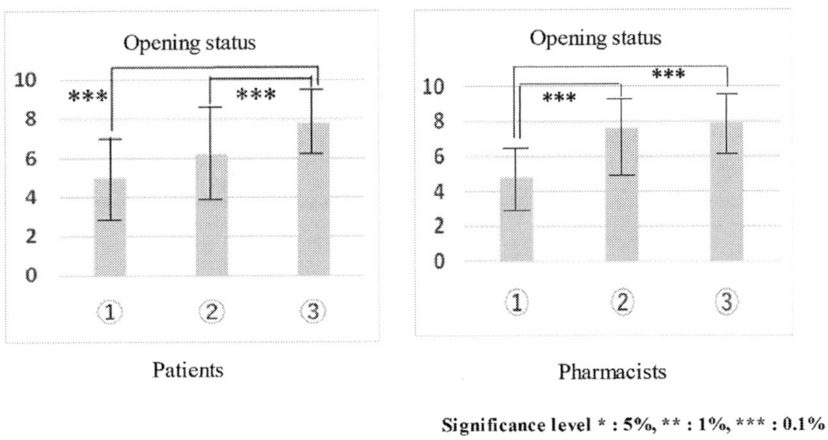

NRS sensual test results by all patients and pharmacists

Fig3.2 shows the opening evaluation of three diffrent packages. Both patients and pharmacists highly evaluated ③. Compare with patients and pharmacists, patients were more evaluated ③. Since pharmacists are no weak pinch power and grip strength, their evaluated ② and ③ almost the same level. For pharmacist, easy vertical opening type of ② could be reasonable to use. On the other hand, patients with weak pinch power had some difficulty to open ② such as pulling.

3.3 Figure

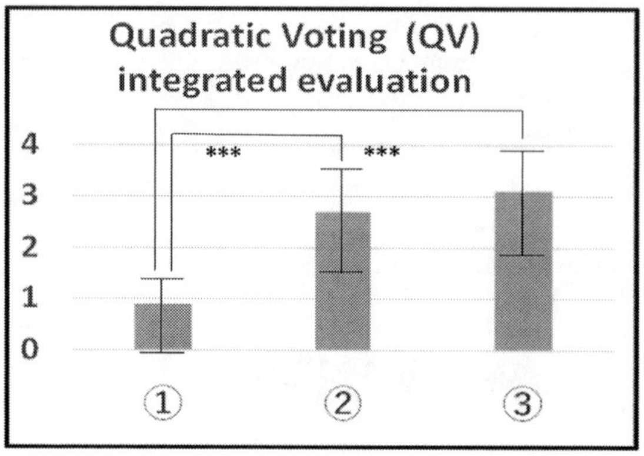

Significancelevel * : 5%, ** : 1%, *** : 0.1%

QV evaluation of pharmacists

3.3 Figure shows the QV integrated evaluation of pharmacists.

Pharmacists evaluate overall ③ was the best. However they evaluate also ②. It means that they evaluate ③ and ② rater than ①, there is significant differences among them. So, both new type packages are effective for their working performances.

4. CONCLUSIONS

From the study of three different pillow packages, new design of two types of packages obtained higher evaluation scores both by patients and pharmacists. So, the reforming the packages could be effective for all users. In addition, a higher evaluation score of ③ packages seems to be not only opening but easier understanding and other elements could be included.

5. ACKNOWLEDGEMENT

We are deeply appreciating for cooperative company, they provide samples. And patients and pharmacists who participated in the study were grateful for their kindness.

6. REFERENCES

Journal:

[1] H. Laetitia Hattingh, Denise Robinson, Alison Kelly: Evaluation of a simulation-based hospital pharmacy training package for pharmacy students, International Journal of Educational Technology in Higher Education volume 15, Article number: 39 (2018)

[2] Sefan Balkanski, Joana Simeonova, Ivan Gitev, Ilko Getov: Evaluation of the current status of the value-added pharmacy services and pharmacists' attitude in Bulgaria, Pharmacia 66(2)79-83 2019

Book:

[3] Lachman, L., Lieberman, H.A. and Kanig, J.L. (1974). The Theory and Practice of Industrial Pharmacy. Philadelphia: Lea and Febige

MD-O02

Machine Learning Modeling of Patients' Attention of Over-the-Counter Medication Label Design

Shiva Esfahanian[1]*, Euihark Lee[1], Lanqing Liu[1] and Laura Bix[1]

[1]*Packaging School, Michigan State University, East Lansing, 48824, USA*

***Correspondence to:** Packaging School, Michigan State University, 448 Wilson Rd, East Lansing, 48824, USA. Esfahan6@msu.edu*

ABSTRACT: In the US, medication is legislated into two categories, prescription, which need doctor authorization, and Over-the-Counter (OTC), that does not need specialist permission. Researchers showed for every dollar expended on OTC medication, healthcare systems save 7.20 dollars. Therefore, OTC plays an important role in the American healthcare system. On the other hand, OTC medication has its own risks like negative effect or Adverse Drug Reaction (ADR) due to drug misuse. Therefore, one of the important sources to buy OTC medication is the label of the product. But it is important for patients to read the label before buying it. Previous studies tried different approaches to increase patients' attention to the label. For instance, putting a small box in Front of the Package (FOP) and highlighting important information. These studies need actual participants which can be costly and time consuming. In order to remedy that, this study proposes a new approach based on Machine Learning (ML) techniques that predict the pattern in new features without the need for new data gathering. This paper is evaluating a prediction of patients' attention for OTC label design by using different ML algorithms. Three ML models- decision tree (DT), random forest (RF), and k-nearest neighbors (KNN)- are implemented to compare accuracy. The result indicated the classification accuracy for DT, RF, and KNN are 79%, 92.6%, and 91.4% respectively.

Keyword: Medical Packaging, Decision Tree, Over-the-Counter (OTC) Labeling, Machine Learning

1. Introduction

Medications in the US in 1951, with the passage of the Durham-Humphrey Amendment to the Federal Food Drug and Cosmetic Act (FEDCA), were legislated into two categories, over-the-counter (OTC) and prescriptions [1, 2]. OTC medication, or non-prescription medications, refers to a safe and effective drug that does not need physician's oversight or prescription [3]. Moreover, for every dollar expended on OTC, the health care system saves approximately $7.20, which is an estimated total of $146 billion savings annually [4]. Furthermore, OTC has other advantages like privacy, convenience, flexibility, and quick access. Consequently, OTC found a vital role in America's health system [5].

Beside all of these advantages, OTC medication has its own risks. Like negative effects or Adverse Drug Reaction (ADR) due to drug misuse. Drug misuse can be a consequence of drug-drug interactions or drug diagnosis interactions. ADR can be defined as "an appreciably harmful and unpleasant reaction resulting from an intervention related to the use of a medicinal product" [6]. A meta-analysis showed annually 106,000 US deaths occur because of ADRs [7]. One way to prevent ADR is to read the label of medication. Obviously, the label of OTC medication has an important role, and it is important for patients to read it before purchasing. Therefore, it is imperative to increase the patients' attention to the label of OTC drugs.

There are many studies that have tried to increase patients' attention by changing the label's design. For example, bigger and bolder fonts, placement of vital information on the front of the package, and highlighting warning instructions [8, 9]. One of these studies, which is the basis of our paper, suggests putting a small box on Front of the Package (FOP) and highlighting important information. In that study, the researchers collected data from ninety-two participants by using a change detection method. Participants should come to Michigan State University and do the pre-test (containing features of the participant), which involves sitting in front of a computer for two hours and answering some questions. This process is very costly and time consuming for both participants and researchers. On top of that, new in-person tests are required whenever new label design is proposed, or new groups of participation are required.

To overcome the disadvantages of in-person tests, this paper proposed a Machine Learning (ML) modeling of patients' attention of label design. ML is a branch of artificial intelligence (AI), and is the analysis of computational algorithms [10]. The crucial role of ML is obvious in various areas of engineering and science [11-15], especially in medical science like disease discovery [16], medical diagnosis [17], and medical imaging [18]. Packaging design is not excluded from this growth. For instance, Knoll et al. automate packaging planning by using different machine learning models. The researchers use a combination of regression and classification models to find the fill rate based on packaging characteristics [19].

This paper focuses on the effectiveness of different ML model implementation on the data from previous labeling design. The chosen models for this study are random forest (RF), decision tree (DT) and k-nearest neighbors (KNN) as they are most commonly used for classification modeling. These methods will be explained in section 2. As a result, the accuracy, area under the curve (AUC), and confusion matrix of these three models are compared with each other.

The remainder of this paper is organized as follows. Section 2 explains the method of different models on the data. Sections 3 and 4 present the results and conclusion, respectively.

2. Method

The objective of the proposed method is to evaluate and predict accuracy for medical packaging of OTC drugs by using different ML models.

2.1 Review of label recognition in-person test

In this study, we used data collected by the Healthcare, Universal Design, Biomechanics (HUB) research group at Michigan State University (MSU) for training and testing our models. The data was collected through an in-person test. The participants were all between 18-65 years old. Each participant came to MSU and completed the pre-test. The pre-test contains questions regarding the participant's features, sex, age, ethnicity, education, and language. Then, they should do three tests regarding their visual acuity, literacy, and color differentiation ability. Each of these participants should answer 168 different trials. At the beginning, researchers explain and demonstrate the procedure. A total of 92 participants participated in this study using the label change detection method.

In change detection, or flicker task, two images (original and modified) as well as a grey one is shown to participants intermittently as demonstrated in Figure 1. This loop of images will appear for nineteen seconds and, if during this time the participants can find the difference between the original and the modified image, they click on the difference using a mouse; otherwise, the computer goes to the next trial.

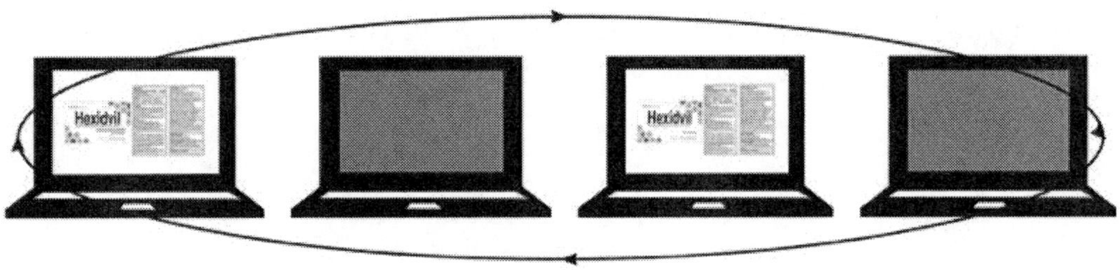

Figure 1 Change detection process [20]

There are three mock brands used in the designing of this study. Hexidvil (pain relief/fever reducer), Circussin (antitussive), and Recantan (acid reducer), with their three different active ingredients being Ibuprofen (IBU), Dextromethorphan (DEX) and Ranitidine (RAN), respectively. Each of these brands have 56 trials in the test which, in total is 168 trials ($3 \times 56 = 168$). Each of these brands are divided into front of package (FOP), which means a small box in front of package or standard (STD) without box. Then, depending on the change that occurs in the front or side of the package, these label designs are further divided into principal display panel (PDP) or a drug fact label (DFL), respectively. If the changes in information are related to the safety and effectiveness of the products, then they will be classified as critical. Also, each label can be highlighted or none highlighted. Final classification depends on change in content and are as follows: AI has any change in active ingredient, DD1 means drug-drug interaction and DD2 means drug- diagnosis. Each brand was created in four treatments: FOP present/ highlight, FOP present / non highlight, STD (FOP absent)/ highlight, and STD (FOP absent)/non highlight (Figure 2).

(a) (b)

Figure 2 Example of trial, image (a) shows FOP/non-highlight and image (b) shows FOP/highlight [20]

This study focuses on the first treatment, FOP, and presents highlights with these features; critical, IBU, and PDP. This work only considers change on one of trial features which is change content because it focuses on participants features with response trial as a target. Therefore, the machine predicts the response trial without collecting new data from in person participants. For training, we used different ML models such DT, RF and KNN explained as follows:

2.2 Decision Tree

DT is a supervised machine learning method for regression and classification. It is a classification algorithm that makes a model in the shape of a tree [10]. The aim of this model is training the machine based on data features to predict the value of the target [21]. The tree contains different parts as root node, branches and leaf node. As you can see in Figure 3 root node is the top-level that represents a big decision or objective.

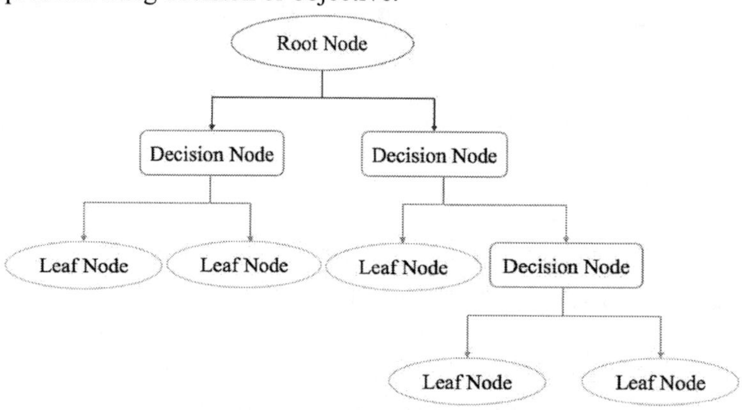

Figure 3 Demonstration of DT

Branches are using to show different options and mostly represented by an arrow. Finally, the leaf node is the result of the decision [22]. DT is dividing a node to different sub nodes and at the end to leaf node. To find the root node, the first step is to measure the impurity of each features. There are different methods for the measurement like entropy and Gini index/ Gini impurity. Entropy is the first step of the decision tree to find randomness in data. The formula of entropy is as follow:

$$\text{Entropy} = -\sum_{i=1}^{n} pi \times log(pi) \qquad (1)$$

Which i is total number of samples, and p is the probability of i. At the beginning the entropy for the entire data set should be calculated. Then, the information gain should be calculated based on Equation 2 [23].

$$\text{Information Gain (T, X)} = \text{Entropy(T)} - \text{Entropy (T, X)} \qquad (2)$$

T is the target variable and X is the feature and Entropy (T, X) is entropy after the data point is split on feature X. At the end, the feature with highest information gain is the root node [23].

The advantage of DT is that it is suitable for large data sets and works for continuous and categorical data. There are some disadvantages with DT such as, usually it has an overfitting problem, and although it works well in training data, it has a low accuracy on new data leadings [24].

2.3 Random Forest

RF algorithm is one of the ML algorithms which can be used for different tasks, like classification and regression. RF is a combination of multiple different DTs. One of the differences between RF and DT is, RF is usually more accurate compared to DT. RF is used in different industries such as banking, healthcare, and marketing.

As shown in Figure 4, it gets prediction results from each tree and the majority of the prediction is the final prediction of RF [25]. RF uses Bagging or Bootstrap Aggregation as ensemble technique. The procedure is: choosing random samples from original data with replacement, then each model (DT) created based on samples (generate DT based on entropy or Gini impurity). At the end, each model has its own result and final output come from the majority voting of samples [23].

The advantage of RF is having low risk of overfitting and its ability to handle large data sets. The downsides of using RF are that it is very slow in training the data and sometime working biased with categorical variables [24].

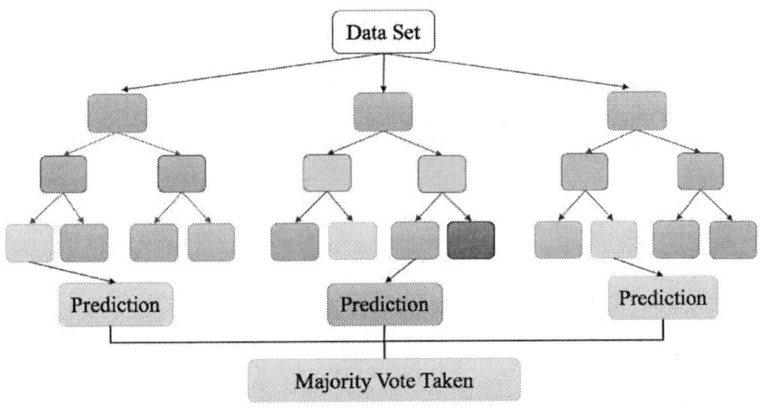

Figure 4 Demonstration of RF

2.4 K-nearest neighbours

KNN is a supervised machine learning algorithm used for both regression and classification prediction issues. But it is more used for classification problems. KNN algorithm works by estimating the likelihood that a data point will become a member of one group, or another based on what group the data points is closest to. In essence KNN works on the assumption that similar data points are close to each other [26]. The first step is to find the K value, which does not have any role, mostly chosen as 5. Second, the similarity among all data points and new data should be calculated by using the Euclidean distance (Equation 3 [27]). Equation 3 measures the distance between two points of p and q. Then, by using this equation the nearest K amount is selected in training data [27]. At the end, the nearest point to the new data is the prediction. If k nearest points are different as shown in Figure 5, for k= 4, the result is the majority vote which will be the stars in the figure 5.

$$d(p,q) = \sqrt{(q_1 - p_1)^2 + (q_2 - p_2)^2 + \cdots + (q_n - p_n)^2} = \sqrt{\sum_{i=1}^{n}(q_i - p_i)^2} \quad (3)$$

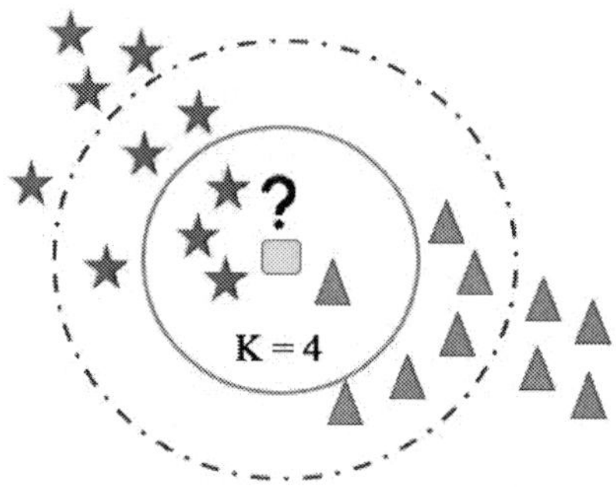

Figure 5 Demonstration of KNN

As you can see it is very easy to work with KNN, but this algorithm is slow and does not work very well on imbalanced data [28].

In this study, patients' attention has been classified and predicted using the three methods of DT, RF, and KNN. The comparison between these methods is done by using the accuracy, area under the curve (AUC), and the confusion matrix of each of them. The AUC is a measure of the classifier's capability to differentiate between classes [29]. The confusion matrix indicates a summary of results of prediction with their values for each model. The results of this study are presented in the next section.

3. Results

The labeling design study has 14 different features. Eight of features refer to the participants' demographic information in terms of, age, color differentiation, education, ethnicity, language, literacy, visual acuity, and sex. Remaining 6 are dependent on the trial features. The way that the model is trained is as follows, 5 features (critical, FOP, highlight, IBU, PDP) are considered as constant and the remaining one (change of content) is able to alternate. Note that change of content by itself consists of AI, DD1, and DD2. So, the number of data points with these features are 92 × 3 = 276. The model is trained over 70% of these data points (195 trials) and tested over the remaining 30% (81 trials). Furthermore, the target of this study is a response trial, which has three answers, hit (correct response), miss (wrong response), and time-out (the time of trial finished without an answer). The results can be categorized into three sections: DT, RF, and KNN.

The first model that we used is DT. The tree of this data is very large, as it has 61 nodes and 36 leaves. The classification accuracy of this model is 79% with AUC of 0.607. The second model is RF, which is a combination of different DTs. The number of trees in this model are 10 and each of these trees has a different number of nodes and leaves. The classification accuracy of RF is 92.6% with AUC of 0.818. The last model that is discussed here is KNN with the number of neighbors of 5. The classification accuracy for this model is 91.4% with AUC of 0.555. Figure 6 shows the accuracy comparison of three models. As you can see RF has the highest accuracy (also highest AUC), then KNN has higher accuracy compared to DT.

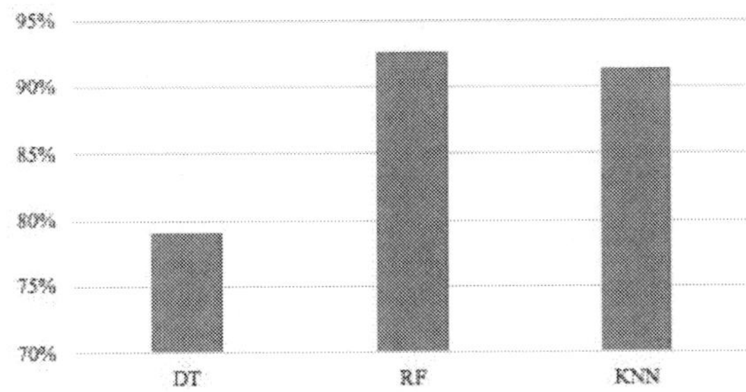

Figure 6 Accuracy of three models

Figure 6 indicate the classification accuracy, but it does not to indicate the result of prediction. As mentioned in previous section, the confusion matrix is used to indicate the summary of result prediction. Table 1 shows the comparison of the confusion matrix between these three models. As expected from accuracy and AUC, RF has the highest number of correct responses to hit. In RF and KNN, there were 74 total hit numbers (all predicted correctly), while DT were able to predict only 64 of them. Moreover, for the time-outs only RF can predict one of them and for the misses all three models fail to perform as expected.

Actual \ Predicted	Hit	TimeOut	miss	Σ
Hit	64	6	4	74
TimeOut	4	0	1	5
miss	1	1	0	2
Σ	69	7	5	81

DT

Actual \ Predicted	Hit	TimeOut	miss	Σ
Hit	74	0	0	74
TimeOut	4	1	0	5
miss	1	1	0	2
Σ	79	2	0	81

RF

Actual \ Predicted	Hit	TimeOut	miss	Σ
Hit	74	0	0	74
TimeOut	5	0	0	5
miss	2	0	0	2
Σ	81	0	0	81

KNN

Table 1 Confusion matrix of three models

As it can be seen from the confusion matrix of these models, although RF and KNN have good accuracy for the data, and predicting hits, they cannot predict time-outs or misses correctly.

The reason for this discrepancy is that accuracy (as indicated in equation (4)) only shows the total of correct predictions over total predictions. Therefore, if the data are imbalanced and the number of one of them is near the total number of predictions, it can decide the outcome of accuracy alone. As it is the case here. Number of the hits (175 trials from 195) are far greater than the other two features (20 trials) and therefore they are the deciding factors. That is why we needed another criteria to compare the models with each other (AUC). It should be noted that low AUC is also another indication of this problem.

$$\text{Accuracy} = \frac{Number\ of\ correct\ predictions}{Total\ number\ of\ predictions} \quad (4)$$

There are many methods to tackle imbalanced data, such as by up sampling minority classes, down sampling majority class, or a combination of over and under sampling. These methods are beyond the scope of this study.

4. Conclusions

This paper proposed three different ML algorithms models on OTC labeling design data: DT, RF, and KNN. ML is implemented in this study to predict the pattern in new features without the need for new data gathering. Although the models have acceptable classification accuracy as 79%, 92.6%, and 91.4%, respectively, their low AUC value indicates that they cannot predict well when one target data (hit) is dominating other targets (time-out and miss). There are different approaches with such imbalanced data such as by up sampling minority classes, down sampling majority class, or a combination of over and under sampling. We are planning to work on these approaches in this study to solve the imbalanced data problem.

5. References

[1] US Food and Drug Administration. (2018, Jan 2). Part III: Drugs and Foods Under the 1938 Act and Its Amendments.https://www.fda.gov/about-fda/changes-science-law-and-regulatory-authorities/part-iii-drugs-and-foods-under-1938-act-and-its-amendments

[2] Mudambi, S. M., & Schuff, D. (2010). Research note: What makes a helpful online review? A study of customer reviews on Amazon. com. MIS quarterly, 185-200.

[3] Albert, S. M., Bix, L., Bridgeman, M. M., Carstensen, L. L., Dyer-Chamberlain, M., Neafsey, P. J., & Wolf, M. S. (2014). Promoting safe and effective use of OTC medications: CHPA-GSA National Summit. The Gerontologist, 54(6), 909-918.

[4] Consumer Healthcare Products Association (CHPA). (2019, March 18). New Study: Over-the-Counter (OTC) Products Used by Millions of Americans Saves Healthcare System Billions Annually.
https://apnews.com/press-release/business-wire/business-health-58f7099d41a6445382dac7c361455083

[5] Consumer Healthcare Products Association (CHPA). (1964-2020). OTC Sales Statistics. https://www.chpa.org/OTCRetailSales.aspx

[6] Aronson, J. K., & Ferner, R. E. (2005). Clarification of terminology in drug safety. Drug safety, 28(10), 851-870.

[7] Lazarou, J., Pomeranz, B. H., & Corey, P. N. (1998). Incidence of adverse drug reactions in hospitalized patients: a meta-analysis of prospective studies. Jama, 279(15), 1200-1205.

[8] Shiyanbola, O. O., Meyer, B. A., Locke, M. R., & Wettergreen, S. (2014). Perceptions of prescription warning labels within an underserved population. Pharmacy practice, 12(1).

[9] Shiyanbola, O. O., Smith, P. D., Mansukhani, S. G., & Huang, Y. M. (2016). Refining prescription warning labels using patient feedback: a qualitative study. PLoS One, 11(6), e0156881.

[10] Abdulkareem, N. M., & Abdulazeez, A. M. (2021). Machine learning classification based on Radom Forest Algorithm: A review. International Journal of Science and Business, 5(2), 128-142.

[11] Wu, Y., Schuster, M., Chen, Z., Le, Q. V., Norouzi, M., Macherey, W., ... & Dean, J. (2016). Google's neural machine translation system: Bridging the gap between human and machine translation. arXiv preprint arXiv:1609.08144.

[12] Quanz, B., Sun, W., Deshpande, A., Shah, D., & Park, J. E. (2020). Machine learning based co-creative design framework. arXiv preprint arXiv:2001.08791.

[13] Butler, K. T., Davies, D. W., Cartwright, H., Isayev, O., & Walsh, A. (2018). Machine learning for molecular and materials science. Nature, 559(7715), 547-555.

[14] Wei, J., Chu, X., Sun, X. Y., Xu, K., Deng, H. X., Chen, J., ... & Lei, M. (2019). Machine learning in materials science. InfoMat, 1(3), 338-358.
[15] Chaabene, W. B., Flah, M., & Nehdi, M. L. (2020). Machine learning prediction of mechanical properties of concrete: Critical review. Construction and Building Materials, 260, 119889.
[16] Esteva, A., Kuprel, B., Novoa, R. A., Ko, J., Swetter, S. M., Blau, H. M., & Thrun, S. (2017). Dermatologist-level classification of skin cancer with deep neural networks. nature, 542(7639), 115-118.
[17] Kononenko, I. (2001). Machine learning for medical diagnosis: history, state of the art and perspective. Artificial Intelligence in medicine, 23(1), 89-109.
[18] Erickson, B. J., Korfiatis, P., Akkus, Z., & Kline, T. L. (2017). Machine learning for medical imaging. Radiographics, 37(2), 505-515.
[19] Knoll, D., Neumeier, D., Prüglmeier, M., & Reinhart, G. (2019). An automated packaging planning approach using machine learning. Procedia Cirp, 81, 576-581.
[20] Esfahanian, S. (2020). A Patient-Centered Approach to Labeling for Over-The-Counter Medications: Using Data to Drive Design Decisions for the Benefit of Older Adults. Michigan State University.
[21] Scikit-learn's development and maintance.(2007-2022). Decision Trees. https://scikit-learn.org/stable/modules/tree.html
[22] Cravit R. (2021, Aug 3). What is a Decision Tree and How to Make One [Templates + Examples]. https://venngage.com/blog/what-is-a-decision-tree/
[23] Ronaghan S. (2018, May 11). The Mathematics of Decision Trees, Random Forest and Feature Importance in Scikit-learn and Spark. https://towardsdatascience.com/the-mathematics-of-decision-trees-random-forest-and-feature-importance-in-scikit-learn-and-spark-f2861df67e3
[24] Trehan D. (2020, July 2). Why choose Random Forest and Not Decision Trees. https://towardsai.net/p/machine-learning/why-choose-random-forest-and-not-decision-trees
[25] Mbaabu O. (2020, Dec 11). Introduction to Random Forest in Machine Learning. https://www.section.io/engineering-education/introduction-to-random-forest-in-machine-learning/
[26] Harrison O. (2018, Sep 10). Machine Learning Basics with the K-Nearest Neighbors Algorithm. https://towardsdatascience.com/machine-learning-basics-with-the-k-nearest-neighbors-algorithm-6a6e71d01761
[27] Al-Serw N. (2021, May 17). K-nearest Neighbor: The maths behind it, how it works and an example. https://medium.com/analytics-vidhya/k-nearest-neighbor-the-maths-behind-it-how-it-works-and-an-example-f1de1208546c
[28] Genesis. (2018, Sep 25). Pros and Cons of K-Nearest Neighbor. https://www.fromthegenesis.com/pros-and-cons-of-k-nearest-neighbors/
[29] Aniruddha. (2020, Jun 16). AUC-ROC Curve in Machine Learning Clearly Explained. https://www.analyticsvidhya.com/blog/2020/06/auc-roc-curve-machine-learning/

CONFREENCE PROCEEDINGS :

PACKAGING DESIGN, ERGONOMICS & HUMAN FACTORS

• ORAL PRESENTATION

• GENERAL STREAM

DG-O01 Future-Driven Packaging Design: A Foresight Method to Aid in Designing Solutions for Future Challenges

DG-002 Tool Design on Sustainable Consumer Behaviour for Food Packaging Companies

DG-O01

Future-Driven Packaging Design: A Foresight Method to Aid in Designing Solutions for Future Challenges

Thomas G. J. Goudsblom[1]*, Bjorn de Koeijer[1], Mauricy Alves da Motta-Filho[1]

[1]Department of Design, Production and Management, Faculty of Engineering Technology, University of Twente, Drienerlolaan 5, 7522 NB Enschede, The Netherlands

***Correspondence to:** Thomas Goudsblom, Department of Design, Production and Management, Faculty of Engineering Technology, University of Twente, Drienerlolaan 5, 7522 NB Enschede, The Netherlands.*
E-mail: t.g.j.goudsblom@student.utwente.nl, or Bjorn de Koeijer: b.l.a.dekoeijer@utwente.nl

ABSTRACT: The task of designers is becoming ever more complex. Volatile challenges in today's world ask for a responsible design approach, and the social consequences of the design decisions need to be analysed during the design process itself. However, foreseeing how these design decisions will unfold in future scenarios is difficult and often allows for multiple ends. The growth of e-commerce and the sizable negative environmental influences created by related (secondary) packaging, impact the social consequences caused by the design and complicate the creation of future scenarios. Moreover, one could wonder to what degree the challenges we currently face could have been anticipated in the design process itself. This problem asks for a design method that can aid in designing packaging solutions whilst considering future challenges, reducing uncertainties, and increasing the resilience of design solutions. This research aims to broaden the knowledge on foresight and future-driven design and proposes a design method for the context of packaging design. A literature review on foresight methods is used as input for the development of the proposed method. The method is evaluated through user tests, after which a design iteration on the working principles of the method is applied to illustrate how the method works. The resulting Bullet Train Model is intended to be used in the preliminary phases of the design process. This model structures the foresight process for the application in packaging design processes. Based on trend research, the model aids in investigating possible and preferred future scenarios. Through the process of backcasting, the developments and decisions required to achieve these future scenarios can be identified. These can be used as input to make strategic design decisions for packaging design.

Keyword: Forecasting, Packaging design, Responsible design, E-commerce, Sustainability

1. INTRODUCTION

The process of designing and the tasks of designers are becoming more complex. In previous centuries, the primary task of the designer could be recognised in decorating and giving shape to products and thinking about their mechanical function. Nowadays, considering the social and environmental consequences of the designed artefacts and the political power these artefacts can hold are part of designers' portfolios [1]. Anticipating how design-decisions taken will unfold in the future is therefore a central question that needs to be answered. However, VUCA (volatile, uncertain, complex, and ambiguous) [2] challenges in today's world impede these evolved tasks of designers even more and make it more difficult to foresee how design-decisions will unfold in the future.

With over 320.000 packages being opened per second worldwide [3], packaging is completely integrated into our daily lives. Moreover, with the growth of e-commerce and its related consequences – the sizable negative environmental impact created by related (secondary) packaging – has changed the role of packaging and provides opportunities to rethink the way packaging is designed and used. Also, when viewing the product life-cycle of packaging from a human-centred perspective, it provides significant opportunities for improving current packaging design solutions [4]. This makes the topic very interesting to be analysed for its future impact, accompanying challenges and opportunities for design.

Having a long term vision in design can help to make design-decisions required to achieve the (preferred) future situation. As such, 'seeing into the future' can help us to identify these new or overlooked opportunities, the challenges of tomorrow and reduce risks in design proposals [1, 4, 5]. When discussing this practice of 'seeing into the future', people are often reminded of the mythical seer by whom the future is predicted. However, foresight or forecasting is not the same as prediction making. It is only possible to make predictions when events are preordained and no actions in the present can influence the future. Hence, prediction concerns itself with future certainty [7], yet, our world is filled with uncertainty [2]. Foresight is a structured approach to think about the future and concerns itself with identifying the full range of future possibilities and mapping out what is yet to be [5, 6, 8]. Whether the forecast turns out to be true is only part of the picture, as its main value lies in answering questions such as "where do we want to be?" [6] and 'what should we do next?'. Effective planning in the present asks for a vision of what the future should or could be [1].

To answer these questions within the application of a (secondary) packaging design process, a new and responsible design approach is needed. The societal impact and consequences of the design should be considered during the design process itself. This leads to the central question of this research: *How can a foresight method be developed to aid the design of (secondary) packaging solutions in e-commerce for future challenges?* In this paper, the practice of foresight is analysed and a new foresight method for the specified application will be discussed.

1.1 METHOD

This research follows an abductive style of reasoning as described by Dorst [9]; first, it focusses on framing the problem, and only then it seeks to solve it and create value. Figure 1 shows the research process. To gain a better understanding of future studies and contemporary methods, a broad spectrum of related literature was reviewed. Secondary packaging for e-commerce was chosen as the scope of the research. Hence, the literature reviewed here focused on investigating the functions, implementation, and the role of packaging within logistic systems.

The literature-based analysis is used to design a concept of the foresight method – the Bullet Train

Model. Next, this concept design is evaluated in a user tests to obtain qualitative insights about the functionality and usability, and to provide an indication of the validity of the method. The results of the user tests together with the concept design are used to create the final design of the foresight method. Finally, by means of a design iteration on the working principles of the method, we illustrate how the method works, and form the conclusion of the research.

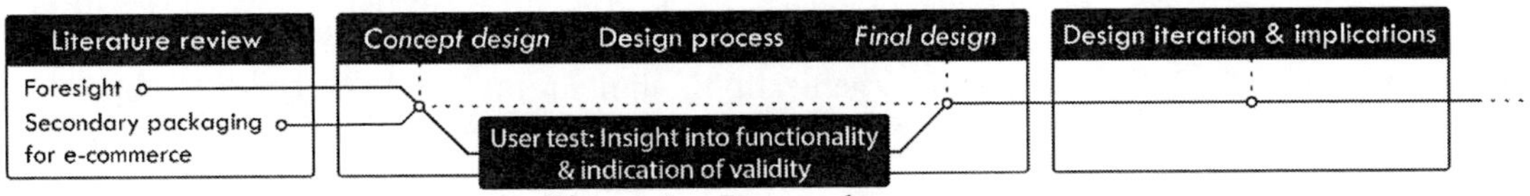

Figure 1 – Research process.

2. LITERATURE REVIEW

2.1 FORESIGHT AND DESIGN

As there are different purposes for using foresight, a variety of methods are available, each specialized for different goals. Table 1 shows different foresight methods. The methods presented in one row can be regarded as similar, despite the use of different terminology. For this research, investigating which methods can be used for designerly purposes is most interesting.

Table 1 – The six main categories of foresight methods compared from different literature.

By Cole [10] and Petermann [5]		By Chambers et al. [11]	
Category	*Description*	*Category*	*Description*
Forecasting	Data-based projections using economic, demographic, or environmental data, etc.	**Time series analysis and projection**	Focus entirely on patterns and pattern changes, and thus relies entirely on historical data
Envisioning	Theory, current or historical events, issues or imagination	**Qualitative techniques**	Qualitative data, information about special (historical) events are brought together in a logical, unbiased and systematic way
Polling	Representative viewpoints solicited from participants		
		Causal models	Uses highly refined and specific information about relationships between system elements

According to Chambers et al. [11], qualitative techniques are most often used in the product development stage. This statement is supported by Evans [8, 12], where case studies discussed in the research focus on applying scenario development and trend research (qualitative techniques or envisioning methods) for (1) shaping futures for their respective design industry, and (2) making an effective design brief to evaluate and inform design activities. To this, Petermann [5] adds how trend forecasting can support the decision-making process. Moreover, research by Kishita et al. [13] applies backcasting (i.e. describing the pathway to an envisioned future) scenarios shaped by causal models, envisioning methods and qualitative techniques to envision sustainable manufacturing industries. Overall, it can be recognized that qualitative techniques or envisioning methods are most often used for designerly purposes. Commonly, these methods are most often applied to develop long-term (10+ years) visions.

Within the design process, no decisive consensus can be found on where applying foresight in the process is most valuable. Evans [8, 14] describes applying foresight in the embodiment phase or as part of the creative process as most relevant, as understanding where trends, values, beliefs and lifestyles are heading aids in identifying possible opportunities for design. Following Grinyer et al. [15], “Both design and foresight have the greatest potential when used during the early stages of a project”. As this is also the moment a design strategy needs to be defined [16], the preliminary phase in the design process seems most valuable for the application of foresight.

Between the different foresight methods that make use of envisioning methods and qualitative techniques, variations can be found in their structural approach. Some of these methods focus on the process as a whole, whereas others focus on a specific part of the foresight process. Based on a literature review examining different structural approaches of the foresight process [12–14, 17–22], it can be seen that most structural approaches follow the foresight process as discussed by Evans [12]: (1) Analysis – Environmental analysis and problem identification; (2) Projection – Scenario building; (3) Synthesis – Developing strategies for the potential challenges the organization may face, considering the organization's limited internal resources. To support the environmental analysis of the foresight process, the PESTEL framework can be used [19, 23].

2.2 (SECONDARY) PACKAGING FOR E-COMMERCE

To design the foresight method, it is important to obtain a sufficient understanding of the scope of the research. Similarly, when applying foresight, a proper understanding of the context of application is required. Hence, the functions of (secondary) packaging, their implementation, and the role of secondary packaging in the logistic system of e-commerce are investigated.

Examining and comparing research journals and books [24–28], and white papers [29, 30] – both on packaging in general and e-commerce specific – shows there is a significant focus on the following packaging functions; protecting the product, communication, providing consumer and logistic convenience, and containing products. Interestingly, throughout the literature different terminology is used. However, when this terminology is explained, the high similarity in the packaging functions described by the different literature can be seen. Moreover, also specific differences between research journals and books, and white papers can be recognized. For example, in white papers, the packaging functions are explained as specific goals the client wants to reach (e.g., "convey brand impression" [30]), instead of describing the packaging functions with a more overarching expression or goal (e.g., "communication" [27, 28]).

When reviewing possible implementations of the functions, it should be considered that the combined performance of all levels of packaging is what matters. E.g., a weak primary packaging needs a strong secondary packaging for physical strength, whereas a strong primary packaging does not require a secondary packaging [26]. Within the context of packaging for e-commerce, the collaboration between these different levels of packaging is very characteristic.

The logistic system of e-commerce describes all actions that are taken from the moment a customer orders a product online to the arrival of the package at the customer (by delivery or collection) [29, 31, 32]. The role of secondary packaging in the logistic system of e-commerce can be viewed as keeping products belonging to the same order together, protecting the product, communicating about the state of the package, its disposal or identity, facilitating dispatch and consumer handling [29, 31, 32]. Logically, these roles can be recognised in the packaging functions as summarized earlier. However, through describing the role of secondary packaging in the logistic system of e-commerce, the differences in different parts of the logistic system can be seen.

The insights obtained from this literature review can help to investigate future visions by using foresight. Moreover, the insights can be used to apply sufficient focus when designing a foresight method.

3. DESIGN PROCESS

Based on the literature review, a concept design has been created: the Bullet Train Model. This model aims to present a visualization and structure of the foresight process within the application of a (secondary) packaging design process. As such, the main goal is to generate future scenarios and use these to develop a strategic path and plan of action. The model builds onto the structural approach of the foresight process as described by Evans [12] and the literature review as addressed. The concept design is used as input for the user tests and the final design.

3.1 USER TEST

The user test is conducted to assess the designed method for its usability and functionality, with the goal to obtain possible improvements for the design and the ability to provide an indication of the validity of the method. This is done by having three participants from different professional backgrounds related to design, packaging and/or e-commerce, apply the method in the context of a small assignment. After finishing the assignment, the participants filled in a questionnaire. The results of the user test show that the method functions as a whole and can achieve its main goal: facilitate in exploring the future to aid in designing (secondary) packaging solutions in e-commerce for tomorrow's challenges. The method supports the user in working with abstract tasks of developing future visions from research and translate this into a strategy draft. However, it can also be recognized some parts of the method are not clear in their use and need improvement. From the user test, the following implications have been obtained: (1) The first three steps of the method are redesigned for a clearer use and better understandability, (2) The usability of the model is increased by combining the functionality of the template, that was provided to the participants during the user test, and the Bullet Train Model Canvas (a more detailed explanation of the model).

4. THE BULLET TRAIN MODEL

The Bullet Train Model (Figure 2), a name is inspired by the futuristic and high-speed bullet train, is designed to assist in the foresight process of generating future scenarios and using these to develop a strategic path and action plan. The analogy describes a train that is driving through a portal leading to the future. Each wagon on the train supports the train to reach its future destinations (future scenarios). Once the possible future scenarios are known, the route and track switches (developments, events, and decisions) the train takes between today and the described future scenarios, can be defined. These are used to determine a strategic path. The model is explained through the following six steps: (1) Navigation – the context, scope, main actor and the purpose of the foresight process are defined; (2) Analysis – present problems and wishes for the future are identified, and PESTEL trend analysis is executed; (3) Loading the train – main drivers of change are defined and evaluated for their uncertainty and impact; (4) Scenarios – future scenarios are written based on the previously executed analyses; (5) Backcasting – the events, developments and decisions needed to arrive at the chosen future scenario are described on a timeline, and (6) Determine strategy path – through using the backcasting process and navigation step, the strategy needed to prepare for future possibilities is defined. The Bullet Train Model is supported through the Bullet Train Model Canvas (view here), which provides more detail on each of the steps and supports the process by providing a template.

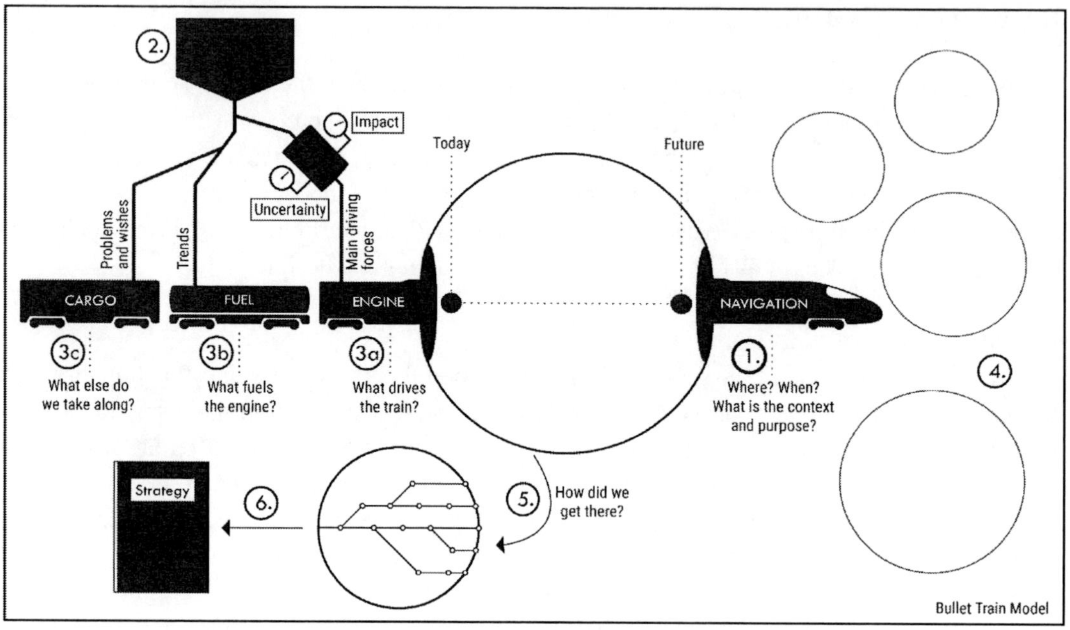

Figure 2 - Visual representation of the Bullet Train Model.

The model is intended to be used in the preliminary phase of the design process, as this is the moment where initial research is conducted, a design strategy is made, and must be decided if the new product will have a chance of success [16]. Moreover, this is also the design phase where foresight is of most value [14, 15]. Finally, the model is intended to be used in a team with a multidisciplinary approach, as input from experts with different backgrounds is valuable [8, 14]. When applying the method in the design iteration, the following can be recognised: the process and the outcome of the method mainly provide strategic implications for both the design process and the implementation of (secondary) packaging within the logistic system for e-commerce. Furthermore, it provides insights that could be used in the corporate strategy related to (secondary) packaging for e-commerce. This aligns with the initial experiences of participants in the user test.

5. CONCLUSION

This research has made a first step in integrating foresight more into the design process of (secondary) packaging, to aid in designing solutions for future challenges. The Bullet Train Model aims at using insights obtained from sketching future possibilities in development or decision making. Using foresight methods shows to have the most value when used during the preliminary phase of the design process. Here, foresight can assist designers in strategy determination and evaluating the possible success of a new product. Moreover, it helps to analyse the societal impact and consequences the design could have, already during the design process itself.

As the method is primarily based on envisioning methods, a creative contribution of the user is required to explore what could happen in the future and handle the associated uncertainties. However, this also makes the method prone to biases. Hence, it is recommended to investigate how artificial intelligence, machine learning and systematic technological assessment can be incorporated into the method and trend research, to possibly reduce the biases. Although the user test provided insights into the usability of the method, it is difficult to describe the effectiveness of the method at this moment. More extensive research is required to evaluate the effectiveness of applying the method within a real-life (secondary) packaging design process. Next to this, the research should be continued to further develop the method, with a specific focus to provide more structure in determining a strategy path.

Although the scope of this research was limited to (secondary) packaging for e-commerce, its relevance also proposes applications beyond this field, as foresight can provide aid in making design decisions. Hence, it is recommended to investigate the application of the research in other domains, such as strategy development or policymaking.

ACKNOWLEDGEMENT

This paper builds on the thesis 'The future of secondary packaging in e-commerce: a forecasting method to aid in designing solutions to future challenges' by T.G.J. Goudsblom (2020).

REFERENCES

[1] Margolin, V. (2007). Design, the Future and the Human Spirit. Design Issues, vol. 23, no. 3, pp. 4–15, DOI: 10.1162/desi.2007.23.3.4.

[2] Kim, E., Beckman, S. L., and Agogino, A. (2018). Design roadmapping in an uncertain world: Implementing a customer-experience-focused strategy. California Management Review, vol. 61, no. 1, pp. 43–70. DOI: 10.1177/0008125618796489.

[3] NVC Netherlands Packaging Centre. (2020). The PUMA Manifesto. Retrieved on March 13, 2022, from: https://www.en.nvc.nl/puma-the-end-of-packaging-as-an-environmental-issue/.

[4] Norman, D. (2020). Waste is an enormous problem. But recycling is the wrong solution. Fast Company. Retrieved on March 13, 2022, from: https://www.fastcompany.com/90463116/waste-is-an-enormous-problem-but-recycling-is-the-wrong-solution.

[5] Petermann, E. (2014). Archaeology of the Future - Reconsidering the Place and Nature of Trend forecasting in Design Discourse. In Proceedings of DRS Conference, pp. 645-662. June 16-19, 2014, Umeå, Sweden.

[6] Prosser, Z. and Basra, S. (2019). Future Thinking; A Mind set, not a method. Touchpoint, vol. 10 N.2 - Designing the Future. Retrieved on March 8, 2022, from: https://medium.com/touchpoint/futures-thinking-a-mind-set-not-a-method-64c9b5f9da37.

[7] Saffo, P. (2007). Six rules for effective forecasting. Harvard Bussiness Review, vol. 85, 7/8, pp. 122–131. Retrieved on May 6, 2020, from: https://hbr.org/2007/07/six-rules-for-effective-forecasting?

[8] Evans, M. (2003). Trend Forecasting for Design Futures. In Proceedings of EAD 5th International Conference, pp. 1–10, April, 2003, Barcelona, Spain.

[9] Dorst, K. (2011). The core of 'design thinking' and its application. Design Studies, vol. 32, no. 6, pp. 521–532, DOI: 10.1016/j.destud.2011.07.006.

[10] Cole, S. (2001). Dare to Dream: Bringing Futures into Planning. Journal of American Planning Association, vol. 67, no. 4, pp. 372–383, DOI: 10.1080/01944360108976246.

[11] Chambers, J. C., Mullick, S. K. and Smith, D. D. (1971). How to choose the right forecasting technique. New York: Wiley.

[12] Evans, M. (2005). I-SPY: Utilising Forecasting and Scenario Planning for Design Futures. In Proceedings of EAD Conference, March 29-31, 2005, Bremen, Germany, DOI: 10.13140/RG.2.1.4800.9763.

[13] Kishita, Y., Mizuno, Y. and Umeda, Y. (2016). Scenario Design Approach to Envisioning Sustainable Manufacturing Industries to 2050. Procedia CIRP, vol. 48, pp. 407–412, 2016, DOI: 10.1016/j.procir.2016.04.103.

[14] Evans, M. (2004). A Design Approach to Trends and Forecasting. In Proceedings of Future Ground: DRS International Conference, November 17-21, 2004, Melbourne, Australia.

[15] Grinyer, C. et al. (2002). Design for Future Needs. In Proceedings of Future Perfect Conference, December 10-11, 2002, Brussels, Belgium. Retrieved on March 3, 2022, from: https://web.archive.org/web/20030622162826/http://www.dffn.org/Research.htm.

[16] Eger, A., Bonnema, M., Lutters, D. and Voort, M. van der. (2013). Product Design, 4th ed. Den Haag: Eleven International Publishing.

[17] Ziegler, W. (1991). Envisioning the future. Futures, vol. 23, no. 5, pp. 516–527, 1991, DOI: 10.1016/0016-3287(91)90099-N.

[18] Robinson, J. B. (1990). Futures Under Glass: A Recipe for People Who Hate to Predict. Futures, vol. 22, no. 9, pp. 820–841, 1990, DOI: 10.1016/0016-3287(90)90018-D.
[19] Burt, G., Wright, G., Bradfield, R., Cairns, G. and Heijden, K. van der. (2006). The Role of Scenario Planning in Exploring the Environment in View of the Limitations of PEST and Its Derivatives. Int. Stud. of Manag. & Organ., vol. 36, no. 3, pp. 50–76, 2006, DOI: 10.2753/IMO0020-8825360303.
[20] Kuzmina, K., Prendeville, S., Walker, D. and Charnley, F. (2018). Future scenarios for fast-moving consumer goods in a circular economy. Futures, pp. 74–88, vol. 107, no. April, DOI: 10.1016/j.futures.2018.12.001.
[21] Eggink, W., Reinders, A. and Meulen, B. van der. (2009). A practical approach to product design for future worlds using scenario development. In Proceedings of the 11th International Conference on Engineering and Product Design Education, pp. 538–543, September 10-11, 2009, Brighton, United Kingdom.
[22] Nathan, L. P., Klasnja, P. V. and Friedman, B. (2007). Value scenarios: A technique for envisioning systemic effects of new technologies. In Proceedings of Conf. Hum. Factors Comput. Syst. - Proc., pp. 2585–2590, 2007, DOI: 10.1145/1240866.1241046.
[23] Johnson, G., Scholes, K. and Whittington, R. (2009). Fundamentals of strategy. Harlow: Pearson Education Limited.
[24] Klooster, R. ten, Dirken, J. M., Lox, F. and Schilperoord, A. A. (2015). Onderscheidenen en ordenen van verpakkingsfuncties. Zakboek Verpakkingen, Second., Delft: Plato Product Consultants, pp. 20–28.
[25] Lutters, D. and Klooster, R. ten. (2008). Functional requirement specification in the packaging development chain. CIRP Ann. - Manuf. Technol., vol. 57, no. 1, pp. 145–148, 2008, DOI: 10.1016/j.cirp.2008.03.052.
[26] Emblem, A. (2012). Packaging Technology, London: Woodhead Publishing Limited, pp. 24–49.
[27] Yam, K. L. (2010). The Wiley encyclopedia of packaging technology, vol. 47, no. 11.
[28] Koeijer, B. de, Wever, R. and Henseler, J. (2017). Realizing product-packaging combinations in circular systems: Shaping the research agenda. Packag. Technol. Sci., vol. 30, pp. 443–460, DOI: 10.1002/pts.2219.
[29] Schumacher Packaging Group. (n.d.). Everything that you need to know about packaging in e-commerce. Retrieved on May 16, 2020, from: https://schumacher-packaging.com/fileadmin/downloads/Schumacher_Packaging_ebook_eCommerce_ENG_1806.pdf.
[30] Noissue. (2019). E-commerce and Sustainable Packaging. Retrieved on March 6, 2022, from: https://www.noissue.co/media/2019SustainabilityinPackagingStudy.pdf.
[31] Amcor Limited. (2016). Packaging for a New Era of E-Commerce. Retrieved on April 28, 2020, from: http://www.bemis.com/Bemis/media/Library/pdf/restricted/amcor-ebook-ecommerce-na.pdf.
[32] Billerudkorsnäs. (n.d.). How Packaging can add Value in E-commerce. Retrieved on March 4, 2022, from: https://www.billerudkorsnas.com/globalassets/billerudkorsnas/about-us/stories/five-ways-to-boost-online-sales-with-packaging/how_packaging_can_add_value_report.pdf.

DG-002

Tool Design on Sustainable Consumer Behaviour for Food Packaging Companies

Tao L.F. Heslenfeld[1], Bjorn de Koeijer[1]*

[1]*Department of Design, Production and Management, Faculty of Engineering Technology, University of Twente, Drienerlolaan 5, 7522 NB Enschede, The Netherlands*

***Correspondence to:** Bjorn de Koeijer, Department of Design, Production and Management, Faculty of Engineering Technology, University of Twente, Drienerlolaan 5, 7522 NB Enschede, The Netherlands. E-mail: b.l.a.dekoeijer@utwente.nl*

ABSTRACT: In the design of new packaging, sustainability and consumer behaviour are two important points of focus. There is a clear gap to combine these two points of focus, since companies struggle how to motivate consumers to behave more sustainably with the help of packaging cues. Therefore, this study explores the possibilities of a tool to support food packaging companies in dealing with sustainable consumer behaviour. By means of a theoretical model, consumer psychology, design cues, and brand management are researched. Furthermore, a qualitative user study (n=22) has been conducted to explore the needs of food packaging companies. Consequently, three points of limitation for companies have been discovered: (1) a gap in consumers' knowledge, (2) communicating sustainability-related elements and features on the packaging, and (3) alignment of challenges between companies' partners, clients, and departments. The user study and theoretical model provided input for a tool pack design, where three tools delegate insights on different levels of information and interaction. The tool pack provides companies with a centralized place to find practical insights about sustainable consumer behaviour in relation to food packaging. With that, the tool aims to assist companies with overcoming the three discovered points of limitation. To validate the tool pack, a user test (n=24) was executed. This test consisted of qualitative semi-structured interviews and an Attrakdiff questionnaire. Based on the results, it is expected that the tool pack can help and provide companies with new insights into sustainable consumer behaviour, in relation to potential packaging design interventions. In conclusion, the theoretical model provides a new perspective on summarising the sustainable consumer behaviour theories relevant for food packaging. The tool pack contributes to the packaging industry by introducing new ways to retrieve the latest sustainable consumer behaviour insights in a simple, playful, and detailed manner.

Keyword: sustainability, consumer behaviour, packaging design, tool design, co-creation

1. INTRODUCTION

When assessing a packaging by a lifecycle analysis (LCA) and finding the packaging has a relatively low environmental impact, it can be called a sustainable packaging [1]. It is beyond doubt that consumers are more concerned about sustainable issues [2] and demanding more eco-friendly products [3]. For these reasons, it is no surprise that consumers hold positive attitudes towards sustainable packaging [4, 5]. Packaging can influence behaviour, for instance the environmental aspects of packaging can affect the consumers' choice behaviour [6]. Furthermore, the packaging can influence purchase intention [7,8], determine price and (food) quality expectations [7], taste impressions [9] and even influence the brand evaluation [10]. Packaging design is of great importance for a company, since it is a valuable communication attribute [11]. Consumers play an important role, since they are the ones who buy, use, and dispose products (including packaging). For this reason, consumer behaviour is one of the key factors in the five packaging perspectives model by the Netherlands Institute for Sustainable Packaging's (KIDV) [12]. Consumer behaviour is a complex topic, let alone the combination of sustainability and consumer behaviour. Most of the time companies do not have the time to keep up to date with scientifical knowledge. Furthermore, scientific literature mainly provides abstract information which can not immediately be used by companies. Therefore, the information needs to be more practical in order to be applied by companies. Consequently, the need to design a tool to support companies in this topic is in the interest of society and may help in the transition towards (more) sustainable packaging.

The study focuses on (Western) food packaging companies. Packaging experts are the prospected end users. 'Packaging experts' is used as an overarching term covering different roles, from packaging designers, packaging technologist to marketers and sustainability managers. The aim of the assignment was to create an overview of existing knowledge and to translate that into a tool which can support companies. As a reason the study aims to explore the following research question: "How to provide practical knowledge through a tool which can support companies to deal with sustainable consumer behaviour, spark them to work on the topic and implement the tool and insights in the packaging development process?"

2. LITERATURE REVIEW

Three main topics that have been studied: (1) sustainable consumer behaviour (psychology), (2) how to influence consumers via the packaging (design cues), and (3) how to convince companies to be part of this sustainability transition (brand management). These topics have been chosen to design a tool which is both useful and valuable (in presenting insights). A theoretical model was created to summarise the literature and serve as the theoretical backbone for the tool design (Figure 1)

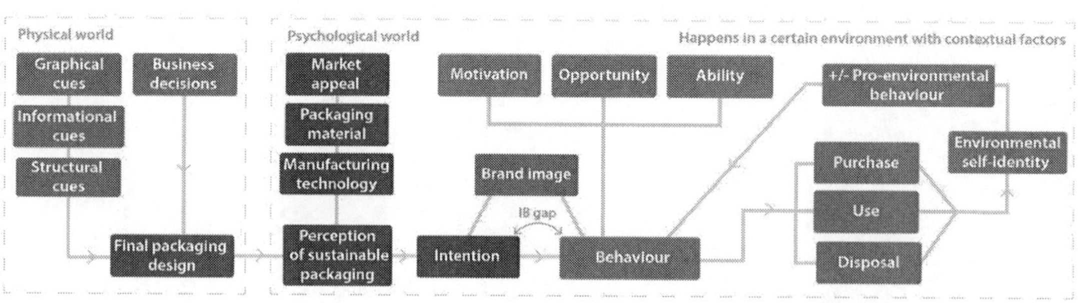

Figure 1 Theoretical model

Knowing the consumer is one of the first things to consider when designing packaging, since various consumer groups with different levels of interest in sustainability reacts differently on sustainable communication efforts on packaging [4,13]. Consumers have a biospheric value, which reflect how much they care about the environment, and it is a predictor for pro-environmental behaviour [14]. Consumers evaluate a packaging's sustainability based on three characteristics: market appeal, manufacturing technology and packaging materials [8]. The market appeal concerns the eco-friendly expectations regarding graphic design, price, and performance [8]. The second characteristic, consumers evaluate the packaging material to decide what the environmental impact of the packaging will be [15]. However, consumers have little knowledge about packaging materials and manufacturing technologies, there is a gap between LCA outcomes and what people believe to know [16]. Finally, the manufacturing technology reflects the manufacturing process evaluation, for instance the use of energy or new technologies being used [8]. Knowledge could be a trigger for sustainable behaviour [17], however, others have found discrepancy for this trigger [18, 19].

This study uses the Motivation, Opportunity, and Ability model [20]. Ability concerns the knowledge, the skills and tools needed to reach a certain goal [20]. The purchase process of sustainable packaging may be stressful, expensive, and time-consuming [21], and habits, routines and cues create cognitive boundaries for consumers [22]. Motivation refers to beliefs, attitudes, and social and personal motivations [20]. Here, emotions play a role in predicting pro-environmental purchase behaviour of packaging [23, 24]. Opportunities concern the external factors outside the person itself [25]. In the past decades, the importance of the environment on user perception and behaviour has been extensively studied [26, 27].

A big problem for pro-environmental behaviour is the intention-behaviour gap. The gap reflects the fact that consumers express to have certain intentions (e.g., environmental), however their behaviour might not reflect these expressed intentions [28]. After a certain behaviour, the self-identity is influenced [29], the way someone describes oneself [30]. Someone who has a high environmental self-identity is more likely to see themselves acting environmentally friendly and showing pro-environmental behaviour [31].

There are three different cues for consumers which a company can use for its packaging design regarding sustainability: informational, graphical, and structural cues [32]. Firstly, informational cues inform consumers about the product's and or packaging's features. For instance, verbal cues can be used to explicitly communicate the packaging's sustainability [33]. In addition, labels can be helpful as well, however these can be confusing for consumers due to the variety of labels [34]. Secondly, graphical cues are cues such as colour, logos, and images. For example, the colour green is associated to sustainability [33]. Also, nature imagery can be a tactful cue, because the positive emotional effect of nature can be beneficial for the advertising effect [35]. Thirdly, structural cues such as shape can play an important role in the consumer perception. For instance, users choose a smaller portion of food when it came from a squeeze tube in comparison to a traditional container [36]. In addition, research showed that people tend to see altered objects as waste, for example torn objects were more likely to be littered [34]. Keep in mind, consumers can be misled by cues which do not communicate something about the real sustainability consequences [16].

It is important to convince companies to be part of this transition towards (more) sustainable packaging, since sustainability has not always been directly acknowledged during the packaging development process [37]. Sustainable packaging can help a company to differentiate itself from competitors [38] and stand out, since consumers see packaging as an environmental burden [32]. A consumer's perception of a brand is called the brand image, and in case of green brands, how sustainable a brand is in the eyes of a consumers [39].

3. METHODOLOGY

3.1 User study

To explore the needs of potential users, semi-structured interviews have been conducted as a user study. Through a semi-structured discussion, identical topics can be covered, but there is still room left for other interesting points proposed by the participant [40]. The interview contained the following discussion topics: packaging process, consumer behaviour, sustainability, and tools used during the packaging development process. Professionals working for a food packaging company and working on sustainable consume behaviour have been interviewed. Both small and large companies were welcome to participate, to keep the exploration phase as open as possible. The interviews were held online and a template with visuals was used to support the discussion. The interviews were recorded and transcribed for providing an accurate review of what had been said. The transcriptions led to codes which were categorised in themes. The interview templates provided semi-quantitative answers as well (e.g., insights in where consumer behaviour is applied throughout the development process).

3.2 User test

A user test was conducted to explore the user's thoughts, ideas, and experiences on the tool design prototype. The user tests applied both a semi-structured discussion as a qualitative evaluation method and the Attrakdiff questionnaire as a quantitative method. The semi-structured discussion contained basic questions and tool-dependent questions. The questionnaire had to support the qualitative data and was used to compare the tools in a more objective way, with less room for interpretations made by the researcher (in comparison to the discussion outcomes). Participants for the first user test were arranged in two groups and spread over two weeks based on their availability, the first tool was tested in the first week, the second tool was tested in the second week. Participants received the assigned tool in advance and were asked to review the tool before the user test. The Attrakdiff questionnaire evaluates the attractiveness of an interactive product based on pragmatic quality (usability and utility) and hedonic qualities (user's psychological needs) [41, 42]. The importance of the hedonic qualities next to the functionality qualities of a product, is highlighted by previous research [41]. With the input and active collaboration of prospected end users throughout the design process, co-creation helped forming the designed tool where designers and end users pursue collective creativity [43].

4. RESULTS

4.1 User study

22 participants have been interviewed, 45% of the participants were men and 55% women, they had a mean age of 39,9 years old, ranging from 25 till 68 years old. In total 29 codes were collected from the transcriptions, they were spread over four main themes and five additional subthemes. The outcomes led to user requirements and a better understanding of the company's circumstances. During the interviews companies talked about wishes. The most prominent wishes were: not time consuming; user friendly; meaningful; trustworthy; provides perspective; for multiple users; convincing arguments; providing a summary and a preference for quantified information. Regarding consumer behaviour, companies are mainly focused on the creation of a sustainable perception for the purchase phase. A big recurring topic is the wrong perception of plastic. Consumers are not always aware of how companies make packaging more sustainable, because companies may not communicate their sustainability efforts. Most companies already use icons, claims, certification marks, or advise their clients to use these cues. Companies explained their wishes for advice how to properly communicate information without risking possible greenwashing. The need for more knowledge on sustainable consumer behaviour was a big

discussion point. There is a knowledge gap between the different stakeholders where people make different claims, for which more independent knowledge can be useful. All participating companies expressed that the biggest barrier for a more sustainable packaging is higher costs. At the moment, most tools were used in the beginning of the development process and are used to gain insights about a certain topic, to communicate information among others and as validation of requirements. In conclusion, three big problems for companies are captured: (1) the knowledge gap of consumers, (2) communicating sustainability, and (3) alignment problems between partners, clients, and departments.

4.2 Tool design

Following the literature review, the user study, and an iterative human-centred design process, a concept was created called the tool pack. Requirements both from the users and KIDV served as boundaries for the tool design. The tool pack can delegate information in three steps, since the topic is complex and can be too overwhelming when one presents it at once. The first step is a playful and physical cube with minimal information about cues. This 'Cues Cube' can help companies in stimulating sustainable consumer behaviour, provide food for thought, and can redirect companies towards the KIDV website as a knowledge hub. The second tool, an interactive infographic, is the central tool where the level of information and interaction is balanced, it provides an overview of the main factors in sustainable consumer behaviour. The expectation is that most users will use the infographic, due to its accessibility and in combination with the amount of information that are provided. The infographic prototype can be found on https://xd.adobe.com/view/8d6eb29b-89bf-4d86-b301-df2d80101e87-d83d/?fullscreen. Finally, a theoretical report acts as the third tool. It is the most static tool and provides more in-depth information on the topic. The complete tool pack (Figure 2) helps the user in gaining knowledge based on their own learning pace and intentions, it empowers the user to make their own learning curve. Delegating information in steps may result in reaching different people within the company that have different levels of interest in the topic. Users can follow a funnel, where they dive deeper into the topic and experience different interactions based on their goal and limitations, like available time.

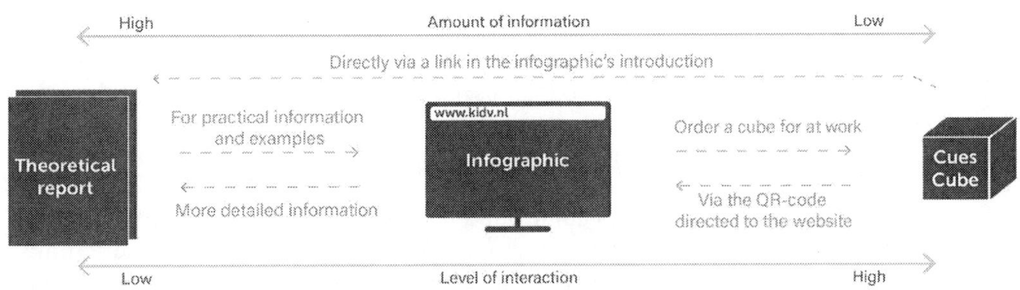

Figure 2 The tool pack and its corresponding tool interactions

4.3. User test

In total 24 participants have been interviewed in two test rounds (for the first-round n=22 and the second-round n=2), gender among participants was equally balanced and they had a mean age of 39,8 years old, ranging from 24 till 58 years old. Half of the respondent group consisted of participants who also participated in the user study. In general, the user test showed that participants perceived the tool pack as helpful, that it may help participants to gain insights about the topic. The Cues Cube was a less meaningful tool in comparison to the infographic, since it provided less information than participants expected and wanted. Participants could still see the purpose of the cube, they called the cube "a starting point" and "a reminder". The infographic was well received, especially the visually attractive way of presenting insights was welcome. The infographic was too long (amount of information), however almost all participants expressed they would like to have more information or examples on certain topics. Participants also expect that different people within the company will use different tools, based on whether packaging is a core activity for their job. The theoretical report was evaluated as a useful tool, since it will provide a summary with details of sustainable consumer behaviour, and it is easy to save and store. The Attrakdiff questionnaire showed that the infographic is perceived as slightly more attractive. The infographic is perceived to be slightly more desirable than the Cues Cube, due to a better balance in hedonic versus pragmatic qualities. The quantitative outcomes provide an indication what qualities can be improved and showed that the tools only need little improvements to be 'desirable'. A small secondary user test was conducted for the theoretical report on a later moment. The process for this test was similar to the other user test. The participants (n=2) found the report useful, it contained interesting topics based on scientific research which the users appreciated. The facts could provide new insights and a new way of looking at the topic, however they missed concrete examples in the theoretical report.

5. DISCUSSION AND CONCLUSIONS

In a design process, results from user studies and user tests where interviews are conducted, always come with a certain extent of subjectivity. The implementation of these results in the design process is also subjective to authors' interpretations. The urgency of certain topics and the feasibility of the implementation of the design, were important factors for the researchers. The user study outcomes are in line with the literature review, participants experienced the same problems as mentioned in literature, such as the consumers' knowledge gap.

The user study and user tests provided rich data, thanks to a large group of participants. Luckily, participants came with overarching discussion topics and problems, which says something about the data saturation. The Attrakdiff questionnaire does not provide detailed feedback about the evaluated tool, one cannot know which semantic differential pair was the most important for users for future improvements. Also relevant, the outcomes of the small user test of the theoretical report gave an indication of how future users could perceive the report, but no concrete points of improvements and conclusions can be made.

The designed tool pack can have positive implications for companies by providing and sharing insights, the tool can reach a substantial group of users, especially when companies share the tool with their clients (as expressed by participants). The tool pack may have negative implications for society as well, since the tool may stimulate misuse by greenwashing. The application of the tool is meant for a broad target group, where no job-specific implications can be assigned to individuals. KIDV also attracts non-food companies, it may come about that non-food companies find the tool pack as well. Application-wise, it is questionable to what extend the tool may be useful for non-food companies. The results contribute to the scientific field by providing a summary of sustainable consumer behaviour in relation to packaging (the theoretical model). Finally, the designed tool contributes to the packaging industry by providing tools which can help in making the 'right' sustainable packaging design choice. In addition, the tool helps companies to pursue sustainability goals as well, without overlooking its commercial and strategic goals. By creating such a practical tool set, the complex and abstract literature will be available to and applicable for a wider audience. The literature research showed some research gaps where more work needs to be done. For instance, more guidance about bridging the intention-behaviour gap focused on sustainable packaging is required to provide companies with advice and inspiration.
The designed tool provides insights in sustainable consumer behaviour factors for (food) packaging based on the literature review. The tool pack delegates information in three different levels, which empowers the user to choose which one is the most suitable for themselves

depending on the available time and the questions they have. To sum up, the tool pack can be used in the process as a summary source for knowledge, to find inspiration, to share insights with colleagues and clients, and to spark the interest in the topic among others. Finally, insights can be used as argumentation and to gain empathy with the consumer to understand their problems as well. The tool can help in a better understanding and designing for sustainable consumer behaviour through packaging.

6. ACKNOWLEDGEMENT

The authors would like to thank the Netherlands Institute for Sustainable Packaging (KIDV) for providing this thesis assignment, and for publication of the tools. KIDV's knowledge and network helped in the creation of this research and development of the tool design.

7. REFERENCES

[1] Glavič, P. and Lukman, R. (2007). Review of sustainability terms and their definitions. Journal of Cleaner Production, 15(18), 1875–1885. DOI: 10.1016/j.jclepro.2006.12.006.

[2] Olsen, M. C., Slotegraaf, R. J. and Chandukala, S. R. (2014). Green Claims and Message Frames: How Green New Products Change Brand Attitude. Journal of Marketing, 78(5), 119–137. DOI: 10.1509/jm.13.0387.

[3] Deloitte. (2021). Sustainability & Consumer Behaviour 2021. Deloitte United Kingdom. Retrieved 27 May 2021, from https://www2.deloitte.com/uk/en/pages/consumer-business/articles/sustainable-consumer.html

[4] Martinho, G., Pires, A., Portela, G. and Fonseca, M. (2015). Factors affecting consumers' choices concerning sustainable packaging during product purchase and recycling. Resources, Conservation and Recycling, 103, 58–68. DOI: 10.1016/j.resconrec.2015.07.012.

[5] Prakash, G. and Pathak, P. (2017). Intention to buy eco-friendly packaged products among young consumers of India: A study on developing nation. Journal of Cleaner Production, 141, 385–393. DOI: 10.1016/j.jclepro.2016.09.116.

[6] Rokka, J. and Uusitalo, L. (2008). Preference for green packaging in consumer product choices - Do consumers care? International Journal of Consumer Studies, 32(5), 516–525. DOI: 10.1111/j.1470-6431.2008.00710.x.

[7] Magnier, L., Schoormans, J. and Mugge, R. (2016). Judging a product by its cover: Packaging sustainability and perceptions of quality in food products. Food Quality and Preference, 53, 132–142. DOI: 10.1016/j.foodqual.2016.06.006.

[8] Nguyen, A. T., Parker, L., Brennan, L. and Lockrey, S. (2020). A consumer definition of eco-friendly packaging. Journal of Cleaner Production, 252, 119792. DOI: 10.1016/j.jclepro.2019.119792

[9] van Rompay, T. J., Finger, F., Saakes, D. and Fenko, A. (2017). "See me, feel me": Effects of 3D-printed surface patterns on beverage evaluation. Food Quality and Preference, 62, 332–339. DOI: 10.1016/j.foodqual.2016.12.002.

[10] Underwood, R. L. and Klein, N. M. (2002). Packaging as Brand Communication: Effects of Product Pictures on Consumer Responses to the Package and Brand. Journal of Marketing Theory and Practice, 10(4), 58–68. DOI: 10.1080/10696679.2002.11501926.

[11] Rettie, R. and Brewer, C. (2000). The verbal and visual components of package design. Journal of Product & Brand Management, 9(1), 56–70. DOI: 10.1108/10610420010316339.

[12] KIDV. (n.d.). KIDV-model Vijf perspectieven op duurzaam verpakken® – Kidv Site. Retrieved 11 May 2021, from https://kidv.nl/kidv-model-vijf-perspectieven-op-duurzaam-verpakken.

[13] Otto, S., Strenger, M., Maier-Nöth, A. and Schmid, M. (2021). Food packaging and sustainability – Consumer perception vs. correlated scientific facts: A review. Journal of Cleaner Production, 298, 126733. DOI: 10.1016/j.jclepro.2021.126733.

[14] Steg, L., Berg, D. A. V. E. and Groot, J. D. I. M. (2012). Environmental Psychology: An Introduction (1st ed.). Wiley-Blackwell.

[15] Lindh, H., Olsson, A. and Williams, H. (2015). Consumer Perceptions of Food Packaging: Contributing to or Counteracting Environmentally Sustainable Development? Packaging Technology and Science, 29(1), 3–23. DOI: 10.1002/pts.2184.

[16] Steenis, N. D., van Herpen, E., van der Lans, I. A., Ligthart, T. N. and van Trijp, H. C. (2017). Consumer response to packaging design: The role of packaging materials and graphics in sustainability perceptions and product evaluations. Journal of Cleaner Production, 162, 286–298. DOI: 10.1016/j.jclepro.2017.06.036.

[17] Peattie, K. (2010). Green Consumption: Behavior and Norms. Annual Review of Environment and Resources, 35(1), 195–228. DOI: 10.1146/annurev-environ-032609-094328.

[18] Pedersen, E. R. and Neergaard, P. (2005). Caveat emptor - let the buyer beware! environmental labelling and the limitations of 'green' consumerism. Business Strategy and the Environment, 15(1), 15–29. DOI: 10.1002/bse.434.
[19] Jensen, B. B. (2002). Knowledge, Action and Pro-environmental Behaviour. Environmental Education Research, 8(3), 325–334. DOI: 10.1080/13504620220145474.
[20] Ölander, F. and ThØgersen, J. (1995). Understanding of consumer behaviour as a prerequisite for environmental protection. Journal of Consumer Policy, 18(4), 345–385. DOI: 10.1007/bf01024160.
[21] Barbarossa, C. and Pastore, A. (2015). Why environmentally conscious consumers do not purchase green products. Qualitative Market Research: An International Journal, 18(2), 188–209. DOI: 10.1108/qmr-06-2012-0030.
[22] Jackson, T. (2005). Motivating sustainable consumption: a review of evidence on consumer behaviour and behavioural change: a report to the Sustainable Development Research Network. Centre for Environmental Strategy, University of Surrey.
[23] Kollmuss, A. and Agyeman, J. (2002). Mind the Gap: Why do people act environmentally and what are the barriers to pro-environmental behavior? Environmental Education Research, 8(3), 239–260. DOI: 10.1080/13504620220145401.
[24] Koenig-Lewis, N., Palmer, A., Dermody, J. and Urbye, A. (2014). Consumers' evaluations of ecological packaging – Rational and emotional approaches. Journal of Environmental Psychology, 37, 94–105. DOI: 10.1016/j.jenvp.2013.11.009.
[25] Langeveld, G., Pietersma, S., de Goede, K., Boomsma, M., van Rhee, M., van Soest, D., Midden, C., Mastop, J. and Weenk, A. (2020). Verbetering Afvalscheiding in de Hoogbouw. VANG-HHA. https://www.vang-hha.nl/kennisbibliotheek/%40235221/verbetering-afvalscheiding-hoogbouw/.
[26] van Rompay, T. J., Deterink, F. and Fenko, A. (2016). Healthy package, healthy product? Effects of packaging design as a function of purchase setting. Food Quality and Preference, 53, 84–89. DOI: 10.1016/j.foodqual.2016.06.001.
[27] Geiger, J. L., Steg, L., van der Werff, E. and Ünal, A. B. (2019). A meta-analysis of factors related to recycling. Journal of Environmental Psychology, 64, 78–97. DOI: 10.1016/j.jenvp.2019.05.004 .
[28] Darby, S. (2000). Making it obvious: Designing feedback into energy consumption. 2nd International Conference on Energy Efficiency in Household Appliances and Lighting. Italian Association of Energy Economists/ EC-SAVE Programme., 26–28.
[29] Lee, L., Piliavin, J. A. and Call, V. R. A. (1999). Giving Time, Money, and Blood: Similarities and Differences. Social Psychology Quarterly, 62(3), 276. DOI: 10.2307/2695864.
[30] Cook, A., Kerr, G. and Moore, K. (2002). Attitudes and intentions towards purchasing GM food. Journal of Economic Psychology, 23(5), 557–572. DOI: 10.1016/s0167-4870(02)00117-4.
[31] van der Werff, E., Steg, L. and Keizer, K. (2013). The value of environmental self-identity: The relationship between biospheric values, environmental self-identity and environmental preferences, intentions and behaviour. Journal of Environmental Psychology, 34, 55–63. DOI: 10.1016/j.jenvp.2012.12.006.
[32] Magnier, L. and Crié, D. (2015). Communicating packaging eco-friendliness: an exploration of consumers' perceptions of eco-designed packaging. International Journal of Retail & Distribution Management, 43(4/5), 350–366. DOI: 10.1108/ijrdm-04-2014-0048.
[33] Pancer, E., McShane, L. and Noseworthy, T. J. (2015). Isolated Environmental Cues and Product Efficacy Penalties: The Color Green and Eco-labels. Journal of Business Ethics, 143(1), 159–177. DOI: 10.1007/s10551-015-2764-4.
[34] Baxter, W., Aurisicchio, M. and Childs, P. R. (2016). Tear here: The impact of object transformations on proper disposal.
[35] Hartmann, P., Apaolaza, V. and Alija, P. (2013). Nature imagery in advertising: Attention restoration and memory effects. International Journal of Advertising, 32(2), 183–210. DOI: 10.2501/ija-32-2-183-210.
[36] Huyghe, E., Geuens, M. and Vermeir, I. (2017). To squeeze or not to squeeze: How squeeze tubes affect consumers' serving sizes. Appetite, 111, 56–62. DOI: 10.1016/j.appet.2016.12.034.
[37] de Koeijer, B., de Lange, J. and Wever, R. (2017). Desired, Perceived, and Achieved Sustainability: Trade-Offs in Strategic and Operational Packaging Development. Sustainability, 9(10), 1923. DOI: 10.3390/su9101923.
[38] Jones, P., Clarke-Hill, C., Comfort, D. and Hillier, D. (2008). Marketing and sustainability. Marketing Intelligence & Planning, 26(2), 123–130. DOI: 10.1108/02634500810860584.
[39] Chen, Y. S. (2009). The Drivers of Green Brand Equity: Green Brand Image, Green Satisfaction, and Green Trust. Journal of Business Ethics, 93(2), 307–319. DOI: 10.1007/s10551-009-0223-9.
[40] Preece, J., Rogers, Y. and Sharp, H. (2015). Interaction Design (4th ed.). John Wiley & Sons Ltd.
[41] Hassenzahl, M. (2003). The thing and I: understanding the relationship between user and product. In Funology (pp. 31-42). Springer, Dordrecht.
[42] Hassenzahl, M. (2004). The Interplay of Beauty, Goodness, and Usability in Interactive Products. Human-Computer Interaction, 19(4), 319–349. DOI: 10.1207/s15327051hci1904_2.
[43] Sanders, E. B. N. and Stappers, P. J. (2008). Co-creation and the new landscapes of design. Co-design, 4(1), 5-18.

CONFREENCE PROCEEDINGS :

PACKAGING FOR FOOD & AGRICULTURE

• ORAL PRESENTATION

• GENERAL STREAM

FG-O01 Ultra-accelerated Shelf-life Test for Shelf-Life Determination of Oxygen-Sensitive Foods in Polymeric Packaging Materials

FG-O02 Predicting the Compression Strength in Ventilated Corrugated Paperboard Packages Using Machine Learning Models

FG-O03 Poly (Butylene Adipate-co-Terephthalate)/Thermoplastic Starch Blended Phosphate via Blown Film Extrusion as Functional Biodegradable Food Packaging

FG-O04 Enhanced Preservation Functionality of Bioplastic Materials for Sustainable Food Packaging

FG-O05 Roles of Titanium Dioxide, Zinc Oxide and Silicon Dioxide on Morphology, Properties and Migration of Thermoplastic Cassava Starch and Poly (Butylene Adipate-co-Terephthalate) Nanocomposite Films

FG-O06 Investigating the Effects of Packaging Design Factors on Customer Purchasing Decisions for Goat Manure Fertilizer

FG-O07 Measuring the Hexane Vapor Transmission Rate of Barrier Paper Materials

• PEER REVIEWED STREAM

FP-O01 Application of Biodegradable Film as Modified Atmosphere Packaging for Red Chili (Capsicum annuum cv. Jinda)

• POSTER PRESENTATION

FG-P01 Evaluation of the quality of fresh broccoli for non-refrigerated transportation using a high-performance thermal insulation box

FG-O01

Ultra-accelerated Shelf-life Test for Shelf-Life Determination of Oxygen-Sensitive Foods in Polymeric Packaging Materials

Anbuhkani Muniandy[1], Patnarin Benyathiar[2], Ferhan Ozadali[1,3], Dharmendra Mishra[1]*

[1]*Department of Food Science, Purdue University, West Lafayette, 47907, USA*
[2]*Department of Food Technology, Mahidol University, Kanchanaburi, 71150, Thailand*
[3]*Mead Johnson Nutrition, Reckitt Benckiser, Evansville, 47712, USA*

***Correspondence to:** Department of Food Science, Purdue University, 745 Agriculture Mall Dr, West Lafayette, 47907, USA.*
E-mail: mishradh@purdue.edu

ABSTRACT: An effective analysis method with multiple accelerant factors is crucial for shelf-life determination and prediction for food products with low errors and reduced analysis time. Temperature is the most common accelerant used in the shelf-life analysis of oxygen-sensitive food. Although the rate-limiting step in oxidation is independent of the oxygen concentration, literatures have shown the partial pressure of oxygen at low levels (<0.21 kPa) influences the product oxidation significantly when coupled with moderate temperatures. This approach has the potential to decrease the shelf-life analysis time, but little data is available. Mango smoothie containing whey protein fortified with vitamins A, B1, C and D3 was developed as the model food to investigate the effect of environmental oxygen pressure on product shelf life in polyethylene terephthalate (PET). PET was chosen due to its widespread use in the beverage industry and is a highly recyclable material, providing an opportunity for packaging sustainability. Bottles filled with model food were placed in a high-pressure (138 kPa) environment with 100% oxygen using custom-made chambers at 4°C (UASLT). Control samples (ASLT) were subjected to the same temperature without additional pressure. The changes in color and vitamin degradation were monitored for both storage conditions. In 40 days, UASLT treatment showed excessive browning of the samples compared to ASLT with ΔE of 7.8±0.5 and 1.8±0.2, respectively, due to rapid degradation of ascorbic acid. The degradation (%) of vitamins A, B1, C and D3 were 20.0±3.5, 13.6±1.6, 27.2±2.8, 26.3±0.9, respectively. The degradation (%) in ASLT were 11.8±4.4, 5.1±1.7, 13.3±1.9, 17.0±2.0, respectively. The addition of oxygen pressure significantly increased the degradation reaction rates of the vitamins due to the rapid influx of oxygen. The results indicated that elevated external oxygen pressure can be used as an accelerant along with moderate temperatures for rapid shelf-life testing of products in polymeric packaging. This new approach has potential application in the food industry for faster shelf-life analysis of food and selection of packaging materials.

Keyword: shelf life, polyethylene terephthalate, multi-accelerant, oxygen pressure

1. INTRODUCTION

Shelf-life of a food product indicates if it has acceptable sensory characteristics and has all the nutrients that are compliant with the label. The shelf-life prediction of foods with long-term stability is often subjected to accelerated shelf-life testing (ASLT) under environmental conditions that are intended to speed up the degradation reactions [1, 2]. Temperature is the commonly used accelerant in ASLT studies. Mathematical models are then applied to extrapolate the stability of the product at milder conditions based on results obtained from ASLT. Though ASLT has been utilized by the food industry for more than 40 years, this method can be considered slow, especially for shelf-stable products that have a long shelf-life. ASLT studies of shelf-stable products could take up to 4-5 months depending on the stability of the product. Increasing the temperature of the ASLT method beyond 40-45 °C is not advisable since it can induce reactions that may not naturally occur in food products leading to an unrealistic aging process [1, 3].

Although oxygen has been considered as an accelerant, less emphasis has been given because the oxidation reaction is independent of oxygen concentration > 0.21 kPa. At high temperatures, the impact of oxygen on stability is negligible due to the reduced solubility of oxygen in food. However, at lower partial pressures (<0.21 kPa) and at moderate temperatures (<45 °C), oxygen becomes the primary factor that drives the oxidation reaction compared to temperature [4-6]. A decrease in shelf life with increasing partial pressure of oxygen (<0.21 kPa) has been reported in the literature [5, 7]. These studies were conducted by maintaining the partial pressure of the oxygen throughout the study with direct exposure of oxygen to food. However, the partial pressure of oxygen inside a sealed food packaging changes constantly due to the oxidation reaction and the diffusion of oxygen over time [8, 9]. The impact of the partial pressure of oxygen at a lower level on the stability of food inside the packaging with an influx of oxygen from the environment is not well-known. Such a time-consuming study can be accelerated with the use of high external oxygen pressure which has not been investigated so far. This multi-accelerant approach has the potential to be used for the rapid shelf-life analysis of oxygen-sensitive products in polymeric packaging. Hence, the study aims to investigate the effect of environmental oxygen pressure on the shelf life of a shelf-stable food in a polyethylene terephthalate (PET) package. PET was chosen because it is a highly recyclable material that is widely used in the beverage industry.

2. MATERIALS AND METHODS

2.1 Materials

Mango puree, sugar and whey protein were purchased from local grocery stores. Pectin stabilizer (Dairyblend acidified beverage 120) was obtained from TIC Gums, Inc. (Ingredion, Westchester, IL, USA). Sunflower lecithin (E 322) was obtained from Lasenor USA, LLC (Salem, OH, USA). Citric acid was obtained from Cargill (Minneapolis, MN, USA). Retinyl palmitate (vitamin A) mix and cholecalciferol (vitamin D3) were donated by Mead Johnson Nutrition (Reckitt Benckiser Health, Evansville, IN, USA). Thiamin hydrochloride (vitamin B1) and ascorbic acid (vitamin C) were obtained from Thermo Fisher Scientific (Waltham, MA, USA). All HPLC standards and chemicals were obtained from Thermo Fisher Scientific.

2.2 Sample preparation

Mango smoothie containing whey protein that was fortified with vitamins A, B1, C and D3 was developed as the model food according to the recipe presented in Table.1. The whey protein and pectin stabilizer were mixed with water and left to hydrate with agitation for 20 min. Then, the rest of the ingredients were mixed, and the pH was verified and adjusted to be

4.00 ± 0.03 with citric acid. The product was aseptically processed and packaged using the MicroThermics® 25 HV lab scale UHT/HTST unit (MicroThermics Inc., Raleigh, NC, USA) and Scholle aseptic bag-in-box packaging system, Scholle A.F. 900 Low Acid Filler (Scholle Ipn Corporation, Northlake, IL, USA), respectively. Both MicroThermics® and filler units were sterilized before introducing the food product to the respective units. The MicroThermics® unit was sterilized with hot water at 121.1°C for 30 min. The system was brought to production mode after sterilization. The Scholle filler system was sterilized with steam at a minimum temperature of 121.1 °C for 30 min. The sterility of the filler head was maintained by keeping the temperature above 60°C and sterile overpressure between sterilization cycles with a spray of 140 ppm chlorine into the filler head. The product was processed at a flow rate of 1 L/min with a holding time of 36 s. The temperature at the end of the hold tube was set to achieve 100 °C to deliver a minimum lethality of 1 min at a reference temperature of 93.33 °C and a z-value of 8.89 °C. The product was cooled down to 24 °C before aseptically filling into a pre-sterilized 1-gallon bag-in-box package (Scholle IPN Corporation, Northlake, IL, USA). The spout on each bag was sterilized with steam at 146.1 °C for 8 s before filling the product.

The aseptic mango smoothie was transferred from the sterile bag to polyethylene terephthalate (PET) bottles in a laminar hood that was sterilized with UV for 45 min. The bottles and closures were also sterilized during the laminar hood sterilization. Product (244.5 ± 0.5 g) was filled into the PET bottles. Prior to sealing, the bottles were flushed with nitrogen for 12 min to reduce the headspace and dissolved oxygen below 2%. The nitrogen line was fitted with a 0.22 μm air filter to supply sterile air. The aerobic plate count method was used to verify the sterility of the hood during the filling processing.

Table 1 Mango smoothie formulation

Ingredient	**Percentage (w/w)**
Water	57.7%
Mango puree	28.0%
Sugar	9.0%
Whey protein	1.0%
Pectin stabilizer	0.3%
Sunflower lecithin	0.2%
Citric acid	0.3%
Ascorbic acid (vitamin C)	1.0%
Thiamin hydrochloride (vitamin B1)	1.0%
Retinyl palmitate mix (vitamin A)	0.5%
Cholecalciferol mix (vitamin D3)	1.0%

2.3 Storage study

Bottles filled with model food were placed in a custom-made high-pressure chamber supplied with 100% oxygen gas at 138 kPa and 40 °C. This storage condition is referred to as the ultra-accelerated shelf-life test (UASLT). Control samples were placed in an incubator at 40 °C without additional pressures that follows the standard ASLT method. Both UASLT and ASLT samples were kept in their respective storage conditions for up to 40 days while monitoring the changes in the color and vitamin levels over time. Triplicates samples were used for each storage condition.

2.3 Color measurement

The color of the samples was determined using a Hunter LabScan XE Colorimeter (HunterLab, Reston, VA, USA) that was controlled by software and was calibrated with a black and white tile. The color measurements were recorded using the CIELab uniform color space with reference illuminant D65 and a visual angle of 10°. The $L*$, $a*$, $b*$ values were recorded and the total color difference, ΔE, was calculated using eq.1. The color of the aseptically processed mango smoothie before the storage study was used as the reference. All measurements were performed in triplicates

$$\Delta E = \sqrt{\Delta L^{*2} + \Delta a^{*2} + \Delta b^{*2}} \quad (1)$$

2.4 Extraction and quantification of vitamins A, D3, B1 and C

Extraction of vitamin A (retinyl palmitate) and D3 (cholecalciferol) was performed simultaneously according to AOAC Official Method 2012.09. [10]. Briefly, 4g of sample was added to a 50 mL centrifuge tube, mixed with 25 mL methanol, and let to stand for 10 min. 10 mL of isooctane was added followed by a vortex treatment for 45 s. 5 mL of distilled water was added and vortexed for 20 s. The sample was then centrifuged at 5000xg for 10 min at 4 °C. 5 μL aliquot of the clear-yellow supernatant was injected into an Agilent 1260 high-performance liquid chromatography (HPLC) system after filtration with 0.45 μm PVDF syringe filter. The isocratic elution of vitamin A (24.4 min) and D3 (6.3 min) using an Infinity Lab Poroshell 120 EC-C18 column (3.0mm x 150 mm, 2.7μm) was achieved in 30 min using methanol at a flow rate of 0.45 mL/min with detection at 325 nm and 266 nm, respectively.

Extraction of vitamin B1 (thiamin hydrochloride) and C (ascorbic acid) was performed simultaneously. In a 15 mL centrifuge tube, 1g of sample was added and mixed with 2 mL of 4.5 % aqueous *m*-phosphoric acid. The sample was sonicated for 15 min before centrifuging at 10000 rpm for 10 min at 4 °C. After filtering the supernatant with a 0.45 μm PVDF syringe filter, 6 μL aliquot was injected into the same HPLC system fitted with an Agilent Poroshell 120 Sb-Aq column (4.6mm x 150mm, 4 μm). Isocratic elution of vitamins B1 (2.1 min) and C (2.6 min) was achieved in 10 min using 0.25 % aqueous *o*-phosphoric acid at a flow rate of 1 mL/min with detection at 245 nm.

All vitamins were quantified based on a calibration curve developed using retinyl palmitate, cholecalciferol, thiamin hydrochloride, and ascorbic acid as standards. Both the extraction and HPLC quantification were performed in triplicates.

2.5 Statistical analysis

Statistical analysis was performed using SAS 9.4 (SAS Institute, Cary, NC, USA). Means were analyzed using ANOVA and significant differences between the means were determined using Tukey's pairwise comparison at α=0.05.

3. RESULTS AND DISCUSSION

Figure 1 shows the extent of browning in mango smoothie products kept at the ASLT and UASLT conditions for 40 days. The samples kept at the UASLT condition show significant (p <0.05) browning compared to the ASLT sample. The ΔE value for the samples at the ASLT and UASLT conditions were 1.84 ± 0.32 and 8.05 ± 0.36, respectively.

Figure 1 Browning of mango smoothie samples at room temperature (left), ASLT (middle) and UASLT (right) conditions for 40 days

Table 2 Change of color in mango smoothie kept at the UASLT and ASLT conditions

	L*	a*	b*	ΔE
Reference	$68.39^{a} \pm 0.03$	$16.63^{a} \pm 0.03$	$65.39^{a} \pm 0.26$	-
ASLT	$65.90^{b} \pm 0.30$	$16.53^{ab} \pm 0.30$	$63.84^{b} \pm 0.26$	$1.84^{a} \pm 0.32$
UASLT	$61.35^{c} \pm 0.30$	$16.98^{ac} \pm 0.09$	$59.20^{c} \pm 0.23$	$8.05^{b} \pm 0.36$

Values represent the average ± standard deviation of triplicate measurements. Values with the same letter in the same column are not significantly different at α=0.05.

Figure 2 shows the amount of vitamins A, B1, C and D3 remaining in the samples after being stored under ASLT and UASLT conditions. Signification reduction in vitamin quantity was observed after 40 days (p <0.05) in the UASLT sample compared to ASLT. The vitamins values were normalized with the initial amounts. The degradation (%) of vitamins A, B1, C and D3 in UASLT samples were 19.6±3.6, 13.1±1.6, 26.7±2.8, 26.9±2.8, respectively. The degradation (%) in ASLT were 11.2±4.4, 4.2±1.8, 12.4±2.0, 16.9±2.0 for vitamins A, B1, C and D3, respectively.

Results also shows that the susceptibility of vitamins to degradation were in the order of C > D3 > A > B1 for UASLT and D3 > C > A > B1 for ASLT. The degradation reaction in samples subjected to ASLT was driven initially by the headspace and dissolved oxygen. Once the oxygen was depleted, the degradation was driven by the elevated temperature since the diffusion of oxygen to the internal environment is slow. On the other hand, the degradation reaction in UASLT samples was driven by the constant influx of oxygen due to the applied pressure as well as the elevated temperature. This resulted in more degradation compared to samples that were not exposed to additional pressure. Rapid degradation of vitamin C in samples kept at the UASLT condition most likely led to the formation of furfural which may undergo polymerization or combine with amino acids to form brown melanoidin pigments [11]. This is consistent with the extensive browning observed with the samples kept at the UASLT conditions.

The levels of headspace and dissolved oxygen were below 2 % throughout the study for both UASLT and ASLT samples (data not shown). This is due to the constant consumption of any residual oxygen that is present in the internal environment of samples at the UASLT condition, and the slow migration of oxygen in ASLT samples. Another study has shown increased oxygen transfer to the PET bottles kept at the UASLT conditions compared to ASLT when water was used as the model food (unpublished). This study also showed increased diffusivity of oxygen that was linear to the pressure level applied.

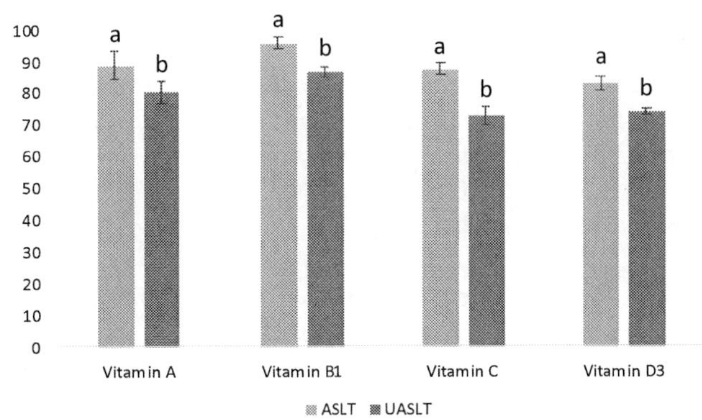

Figure 2 Quantity of remaining vitamins A, B1, C and D3 in mango smoothie after storage at the UASLT and ASLT conditions for 40 days. Values represent the average ± standard deviation of triplicate measurements. Values with the same letter in the same category are not significantly different at α=0.05.

4. CONCLUSIONS

The application of oxygen pressure enhanced the diffusion of oxygen to the internal environment of the PET bottles. The food sample experienced a realistic aging process since the oxygen level inside the packaging was maintained below < 0.21 kPa (atmospheric level) and only moderate temperature (40 °C) was used. The constant influx of oxygen in the samples kept at the UASLT conditions increased the rate of reactions in the food products and resulted in faster degradation of color and vitamins A, B1, C and D3. Results from this study indicate that the elevated external oxygen pressure can be used as an accelerant along with moderate temperatures for rapid shelf-life testing of products in polymeric packaging. This new multi-accelerant approach has potential application in the food industry for faster shelf-life analysis of food and package selection. Comparative analysis for package selection can also be done faster using this rapid method.

5. ACKNOWLEDGEMENT

This work was supported by the USDA National Institute of Food and Agriculture, Hatch project 1023114.

6. REFERENCES

1. Labuza, T.P. and D. Riboh, *Theory and application of arrhenius kinetics to the prediction of nutrient losses in foods.* Food Technology, 1982. **36**(10): p. 66-74.
2. Labuza, T.P. and M.K. Schmidt, *Accelerated shelf life testing of foods.* Food Technology, 1985. **39**(9): p. 57-64.
3. Labuza, T.P., S.R. Tannenbaum, and M. Karel, *Water content and stability of low moisture and intermediate moisture foods.* Journal of Food Technology, 1970. **24**: p. 543-550.
4. Labuza, T.P. and L.R. Dugan, *Kinetics of lipid oxidation in foods.* C R C Critical Reviews in Food Technology, 1971. **2**(3): p. 355-405.
5. Cardelli, C. and T.P. Labuza, *Application of Weibull Hazard Analysis to the Determination of the Shelf Life of Roasted and Ground Coffee.* LWT - Food Science and Technology, 2001. **34**(5): p. 273-278.
6. Poulsen, L. and P.R. Ogilby, *Oxygen Diffusion in Glassy Polymer Films: Effects of Other Gases and Changes in Pressure.* The Journal of Physical Chemistry A, 2000. **104**(12): p. 2573-2580.
7. An, D.S., J.H. Lee, and D.S. Lee, *Shelf life model of powdered infant formula as function of temperature and oxygen concentration.* Food Packaging and Shelf Life, 2018. **15**: p. 130-133.
8. Moufle, A.-L., J. Jamet, and R. Karoui, *Multifactors accelerated aging of sterilized acidic pudding determined by traditional and fluorescence techniques.* Journal of Food Process Engineering, 2018. **41**(6): p. e12807.
9. Soares, N.F.F. and J.H. Hotchkiss, *Comparative effects of de-aeration and package permeability on ascorbic acid loss in refrigerated orange juice.* Packaging Technology and Science, 1999. **12**(3): p. 111-118.
10. AOAC 2012.09, *Simultaneous Determination of Vitamins A and E in Infant Formula and Adult/Pediatric Nutritional Formula by HPLC and Column Switching.* J. AOAC Int, 2013. **96**.
11. Yuan, J.-P. and F. Chen, *Degradation of Ascorbic Acid in Aqueous Solution.* Journal of agricultural and food chemistry, 1998. **46**(12): p. 5078-5082.

FG-O02

Predicting the Compression Strength in Ventilated Corrugated Paperboard Packages Using Machine Learning Models

Khadijeh Shirzad[1], Amin Joodaky[1]*

[1]*School of Packaging, Michigan State University, East Lansing, 48824, United States*

***Correspondence to:** School of Packaging, Michigan State University, 448 Wilson Rd, East Lansing, 48824, United States. joodakya@msu.edu*

ABSTRACT: In the food packaging industry, ventilated corrugated paperboard boxes are commonly used for packaging and transporting fresh products. The ventilation holes reduce the compression strength of the paperboards. In order to optimize the ventilated corrugated paperboard box design, the effects of geometry and location of the holes on the box compression strength are investigated. In this study, we applied four machine learning models of multiple linear regression, support vector regression, decision tree regression, and random forest regression to provide a tool for predicting the compression strength of a ventilated corrugated paperboard with one circular hole. The size and location of the vent holes are varied for each sample. Machine learning is a subfield of artificial intelligence that develops a mathematical approach to predict complex patterns of sample data. In this paper, the sample data for training and testing the machine learning models are provided from experimental tests. The random forest regression model showed 91.26% accuracy. With the presented accuracy, the proposed method predicts the compression strength of the ventilated corrugated paperboards for a wide range of radiuses and locations of the vent hole.

Keyword: Food packaging, Ventilated package, Machine learning, Compression behavior, Package design

1. INTRODUCTION

The fundamental function of packaging is to protect food from mechanical damage [1]. In fresh food product packaging, ventilated corrugated paperboard (VCP) boxes have broad applications. The widespread use of VCP is owing to its cost-effective material properties, recyclability, and effective protection of food products [2]. The main objective of VCP design is to preserve the freshness of the food products and to provide adequate compression strength for the package [3]. The heat exchange of VCP packaging is critical to provide the consistent cooling of fresh products in a transportation system. The heat transfer between the food and cooling air is facilitated by designing vent holes in the box walls. However, the compression strength is reduced by increasing the hole area. Therefore, the structural design of VCP is optimized to maximize the ventilation area and to minimize the compression strength loss.

Several researchers studied the design of the vent holes in ventilated packages for fresh products [4–7]. Berry et al.[4] analyzed the package with four different package designs including three vent hole areas and three corrugated fiberboard grades. The authors reported a negative correlation between the vent area and package strength and a positive relationship between corrugated fiberboard grade and compression strength of the package. Opara and Fadiji [5] studied the effect of the vent area on the compressive strength of apple VCP boxes. They investigated two different boxes, each with a different length-to-height ratio and vent area. They concluded that the mechanical damage was more significant in the package with a lower vent area and length-to-height ratio. Singh et al.[6] conducted a study to find the impact of varying vent size, shape, and location of the hole on the loss of compression strength in the package of perishable products. This study demonstrated that the horizontal holes in corrugated boxes with a rectangular or parallelogram shape retain their strength better than holes with circular shapes. In addition, the venting or handling holes areas and the loss of strength were linearly correlated.

An analysis on the effects of vent shape, orientation, area, and vent hole numbers was conducted by Fadiji et al. [7]. The authors reported a reduction in the compression strength of the box by increasing the height and area of the vent hole. In terms of vent shape, they showed that the rectangular vent holes better preserved the compression strength of the box compared to the circular and oval ones. The compression strength decreased in the packages with two and four vent holes. Moreover, the vent holes placed at 45 degrees enhanced the compression strength of the box.

In the fresh food industry, the majority of VCP package designs are primarily dependent on basic approaches developed from trial-and-error tests. Performing the experimental test is time and cost-consuming. Many computer simulation models are developed to provide more efficient solutions [3]. VCP packages involve complicated structural properties (e.g., the composite structure and nonlinear deformation behavior of paperboards). Hence, the computational VCP package mechanics models are usually based on simplified assumptions such as representing a nonlinear or uncertain behavior by a linear function. Furthermore, the computer simulation models are usually computationally expensive [8]. In the following, the machine learning approach is introduced as a solution that learns from a limited number of actual test data and predicts the results for a wide range of unseen cases without requiring to apply a complex mechanics model.

Machine learning (ML) is a subfield of artificial intelligence that develops a mathematical approach to predict complex patterns of sample data. ML models have been applied in various science and engineering applications such as predicting the stress-strain plots in material design [9], predicting the mechanical properties of concrete [10], detecting road shock [11], reducing packaging costs [12], analyzing the freshness of packaged meat products [13], classifying and selecting the fish gelatin packaging film product [14].

In this paper, we applied four ML methods to predict the compression strength of paperboard plates with a circular vent hole for a wide range of radiuses and locations. The applied ML procedure is coded in an open-source programming language. The computational models with complex mechanics are not involved in the applied ML algorithms. The models learn from a limited number of actual experimental test results. The compression behavior of the box is directly related to the compression of vertical side plates. The obtained results for predicting compressions of ventilated corrugated plates in this study are applied to design VCP boxes. A comparative analysis of the four ML models is presented. The remainder of this study is structured as follows: Section 2 explains the sample data set collected from the experimental test and describes the ML methods applied to the collected data. Section 3 discusses the performance of each ML model. Section 4 presents the conclusions and findings of this study.

2. MATERIALS AND METHODS

In this section, the test sample characteristics and experimental procedure for the data collection step are explained. In the second part, the applied ML methods are discussed.

2.1 Data Acquisition

There are three general steps in an ML analysis. The first step is the acquisition of the sample data. The data set in this study is obtained from experimental compression tests. A C-flute paperboard with a random dimension of 15 cm × 15 cm and one circular hole is considered. The samples had been conditioned for 24 hours at 72°F and 50% RH as required in the test standard ASTM D642. The hole circles with radiuses of 0.5, 1, 1.5, and 2 centimeters are made in the plate using a Kongsberg digital cutting table machine. The lower left corner of the plate is set as the origin (0,0). The location of the hole is varied with a 2.5 cm increment, starting from the point (2.5, 2.5) as shown in Figure 1. The samples are placed vertically inside a compression tester machine (Instron 5565 Universal testing machine). Two blocks are used to ensure the vertical positioning of the plate before the compression (see Figure 2a). Once the top platen starts contacting the sample plate, the blocks are removed. The plates are compressed for 4% to 5% strains (6 to 7.5 mm). Figure 2b shows the collapse trace of the sample after performing the compression test.

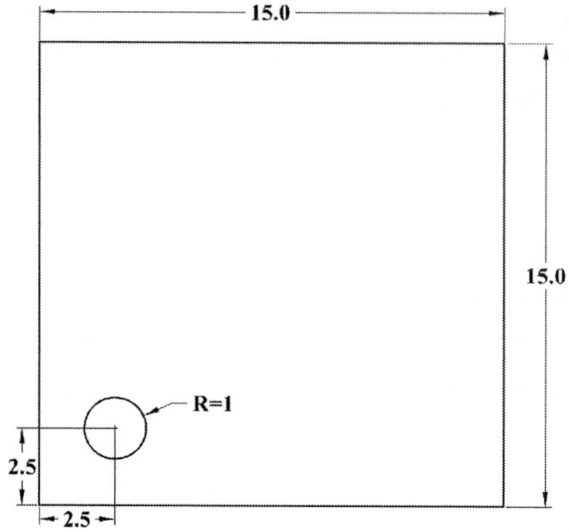

Figure 1 Sketch of a sample plate for experimental tests. The sizes are in centimeters

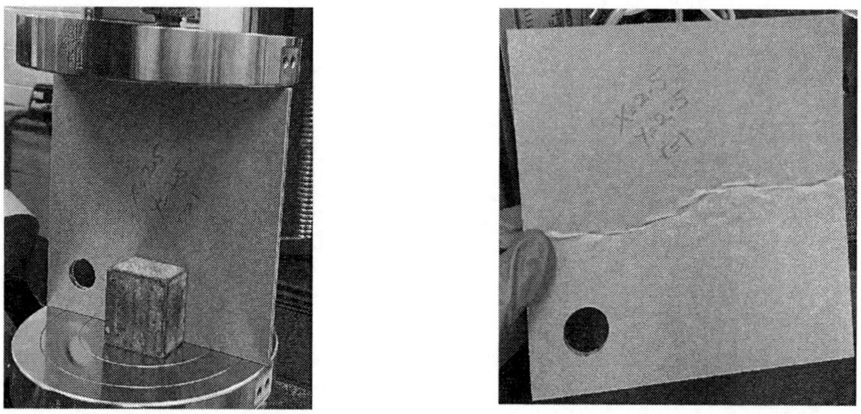

(a) (b)

Figure 2 Experimental test specimen (a) before starting the test, (b) after performing the test

Figure 3 shows the compression behavior of the plates for the holes located on the plate center at (7.5,7.5) with the four radiuses of 0.5, 1, 1.5, and 2 cm. The compression maximum value for each case is collected as the VCP compression strength. The compression strength is reduced from 476.74 kPa to 348.81 kPa when the hole size is changed from 0.5 cm to 2 cm. In other words, the compression loss is higher for the plates with larger hole radiuses.

The test is repeated four times for a case, each time on a new sample. The mean of four maximum compression stress data for each case is collected and stored for postprocessing by the ML models. The plate is compressed symmetrically from the top edge. Therefore, studying a vertical half of the plate is sufficient; the results are copied symmetrically for the other half.

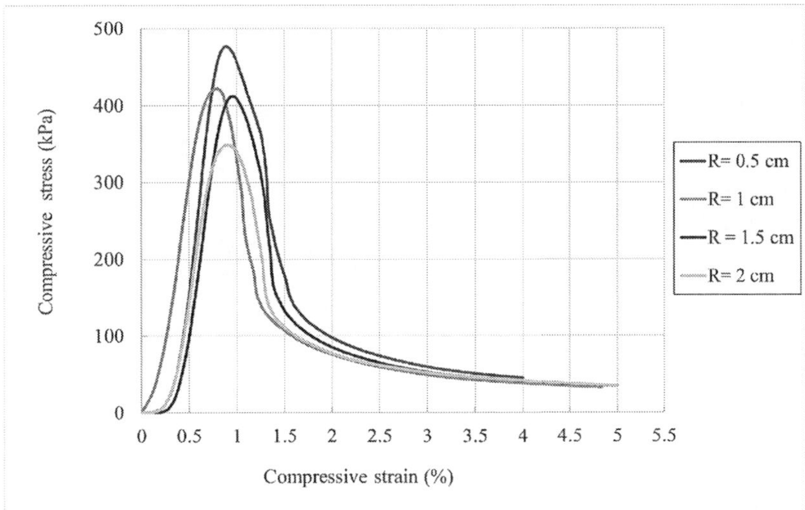

Figure 3 Compression of four samples with different hole radius sizes located on the center of plate

2.2 Postprocessing using ML

In the second step, the sample data is used to design the input data frame for the ML models. The data set is split into two groups, the training and testing sets. The ML models in this study include multiple linear regression, support vector regression, decision tree regression, and random forest regression. Figure 4 illustrates the workflow of the proposed ML method in this study. Figure 5 presents the compression strength (kPa) as the heatmap of a sample data for different hole locations when the radius is equal to (a) 0.5 cm, (b) 1 cm, (c) 1.5 cm, and (d) 2 cm. Only for the 0.5 cm radius, designing the hole in the center results in higher compression strength. For the other hole sizes, the top and lower locations provide higher strengths. One of the advantages of applying the ML approaches is finding a highly accurate prediction for such problems with uncertain trends.

In the last step, the ML models are trained by the sample data (70% of data) to enable them to predict the compression strength for the unseen cases (30% of data). Then the prediction from the ML model is compared with the testing set. The Python programming language version 3.9.7 is used to implement the ML models. The next section provides a summary of the ML models applied in this study.

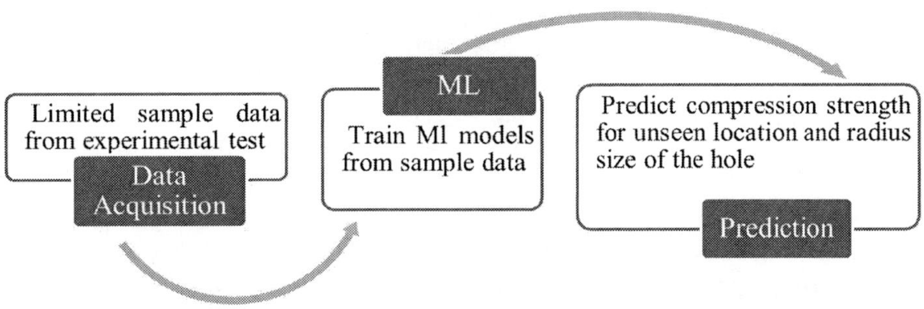

Figure 4 Schematic of proposed ML workflow in this paper

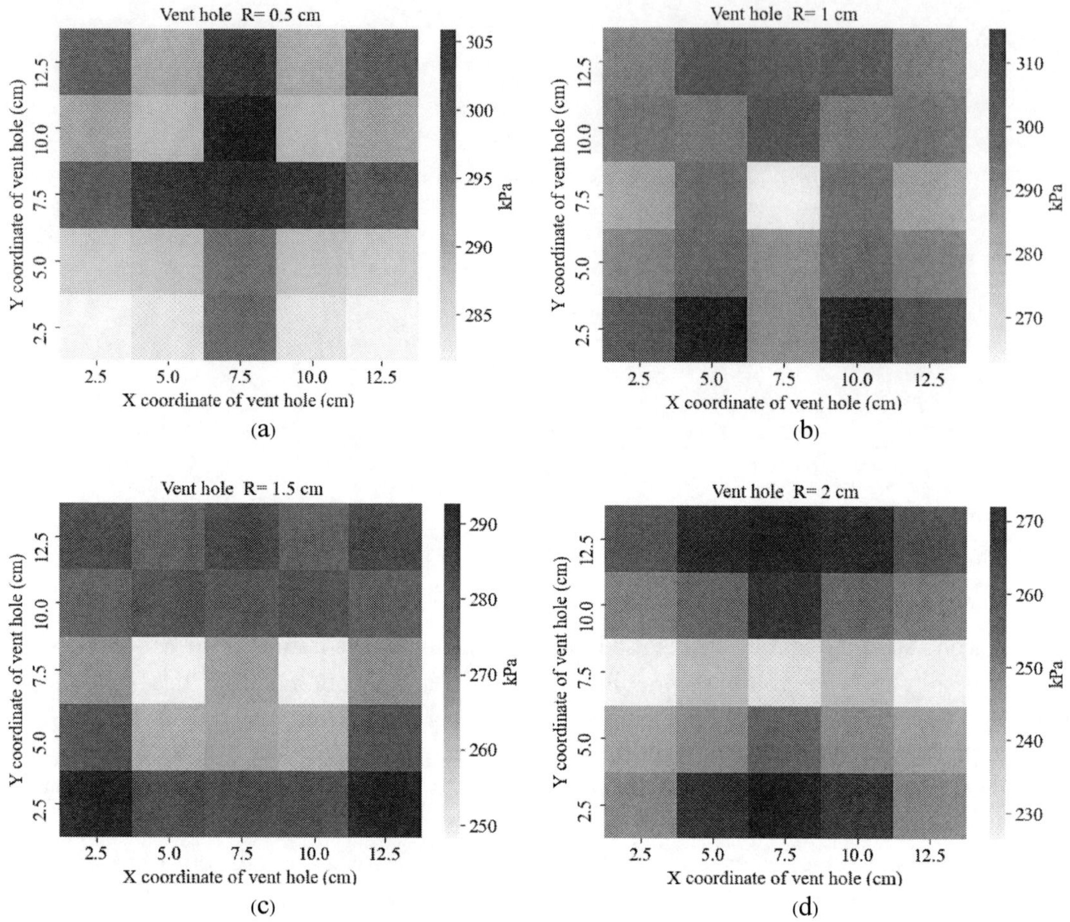

Figure 5 Heatmap of compression data for the hole sizes (a) R=0.5 cm, (b) R=1 cm, (c) R=1.5 cm, and (d) R=2 cm

2.2.1 *Multiple Linear Regression (MLR)*

Prediction of the compression strength value is a type of regression problem. The regression method is used to analyze the relationship between one or more independent variables and a dependent variable. The MLR is a regression model with more than one regressor variable. The MLR model with □ regressor is defined as follows [15].

$$□_□ = □_0 + □_1□_1 + \cdots + □_□□_□ + □, \quad (□ = 1,2,\ldots,□) \qquad (1)$$

Where $□_□$ denotes the dependent variable, $□_□$ is the independent variable, and □ is the model error term. $□_0$ is the constant term and $□_□$ is the coefficient for each independent variable. In this study, the compression strength is selected as the dependent variable. The location of the vent hole (i.e., x and y) and the radius of the vent hole are selected as the independent variables. The MLR model is trained on the data set including independent and dependent values. The trained MLR can predict the dependent variable for a new set of independent variables.

2.2.2 *Support Vector Regression (SVR)*

The SVR method is an established method for predicting linear and non-linear regression problems. SVR is applicable for the small amount of sample data. The SVR basics are developed by Vapnik and Chervonenkis and follow the same methodology as a support vector machine in classification methods [16]. In a support vector machine classification algorithm, each labeled sample data in the training set is considered a data point in a multidimensional feature space. In this feature space, a hyperplane classifies the training sample data. The SVR model develops an ε-insensitive loss function to compute a hyperplane in which there is a ε deviation of the predicted values from actual values in the training sample data [17].

2.2.3 *Decision Tree Regression (DTR)*

DTR is a variation of a decision tree classifier that is appropriate for a type of regression problem. The principle of DTR is based on an iterative process in which the training sample data divide into a binary partitioning to structure the regression tree. The method splits the training data set into two parts and chooses the one that minimizes the total of the squared deviations from the mean in the two parts. The DTR method employs a divide-conquer greedy algorithm to split the data [18].

2.2.4 *Random Forest Regression (RFR)*

An ensemble of decision trees can be described as an RFR. In an RFR algorithm, various decision trees are averaged to build a more robust (i.e., generalization) model. Generality can handle more data variations and is less vulnerable to overfitting. The randomness in an RFR reduces the variance of the model, in which it outperforms an individual decision tree in terms of generalization [19].

3. RESULTS AND DISCUSSION

In this section, the results of the ML models are presented, and the applicability of the ML approaches is validated. The results of the ML prediction models are shown in figures 6a-d. Various statistical measures have been applied to evaluate the results of ML prediction when compared to the actual data. In this study, the coefficient of determination (R^2), mean squared error (MSE), root mean squared error (RMSE), mean absolute error (MAE), and mean absolute percentage error (MAPE) measures were calculated. All the statistical evaluation results of the ML models are shown in Table 1.

The DTR and RFR models have demonstrated high accuracy in the prediction of the compression strength and showed a significant correlation with the actual data. Regarding the accuracy of the results, the SVR and MLR models are in third and fourth place. The R^2 value for the MLR, SVR, DTR, and RFR models are calculated as 0.7465, 0.7973, 0.8956, and 0.9126, respectively. The DTR and RFR models reported the least MSE value among all four ML models. The RFR model with R^2= 0.9126, MSE= 42.0771, RMSE= 6.4860, MAE= 5.0522, and MAPE= 0.0176 obtained the most accurate compression predictions. The DTR demonstrated close results to the RFR model. The MLR model showed a fairly accurate predictions and received the lowest accuracy rank among the four models.

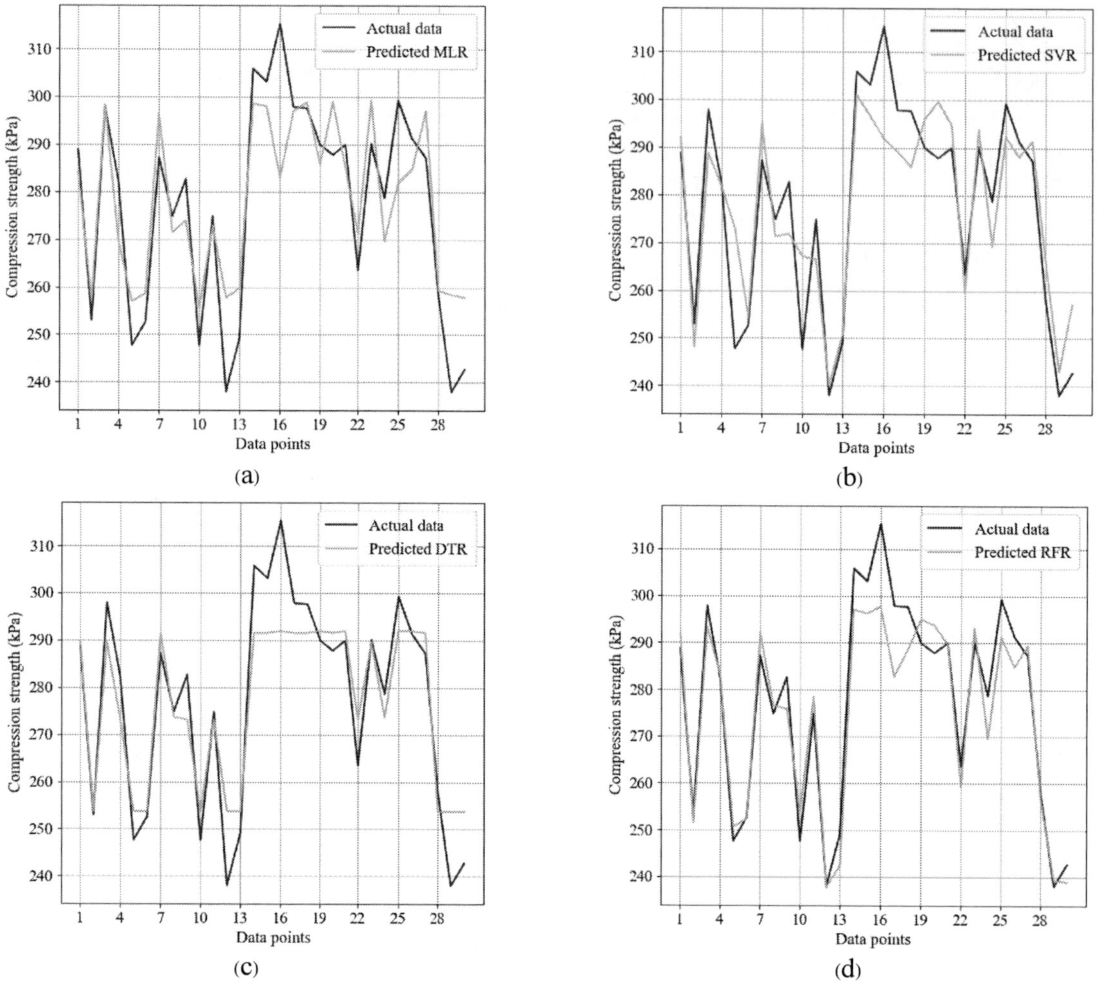

Figure 6 Compression strength predictions by (a) MLR, (b) SVR, (c) DTR, and (d) RFR models

Table 1 A Summary of the results from machine learning models

Method	R^2	MSE	RMSE	MAE	MAPE [%]
MLR	0.7465	122.0707	11.0485	8.7335	0.0320
SVR	0.7973	97.5930	9.8789	7.8509	0.0285
DTR	0.8956	50.2555	7.0891	5.4147	0.0196
RFR	0.9126	42.0771	6.486	5.0522	0.0176

4. CONCLUSIONS

The ML approach for predicting the compression strength of the VCPs is an alternative solution to the current time and cost-consuming experimental tests and designs. In this paper, we used four ML models to predict the compression strength of the VCP packages. A limited sample data set from the experimental tests are collected to train the ML algorithms. The location and size of the vent hole as two parameters that affect the compression strength of the VCPs are considered as the input data for the ML models. Five statistical evaluation measures have been applied to validate the results of the ML models. All ML models presented reasonably accurate results in predicting the compression strength of the VCPs. The statistical measurements indicated that the RFR approach provided remarkably accurate predictions for the compression strength of the VCPs.

5. REFERENCES

[1] P.B. Pathare, U.L. Opara, Structural design of corrugated boxes for horticultural produce: A review, Biosyst. Eng. 125 (2014). https://doi.org/10.1016/j.biosystemseng.2014.06.021.

[2] T. Fadiji, C.J. Coetzee, T.M. Berry, U.L. Opara, Investigating the role of geometrical configurations of ventilated fresh produce packaging to improve the mechanical strength – Experimental and numerical approaches, Food Packag. Shelf Life. 20 (2019). https://doi.org/10.1016/j.fpsl.2019.100312.

[3] M. Mukama, A. Ambaw, U.L. Opara, Advances in design and performance evaluation of fresh fruit ventilated distribution packaging: A review, Food Packag. Shelf Life. 24 (2020). https://doi.org/10.1016/j.fpsl.2020.100472.

[4] T.M. Berry, T.S. Fadiji, T. Defraeye, U.L. Opara, The role of horticultural carton vent hole design on cooling efficiency and compression strength: A multi-parameter approach, Postharvest Biol. Technol. 124 (2017) 62–74. https://doi.org/10.1016/j.postharvbio.2016.10.005.

[5] U.L. Opara, T. Fadiji, Compression damage susceptibility of apple fruit packed inside ventilated corrugated paperboard package, Sci. Hortic. (Amsterdam). 227 (2018) 154–161. https://doi.org/10.1016/j.scienta.2017.09.043.

[6] J. Singh, E. Olsen3, S.P. Singh, J. Manley2, F. Wallace, The Effect of Ventilation and Hand Holes on Loss of Compression Strength in Corrugated Boxes, 2008.

[7] T. Fadiji, C. Coetzee, U.L. Opara, Compression strength of ventilated corrugated paperboard packages: Numerical modelling, experimental validation and effects of vent geometric design, Biosyst. Eng. 151 (2016) 231–247. https://doi.org/10.1016/j.biosystemseng.2016.09.010.

[8] T. Fadiji, C.J. Coetzee, T.M. Berry, A. Ambaw, U.L. Opara, The efficacy of finite element analysis (FEA) as a design tool for food packaging: A review, Biosyst. Eng. 174 (2018) 20–40. https://doi.org/10.1016/j.biosystemseng.2018.06.015.

[9] S. Pal, K. Naskar, Machine learning model predict stress-strain plot for Marlow hyperelastic material design, Mater. Today Commun. 27 (2021) 102213. https://doi.org/10.1016/j.mtcomm.2021.102213.

[10] W. Ben Chaabene, M. Flah, M.L. Nehdi, Machine learning prediction of mechanical properties of concrete: Critical review, Constr. Build. Mater. 260 (2020) 119889. https://doi.org/10.1016/j.conbuildmat.2020.119889.

[11] J. Lepine, V. Rouillard, M. Sek, On the use of machine learning to detect shocks in road vehicle vibration signals, Packag. Technol. Sci. 30 (2017) 387–398.

[12] C. Zhao, M. Johnsson, M. He, Data mining with clustering algorithms to reduce packaging costs: A case study, Packag. Technol. Sci. 30 (2017) 173–193.

[13] Y. Sun, M. Zhang, B. Adhikari, S. Devahastin, H. Wang, Double-layer indicator films aided by BP-ANN-enabled freshness detection on packaged meat products, Food Packag. Shelf Life. 31 (2022) 100808. https://doi.org/10.1016/j.fpsl.2021.100808.

[14] N. da Silva e Silva, F. de Souza Farias, M.M. dos Santos Freitas, E.J.G. Pino Hernández, V.V. Dantas, M. Enê Chaves Oliveira, M.R.S.P. Joele, L. de Fátima Henriques Lourenço, Artificial intelligence application for classification and selection of fish gelatin packaging film produced with incorporation of palm oil and plant essential oils, Food Packag. Shelf Life. 27 (2021). https://doi.org/10.1016/j.fpsl.2020.100611.

[15] D.C. Montgomery, E.A. Peck, G.G. Vining, Introduction to linear regression analysis, John Wiley & Sons, 2021.

[16] A.J. Smola, B. Schölkopf, A tutorial on support vector regression, Stat. Comput. 14 (2004) 199–222.

[17] F. Zhang, L.J. O'Donnell, Support vector regression, Elsevier Inc., 2019. https://doi.org/10.1016/B978-0-12-815739-8.00007-9.

[18] W.Y. Loh, Classification and regression trees, Wiley Interdiscip. Rev. Data Min. Knowl. Discov. 1 (2011) 14–23. https://doi.org/10.1002/widm.8.

[19] S. Raschka, V. Mirjalili, Python machine learning: Machine learning and deep learning with Python, scikit-learn, and TensorFlow 2, Packt Publishing Ltd, 2019.

FG-O03

Poly (Butylene Adipate-co-Terephthalate)/Thermoplastic Starch Blended Phosphate via Blown Film Extrusion as Functional Biodegradable Food Packaging

Phanwipa Wongphan[1], Nathdanai Harnkarnsujarit[1,2]*

[1]*Department of Packaging and Materials Technology, Faculty of Agro-Industry, Kasetsart University, Bangkok 10900, Thailand*
[2]*Center for Advanced Studies for Agriculture and Food, Kasetsart University, Bangkok 10900, Thailand*

***Correspondence to:** Department of Packaging and Materials Technology, Faculty of Agro-Industry, Kasetsart University, Bangkok 10900, Thailand.*
E-mail; nathdanai.h@ku.ac.th

ABSTRACT: Thermoplastic starch (TPS) and polybutylene adipate-co-terephthalate (PBAT) are bioplastic materials that has potential to convert into biodegradable films which supports global policies on sustainability. Blending of these polymers improves processing efficiency in TPS production, while modifying flexibility of the films. Phosphate compounds have high numbers of reactive groups which possibly interacted and provide functionality to the bioplastic polymers. This research aimed to investigate effects of phosphate derivatives namely sodium hexametaphosphate ($Na_6((PO_3)_6)$) and sodium tripolyphosphate ($Na_5P_3O_{10}$) on mechanical, barrier and morphological properties of bioplastic TPS/PBAT blown films. Infrared spectra indicated hydrogen bonding and carbonyl group of ester bonds between polymer with $Na_5P_3O_{10}$ and ($Na_6((PO_3)_6)$) leading to modified properties of the films. TPS/PBAT films containing ($Na_6((PO_3)_6)$) had more homogenous microstructure as shown by scanning electron microscopy and slightly increased elongation of the films, indicating improved compatibility. Conversely, $Na_5P_3O_{10}$gave non-homogenous structure and dispersion of clumps within film matrices, giving decreased mechanical properties. Both phosphate compounds modified thermal stability and barrier properties. ($Na_6((PO_3)_6)$) enhanced compact structure due to decreased melting of starch granules giving higher barrier property than $Na_5P_3O_{10}$. In addition, films containing ($Na_6((PO_3)_6)$) and $Na_5P_3O_{10}$ effectively preserved quality of packaged processed meat during storage than control. Accordingly, both phosphate compounds provided extra functionality to biodegradable TPS/PBAT film, producing shelf-life extension films for sustainable food packaging.

Keyword: Phosphate, Thermoplastic starch, Poly (butylene adipate-co-terephthalate), Bioplastic, Food packaging

1. INTRODUCTION

Petroleum-based plastic packaging is a major cause for the environmental impact due to petroleum plastics are poorly biodegradable, taking more than hundred years to degrade. The environmental concern has driven the development of bio-based plastic packaging. Bio-based plastic packaging has been widely interested in food industry. Bio-based plastic is a major global policy for sustainable economy. The bio-circular-green (BCG) economic model has the potential to improve competition, increasing value and meet consumer demand for biodegradable and environmentally friendly plastics. The bio-based plastic packaging is used as flexible packaging. Polybutylene adipate terephthalate (PBAT) and thermoplastic starch are the most widely used biodegradable polymers. However, major drawback of bio-based plastic includes poor mechanical and barrier properties with low thermal resistance, giving disadvantage when compared with petroleum plastics. The bio-based packaging has been developed to improve the properties by incorporate additive compound, composite polymer and polymer blends [1-3].

Thermoplastic starch (TPS) is often applied in polymer blends because of its poor mechanical properties, moisture sensitivity, poor stability during processing and storage. Blending with more hydrophobic polymer improves moisture resistance and processing efficiency. TPS is mostly blended with PBAT because the polyester has more hydrophobicity, good mechanical and high processability [4-5]. However, incompatibility between polymers gives poor mechanical and barrier properties. Therefore, compatibility between polymers is necessary to improve the properties of final product [6-7] and enhance interfacial adhesion of polymer blend for homogenous structure, strong mechanical properties, high barrier and thermal resistance. Several compatibilizers have been used to improve the interfacial adhesion between PBAT and TPS such as tartaric acid, titanium dioxide, nisin, sodium nitrite, sodium benzoate and potassium sorbate [8-12]

Phosphate compounds have been used in food products as food preservative according to U.S. Food and Drug Administration (FDA) to improve the functionality, especially to increase the water holding capacity. Phosphate compounds have high numbers of reactive groups which possibly interacted and provide functionality to the bioplastic polymers. The effectiveness of phosphates on functional properties of meat products depends on the type of phosphate, the amount used, and the specific food products [13]. Quality and shelf life of foods are important parameters affecting consumer perception and safety. The addition of functional compounds into packaging produces "active packaging" which potentially improves quality and extends the shelf life of food products. The development of food packaging considered quality and safety as primary factors. Consequently, the development of bio-based food packaging needs to consider interaction between food and environment because packaging effect food quality and consumer safety.

This research aimed to produce functional biodegradable food packaging from bioplastic TPS/PBAT incorporated phosphate derivatives namely sodium hexametaphosphate ($Na_6((PO_3)_6)$) and sodium tripolyphosphate ($Na_5P_3O_{10}$) via blown films extrusion. The effects of sodium hexametaphosphate ($Na_6((PO_3)_6)$) and sodium tripolyphosphate ($Na_5P_3O_{10}$) on mechanical, barrier and morphological properties of bioplastic TPS/PBAT blown films were investigated.

2. MATERIALS AND METHODS

2.1 Preparation of poly (Butylene Adipate-co-Terephthalate)/thermoplastic starch films

Native cassava starch (Siam Modified Starch Co., Ltd., Pathumthani, Thailand) was dried in hot air oven overnight at 50±2 °C. The starch powder was mixed with 35% glycerol (%w/w starch) (Patum Vegetable Oil Co., Ltd, Phthumthani, Thailand) using a dough mixer (SC-236A, Stelang, China) for 10 min. Phosphate derivatives was mixed at 3% namely sodium tripolyphosphate ($Na_5P_3O_{10}$, Thai Food and Chemical Co., Ltd., Thailand) and sodium hexametaphosphate ($Na_6((PO_3)_6$), Thai Food and Chemical Co., Ltd., Thailand). The mixtures were transformed into TPS pellets using a twin-screw extruder (Labtech Engineering, Samut Prakan, Thailand). The barrel temperature range from the barrel to the die was 90-145 °C with screw speed 180 rpm. The extrudate TPS was cooled in air and then cut into 2.5 mm pellets.

The polybutylene adipate-co-terephthalate (PBAT, Ecoflex®F Blend C1200, BASF, Germany) and TPS and TPS containing functional compound pellets were compounded at weight ratios between PBAT/TPS of 50/50 (PBAT/TPS) and PBAT/TPS containing functional compound of 50/50/1, 50/50/3 and 50/50/5. The temperature of pellets compounds was 95-150 °C using a twin-screw extruder with a screw speed 220 rpm. The blend materials were cut into 2.5 mm pellets and kept in Ziploc bags.

The PBAT/TPS pellets were dried at 50 °C overnight using a hot air oven and were prepared PBAT/TPS films using a single-screw blown-film extruder (Labtech Engineering, Samut Prakan, Thailand) using temperature profile range from the hopper to the die of 140-155 °C. The screw and nip roll speeds were 28-50 and 2.2–2.5 rpm, respectively. The blown-films were stored in a Ziploc bag at room temperature.

2.2 Fourier transform infrared spectroscopy

The infrared absorption spectra of the film samples were determined using a Bruker Tensor 27 FT-IR Spectrometer (Bruker OPTIK GmbH, Germany) with attenuated total reflection (ATR) mode. IR absorbance spectra were recorded from 500 to 4,000 cm^{-1} at a resolution of 4 cm^{-1} (co-added 64 scans).

2.3 Morphology

Microstructures of the samples were determined using a scanning electron microscope (SEM, FEI Quanta 450, Thermo Fisher Scientific, Massachusetts, USA) with an accelerating voltage of 10 kV. Specimens were freeze-cracked and placed on a metal stub. Samples were coated with gold using a sputter coater (Quorum Technology Polaron Range SC7620, UK). The samples were determined for surface and cross-section structures with magnification of 500X and 3000X, respectively.

2.4 Mechanical properties

The films were determined for mechanical properties namely tensile strength and elongation at break following ASTM D882-88 using universal testing machine (Model 5965, Instron, USA). Distance between the handle and the speed was 5 cm and 500 mm/min, respectively. The films were cut into 2.5×10 cm^2 and determined for thickness using a digital micrometer. All film samples were kept at 50%RH for 2 days prior to measurement. The data were averaged from six specimens.

2.5 Water vapor permeability

Water vapor permeability (WVP) was determined using a cup method (ASTM E96-80). The films were cut into 7 cm diameter sample and placed on metal cup containing dried silica gels. The O-ring was placed on the film edge and sealed using hot paraffin wax. The samples were placed in a humidity chamber at 25°C and 50%RH and weighed daily until they recorded constant weight. Water vapor transmission rate (WVTR) was derived from the slope of the linear weight increase. WVP was calculated as follows:

$$WVP = \frac{WVTR \times thickness}{\Delta P} \quad (1)$$

where, ΔP is the difference in water vapor pressure between each side of the films.

2.6 Oxygen permeability

Oxygen transmission rate (OTR) was determined using an oxygen permeation analyzer (model 8500, Illinois Instrument, USA) following ASTM D3985-81. The oxygen permeability (OP) was calculated as follows:

$$OP = \frac{OTR \times thickness}{\Delta P} \quad (2)$$

where, ΔP is the difference in water vapor pressure between each side of the films.

2.7 Meat packaging

Ham steaks were purchased from a supermarket (Bangkok, Thailand) and used for experiments on the same day of purchase stored at 4 °C. Ham steaks were placed between PBAT/TPS films (150 mm × 150 mm) with and without phosphate compounds. The surface of ham was directly contacted with the films. Samples were stored 6 days at 4 °C in a glass door refrigerator. Ham samples were determined quality during storage using water activity (a_w) value. Samples were determined using a Testo 650 water activity meter (Testo Inc., USA)

2.8 Statistical analysis

One-way analysis of variance (ANOVA) was performed using SPSS version 22 (SPSS Inc., USA). Significant difference among samples were determined using Duncan's multiple range tests at a confidence interval of 95% ($P \leq 0.05$).

3. RESULTS AND DISCUSSION

3.1 Fourier transform infrared spectroscopy (FTIR)

Figure 1 shows the FTIR spectra of the film. The peak between 500-1500 cm^{-1} shows fingerprint regions of the polymer. Adding both phosphate compounds had an insignificant effect on PBAT/TPS film. However, the intensity of the peak of PBAT/TPS containing phosphate compounds at 1296 and 1712 cm^{-1} slightly increased attributed to C–O stretching vibration in ester bonds and C=O bond of carbonyl groups, respectively. The absorption peak between 3000 and 3700 cm^{-1} indicated the O-H stretching due to hydrogen bonds in the films. Both phosphate compounds modified intensity of H-bonding leading to modified properties of the films.

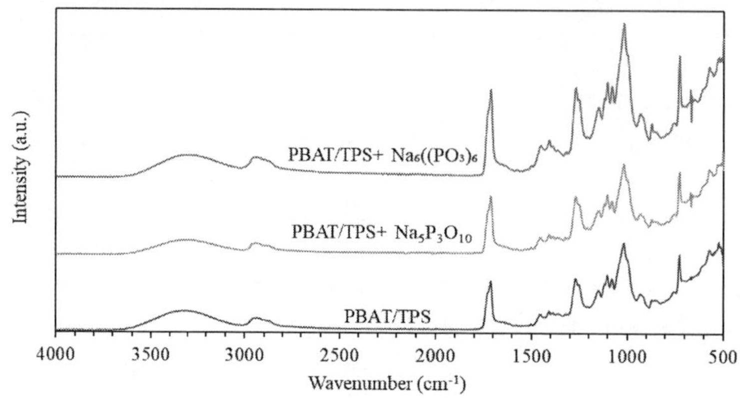

Figure 1 FTIR absorption of PBAT/TPS blend films containing sodium tripolyphosphate ($Na_5P_3O_{10}$) and sodium hexametaphosphate ($Na_6((PO_3)_6)$).

3.2 Morphology

Microstructure of surface and cross-section of the film are shown in Figure 2. PBAT/TPS film had rough surface due to incompatibility between polymer blend. Incorporated tripolyphosphate ($Na_5P_3O_{10}$) had rougher surface than control. Cross-section images shows fine pores that indicated incompatibility between polymers the film metrics. Conversely, hexametaphosphate ($Na_6((PO_3)_6)$) gave higher homogenous structure. The $Na_6((PO_3)_6)$ reduced the melting temperature leading to higher melting of starch granules and subsequently smoother region in cross section. The results indicated $Na_6((PO_3)_6)$ improved compatibility and possibly acted as compatibilizer.

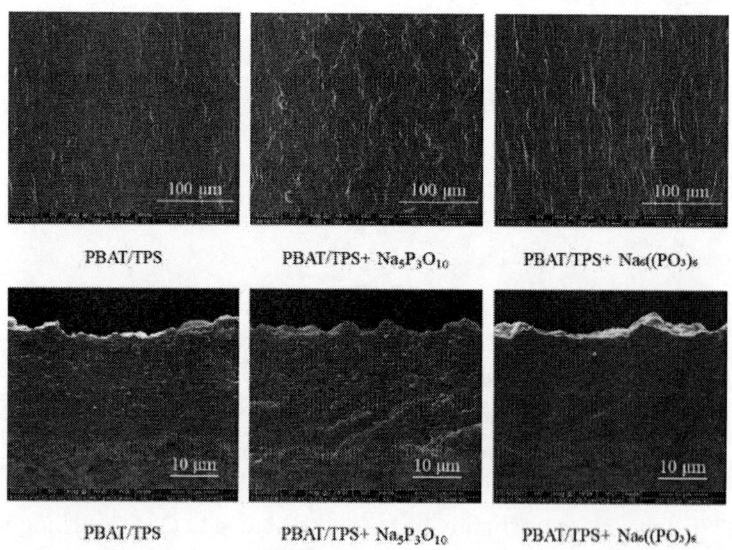

Figure 2 Microstructure of PBAT/TPS blend films containing sodium tripolyphosphate ($Na_5P_3O_{10}$) and sodium hexametaphosphate ($Na_6((PO_3)_6)$) in surface (above) and cross-section (below) structure.

3.3 Mechanical properties

Mechanical properties including tensile strength and elongation of the films are shown in Figure 3a and 3b, respectively. Films in the machine direction (MD) had higher mechanical properties than cross direction due to molecular orientation of the polymers. Phosphate reduced tensile strength and elongation due to non-homogenous structure. Moreover, the intensity of O-H stretching decreased, suggesting amount of H-bonding decreased leading to reduced mechanical properties. However, film containing hexametaphosphate ($Na_6((PO_3)_6)$) gave higher elongation than containing tripolyphosphate ($Na_5P_3O_{10}$) due smooth structure leading to improved compatibility and flexibility of the film (as suggested by SEM image). The results indicated $Na_6((PO_3)_6)$ improved flexible properties, possibly due to enhanced polymer compatibility.

3.4 Water vapor permeability

Water vapor permeability is an important factor to maintain quality for packaged food. Figure 4 shows water vapor permeability (WVP) of the films. Adding phosphate compounds reduced the WVP. The PBAT/TPS containing hexametaphosphate ($Na_6((PO_3)_6)$) had the lowest WVP value due to more homogenous and denser structure leading to reduced diffusivity. Results suggested that WVP of the films was influenced by microstructure and compatibility of the polymer blends had major effects on permeability.

3.5 Oxygen permeability

Adding phosphate compounds were adjusted the oxygen permeability (OP). Figure 4 shows the PBAT/TPS films incorporated tripolyphosphate ($Na_5P_3O_{10}$) had the highest OP value which was coincident with fine pores in cross-section SEM image that led to high diffusivity. Conversely, the PBAT/TPS incorporated hexametaphosphate ($Na_6((PO_3)_6)$) had the lowest OP because the homogenous structure of the film matrices limited oxygen diffusion, suggesting that the major impact of microstructure on oxygen permeability.

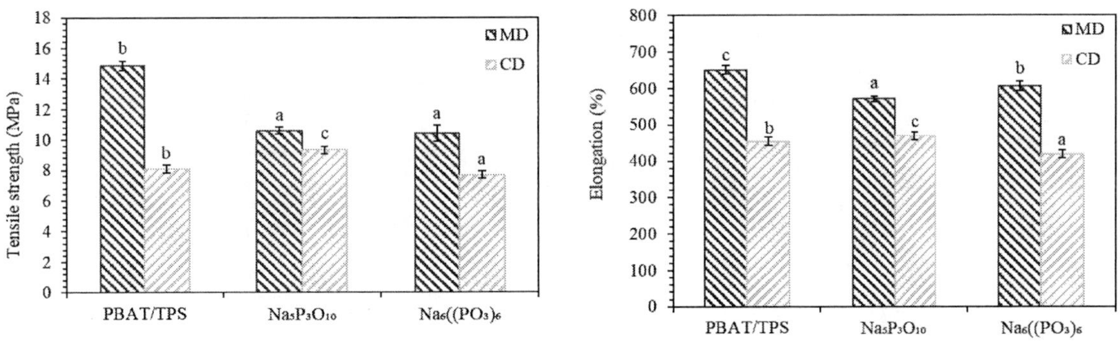

(a) (b)

Figure 3 Mechanical properties namely (a) tensile strength and (b) elongation of PBAT/TPS blend films containing sodium tripolyphosphate ($Na_5P_3O_{10}$) and sodium hexametaphosphate ($Na_6((PO_3)_6)$) in machine (MD) and cross (CD) direction

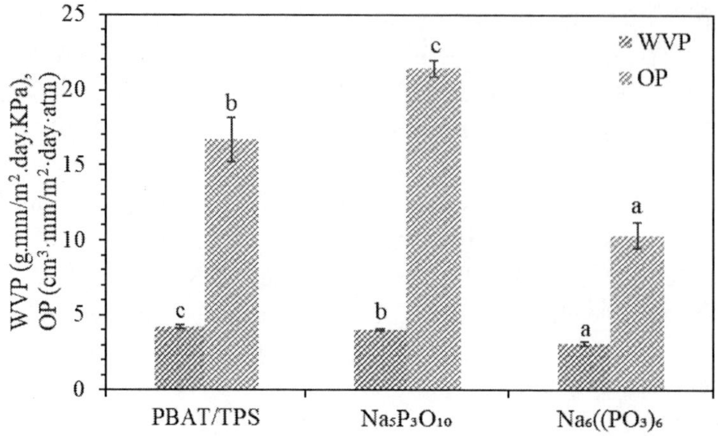

Figure 4 Barrier properties including water vapor permeability (WVP) and oxygen permeability (OP) of PBAT/TPS blend films containing sodium tripolyphosphate ($Na_5P_3O_{10}$) and sodium hexametaphosphate ($Na_6((PO_3)_6)$)

3.6 Meat packaging

Figure 5a shows the appearance of packaged ham. Water activity (a_w) of fresh ham was 0.8437 and slightly decreased during storage (Figure 5b). The water loss from the ham caused reduction of a_w. The PBAT/TPS with phosphate compounds effectively maintained quality of the ham during 3 days than control. Both phosphate compounds have potential to preserve the water holding capacity. $Na_6((PO_3)_6)$ effectively maintained water in the sample during 6 days like as fresh ham, while film containing $Na_5P_3O_{10}$ tended to slightly decrease water activity due to microstructure of the PBAT/TPS+ $Na_5P_3O_{10}$ film had fine pores and the highest WVP leading to high water diffusion and loss from food matrices.

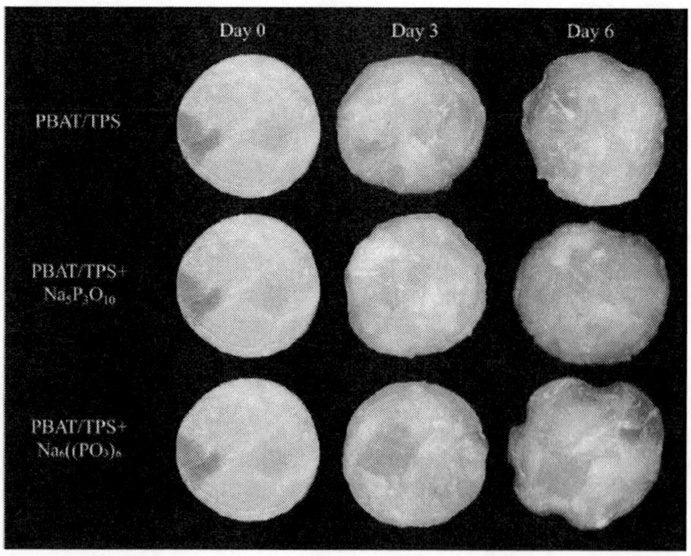

(a)

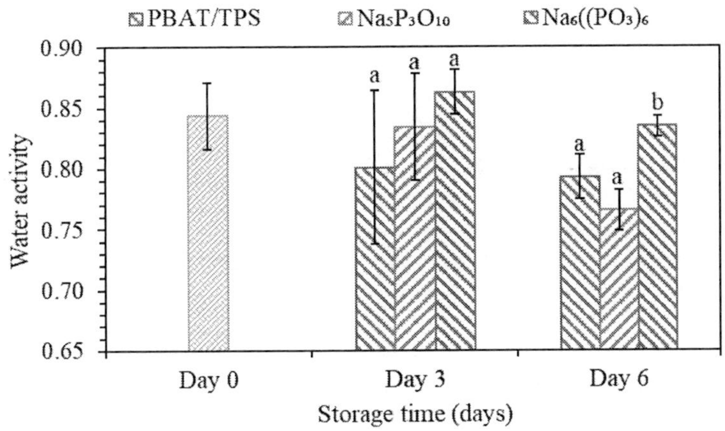

(b)

Figure 5 (a) Appearance and (b) Water activity of packed ham stored in PBAT/TPS blend films containing sodium tripolyphosphate ($Na_5P_3O_{10}$) and sodium hexametaphosphate ($Na_6((PO_3)_6)$)

4. CONCLUSIONS

Bioplastic blends between PBAT and TPS films containing sodium tripolyphosphate ($Na_5P_3O_{10}$) and sodium hexametaphosphate ($Na_6((PO_3)_6)$) significantly affected mechanical, barrier and morphological properties of bioplastic TPS/PBAT blown films. Both phosphate compounds modified intensity of H-bonding and microstructure leading to modified properties of the films. Microstructure of the film was the major factor affecting mechanical strength, water vapor permeability and oxygen permeability. However, $Na_6((PO_3)_6)$ improved flexibility, possibly due to enhanced compatibility between polymer blends. $Na_5P_3O_{10}$ gave non-homogenous structure and fine pores in the film metrics caused decreased mechanical strength and barrier properties of the film. Moreover, both phosphate compounds effectively preserved quality of packaged processed meat during storage than control. Accordingly, both phosphate compounds provided extra functionality to biodegradable TPS/PBAT film, producing shelf-life extension for sustainable food packaging.

5. ACKNOWLEDGEMENT

This research project is supported by National Research Council of Thailand (NRCT): (Contact No. N41A640190).

6. REFERENCES

[1] Harnkarnsujarit, N., Wongphan, P., Chatkitanan, T., Laorenza, Y., & Srisa, A. (2021). Bioplastic for sustainable food packaging. In Sustainable food processing and engineering challenges (pp. 203-277). Academic Press. DOI: 10.1016/B978-0-12-822714-5.00007-3.

[2] Jariyasakoolroj, P., Leelaphiwat, P., & Harnkarnsujarit, N. (2020). Advances in research and development of bioplastic for food packaging. Journal of the Science of Food and Agriculture, 100(14), pp. 5032-5045. DOI: 10.1002/jsfa.9497.

[3] Bumbudsanpharoke, N., Wongphan, P., Promhuad, K., Leelaphiwat, P., and Harnkarnsujarit, N. (2022). Morphology and permeability of bio-based poly (butylene adipate-co-terephthalate) (PBAT), poly (butylene succinate) (PBS) and linear low-density polyethylene (LLDPE) blend films control shelf-life of packaged bread. Food Control, 132, 108541, DOI: 10.1016/j.foodcont.2021.108541.

[4] Garcia, P. S., Baron, A. M., Yamashita, F., Mali, S., Eiras, D., & Grossmann, M. V. E. (2018). Compatibilization of starch/poly (butylene adipate□co□terephthalate) blown films using itaconic acid and sodium hypophosphite. Journal of Applied Polymer Science, 135(33), 46629. DOI: 10.1002/app.46629

[5] Olivato, J. B., Grossmann, M. V. E., Bilck, A. P., & Yamashita, F. (2012). Effect of organic acids as additives on the performance of thermoplastic starch/polyester blown films. Carbohydrate Polymers, 90(1), pp. 159-164. DOI: 10.1016/j.carbpol.2012.05.009

[6] Wadaugsorn, K., Panrong, T., Wongphan, P., & Harnkarnsujarit, N. (2022). Plasticized hydroxypropyl cassava starch blended PBAT for improved clarity blown films: Morphology and properties. Industrial Crops and Products, 176, 114311. DOI: 10.1016/j.indcrop.2021.114311.

[7] Wongphan, P., Panrong, T., & Harnkarnsujarit, N. (2022). Effect of different modified starches on physical, morphological, thermomechanical, barrier and biodegradation properties of cassava starch and polybutylene adipate terephthalate blend film. Food Packaging and Shelf Life, 32, 100844. DOI: 10.1016/j.fpsl.2022.100844.

[8] Olivato, J. B., Nobrega, M. M., Müller, C. M. O., Shirai, M. A., Yamashita, F., & Grossmann, M. V. E. (2013). Mixture design applied for the study of the tartaric acid effect on starch/polyester films. Carbohydrate Polymers, 92(2), pp. 1705-1710. DOI: 10.1016/j.carbpol.2012.11.024.

[9] Leelaphiwat, P., Pechprankan, C., Siripho, P., Bumbudsanpharoke, N., & Harnkarnsujarit, N. (2022). Effects of nisin and EDTA on morphology and properties of thermoplastic starch and PBAT biodegradable films for meat packaging. Food Chemistry, 369, 130956. DOI: 10.1016/j.foodchem.2021.130956.

[10] Katekhong, W., Wongphan, P., Klinmalai, P., & Harnkarnsujarit, N. (2022). Thermoplastic starch blown films functionalized by plasticized nitrite blended with PBAT for superior oxygen barrier and active biodegradable meat packaging. Food Chemistry, 374, 131709. DOI: 10.1016/j.foodchem.2021.131709.

[11] Phothisarattana, D., Wongphan, P., Promhuad, K., Promsorn, J., & Harnkarnsujarit, N. (2022). Blown film extrusion of PBAT/TPS/ZnO nanocomposites for shelf-life extension of meat packaging. Colloids and Surfaces B: Biointerfaces, 112472. DOI: 10.1016/j.colsurfb.2022.112472.

[12] Wangprasertkul, J., Siriwattanapong, R., & Harnkarnsujarit, N. (2021). Antifungal packaging of sorbate and benzoate incorporated biodegradable films for fresh noodles. Food Control, 123, 107763. DOI: 10.1016/j.foodcont.2020.107763.

[13] Masniyom, P., Benjakul, S., & Visessanguan, W. (2005). Combination effect of phosphate and modified atmosphere on quality and shelf-life extension of refrigerated seabass slices. LWT-Food Science and Technology, 38(7), pp. 745-756. DOI: 10.1016/j.lwt.2004.09.006.

FG-O04

Enhanced Preservation Functionality of Bioplastic Materials for Sustainable Food Packaging

Nathdanai Harnkarnsujarit*, Pattarin Leelaphiwat, Nattinee Bumbudsanpharoke, Phanwipa Wongphan

Department of Packaging and Materials Technology, Faculty of Agro-Industry, Kasetsart University, Bangkok, 10900, Thailand

***Correspondence to:** Department of Packaging and Material, Kasetsart University, 50 Ngam Wong Wan Rd, Latyao, Chatuchak, Bangkok, 10900, Thailand. nathdanai.h@ku.ac.th*

ABSTRACT: Bioplastics are promising materials which are potentially developed into sustainable food packaging. Increasing numbers of global production for bioplastic materials suggests higher demand in several industries. Packaging for food products requires high barrier properties and/or protective function to delay spoilage and deterioration which led to food loss. Bioplastic packaging, however, has poor barrier properties as compared with petroleum-based plastic which typically limits their application for short shelf-life and ready-to-eat foods. Enhanced functionality of the bioplastic, particularly the protective performance with shelf-life extension function possibly compensates for poor barrier and facilitates utilization as food packaging. The production of bioplastic incorporating active compounds as active packaging via extrusion will be demonstrated in this study. Flower-based essential oil active ingredient i.e., Ylang-Ylang was incorporated into bioplastic materials including polybutylene adipate-co-terephthalate (PBAT) and polybutylene succinate (PBS) blends. Ylang-Ylang essential oil (EO) improved compatibility of the PBAT/PBS films that enhanced smooth surface and reduced phase separation. EO acted as compatibilizer and plasticizer leading to modify mechanical properties. Barrier properties depended on surface hydrophobicity of the polymers. Moreover, PBAT/PBS films incorporated Ylang-Ylang essential oil effectively delayed microbial growth in Thai starch-based desserts which were Pandanus pudding and Thai layer dessert, respectively. Accordingly, compounding of active compounds with polymer via extrusion effectively enhanced functionality of the bioplastic as environmentally friendly food packaging which can reduce food loss for sustainability.

Keyword: Food Packaging; Bioplastic; Active Packaging; Shelf-life; Essential oil

1. INTRODUCTION

Functional packaging is a part of active packaging materials which enhances the the package performance to maintain or adjust the quality and safety of the products [1]. The development of functional packaging for foods depends on deterioration of the products during storage. Thai dessert products mostly consist of flour, coconut milk and sugar with intermediate moisture contents. The components of Thai dessert are the main resource for the microbial growth. Moreover, relatively high-water activity supports growth of microorganisms, leading to changes of quality and safety e.g., flavor, color and texture during storage that affects consumer perception and led to food loss [2].

Bioplastics are promising materials which are potentially developed into sustainable food packaging. Increasing numbers of global production for bioplastic materials suggests higher demand in several industries. Packaging for food products requires high barrier properties and/or protective function to delay spoilage and deterioration which led to food loss. Bioplastic packaging, however, has poor barrier properties as compared with petroleum-based plastic which limits their applications as short shelf-life and ready-to-eat foods [3]. Enhanced functionality of the bioplastic, particularly the protective performance with shelf-life extension function possibly compensates for poor barrier and facilitates utilization as food packaging. Several active ingredients either synthetic and natural products have been incorporated into bioplastic materials including thermoplastic starch (TPS), polylactic acid (PLA) polybutylene succinate (PBS) and polybutylene adipate-co-terephthalate (PBAT) blends. Small molecular weight substances e.g., sodium nitrite and essential oils showed plasticization effects to the polymers. Incorporations of active compounds modified morphology of the films which subsequently affected the packaging properties [4,5].

Here, functionality of bioplastic materials was developed by blending PBAT and PBS at different concentration of natural essential oil namely Ylang-Ylang essential oil using single screw cast extrusion. This study aimed to investigate effects of Ylang-Ylang essential oil on morphology, mechanical and barrier properties and antimicrobial capacity against microbial growth in the packaged Thai dessert. Findings can be used to reduce plastic packaging waste and food loss for sustainability.

2. MATERIALS AND METHODS

2.1 Blend film preparation

Polybutylene adipate-co-terephthalate (PBAT, Ecoflex®F Blend C1200, BASF, Germany) and polybutylene succinate (PBS, BioPBS™ FZ91PM, PTT MCC Biochem Company Limited, Thailand) pellets were dried at 50 °C overnight using a hot air oven. Blend pellet at the PBAT/PBS ratio is 70/30 containing Ylang-Ylang essential oil (Thai - China Flavours and Fragrances Industry Co., Ltd., Thai;and) at difference concentration of 0, 3, 4.5 and 6% (w/w) using a single screw cast-extruder (CF-W400, Chareon TUT, Thailand). The temperature ranges from hopper to die was 145-175 °C. The screw and nip roll speeds were 30 and 3.1±0.3 rpm, respectively. Samples were stored in a Ziploc bag at room temperature.

2.2 Microstructure

The samples were immersed in liquid nitrogen and freeze-cracked. Samples were placed on the stub and coated with gold using a sputter coater (SCD040, Balzers Union Ltd., Germany). Microstructures of surface and cross-sectioned structures were determined using FEI Quanta 450 scanning electron microscope (SEM, Thermo Fisher Scientific, USA) at an accelerating voltage of 10 kV and a magnification of 250X (surface and cross-section)

2.3 Mechanical properties

Mechanical properties were characterized using an Instron machine (Model 5965, Instron, USA) at a speed of 500 mm/min and distance between the handle was 5 cm following ASTM D882-88. Films were cut into 15 cm × 2.5 cm and stored at 50 % RH for 2 days before measurement.

2.4 Barrier properties

2.4.1 Contact angle (CA)

Hydrophilic-hydrophobicity of the films surface was determined using a contact angle (OCA 15 EC, Data Physics Instruments GmbH, Germany). Dropped distilled water (3 μL) on the surface using a microsyringe. The images were immediately taken for contact angle using SCA 20 software (Dataphysics Instruments GmbH, Germany).

2.4.2 Water vapor permeability (WVP)

WVP was determined using the cup method following ASTME96-80. The samples (7 cm diameter) were placed on aluminum cups containing dried silica gel and sealed O-ring with paraffin wax. The cups were placed in humidity chamber at 25 °C and 50% RH and weighed until constant weight. Water vapor trans-mission rate (WVTR) was determined from the linear slope between weight and time. Water vapor permeability (WVP) was estimated according to the following (Eq. 1).

$$WVP = \frac{WVTR \times thickness}{Pressure\ difference} \qquad \text{(Eq. 1)}$$

2.4.3 Oxygen permeability (OP)

Oxygen transmission rate (OTR) was determined using oxygen permeation tester (Model 8000, Illinois, USA) following ASTM D3985. The samples were cut into 14 cm diameter and OP was calculated from Eq. (2).

$$OP = \frac{OTTR \times thickness}{Pressure\ difference} \qquad \text{(Eq. 2)}$$

2.5 Packaged Thai desserts

Thai desserts without food preservatives namely Thai layer cake and Pandanus pudding were purchased from local market (Bangkok, Thailand). Desserts were placed between PBAT/TPS films (150 mm × 150 mm) with and without Ylang-Ylang essential oil. The samples were and determined quality by appearance during storage at 4 °C.

2.6 Statistical analysis

Statistical analysis was used to determine statistical difference among samples using Duncan's multiple range test was used for data analysis at 95% confidence intervals by SPSS22 (SPSS Inc., USA).

3. RESULTS AND DISCUSSION

3.1 Microstructure

Figure 1 shows appearance of the PBAT/PBS blend films at different Ylang-Ylang essential oil (EO) concentrations (0, 3, 4.5 and 6%) during single screw cast-extrusion. Addition of EO improved clarity of the films. Film clarity depended on degree of EO concentrations. Microstructure of the PBAT/PBS blend films with and without EO are shown in Figure 2. The PBAT/PBS film shows the roughest and layered surface due to incompatible between polymer blends. PBAT has more hydrophilic than PBS leading to phase separation of the polymer blends. Increased EO concentration improved compatibility of the PBAT/PBS films that enhanced smooth surface and reduced phase separation. Results indicated that EO acted as compatibilizer and enhanced adhesion between polymers giving smoother surface and more homogenous structure.

3.2 Mechanical properties

Mechanical properties were determined using stress-strain curve. Mechanical properties are shown in Figure 3 and Table 1. Tensile strength and elongation were taken from the maximum tensile strength and tensile strain at break, respectively. Mechanical properties in machine direction (MD) have higher mechanical properties than cross direction (CD) due to stretching and arrangement of molecules during extrusion process. Adding EO increased mechanical strength. Elongation of the films was enhanced with EO due to plasticization effect leading to improved flexibility of the PBAT/PBS blends film. Result shows mechanical properties has correlation with microstructure and plasticization effect.

PBAT/PBS　　PBAT/PBS+ 3% EO　　PBAT/PBS+ 4.5% EO　　PBAT/PBS+ 6% EO

Figure 1 Appearance of the films during single screw cast-extrusion of the PBAT/PBS blend films at different Ylang-Ylang essential oil (EO) concentrations (0, 3, 4.5 and 6%).

Figure 2 Microstructures of the PBAT/PBS blend films at different Ylang-Ylang essential oil (EO) concentrations (0, 3, 4.5 and 6%) in surface (top) and cross-section (below).

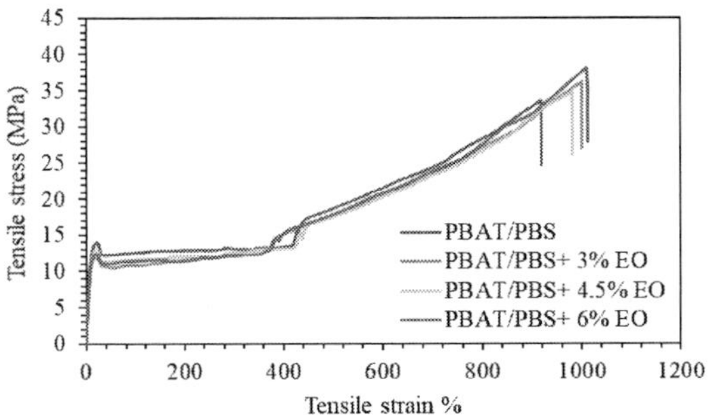

(a)

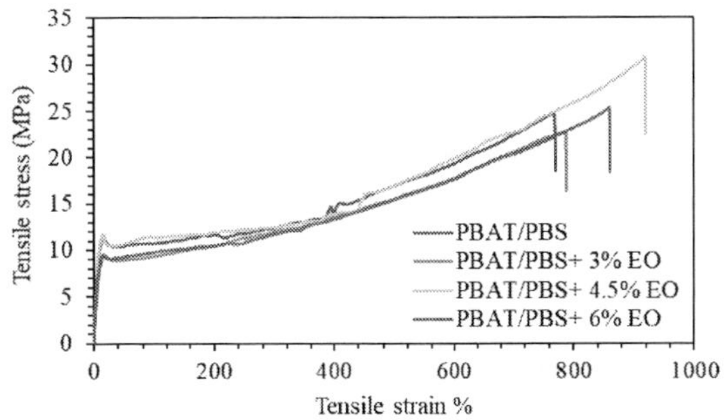

(b)

Figure 3 Stress-strain curve of the PBAT/PBS blend films at different Ylang-Ylang essential oil (EO) concentrations (0, 3, 4.5 and 6%) in (a) machine (MD) and (b) cross (CD) directions.

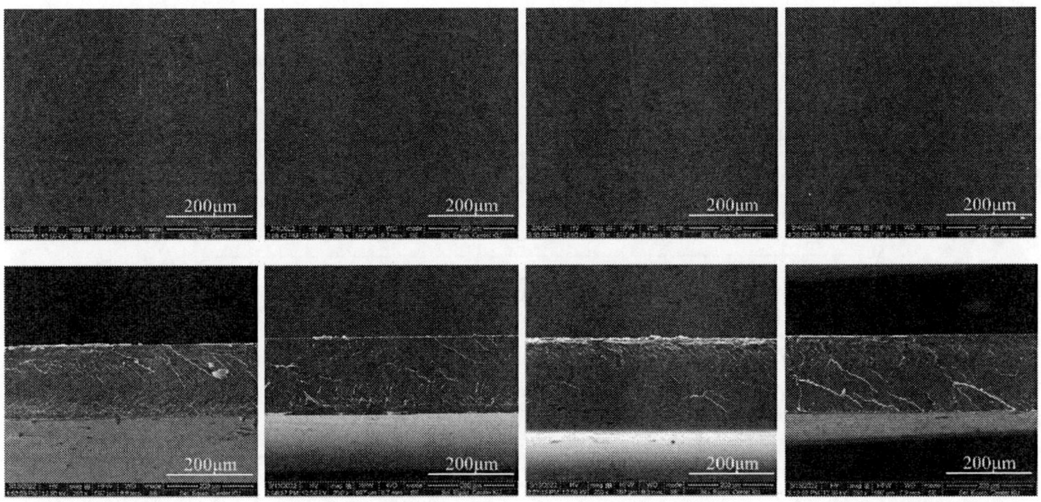

PBAT/PBS PBAT/PBS+ 3% EO PBAT/PBS+ 4.5% EO PBAT/PBS+ 6% EO

Table 1 Mechanical properties of the PBAT/PBS blend films at different Ylang-Ylang essential oil (EO) concentrations (0, 3, 4.5 and 6%) namely tensile strength and elongation in machine (MD) and cross (CD) directions.

Treatments	Tensile strength (MPa)		Elongation (%)	
	MD	CD	MD	CD
PBAT/PBS	37.15±1.79^{ab}	25.59±2.24^{b}	980.78±31.82^{a}	801.48±61.13^{a}
PBAT/PBS+ 3% EO	36.36±3.38^{a}	22.62±0.52^{a}	991.06±55.98^{a}	780.49±18.31^{a}
PBAT/PBS+ 4.5% EO	35.45±1.43^{a}	29.43±1.86^{c}	995.65±28.67^{a}	915.52±39.20^{c}
PBAT/PBS+ 6% EO	38.57±1.53^{b}	25.89±1.46^{b}	1004.24±20.41^{a}	860.73±30.88^{b}

3.3 Barrier properties

3.3.1 Contact angle

Surface hydrophobicity of the film was determined using contact angle (Figure 4). PBAT/PBS with 3% EO film had the highest contact angle value which indicated that addition of EO improved hydrophobicity of the film. However, Film containing 4.5% EO significantly decreased hydrophobicity due to microstructure had smoother surfaces than PBAT/PBS blend films. Reduced roughness of the hydrophobic surface typically led to decreased CA values. Results suggested that the hydrophobicity of EO itself and the surface roughness had impact on wettability of the films.

3.3.2 Water vapor permeability

Addition of Ylang-Ylang essential oil insignificantly affected water vapor permeability (WVP) as shown in Figure 4. SEM images showed that all films with no EO and different levels of incorporated EO had homogenous and dense structure which gave similar water vapor permeation.

3.3.3 Oxygen permeability

Figure 4 shows that adding Ylang-Ylang essential oil significantly increased oxygen permeability (OP). Ylang-Ylang essential oil enhanced hydrophobicity of the film. Oxygen readily dissolved in non-polar matrices, lead to higher oxygen diffusion. Though, the microstructures had similar dense matrices, hydrophobicity of the matrices gave significant impact on oxygen permeation.

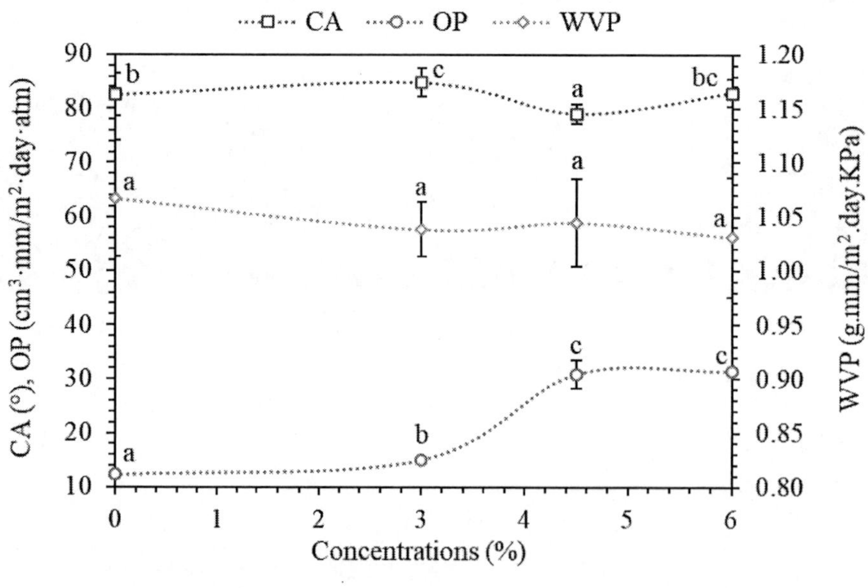

Figure 4 Barrier properties of the PBAT/PBS blend films at different Ylang-Ylang essential oil (EO) concentrations (0, 3, 4.5 and 6%) namely contact angle (CA), water vapor permeability (WVP) and oxygen permeability (OP).

3.4 Packaged Thai desserts

Figure 5 shows the appearance of packaged Thai desserts namely Pandanus pudding and Thai layer dessert during storage at 4 °C. The dessert packaged in PBAT/PBS films (control) showed microbial growth after 18 and 10 days for Pandanus pudding and Thai layer dessert, respectively. Conversely, the PBAT/PBS films incorporated Ylang-Ylang essential oil effectively delayed microbial growth and extended shelf life of Thai desserts for 5 and 4 days in Pandanus pudding and Thai layer dessert, respectively. Results indicated films with Ylang-Ylang essential oil had potential for extended shelf life of dessert products.

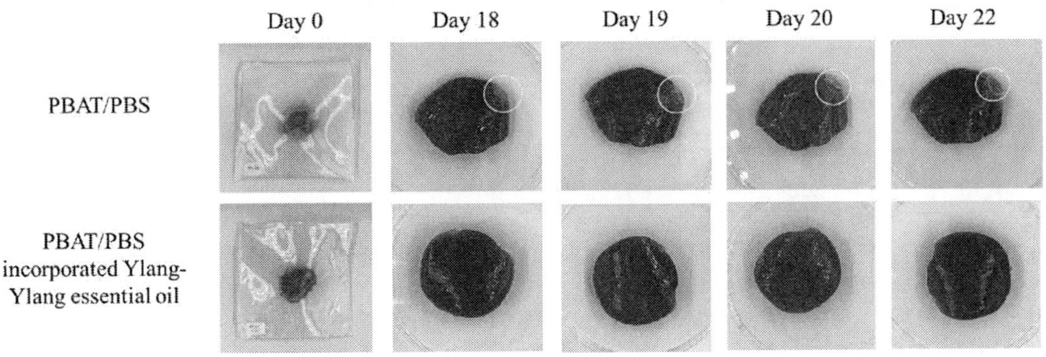

(a)

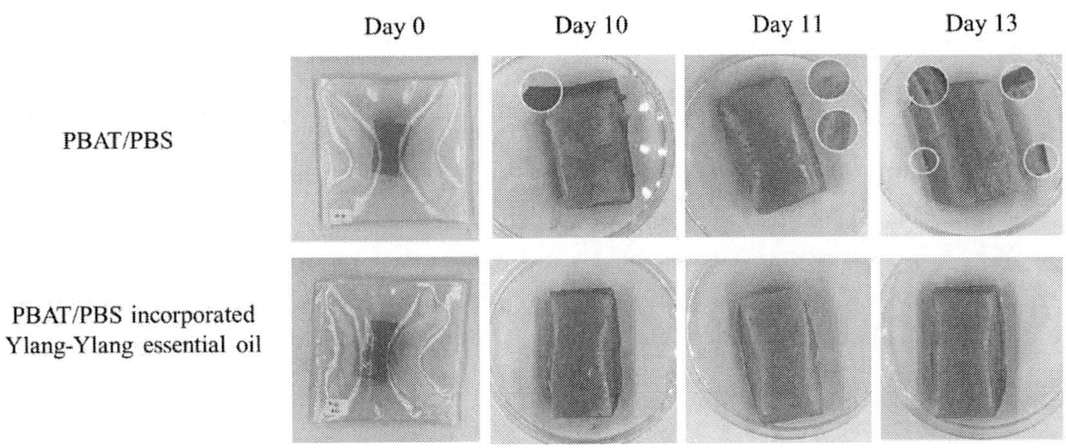

(b)

Figure 5 Appearance of packaged (a) Pandanus pudding and (b) Thai layer dessert during storage at 4 °C

4. CONCLUSIONS

Incorporating Ylang-Ylang essential oil into polybutylene adipate-co-terephthalate (PBAT) and polybutylene succinate (PBS) blends film modified microstructure, mechanical and barrier properties. EO acted as compatibilizer and enhanced adhesion between polymers giving smoother surface and homogenous structure, leading to modify mechanical properties. However, barrier properties depended on surface hydrophobicity and microstructure. In addition, Ylang-Ylang essential oil had potential for extended shelf life of dessert products for 5 and 4 days in Pandanus pudding and Thai layer dessert, respectively. Films with Ylang-Ylang essential oil potentially serves as functional packaging with shelf-life extension and compatible with aroma of Thai dessert products.

5. ACKNOWLEDGEMENT

This research was financially supported by Kasetsart University Research and Development Institute (KURDI) under grant no. FF(KU)25.64.

6. REFERENCES

[1] Katekhong, W., Wongphan, P., Klinmalai, P., and Harnkarnsujarit, N. (2022). Thermoplastic starch blown films functionalized by plasticized nitrite blended with PBAT for superior oxygen barrier and active biodegradable meat packaging. Food Chemistry, 374, 131709, DOI: 10.1016/j.foodchem.2021.131709.

[2] Nopwinyuwong, A., Trevanich, S., and Suppakul, P. (2010). Development of a novel colorimetric indicator label for monitoring freshness of intermediate-moisture dessert spoilage. Talanta, 81(3), pp. 1126-1132, DOI: 10.1016/j.talanta.2010.02.008.

[3] Jariyasakoolroj, P., Leelaphiwat, P., & Harnkarnsujarit, N. (2020). Advances in research and development of bioplastic for food packaging. Journal of the Science of Food and Agriculture, 100(14), pp. 5032-5045, DOI: 10.1002/jsfa.9497.

[4] Bumbudsanpharoke, N., Wongphan, P., Promhuad, K., Leelaphiwat, P., and Harnkarnsujarit, N. (2022). Morphology and permeability of bio-based poly (butylene adipate-co-terephthalate) (PBAT), poly (butylene succinate) (PBS) and linear low-density polyethylene (LLDPE) blend films control shelf-life of packaged bread. Food Control, 132, 108541, DOI: 10.1016/j.foodcont.2021.108541.

[5] Celebi, H., and Gunes, E. (2018). Combined effect of a plasticizer and carvacrol and thymol on the mechanical, thermal, morphological properties of poly (lactic acid). Journal of Applied Polymer Science, 135(8), 45895, DOI: 10.1002/app.45895.

FG-O05

Roles of Titanium Dioxide, Zinc Oxide and Silicon Dioxide on Morphology, Properties and Migration of Thermoplastic Cassava Starch and Poly (Butylene Adipate-co-Terephthalate) Nanocomposite Films

Danaya Phothisarattana[1], Nathdanai Harnkarnsujarit[1,2]*

[1]*Department of Packaging and Materials Technology, Faculty of Agro-Industry, Kasetsart University, Bangkok 10900, Thailand*
[2]*Center for Advanced Studies for Agriculture and Food, Kasetsart University, Bangkok 10900, Thailand*

***Correspondence to:** Department of Packaging and Materials Technology, Faculty of Agro-Industry, Kasetsart University, 50 Ngam Wong Wan Rd., Latyao, Chatuchak, Bangkok 10900, Thailand. E-mail: nathdanai.h@ku.ac.th*

ABSTRACT: Metal oxide are widely used as nanofillers to improve mechanical, thermal and barrier properties in plastics due to several functions as reinforcement, non-toxic and UV-shielding properties. This study aim to investigate effects of different metal oxide nanoparticle namely titanium oxide (TiO_2), zinc oxide (ZnO) and silicon dioxide (SiO_2) on properties of poly(butylene adipate-co-terephthalate) (PBAT) and thermoplastic cassava starch (TPS) blended films produced by blown film extrusion. Microstructures were determined using scanning electron microscope (SEM) and atomic force microscopy (AFM) which indicated micro pore due to incompatibility between polymers. Diverse interaction between nanoparticles and polymers differently modified microstructures and subsequently controlled film properties. Films containing TiO_2 and SiO_2 showed increasing tensile strength due to good dispersion of nanoparticle in polymer matrices, while agglomeration of ZnO gave opposite result. Water vapor permeability (WVP) of the film containing TiO_2 decreased due to good dispersion of TiO_2 leading to tortuous pathway for diffusion of water molecule. Conversely, incompatibility between polymer blends, agglomeration of ZnO and high porosity of SiO_2 containing films increased WVP as similarly found for oxygen permeability of films containing TiO_2 and ZnO. Specific migration of metal releasing from the films into simulants (distilled water, 10% ethanol and 3% acetic acid) were strongly dependent on the simulant and metal types, which had different interaction with polymer matrices. ZnO showed the largest migration in acid simulant due to high dissolution of metal which further enhanced film disintegration. This finding demonstrates different characteristics of the nanofillers on morphology, properties and migration behavior of bioplastic film to produce sustainable packaging via extrusion.

Keyword: Titanium dioxide, Zinc oxide, Silicon dioxide, Biodegradable

1. INTRODUCTION

In present, bio-plastics market trends were increased years to years due to plastic pollution and environmental concerns in both of bio-based and biodegradable, for example poly(butylene adipate-co-terephthalate) (PBAT), thermoplastic starch (TPS) and poly lactic acid (PLA). TPS, bio-based and biodegradable polymer which produced by starch, have an advantage on oxygen barrier, good mechanical properties, but sensitive to water [1,2]. PBAT, aliphatic-aromatic co-polyester polymer, have an advantage on water resistance, ductility, flexibility and biodegradable [2,3,4].

Metal oxides are widely used as fillers with superior reinforcement properties, biocompatibility and non-toxic [2,3]. Incorporating metal oxide into polymer or bio-plastics gives an advantage on mechanical, barrier, thermal and antimicrobial properties, however the solution casting process might not suitable in industrial scale and produces high amount of solvent waste. Instead, extrusion process is more suitable for the commercialization of bioplastic packaging. More investigations on incorporating metal oxide into bio-plastics via extrusion are required. [6,7,8].

This research investigated the effects of metal oxide nanoparticle namely titanium oxide (TiO_2), zinc oxide (ZnO) and silicon dioxide (SiO_2) on morphology, properties and specific migration of PBAT and TPS blended films produced using extrusion process.

2. MATERIALS AND METHODS

2.1. Films preparation

Acetylated cassava starch with degree of substitution 0.01-0.03 (SMS Corp., Pathum Thani, Thailand) and Polybutylene adipate-co-terephthalate (PBAT, Ecoflex®F Blend C1200, polymats co. ltd, Bangkok, Thailand) were dried at 50 °C overnight in hot air oven before using. Starch, glycerol (Sac Sci-Eng Ltd., Bangkok, Thailand) and metal oxide (TiO_2, ZnO and SiO_2) (Prime Nanotechnology Co., Ltd.,Bangkok, Thailand) were mixed at weight ratio 100:35:0 and 100:35:2.5 respectively using dough mixer (SC-236A, Stelang Electric Appliance Co., Ltd., Foshan, China) for 10 min. The compounds were processed using a twin-screw extruder (Labtech Engineering, Samut Prakan, Thailand) at 85 to 150 °C and 180 rpm screw speed. The extrudates were cut into 2.5 cm pellets using a pelletizer (Labtech Engineering, Samut Prakan, Thailand) and manually blended with PBAT at weight ratio 40:60 for the extrudates (TPS-metal oxide) and PBAT respectively before compounding using twin-screw extruder and cut into 2.5 cm pellets.

PBAT/TPS containing metal oxide films were blown using blow film extrusion with single-screw extruder (Labtech Engineering, Samut Prakan, Thailand). Temperature range was 150 to 165 °C using 20 to 27 rpm for screw speed and 2 to 3 rpm for nip roll speed to form a stable vertical tubular bubble. The films were collected in an aluminum bag until use.

2.2. Scanning electron microscopy (SEM)

Films (1 cm × 10 cm) were dipped in nitrogen liquid before broken using forceps. Films were fixed with stub using carbon tape and coated with gold using sputter coater (Quorum Technology Polaron Range SC7620, UK). Microstructures of cross-section area of films were determined using Scanning electron microscope (SEM, FEI Quanta 450, Thermo Fisher Scientific, USA) at 3000x magnification and 10 kV.

2.3. Atomic force microscopy (AFM)

Topographic images (25 μm2) of films were determined using atomic force microscopy (AFM, MFP-3D-Bio, Asylum Research, USA) at 225 μm-length cantilever, 8 nm nominal tip radius curvature and 0.8 Hz scan rate using tapping mode.

2.4. Mechanical properties

Tensile strength (TS), Elongation at break (EB) and Young's modulus (YM) of 2.5 cm × 15 cm films in machine direction were determined using instron universal testing machine (Model 5965, Intron, Norwood, MA, USA) in tension mode. Films were cut and measured the thickness before placed in humidity chamber at 50% RH at 25 °C for 48 h. films were tested following ASTM D882-12. Films were placed at 5 cm distance using 500 mm/min crosshead speed. Samples were tested at ten replications.

2.5. Barrier properties

Water vapor transmission rate (WVTR) of four replicate 7 cm diameter circles films was determined using standard cup method (ASTM E96). Films were conditioned in humidity chamber at 50% RH at 25 °C for 48 h before placed in aluminum cup containing silica gel, covered with O-ring and sealed with paraffin. Samples were placed in humidity chamber at 50% RH at 25 °C. WVTR was calculated from the linear slope between weight gain and storage time and converted to water vapor permeability (WVP) using Eq. (1)

$$\text{WVP} = \frac{(\text{WVTR} \times \text{thickness})}{\text{difference in vapor pressure between two sides of the film}} \qquad (1)$$

Oxygen permeability (OP) of duplicate 13 cm diameter films were determined using an 8500 Model oxygen permeability analyzer (Illinois Instruments, Inc., Johnsburg IL, USA) following ASTM D3985-051. Films were placed in humidity chamber at 50% RH at 25 °C for 48 h before testing. Testing condition was 23 ± 2 °C and 50% RH. OTR was conversed OP following Eq. (2).

$$\text{OP} = \frac{(\text{OTR} \times \text{thickness})}{\text{difference in oxygen partial pressure between two sides of the film}} \qquad (2)$$

2.6. Dynamic mechanical thermal analysis (DMTA)

Thermomechanical properties of the films were determined using using a dynamic mechanical thermal analyzer (DMTA, Mettler Toledo, Greifensee, Switzerland). Films were tested in tension mode with temperatures range from -100 to 100 °C using 2 °C/min scanning rate at 0.5 Hz frequencies.

2.7. Specific migration

Films were cut and immersed in different simulant (distilled water, 10 %v/v ethanol, 3 %w/v acetic acid) in ratio 1dm^2/50 ml following Commission Regulation (EU) No 10/2011 for 10 days at 5 ± 2 ˚C before removing films from simulants. Metal in solutions (titanium, zinc and silicon for film containing TiO_2, ZnO and SiO_2, respectively) were tested using an inductively coupled plasma - optical emission spectrometry (ICP-OES, Agilent 5100 SVDV ICP-OES, Agilent Technologies, Inc., Santa Clara, CA, USA) at 5 s read time, 1.20 kW RF power and 20 s stabilization time using 8 mm viewing height with 0.8 L/min nebulizer flow, 12 L/min plasma flow and 1 L/min aux flow. Specific metal migration was determined for two replications and reported as mg/dm^2.

2.8. Statistical analysis

Samples were determined for significant differences using analysis of variance (ANOVA) by Duncan's multiple range test with $\alpha = 0.05$ using SPSS 17.0 (SPSS Inc., USA).

3. RESULTS AND DISCUSSION

3.1. Microstructure

Cross-section area of TPS/PBAT (control) and TPS/PBAT containing metal oxide (TiO_2, ZnO and SiO_2) blend films are shown in Figure 1A. In control film, the cross-section image presented micro-pore and dispersion of incomplete melting granule. Adding metal oxide decreased numbers of incomplete melting granule, but increased number of micro-pores. These micro-pores indicated incompatibility between polymer matrices which was due to different polarity between TPS (hydrophilic) and PBAT (hydrophobic) [9,10,11,12,13].

Topographic images indicating roughness of films are shown in Figure 1B. Lighter area indicated higher area, while darker color indicated deep area on films surface. Adding TiO_2 and ZnO increased height of the peak and deep of the valley because of micro-pore, attributing to incompatibility between matrices and agglomeration. Films containing SiO_2 showed smoother surface due to better dispersing of SiO_2 in polymer matrices [14]. These micro-pore, incompatibility and agglomeration of metal oxides in the matrices had major effect to films properties.

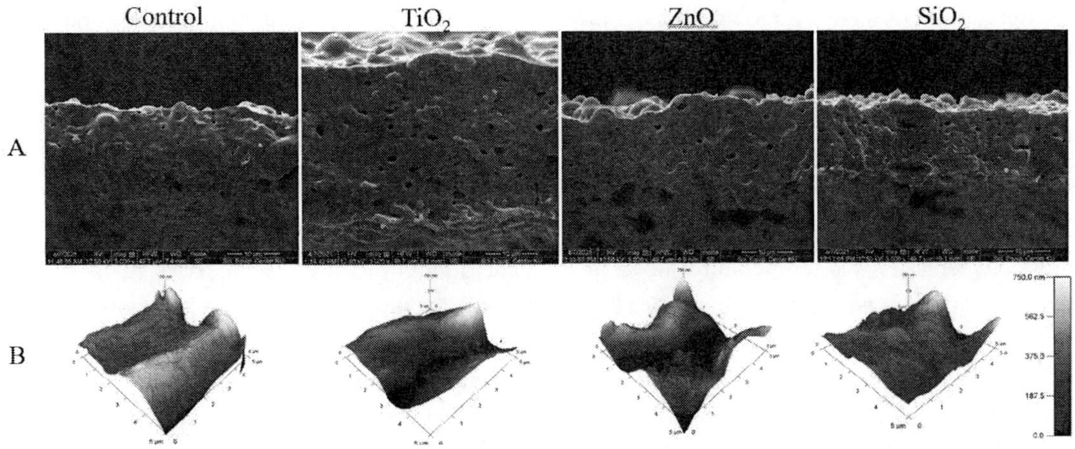

Figure 1 Microstructure of TPS/PBAT (control) and TPS/PBAT containing metal oxide (TiO_2, ZnO and SiO_2) blend films (A) cross-section image (B) Topographic image

3.2. Mechanical properties

Mechanical properties of the films are shown in Table 1. Adding TiO_2 and SiO_2 significantly increased TS due to good dispersing of metal oxide nanoparticle in polymer matrices. Favor interaction between polymer matrices and particle caused a good stress transfer and preventing breakage. Conversely, agglomeration of ZnO in matrices led to poor interaction between ZnO and the matrices. Limited interaction with polymer matrices caused phase separation and defect which decreased TS [4,7,8].

EB of the films containing TiO_2 and ZnO were insignificantly affected by metal oxide. Films containing SiO_2 had decreased EB due to anti-plasticizer effects of SiO_2 [9,15,16]. Adding metal oxide (TiO_2, ZnO and SiO_2) significantly increased YM leading to stiffening and hardening on films. However, no significant difference was found between types of metal oxide [4,10].

Table 1 Mechanical and barrier properties of TPS/PBAT (control) and TPS/PBAT containing metal oxide films

Films	TS (MPa)	EB (%)	YM (MPa)	WVP ($g.mm/m^2.h.atm$)	OP ($ml.mm/m^2.day$)
Control	9.78 ± 0.72^b	516.13 ± 35.04^b	30.73 ± 5.06^a	11.18 ± 0.25^a	26.39 ± 1.23^a
TiO_2	11.17 ± 1.25^c	567.64 ± 55.18^b	57.51 ± 9.30^b	10.53 ± 0.52^a	30.69 ± 1.21^{bc}
ZnO	7.72 ± 0.60^a	514.16 ± 58.68^b	51.29 ± 11.14^b	15.23 ± 0.88^b	32.84 ± 0.75^c
SiO_2	12.83 ± 2.00^d	336.66 ± 74.17^a	48.48 ± 15.27^b	16.52 ± 0.51^c	28.68 ± 1.27^{ab}

$^{a-d}$ indicate significant difference between samples at $\alpha = 0.05$.

3.3. Barrier properties

WVP of the films are shown in Table 1. Films containing TiO_2 and control had insignificant different WVP. Films containing ZnO had higher WVP due to agglomeration of nanoparticles. SiO_2 gave large number of fine micro-pore in the film contributing to the highest WVP value. Similarly, the effect of micro-pore in film matrices and agglomeration of particles also affected OP of the films [3,4,6].

3.4. Dynamic mechanical thermal analysis

Thermomechanical properties of the films were determined using DMTA. Peak of the plot between tan δ and temperature shows phase transition temperature i.e., relaxation, involving glass transition in amorphous phase [17,18,19]. Relaxation temperature of films are shown in Table 2. The main relaxation of films occurred due to PBAT phase which normally about -30 °C [20]. Adding metal oxide (TiO_2, ZnO and SiO_2) increased relaxation temperature of composite films due to interaction between matrices and polymer matrices preventing mobility of the macro molecule polymeric chains [7,8].

Table 2 Relaxation temperature of TPS/PBAT (control) and TPS/PBAT containing metal oxide (TiO_2, ZnO and SiO_2) blend films

Films	Relaxation temperature (°C)
Control	-26.35
TiO_2	-24.25
ZnO	-25.17
SiO_2	-25.67

3.5. Specific migration

Specific migration (SM) of metal (titanium, zinc and silicon) in simulants are shown in Table 3. Films containing TiO_2 and SiO_2 showed the highest metal releasing into distilled water. This was mainly due to high solubility of TPS polymer into the simulants [21]. Water solubility of TPS caused swelling, deformation of the films and releasing of metals into simulant. Moreover, hydrogen bond between water and oxygen atom of metal oxide (TiO_2, ZnO and SiO_2) or hydroxyl group of starch which could replace existing hydrogen bond of metal oxide and starch which increased the migration [21,22].

Films containing ZnO presented the highest migration especially into acid simulants. Ionization accelerated ZnO dissolution in acid media. Conversely, TiO_2 and SiO_2 were more inactive and presented lower migration. SiO_2 was an inert material, while TiO_2 can slightly dissolve in acid and organic compound [21,22,23].

Table 3 Specific migration of titanium, zinc and silicon from film containing TiO_2, ZnO and SiO_2 into different simulants.

Films	Specific migration ($\times 10^{-2}$ mg/dm^2)		
	Distilled water	**10% Ethanol**	**3% Acetic acid**
TiO_2	1.26 ± 0.23^b	0.66 ± 0.36^a	0.60 ± 0.14^a
ZnO	15.28 ± 0.31^c	14.54 ± 1.10^b	38.75 ± 12.01^b
SiO_2	0.66 ± 0.17^a	nd	nd

$^{a\text{-}d}$ indicate significant difference between samples at $\alpha = 0.05$.

nd indicate non-detected

4. CONCLUSIONS

Incorporating metal oxide (TiO_2, ZnO and SiO_2) in TPS/PBAT blended films via extrusion process enhanced melting of the polymer granule and increased film roughness. The dispersing of the nanoparticle is the major factor which affected the morphology and properties of films. Good interaction between metal oxide and polymer matrices enhanced dispersion, leading to good stress transfer and limited mobility of the polymer chains. Agglomeration of the particle decreased mechanical properties and created defect which increased permeability of the films. Specific migration of the films was affected by solubility of polymer which caused releasing of particle into simulant. Types of metal oxide and reaction with simulant also had major effects on degree of migration. Accordingly, the different nanofillers gave a diverse morphology which subsequently impacted properties and migration behavior of bioplastic films produced via extrusion.

5. ACKNOWLEDGEMENT

This study was technical support by Phanwipa Wongphan, Khwanchat Promhuad and Juthathip Promsorn. This research was funded by Kasetsart University through the Graduate School Fellowship Program. This research was partially funded by Kasetsart University Research and Development Institute, KURDI grant no. FF(KU)1.65.

6. REFERENCES

[1] Abdullah, A. H. D., Putri, O. D., Fikriyyah, A. K., Nissa, R. C., Hidayat, S., Septiyanto, R. F., Karina, M. and Satoto, R. (2020). Harnessing the excellent mechanical, barrier and antimicrobial properties of zinc Oxide (ZnO) to improve the performance of starch-based bioplastic. Polymer-Plastics Technology and Materials, 59(12), pp. 1259-1267, DOI:10.1080/25740881.2020.1738466.

[2] Amin, M. R., Chowdhury, M. A. and Kowser, M. A. (2019). Characterization and performance analysis of composite bioplastics synthesized using titanium dioxide nanoparticles with corn starch. Heliyon, 5(8), pp. e02009, DOI:10.1016/j.heliyon.2019.e02009.

[3] Venkatesan, R. and Rajeswari, N. (2016). ZnO/PBAT nanocomposite films: Investigation on the mechanical and biological activity for food packaging. Polymers for Advanced Technologies, 28(1), pp. 20-27. DOI:10.1002/pat.3847.

[4] Venkatesan, R. and Rajeswari, N. (2017). TiO_2 nanoparticles/poly(butylene adipate-co-terephthalate) bionanocomposite films for packaging applications. Polymers for Advanced Technologies, 28(12), pp. 1699-1706. DOI:10.1002/pat.4042.

[5] Shankar, S. and Rhim, J.-W. (2019). Effect of types of zinc oxide nanoparticles on structural, mechanical and antibacterial properties of poly(lactide)/poly(butylene adipate-co-terephthalate) composite films. Food Packaging and Shelf Life, 21. pp. 100327, DOI:10.1016/j.fpsl.2019.100327.

[6] Al-Tayyar, N. A., A. M. Youssef and R. R. Al-Hindi (2020). Antimicrobial packaging efficiency of ZnO-SiO_2 nanocomposites infused into PVA/CS film for enhancing the shelf life of food products. Food Packaging and Shelf Life, 25. DOI:10.1016/j.fpsl.2020.100523.

[7] Cao, C., Wang, Y., Zheng, S., Zhang, J., Li, W., Li, B., Guo, R. and Yu, J. (2020). Poly (butylene adipate-co-terephthalate)/titanium dioxide/silver composite biofilms for food packaging application. Lwt, 132, 109874. DOI:10.1016/j.lwt.2020.109874.

[8] Hassannia-Kolaee, M., Khodaiyan, F., Pourahmad, R. and Shahabi-Ghahfarrokhi, I. (2016). Development of ecofriendly bionanocomposite: Whey protein isolate/pullulan films with nano-SiO_2. International Journal of Biological Macromolecules, 86, pp. 139-144. DOI:10.1016/j.ijbiomac.2016.01.032.

[9] Oleyaei, S. A., Zahedi, Y., Ghanbarzadeh, B. and Moayedi, A. A. (2016). Modification of physicochemical and thermal properties of starch films by incorporation of TiO_2 nanoparticles. Int J Biol Macromol, 89, pp. 256-264, DOI:10.1016/j.ijbiomac.2016.04.078.

[10] Ostafińska, A., Mikešová, J., Krejčíková, S., Nevoralová, M., Šturcová, A., Zhigunov, A., Michálková, D. and Šlouf, M. (2017). Thermoplastic starch composites with TiO_2 particles: Preparation, morphology, rheology and mechanical properties. International Journal of Biological Macromolecules, 101, pp. 273-282, DOI:10.1016/j.ijbiomac.2017.03.104.

[11] Wongphan, P., Panrong, T. and Harnkarnsujarit, N. (2022). Effect of different modified starches on physical, morphological, thermomechanical, barrier and biodegradation properties of cassava starch and polybutylene adipate terephthalate blend film. Food Packaging and Shelf Life, 32, pp. 100844, DOI: 10.1016/j.fpsl.2022.100844.

[12] Katekhong, W., Wongphan, P., Klinmalai, P. and Harnkarnsujarit, N. (2022). Thermoplastic starch blown films functionalized by plasticized nitrite blended with PBAT for superior oxygen barrier and active biodegradable meat packaging. Food Chemistry, 374, pp. 131709, DOI: 10.1016/j.foodchem.2021.131709.

[13] Wadaugsorn, K., Panrong, T., Wongphan, P. and Harnkarnsujarit, N. (2022). Plasticized hydroxypropyl cassava starch blended PBAT for improved clarity blown films: Morphology and properties. Industrial Crops and Products, 176, pp. 114311, DOI: 10.1016/j.indcrop.2021.114311.

[14] Perera, K. Y., Jaiswal, S. and Jaiswal, A. K. (2021). A review on nanomaterials and nanohybrids based bio-nanocomposites for food packaging. Food Chemistry, 376, pp. 131912, DOI: 10.1016/j.foodchem.2021.131912.

[15] Ezati, P., Riahi, Z., & Rhim, J. W. (2022). CMC-based functional film incorporated with copper-doped TiO_2 to prevent banana browning. Food Hydrocolloids 122, 107104, DOI: 10.1016/j.foodhyd.2021.107104.

[16] Dong, X., Liang, X., Zhou, Y., Bao, K., Sameen, D. E., Ahmed, S., Dai, J., Qin, W. and Liu, Y. (2021). Preparation of polylactic acid/TiO_2/GO nano-fibrous films and their preservation effect on green peppers. Int J Biol Macromol, 177, pp. 135-148, DOI: 10.1016/j.ijbiomac.2021.02.125.

[17] Wongphan, P. and Harnkarnsujarit, N. (2020). Characterization of starch, agar and maltodextrin blends for controlled dissolution of edible films. Int J Biol Macromol, 156, pp. 80-93. DOI: 10.1016/j.ijbiomac.2020.04.056.

[18] Phothisarattana, D., Wongphan, P., Promhuad, K., Promsorn, J. and Harnkarnsujarit, N. (2021). Biodegradable Poly(Butylene Adipate-Co-Terephthalate) and Thermoplastic Starch-Blended TiO_2 Nanocomposite Blown Films as Functional Active Packaging of Fresh Fruit. Polymers (Basel), 13(23), pp. 4192, DOI:10.3390/polym13234192.

[19] Phothisarattana, D., Wongphan, P., Promhuad, K., Promsorn, J. and Harnkarnsujarit, N. (2022). Blown film extrusion of PBAT/TPS/ZnO nanocomposites for shelf-life extension of meat packaging. Colloids and Surfaces B: Biointerfaces, 214, pp. 112472, DOI: 10.1016/j.colsurfb.2022.112472.

[20] Deng, Y., Yu, C., Wongwiwattana, P. and Thomas, N. L. (2018). Optimising Ductility of Poly(Lactic Acid)/Poly(Butylene Adipate-co-Terephthalate) Blends Through Co-continuous Phase Morphology. Journal of Polymers and the Environment, 26(9), pp. 3802-3816. doi:10.1007/s10924-018-1256-x.

[21] Bumbudsanpharoke, N., Choi, J., Park, H. J. and Ko, S. (2019). Zinc migration and its effect on the functionality of a low density polyethylene-ZnO nanocomposite film. Food Packaging and Shelf Life, 20, pp. 100301, DOI:10.1016/j.fpsl.2019.100301.

[22] Lian, Z., Zhang, Y. and Zhao, Y. (2016). Nano-TiO_2 particles and high hydrostatic pressure treatment for improving functionality of polyvinyl alcohol and chitosan composite films and nano-TiO 2 migration from film matrix in food simulants. Innovative Food Science & Emerging Technologies, 33, pp. 145-153. DOI:10.1016/j.ifset.2015.10.008.

[23] Chen, J., Dong, X., Zhang, Q. and Ding, S. (2019). Migration of titanium dioxide from PET/TiO_2 composite film for polymer-laminated steel. Food Additives & Contaminants: Part A, 36(3), pp. 483-491, DOI:10.1080/19440049.2019.1577992.

FG-O06

Investigating the Effects of Packaging Design Factors on Customer Purchasing Decisions for Goat Manure Fertilizer

Narucha Tanaiutchawoot*[1], Warangkana Saengtawan[2], Jiraporn Sinlapasiriwat[3], Nadthanicha Sukkeaw[4]

[1,2,3,4] *Industrial Engineering, Suranaree University of Technology, Nakhon Ratchasima, 30000, Thailand*

***Correspondence to:** Industrial Engineering, Suranaree University of Technology, Nakhon Ratchasima, 30000, Thailand. narucha@g.sut.ac.th*

ABSTRACT: Goat manure is one of the best animal manure for fertilizer, and it helps to provide the ideal growth conditions for plants. The majority of packaging in this category is designed to keep products contained rather than appeal to customers. Because competition is fierce both online and on the ground, it's critical to improve competitiveness through package development and design. The goal of this article is to look at 1) consumer behavior while making a fertilizer purchase and 2) the impact of packaging design on customer purchasing decisions by comparing the old and new designs. To create the packaging for the community's product, researchers collaborated with one community enterprise at Petchburi in Thailand. According to the pre-survey, 64.5 percent of 110 participants were more concerned with packaging design than with pricing, market channels, or marketing. The packaging, originality, product feature representation, cleanliness, safety, and the ability to pack and unpack the goods were all important considerations when buying goat dung fertilizer. The packaging is then designed with the Zip Lock plastic bag that helps maintain product quality for longer. Color, image, and information on the packaging show identity of the community. A scaling mark is designed on the packaging to reflect consumer behavior towards the use of products according to the theory of NUDGES. After designing and production, the old and new packages are tested in the market via online and onsite channels. Although the quality and quantity of products remain the same, 87.5 percent of the 32 consumers who purchased fertilizer chose the new packaging design for 60 baht and the rest chose the other option for 50 baht. As a result, packaging design raises the perceived value of a product and promotes customer acceptability.

Keyword: Packaging design, decision-making, customer behaviors, goat manure fertilizer

1. INTRODUCTION

Fertilizer is a crucial component in improving agricultural output. When chemical fertilizers are compared to organic fertilizers, the former is more expensive and can harm the environment and human health [1]. As a result, people have become interested in organic fertilizers. This fertilizer is not only cheap, but it's also good for plants because organic matter that doesn't decompose has the capacity to absorb current plant nutrients [2]. Nong Yangtoi Subdistinct, Si Thep District, Petchabun Province, Thailand, is one of several community companies focused on goat manure-based organic fertilizers. This community enterprise sells a variety of goat and sheep products. These products are derived from natural sources derived from community goats and sheep. Normally, goat manure fertilizer is sold to the community and locals without a market strategy in place. This product's packaging serves merely to contain the product, which is the traditional function of packing. As a result, sales volume is minimal, and the cash generated by this product is insufficient to sustain the community company. Organic fertilizer, on the other hand, is experiencing rapid expansion in the market. Planting trees is also becoming more popular. As a result, if they have a competent marketing plan, such as enhancing product value, growing customer awareness, and expanding market channels, this business should be successful. One possibility to increase product value and customer awareness is to add the packaging design to the product. Packaging is considered a distinguishing technique that assists customers in making a decision among a large range of identical products, as well as inciting consumer purchasing behavior [3]. Despite massive efforts in mass-media advertising to influence consumer purchases, a greater proportion of sales are impacted by the marketplace environment. According to Annon's research, 66 percent of all supermarket purchasing choices are made in-store [4]. According to Rettie and Bruwer, 70% of purchases are made while viewing the products [5]. Packaging is vital not just for product protection, but also for shelf attractiveness, product information, and generating brand image and awareness. It serves as a link between the product and the consumer. The packaging design is the product's initial impression for customers, and it is critical to the product's market placement. To attain a higher position, it is necessary to concentrate on a certain segment of the market in the brand's identity [6, 7]. The right packaging may help a brand's standing in the market and the minds of customers [8]. It is also recognized as the primary mode of communication and has evolved over time [9]. Brand-marking packages can help the product's maker, packer, and shipper get attention [10]. Packaged product components like shape, colors, patterns, and texture can excite the consumer's attention and influence whether or not the product is purchased [11]. The visual qualities of a package contribute to product appreciation by attracting attention and capturing it at a higher rate [12]. Design techniques are thus critical for marketing competitiveness.

Thaler's Nudge Theory is one strategy for changing people's behavior in a predictable way without constraining alternatives and employing any type of decision architecture [13]. As an acronym for "NUDGES," six components are formed: incentive (N), understand mapping (U), defaults (D), give feedback (G), expect error (E), and structure complex choices (S). This method is used to create imagery on food packaging [14]. By depicting a meal portion in a smaller container, the illusion of a greater portion is created, even when the real amount of food stays the same. This hypothesis is also used by Wensing and colleagues to generate green nudges on consumer assessment of bio-based plastic packaging [15]. To assess consumer behavior, four tactics are used: nature pictures, reflection questions, information on bio-based plastic, and normative information. According to the findings, the largest impacts are produced when the nudging approach matches the features of the customers' cognitive style.

Although various strategies and technologies have been offered, the package design of most agricultural products is still insufficient to compete in the local and international markets [16]. Although certain locally created, high-quality items are available, they are not considered decent products due to the appearance of their packaging design. As a result, researchers identify these issues and want to improve the value of agricultural goods in Thailand through package design, beginning with the first community enterprise in Petchabun Province. This study then focuses on 1) customer behavior in making purchase decisions for goat dung fertilizer and 2) the influence of container design on consumer purchasing decisions.

2. RESEARCH OVERVIEW

Researchers cooperated on this research with the community enterprise in Petchabun Province in Thailand. Research procedures are conducted as presented in Figure 1.

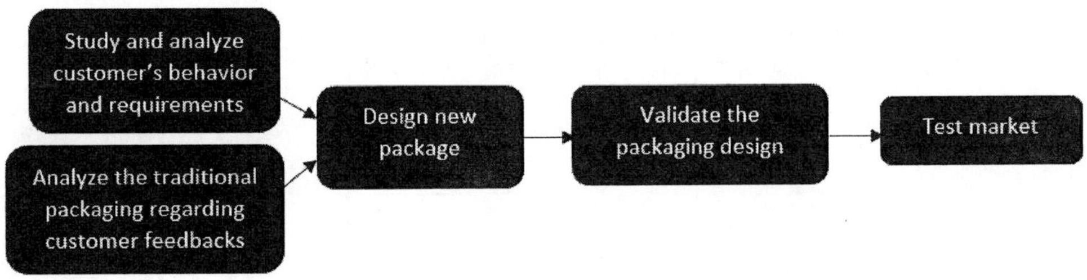

Figure 1 Research procedures

Researchers prepared a pre-survey to acquire information about customer behavior and expectations after reviewing relevant literature on package design, marketing, and goat manure production. Face-to-face interviews with individuals who have consumed the product are conducted to get customer feedback on conventional packaging design. The results of these surveys influence the direction of this community enterprise's new package design. Conceptual designs are confirmed in a post-survey before building a prototype. Due to funding constraints, 20 prototypes were launched for market testing. The market test results are then gathered and examined.

3. STUDY AND ANALYZE THE CUSTOMER'S BEHAVIOR AND REQUIREMENTS

The survey was developed to explore customer behavior and requirements, and it was separated into two sessions: basic information and consumer behavior. Gender, salary, and user group are examples of general information. Consumer behavior is centered on the product's weight and other essential aspects for decision-making. With 100 participants, this survey was done using face-to-face and online interviews.

According to the survey results, 76.36 percent are men, and the average monthly payment for the major participants is less than 20,000 baht. These individuals purchased the fertilizer for personal use rather than commercial purposes, hence the product weight should be approximately 1 kg (73.6%). When it comes to decision-making elements, the most important ones are a package, pricing, marketing channel, and promotion, in that order. These findings suggest that customers prefer a small quantity of fertilizer since their goal is personal usage. Packaging design is the most important component in attracting customers and influencing their purchasing decisions. Customer decisions, on the other hand, are influenced by price, marketing channel, and promotion. As a result, rather than lowering prices, increasing marketing channels, or giving promotions, a new packaging design may be one approach to enhance market opportunities. The outcome is summarized in Table 1.

Table 1 The result of general information and consumer behavior from 100 participants

1 General information						
1.1 Gender	**Male**	**Female**				
%	76.36	23.64				
1.2 Salary/Month	**No**	**<10,000 baht**	**10,001-20,000 baht**	**>20,000 baht**		
%	2.7	40.9	40.0	16.4		
1.3 User group	**Personal use**	**Business owner**				
%	90.9	9.1				
2 Consumer behavior						
2.1 Weight of product	**1 kg**	**5 kg**	**50 kg**			
%	73.6	21.8	4.6			
2.2 Factors for decision-making	**Level of interest**					
	4	**3**	**2**	**1**	**Average**	**SD**
- Package	64.5%	29.1%	6.4%	0%	3.94	0.745
- Price	29.1%	60.0%	8.2%	2.7%	3.47	0.787
- Marketing channel	11.8%	30.0%	42.7%	15.5%	2.62	0.965
- Promotion	10.9%	21.8%	35.5%	31.8%	2.33	1.056

4. ANALYZE THE TRADITIONAL PACKAGING REGARDING CUSTOMER FEEDBACKS

The traditional packing in this community company comprises 50 kilos of goat dung fertilizer with a selling price of 250 baht, as shown in Figure 2.

Figure 2 The traditional packaging for the community enterprise, Nong Yangtoi Subdistinct, Si Thep District, Petchabun Province in Thailand

Researchers gathered user feedback and suggestions from 49 participants who utilized this community enterprise's product. Seventy-eight percent of them were unable to identify the package's brand. Eighty percent of them want to view the product before purchasing it so that they can determine the quality of the product and have faith in it. Another proposal is to provide information such as ingredients, resources, and weight on the container that is not already available (80%). Customers also compliment cleanliness and ease of usage. These two criteria are additional considerations in the purchase choice.

5. DESIGN PACKAGING FOR GOAT MANURE FERTILIZER

Specifications for packaging design are determined based on the results of the consumer behavior study and consumer feedback. The product's weight is decreased to 1 kg, making it acceptable for personal usage by the customer's group. When logistics costs are considered, this size is also appropriate for online marketing. The following information should be included on the package: brand name, logo, product name, kind of fertilizer, weight, ingredient, production name, expiry date, instruction of use, and qualification. The appearance of the new packaging should be distinct, clean, appealing, easy to use, and perhaps present the product. By integrating white and purple into the design, community identity is employed to frame the package design idea. The logo is founded on the idea of a goat, which is a symbol for this product. The package is designed using the nudging principle. A ruler on the packaging is intended to illustrate the appropriate proportion of consumption per time unit. Consumers often utilize a product by estimating the right amount per plant or by utilizing extra measurement instruments such as measuring cups and scoops. This design seeks to assist customers with the simplicity of use and decision-making when utilizing the product without the need for an additional tool. This also reduces the possibility of user mistakes when planting or farming is new to the user. However, this notion is evaluated in the package design validation process to observe the customer's responsibility.

These specifications serve as the foundation for the new packaging design. The plastic bag with a zip lock is chosen as the packing material since it may be reused several times. Figure 3 shows four package designs created with Adobe Illustrator and Photoshop. Design 1 and Design 2 are nearly identical. Design 1 adds only the ruler design on the left side of the packaging, which is comparable to Design 3. Design 4 is similar to Design 3, however, there is no ruler on the packaging.

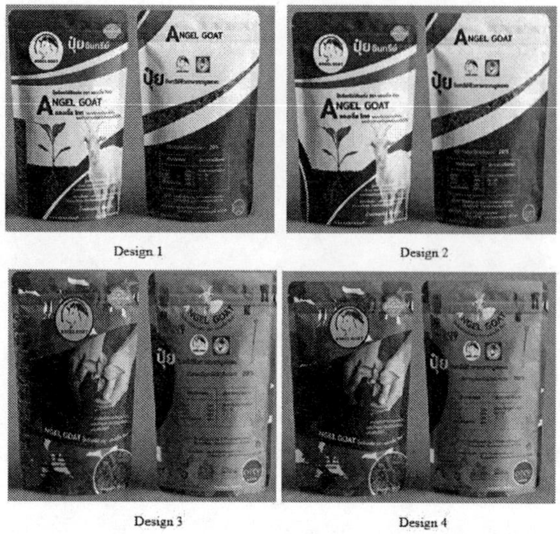

Figure 3 Four types of packaging design for the validation process

6. VALIDATE THE PACKING DESIGN

The post-survey is designed to confirm marketing based on consumer input for each conceptual design and to assess the market opportunity for a new product. This method is essential to provide the owner with more confidence before investing in the costly production process. One hundred and twenty participants responded to the post-survey by picking their preferred design, as shown in Figure 3. According to the results, 53.3% favor Design 1, and 23.2% prefer Design 2. Participants who chose Design 3 and Design 4 account for 16% and 7.5%, respectively. As a result, Design 1 was chosen to create the prototype. Due to budget constraints, only 20 package prototypes are prepared for market testing.

7. MARKET TEST

The marketing experiment is being carried out in conjunction with the annual fair for agricultural goods in Petchabun Province. As shown in Figure 4, the marketing is assessed by comparing the purchase volume of the product with the new packaging design to the product without the packaging design (standard plastic bag). Both groups' items are of comparable quality and weight. The product with the new package design cost 30 baht, which was more expensive than the other product, which cost 10 baht. These selling prices are determined by a cost-benefit analysis. The packing cost is computed on the assumption that the product would be packaged in 1,000 units. The price decreases as the volume of orders increases. All 20 new package design goods were sold within one hour, and 29 people pre-ordered. The other one was not purchased by any of the consumers. They said that the new package appears to be more valuable and dependable than the previous one, although they are the identical product, as explained by the researcher. Buyers are prepared to pay a higher price for a more expensive product; on the other hand, customers prefer to pay a higher price for more efficient packing, despite in agricultural products.

Figure 4 Product testing in the market with and without packaging design

Five factors from traditional packaging are evaluated for five levels (-2 to 2) by interviewing 49 people who are interested in the new package design and plan to purchase the product. The result is presented in Table 2.

Table 2 Level of satisfaction on the new packaging design regarding 5 factors

Factors	**Level of satisfaction**						
	2	**1**	**0**	**-1**	**-2**	**average**	**SD**
1) Uniqueness/ brand awareness	40	9	0	0	0	1.82	0.15
2) Indicating the properties of the product	37	12	0	0	0	1.76	0.19
3) Cleanness and safety	38	11	0	0	0	1.78	0.18
4) Complete information	38	11	0	0	0	1.78	0.18
5) Ease of use	39	10	0	0	0	1.80	0.17

2= very satisfied, 1=satisfied, 0=neutral, -1=unsatisfied, -2 very unsatisfied

The result shows that most customers are very satisfied with the new packaging design in all factors (> 1.5 on average). That means the new design can support the customer's requirements and add product value.

8. DISCUSSION AND CONCLUSION

This article examines the impact of package design on a customer's decision to purchase goat manure fertilizer. The results indicate that customers consider not only the quality of the product but also the appearance of the package. The most crucial component in customer purchasing decisions is the packaging design. Another component that buyers evaluate and are interested in, especially for agricultural products, is a packaging design that allows customers to view the product inside. When creating fertilizer packaging, information that helps identify the product, cleanliness, and convenience of use are all things to consider. Nudging is another key strategy for supporting design thinking in packaging design, since it assists customers when using the product by offering feedback, making it the default option, and presenting a measurable unit. All of these elements can contribute to a prospective package design that raises the product's value.

The amount of additional price acceptability for packaging design, on the other hand, should be explored. If applicable, the link between the amount of fluctuation and the price of the previous and current packing should be shown. This result will assist the designer in balancing the amount of novelty and investment while creating a new package. The influence of nudging theory as incorporated into packaging design (ruler image) should be researched further to determine how this component affects customer behavior when using the product.

9. ACKNOWLEDGEMENT

We would like to thank the teams from the Nong Yangtoi Subdistrict, Si Thep District, Petchabun Province in Thailand. They give assistance and information throughout the whole research process. This research is funded by the Science Park at Technopolis, Suranaree University of Technology (SUT), and the Department of Industrial Engineering, SUT to produce a prototype and perform market tests and travel costs.

10. REFERENCES

[1] Savci, S. (2012). An agricultural pollutant: chemical fertilizer. International Journal of Environmental Science and Development, **3**(1), pp. 73.

[2] Senesi, N. (1989). Composted materials as organic fertilizers. Science of the Total Environment, **81**, pp. 521-542, DOI: 10.1016/0048-9697(89)90161-7.

[3] Wells, L.E., Farley, H. and G.A. Armstrong. (2007). The importance of packaging design for own□label food brands. International Journal of Retail & Distribution Management, 35(9), pp. 677-690, DOI: 10.1108/09590550710773237.

[4] Annan, B.A. (2018) The effects of packaging characteristics on consumer perception about liquid soaps in Ghana. Master thesis. Lappeenranta University of Technology, Institutional School of Business and Management. https://urn.fi/URN:NBN:fi-fe2018121050276

[5] Rettie, R. and Brewer, C. (2000). The verbal and visual components of package design. Journal of product & brand management, 9(1), pp. 56-70, DOI: 10.1108/10610420010316339

[6] Gonzalez, M. P., Thornsbury, S. and Twede, D. (2007). Packaging as a tool for product development: Communicating value to consumers. Journal of food distribution Research, **38**(856-2016-57932), pp. 61-66.

[7] Silayoi, P. and Speece, M. (2007). The importance of packaging attributes: a conjoint analysis approach. European journal of marketing, 41(11/12), pp. 1495-1517, DOI: 10.1108/03090560710821279

[8] Louw, A. (2006). The Power of Packaging. pp. 186-216.

[9] Orth, U.R. and Malkewitz, K. (2008). Holistic package design and consumer brand impressions. Journal of marketing, **72**(3), pp. 64-81, DOI: 10.1509/JMKG.72.3.064

[10] Balakrishnan, M.S. (2015). PepsiCo MEA: the role of packaging in brand activation. Emerald Emerging Markets Case Studies, 5(3), pp. 1-12, DOI: 10.1108/EEMCS-03-2015-0043

[11] Chitturi, R., Londono, J.C. and Amezquita, C.A. (2019). The influence of color and shape of package design on consumer preference: The case of orange juice. International Journal of Innovation and Economic Development, **5**(2), pp. 42-56, DOI: 10.18775/ijied.1849-7551-7020.2015.52.2003

[12] Dobson, P. and Yadav, A. (2012). Packaging in a market economy: The economic and commercial role of packaging communication. Prepared on behalf of the British Brands Group, University of East Anglia.

[13] Thaler, R.H. and C.R. Sunstein. (2008). Nudge: Improving decisions about health, wealth, and happiness. Yale University Press.

[14] Petit, O., Velasco, C. and Spence, C. (2018). Are large portions always bad? Using the Delboeuf illusion on food packaging to nudge consumer behavior. Marketing Letters, **29**(4), pp. 435-449, DOI: 10.1007/s11002-018-9473-6

[15] Wensing, J., Caputo, V., Carraresi, L., and Bröring, S. (2020). The effects of green nudges on consumer valuation of bio-based plastic packaging. Ecological Economics, 178, pp. 106783, DOI: 10.1016/j.ecolecon.2020.106783

[16] Kwaku, A. R., and Fan, Q. (2020). Effect of good product design and packaging on market value and the performance of agricultural products in the Ghanaian market. Open Access Library Journal, 7(9), pp. 1-14, DOI: 10.4236/oalib.1106714

FG-O07

Measuring the Hexane Vapor Transmission Rate of Barrier Paper Materials

Yves Wyser*, Sander Koster, Chantal Pelletier

Nestlé Institute of Packaging Science, Société des Produits Nestlé. 1000 Lausanne, Switzerland

***Correspondence to:** Nestlé Research, rte du Jorat 57, 1000 Lausanne 26, Switzerland, yves.wyser@rdls.nestle.com*

ABSTRACT: Recent developments have yielded paper-based materials with properties fulfilling the protection requirements of many types of food. In addition to moisture and oxygen, the barrier to mineral oils is one of the key performances required from packaging. While the determination of mineral oil migration is complex and time consuming, it has been shown that Hexane Vapor Transmission Rate (HVTR) could be a good proxy for the screening of mineral oil barrier of packaging materials. Methods used to determine barrier properties of current packaging material have shown limitations when applied to novel fiber-based materials, mainly linked to their hygroscopic nature. HVTR is not an exception and several artefacts have been identified when applying current methods to barrier paper. The study aimed at developing an automated gravimetric method for the determination of the HVTR of barrier papers that avoids pitfalls linked to the sample setup and to the hygroscopic nature of paper. It further presents a method that enables the determination of HVTR of pouches and other 3D packages to assess the properties of a complete package, including seals, leaks and other potential defects introduced by the pack-forming process. The results show that by understanding and avoiding common artefacts of paper testing, a method with significantly improved robustness compared to those commonly reported for HVTR testing of paper-based materials could be developed for flat materials and adapted to test complete packages.

Keyword: Barrier paper, Hexane Water transmission Rate, Mineral Oil Migration

1. INTRODUCTION

Paper-based materials have been used in packaging applications since the beginning of the industrial era. While, in the past, paper was preferred over other materials for its mechanical strength and ability to be printed and converted into packages [1], paper has recently gained increasing attention as an alternative to fossil-based packaging materials, as paper is a widely recyclable material made from renewable resources.

Classical paper materials do not provide to food products the level of protection required to maintain their quality over the duration of their shelf life. For food products requiring high protection against water vapor and/or oxygen, multilayer laminates are most often employed, which are typically composed of combinations of different polymers and/or coatings [2]. The use of such materials poses a problem for recycling, due to highly complex structures and inability to efficiently separate layers for re-processing.

To expand the applicability of paper packaging to food products requiring protection against external factors such as moisture uptake or loss or oxygen pickup [3,4], recent research focus has been put on the development of barrier or functional papers.

Barrier papers are typically coated with one or more polymeric layers with an overall thickness of less than 10 μm which are applied as water- or solvent-based polymer dispersions or solutions on paper. Vacuum deposited thin films of metals (aluminium) or metal oxides (silicon and aluminium) can be used to further improve barrier properties of these materials [5,6]. These materials achieve barrier performance close to that of multilayer plastic laminates while remaining recyclable in paper recycling stream as mixed paper waste if designed properly [7].

While oxygen and water permeation are key performance parameters for packaging in general and for barrier papers in particular, they are not sufficient to guarantee the quality and safety of food products. Another property to be assessed is the ability of the packaging to limit or suppress the migration into the food of other undesired substances,

With the increased use of recycled materials in food contact applications, the concern around such issues is increasing [8]. One example of substances being increasingly present are mineral oil components (MOH). MOH mainly enter the recycling stream through newspapers, for which MOH-based printing inks are used. The components of these oils susceptible to migrate into foods are complex mixture of saturated hydrocarbons (MOSH) and unsaturated hydrocarbons (MOAH) [9].

Determining oxygen and water vapor properties of packaging materials is rather straight forward and well established for packaging materials (even though some pitfalls related to the hygroscopic nature of the paper have to be avoided [5]). Conversely, the fast estimation of the migration of mineral oils is still challenging.

Migration testing is rendered difficult and time consuming due to the complexity of the mixture of mineral oil components migrating and the low level of detection needed to comply with safety limits (linked to the fact that many of these components and their toxicity are unknown) [9]. Moreover, the chemical composition of mineral oils used in different applications is variable and methodologies used to analyze in foods are non-harmonized [10].

A fast approach proposed initially by Fiselier &Grob [11] is the determination of the permeability and the definition of a breakthrough period of a component through a barrier material. A donor material spiked with MOH is separated from an acceptor material (typically Low Density Polyethylene, LDPE [11] or silicone paper [12]) by the material to be tested. A strip of the receptor material is removed at different time points and the concentration of MOH is measured. A reduction of a factor 100 in the concentration between the donor and receptor material has been defined as maximum migration threshold by Biederman and Grob [13] based on the concentration of contaminants typically found in recycled paper [11]. In many cases, surrogate components are used to avoid the complexity brought by the multiple components present in mineral oils.

Diehl & Welle [9] proposed a lag-time experiment where the time between the start of the exposure of a barrier material to a contaminated gas flow and the detection of the contaminants behind the barrier is determined. The proposed criteria is that if the lag time is larger than the shelf life, the material can be considered as a functional barrier. This means that the testing time might be rather long unless performed at higher temperature.

While these methods are important and efficient to determine the mineral oil barrier of a packaging material, they are not practical to perform rapid screening of numerous variants of a packaging material produced in a development stage.

A simple semi-quantitative gravimetric method based on hexane permeation has been proposed by Gaudreault et al. [14] in which an evaporation chamber containing hexane and closed by the barrier material is weighed at regular intervals. The weight loss as a function of time yields the Hexane Vapor Transmission Rate (HVTR) of the material. N-Hexane is chosen as a non-polar solvent with a vapor pressure at room temperature susceptible to yield short testing times and fairly good correlation with mineral oil migration [15,16].

Upon application of this method to barrier paper material, our experience has shown that unrealistic (negative) results were sometimes obtained, potentially linked to some uncontrolled parameters in the testing method. The present study therefore aimed at refining the method by identifying potential sources of error to improve its precision. Further, a method is proposed to apply the same principle to formed packages, enabling the evaluation of the effect of potential sealing defects or barrier loss during forming of packages.

2. MATERIALS AND METHODS

An automated Water Vapor Transmission tester is used for all measurements (FX 3180 Cupmaster, Textest Instruments, Zürich, Switzerland). The equipment allows the simultaneous testing of 12 evaporation cells, in controlled climatic conditions (fixed at 23 °C and 50 % Relative Humidity (RH) in this study). The equipment allows the automatic weighing of the evaporation cells at regular intervals (30 minutes in the case of this study) with a precision of 0.1 mg.

The evaporation cells are composed of a metal cup of approximately 90 mm in diameter and 25 mm in depth. The tightness is ensured by a rubber gasket placed on the shoulder of the cup. The sample is placed on the gasket and a Teflon gasket is placed on the sample. The set-up is tightened by a metal screw cap (Figure 1). The exposed surface of the material is 50 cm^2. All tests were performed in triplicate with a blank cell (same setup without hexane).

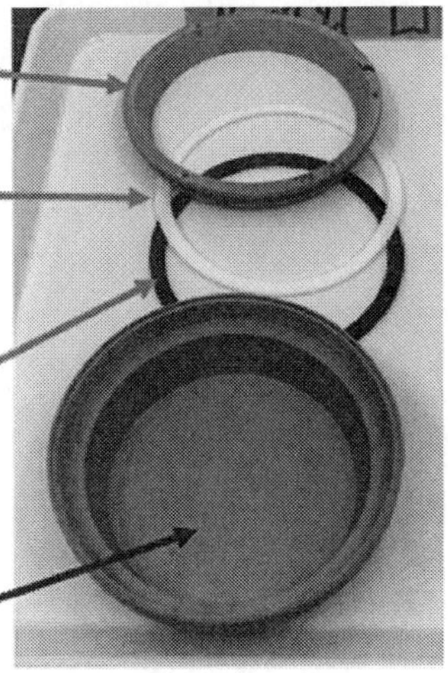

Figure 1: Evaporation cell setup

Two types of rubber gasket materials have been used to assess lag time of permeation through the gasket: a Silicon rubber (Exact composition unknown, supplied by Textest Instruments, Zürich, Switzerland) and a fluoroelastomer polymer (Viton®, supplied by MaagTechnic, Dübendorf, Switzerland). Emplura n-Hexane was supplied by Supelco, (Darmstadt, Germany). The amount of n-Hexane placed in the evaporation cells was 2.5 mL.

Table 1 lists the different barrier materials used. The alu-foil based laminate has been used to optimize the setup and paper-based materials have been chosen from our portfolio. Two metalized papers (MP 1 and MP 2 were in development stage and two commercially available polymer coated papers claiming barrier against mineral oils (CP1 and CP 2) and recyclability in the paper stream.

Table 1. Laminate, barrier paper and packages made thereof.

Sample designation	Sample type	Structure / Dimensions
Alu foil	Plastic laminate with aluminium foil	PET (12 µm) / alu (12 µm) / LDPE (60 µm)
MP 1	Metalized barrier paper	Development
MP 2	Metallized barrier paper	Development
CP 1	Polymer coated commercial paper	81 g/m2
CP 2	Polymer coated commercial paper	89 g/m2
CP 1 SS	4-side seal sachet made from CP 1 on a Vertical Form Fill Seal machine.	14.9 x 5.9 cm2
CP1 Flow	Flow-wrap pouch made from CP 1	17.5 x 8.2 cm2
CP 2 Flow	Flow-wrap pouch made from CP 2	17.5 x 8.2 cm2

In addition, industrially formed empty pouches from some of the base materials listed in Table 1 were used for the evaluation of complete packs. The pouches were cut open on the top and a glass vial containing 0.4 and 6 g of n-Hexane, for the 4-side seal and flow-wraps pouches, respectively, was inserted. The pouches were resealed with a Brugger HSG-CC lab heat sealer (Brugger Feinmemechanik, Münich, Germany) equipped with flat sealing jaws. 5 mm wide jaws were used for CP 1 4-SS while 10mm wide jaws were used for the flow-wraps. For the latter, the lab-made seal was performed in such a way that it does not contain a layer jump, hence avoiding potential leaks created during sample preparation. 4-side seal sachets were resealed as on the industrial line. A sealing temperature of 120°C was used for the CP 1 paper and 180°C for the CP 2 paper. Dwell time and force were kept constant for both materials at 0.5 sec and 200 N.

3. RESULTS AND DISCUSSION

3.1 Selection of rubber gasket material

Figure 2a and b compare the weight loss of the evaporation cells with the 2 types of rubbers gaskets when mounted with the aluminium laminate which has a theoretical HVTR of zero. It is to be noted that the vertical axes of the two graphs are not the same.

Figure 2a shows that for the silicon rubber, a lag-time of approximately 1 day can be observed followed by a high permeation. Figure 2b shows that by using the Viton® rubber gasket increases the lag time to 5 days and significantly decreases the permeation rate thereafter.

It is hypothesized that the hexane permeates trough the rubber gasket and evaporates at the edge. The test on actual samples needs therefore to be completed before the solvent reached the edge of the rubber gasket, i.e. before 1 days with the silicon rubber and before 5 days with the Viton® rubber. Not taking this fact into consideration would yield an error of approximately 13 g/m^2/day on the measured value of a sample with the silicon rubber (taken between day 2 and day 3 of the measurement) and of 0.6 g/m^2/day with the Viton® rubber (taken between day 7 and day 8).

The effect of permeation through the rubber gasket on an actual sample can be seen on Figure 3 where CP 1 samples were tested with the silicon and the Viton® rubber. It can be observed that during the first day of measurement, the slopes of both curves are quasi identical and correspond to the permeation through the material. After this point, the slope of weight loss measured on the sample tested with the silicon rubber drastically increases due to the permeation through the rubber gasket, while the slope of the other curve remains constant over the rest of the 4 days of testing.

Figure 2 suggests that hexane dissolves quite significantly in the rubber, including in the Viton® rubber. This means that before the rubber gasket is used again, sufficient time must be given to aerate the rubber to avoid a bias due to the evaporation of the hexane from the rubber gasket during the test

Figure 4 shows the slope of weight loss (between 1.5 and 2.5 days of testing) that is due to the Viton® rubber gasket as a function of aeration time under a fume hood at 23 °C. The values are divided by the surface of a sample to allow direct comparison. Although the results are relatively scattered, the graph clearly shows that this effect cannot be neglected, and the rubber gaskets need to be aerated for at least 7 days to reduce the effect to 0.1 g/m2/day. With short aeration, the measured weight loss rate corresponds relatively well to the one calculated from Figure 2b and reported above.

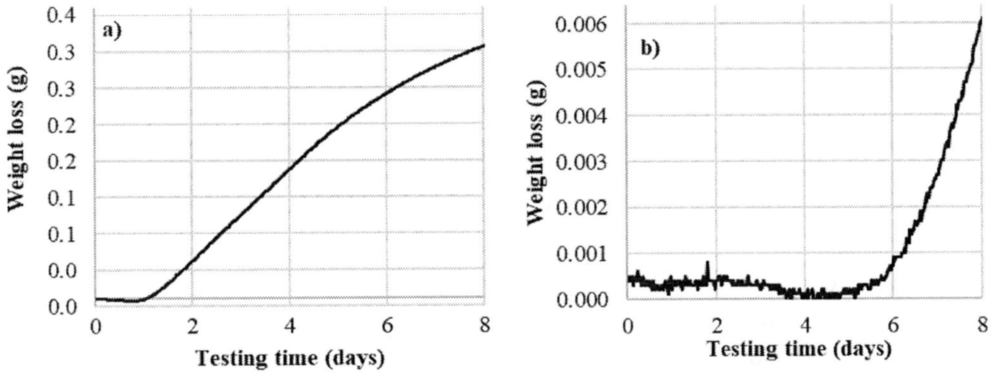

Figure 2: Weight loss measured on the evaporation cell containing 2.5 mL of hexane, sealed with a perfect barrier laminate (Alu foil) and different gaskets materials a: silicon rubber, b: Viton® rubber

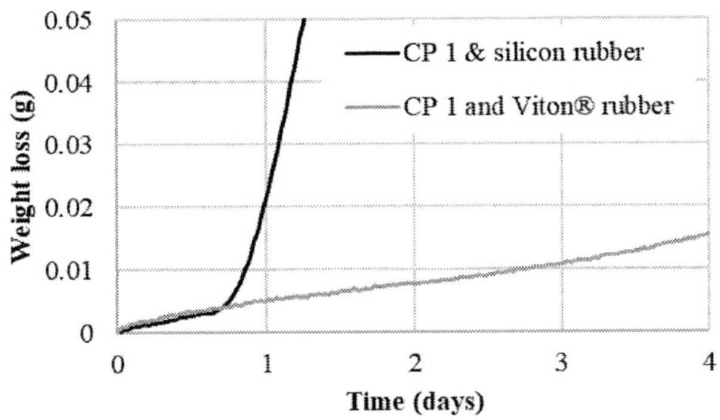

Figure 3: Weight loss measured on the evaporation cells containing 2.5 mL of hexane, sealed with CP 1 and different gaskets materials a: silicon rubber, b: Viton® rubber

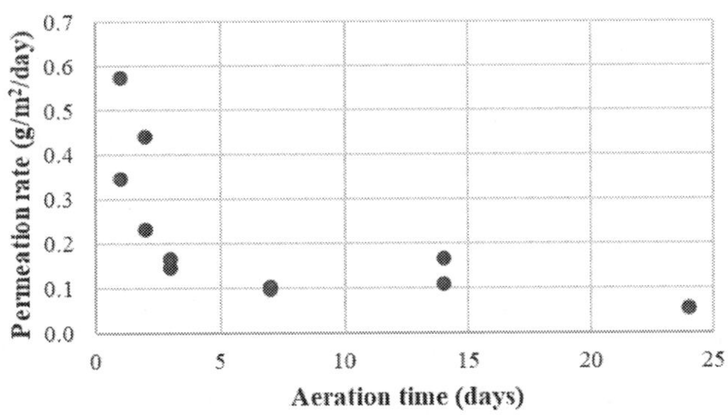

Figure 4: Weight loss rate due to the Viton® gasket as a function of aeration time under a fume hood at 23 °C. The value is divided by the testing surface for comparison purposes.

3.2 Testing of paper

Paper is a hygroscopic material which moisture content depends on the relative humidity surrounding it. When moving from a dry environment to a humid environment, paper can pick up significant amount of moisture.

Figure 5 shows the weight change of a CP 1 sample prepared as above without the introduction of hexane in the evaporation cell (Blank). It shows that during the first day, the sample gains weight (negative weight loss).

Paper materials are in general conditioned at 23 °C and 50 % relative humidity prior to testing in the same conditions to avoid this type of artefact. However, due to the toxicity of hexane, sample preparation must occur under a fume hood which cannot be conditioned in terms of relative humidity. The testing time for such samples must therefore be sufficiently long to allow the paper to equilibrate with its environment, which can be fairly long depending on the paper material used, and a blank sample must be used, whose weight gain must be added to the weight loss of the actual samples.

The time taken by the paper to achieve equilibrium corresponds more or less to the lag time observed for the n-Hexane to start permeating through the silicon rubber (Figure 2a), underlining the importance of using low permeability material to ensure the tightness of the evaporation chamber.

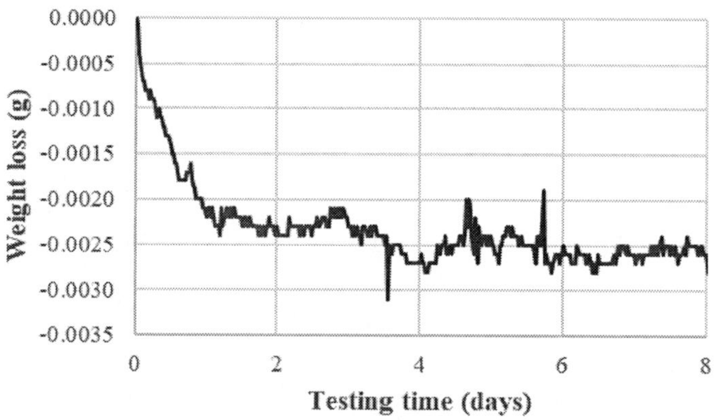

Figure 5: Weight change of CP 1 sample without hexane in the evaporation cell (Blank).

3.4 Optimized method

As mentioned in the introduction, preliminary tests were performed in an external lab with a sub-optimal method. Figure 6 shows the obtained HVTR on two materials with the initial method and the optimized method. This figure shows that by systematically avoiding pitfalls of paper testing, unrealistic results can be avoided, and scatter can be drastically reduced, hence improving the precision of the method.

Table 2 gives the HVTR results obtained on the different samples.

While differences between the flat samples can be seen, they all would be classified as good to medium barrier according to reference [14] who states that materials with HVTR below 10 g/m²/day are high performance barriers, whereas materials with a HVTR higher than 500 g/m²/day is not considered as an effective functional barrier to mineral oil components.

Diehl et al [16] propose a scale of correspondence between HVTR and lag-time of mineral oils in packaging materials where a HVTR of 10 g/m²/day corresponds to a lag-time higher than 3 years, and a HVTR of 100 g/m²/day a lag-time of more than 1 year.

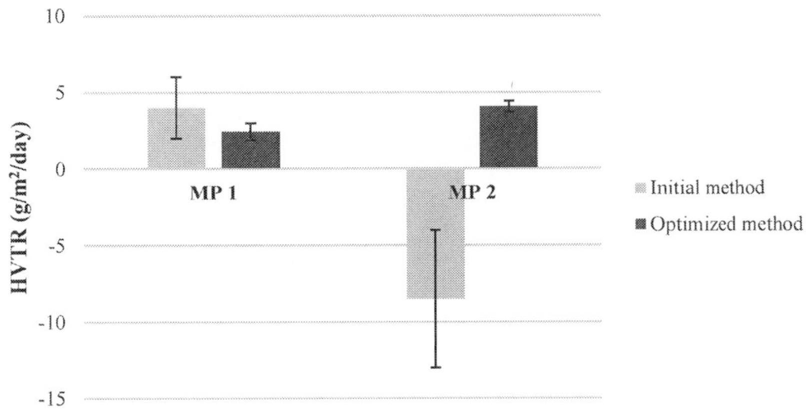

Figure 6: Measured HVTR on two samples with initial and optimized methods

Table 2: HVTR results obtained on the different samples, at 23 C and 50 % RH

Designation	HVTR
	g/m²/day
MP 1	2.4 ± 0.6
MP 2	4.1 ± 0.4
CP 1	0.55 ± 0.14
CP 2	25 ± 6
CP 1 4SS	2.9 ± 0.4
CP 1 Flow	94 ± 4
CP 2 Flow	37 ± 5

The results of tests performed on pouches show that leaks and/or pack-forming processes can have a significant effect on the hexane permeation rate of a formed packaging. The CP 1 paper keeps high barrier performance after forming into CP 1 4SS while it looses most of its properties in CP 1 flow. The consequence is that once formed, the CP 2 Flow sample performs better that the CP 1 Flow, while CP 1 was significantly better than CP 2 when tested as unprocessed material. Both papers have been formed into flow-wraps on the same machine. The root cause of this strong degradation of barrier performance has not been identified at this stage.

These results suggest that, while the determination of HVTR on flat material is a rapid and useful tool to screen the properties of material as barrier to mineral oils, effects of process and sealing must be evaluated once an application and a specific packaging format is considered.

4. CONCLUSIONS

Hexane vapor transmission rate has been shown to be a proxy to rapidly determine the protection level provided by a packaging material towards mineral oil migration.

While methods for the determination of HVTR have been published, experience has shown that when testing barrier paper, biases could be introduced generating unrealistic results.

The study presented here determines the factors affecting the robustness of the method and improvements are proposed to improve the tightness of the testing setup.

The optimized method is shown to yield reliable and reproducible results.

It is further shown that pack-forming processes can have important effects on the performance of the final packaging.

5. REFERENCES

[1] Kirwan M.J. 2012 Handbook of Paper and Paperboard Packaging Technology. ISBN:9781118470930, 2013 John Wiley & Sons, Ltd.

[2] Lange, J.; Wyser, Y. Recent Innovations in Barrier Technologies for Plastic Packaging - A Review. Packaging Technology and Science, 2003, 16 (4), 149.

[3] Söremark C and Tryding J, Packaging, in Paper Products Physics and Technology, Ek M, Gellerstedt G, Henriksson G, Eds, Walter de Gruyter GmbH & Co. KG, Berlin Germany, 2009

[4] Wyser, Y.; Shires, D. Increasing the quality and impact of manuscripts in the field of new materials. Packaging Technology and Science, 2019, 32 (1), 3.

[5] Wyser, Y., Vishtal, A., Giardiello, M.I., Pelletier, C., Deantoni, F., Understanding barrier degradation during water vapour transmission rate testing of high barrier metallized paper, Packaging Technology and Science, 2022, 35(3), pp. 291-299

[6] Field, R., Metallized C1S paper with moisture barrier: The ultimate in sustainabilitv, in Proceeding of PaperCon 2010, Atlanta (Ga), USA, 2010

[7] Lee, T. J.; Yoon, C.; Ryu, J. Y. A new potential paper resource; Recyclability of paper cups coated with water-soluble polyacrylate-based polymer. Nordic Pulp and Paper Research Journal, 2017, 32 (1), 155.

[8] Grob, K., How to make the use of recycled paperboard fit for food contact? A contribution to the discussion (2022) Food Additives and Contaminants - Part A Chemistry, Analysis, Control, Exposure and Risk Assessment, 39 (1), pp. 198-213.

[9] Diehl, H., Welle, F., How to determine functional barrier performance towards mineral oil contaminants from recycled cardboard (2015) Food Packaging and Shelf Life, 5, pp. 41-49

[10] Koster S, Varela J. Stadler RH, Moulin J, Cruz-Hernandez C, Hielscher J., Lesueur C, Roiz J, Simian H Mineral oil hydrocarbons in foods: is the data reliable? (2020) Food Add. Contam Part A 37(1), p.69-83

[11] Fiselier, K., Grob, K. Barriers against the migration of mineral oil from paperboard food packaging: Experimental determination of breakthrough periods (2012) Packaging Technology and Science, 25 (5), pp. 285-301

[12] Biedermann-Brem, S., Biedermann, M., Grob, K. Required barrier efficiency of internal bags against the migration from recycled paperboard packaging into food: a benchmark (2016) Food Additives and Contaminants - Part A Chemistry, Analysis, Control, Exposure and Risk Assessment, 33 (4), pp. 725-740.

[13] Biedermann-Brem, S., Grob, K. Barriers against the migration from recycled paperboard into food: Measuring efficiency by surrogate components (2014) Packaging Technology and Science, 27 (9), pp. 713-726.

[14] Gaudreault, R., Brochu, C., Sandrosck, R, Deglmann, P., Seyffer, H., and Tétreault. A., Overview of practical and theoretical aspects of mineral oil contaminants in mill process and paperboards. In Advances in Pulp and Paper Research, Cambridge 2013, Trans. of the XVth Fund. Res. Symp. Cambridge, 2013, (S.J. I'Anson, ed.), pp 907-925, FRC, Manchester, 2018.

[15] Diehl, H., Pfeiffer, A.-M., Seyffer, H. New testing method for functional barriers (2014) International Paperworld IPW, (8-9), pp. 8-11.

[15] Diehl, H., Seyffer, H., Pfeiffer, A.-M., Barrierewirkung ausgewählter Kunststoffmaterialien gegen die Migration von Mineralölfraktionen in Lebensmittel, BfR Workshop Mineralöle in Lebensmittelverpackungen-Entwicklungen und Lösungsansätze- Berlin 22. / 23. September 2011, https://www.bfr.bund.de/cm/343/barrierewirkung-ausgewaehlter-kunststoffmaterialien-gegen-die-migration-von-mineraloelfraktionen-in-lebensmittel.pdf (accessed 29/04/2022)

FG-P01

Evaluation of the quality of fresh broccoli for non-refrigerated transportation using a high-performance thermal insulation box

Hiroaki Kitazawa[1]*, Naoki Sugino[1], Takashi Watanabe[1]

[1]*Institute of Food Research, National Agriculture and Food Research Organization, Tsukuba, 305-8642, Japan*

***Correspondence to:** Institute of Food Research, National Agriculture and Food Research Organization, Tsukuba, 305-8642, Japan. ktz@affrc.go.jp*

ABSTRACT: In Japan, fresh vegetables are usually transported by refrigerated trucks. However, securing refrigerated trucks has become increasingly difficult due to the lack of refrigerated trucks and drivers for them. In addition, enhancement of the use of non-refrigerated trucks might be recommended from the viewpoint of environmental impact in the near future. Therefore, a transportation system for fresh vegetables using a high-performance thermal insulation box has been proposed. In this system, vegetables are pre-cooled, packaged into high-performance thermal insulation boxes, and then transported by non-refrigerated trucks. Thus, the temperature inside the box gradually increases during transportation. Therefore, it is necessary to identify the quality changes in vegetables under such temperature conditions. In this study, we evaluated the relationship between changes in quality and storage temperature by testing at 2, 5, 10, 15, and 20 °C with 85–90% relative humidity for 21 days. Based on these results, we propose a mathematical model to predict the change in the commercial quality of broccoli under several temperature conditions. This model would contribute to the prediction of the quality change of broccoli in non-refrigerated transportation using high-performance thermal insulation boxes.

Keyword: appearance quality; cumulative temperature; prediction of quality changes

1. INTRODUCTION

Although most fresh vegetables are transported by refrigerated trucks in Japan [1], recently it has become difficult to secure refrigerated trucks due to their scarcity and the lack of drivers for them. Moreover, from an environmental perspective, enhancing the use of non-refrigerated trucks may be recommended in the near future. Considering this background, a transportation system for fresh vegetables using non-refrigerated trucks with a high-performance thermal insulation box (Figure 1) has been proposed. In this system, vegetables on a pallet are pre-cooled, packed into a high-performance thermal insulation box, and then transported. Under such transportation conditions, the temperature inside the box gradually increases. Therefore, it is necessary to determine the quality changes in vegetables under such temperature conditions.

It is well known that the quality of broccoli immediately decreases after harvest [2,3,4]. This produce is usually transported in a cold chain using ice or other ice packs and refrigerated transportation facilities [3,4]. In other words, the commercial quality of broccoli is quickly lost if it is transported without temperature control. Therefore, to successfully transport broccoli using a thermal insulation box with non-refrigerated trucks, we must clarify the relationship between the quality change of broccoli and the temperature conditions after harvest. Moreover, to predict the change in quality following any temperature variations, we must also construct estimation methods, such as a mathematical model.

In this study, we evaluated the relationship between the change in quality and cumulative temperature with a storage test for 21 days.

Figure 1 Image of the thermal insulation box (Fresh Logi™, Asahi Kasei, Tokyo, Japan).
for vegetable transportation
Pre-cooled vegetables on the pallet are surrounded by thermal insulation panels.

2. MATERIALS AND METHODS

The broccoli used (*Brassica oleracea* L. *italica,* cv. Ohayo) was harvested from the experimental field of the Institute of Vegetable and Floriculture Science, NARO (Tsukuba, Japan), on November 16, 2021. The broccoli heads were immediately transported to the Institute of Food Research and were packed into polyethylene bags with a thickness of 0.03 mm and stored in an incubator (KCL-2000A, Tokyo Rikakikai, Tokyo, Japan) for 21 days. The average initial weight (g) was 603. Based on our previous work [5], the storage temperature during the experiment was set to 0, 5, 10, or 20 °C, and the relative humidity inside the chamber was maintained at 85–90%. Six broccoli heads were stored at each temperature.

In terms of commercial quality, we assessed the occurrence of wilt, rot and yellowing on broccoli at 1–3-day intervals. The observed qualities were scored according to the following indices: 5, none; 4, slight or partly; 3, about half; 2, more than half; and 1, whole.

Moreover, we determined a total estimation score considering the minimum value of each quality score on each observation day as follows: 5, same as just harvested; 4, good; 3, fit for sale; 2, unavailable for sale; 1, should be discarded immediately. Thus, when the score was below 3, we assumed that the commercial quality was lost.

3. RESULTS AND DISCUSSION

Wilt occurrence followed an increase in storage temperature. Approximately more than half of each sample wilted on day 3 at 20 °C, on day 4 at 15 °C, and on day 6 at 10 °C (Figure 2a). In contrast, at 5 °C, the samples were observed for 20 days under such conditions. Moreover, at 2 °C, only slight or partial parts of each sample wilted during storage for 21 days.

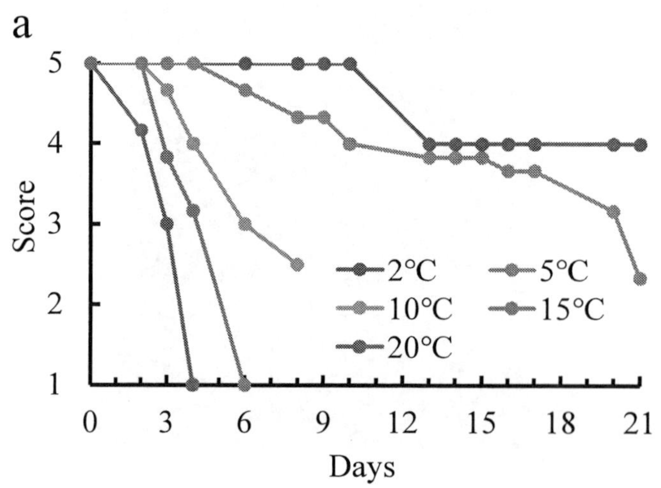

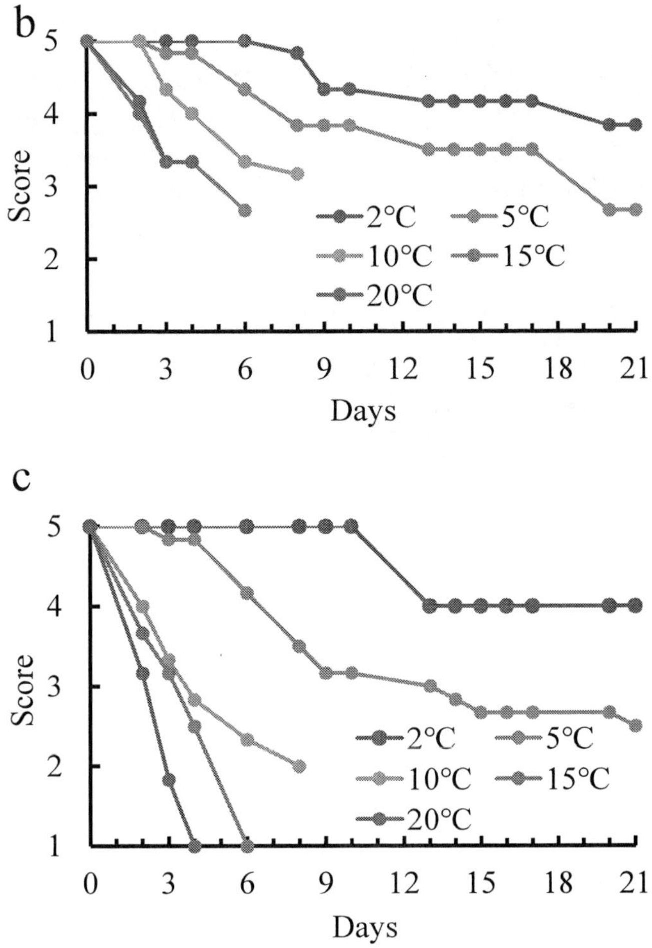

Figure 2 Relationship between the occurrence of wilt (a), rot (b), and yellowing (c), and storage temperature on broccoli during storage. Dots show the average value of the score of each sample.

More than half of each sample rotted on day 3 at 20 and 15 °C, and on day 8 at 10 °C (Figure 2b). However, at 5 °C, the same amount of wilting was observed for each sample after 18 days of storage. Moreover, at 2 °C, only slight or partial parts of each sample rotted during storage for 21 days. Yellowing was found on approximately more than half of each sample on day 2 at 20 °C, on day 4 at 15 and 10 °C, and on day 9 at 5 °C (Figure 2c). Moreover, at 2 °C, only slight or partial yellowing was observed in each sample during storage for 21 days. Thus, the tendency of yellowing (Figure 3) was similar to the occurrence of wilt and rot and supported the results of our previous work [5].

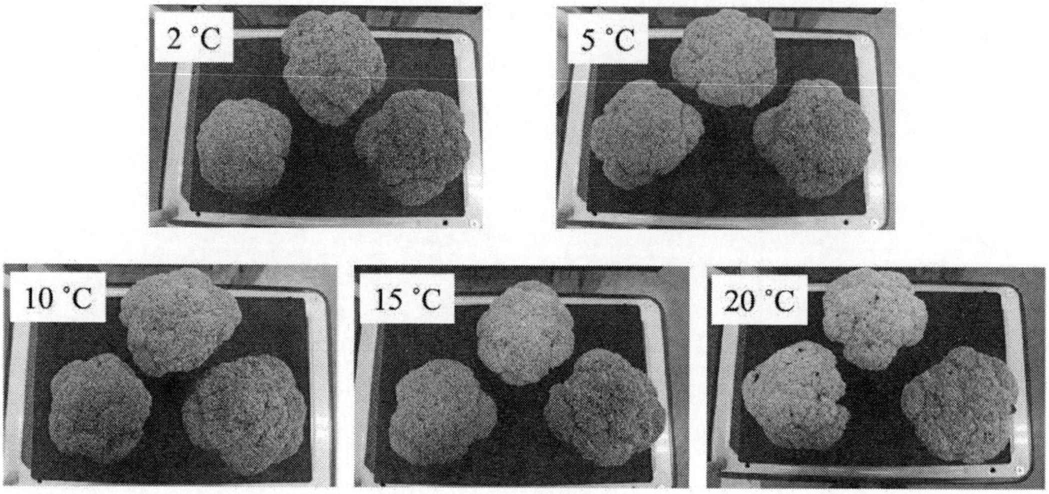

Figure 3 Relationship between storage temperature and the occurence of yellowing on day 3

Figure 4 shows the total estimation score used to determine the commercial quality based on the minimum value of each quality score on each observation day. Compared with the results shown in Figure 3, it was suggested that the main factor in determining the total estimation score was the yellowing score. The relationship between the yellowing score (Y_{Score}) and the total estimation score (T_{Score}) was expressed by the following equation with a high correlation by multiple regression analysis (Figure 5).

$$T_{Score} = 0.9887Y_{Score} + 0.0897 \quad (1)$$

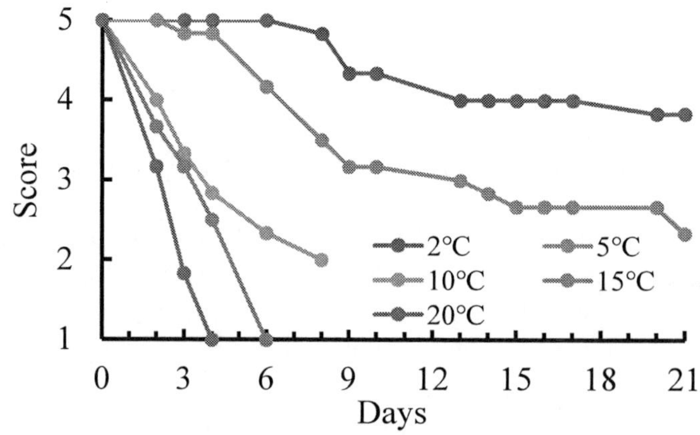

Figure 4 Relationship between storage temperature and the total estimation score on broccoli during storage.
Dots show the average value of the score of each sample.

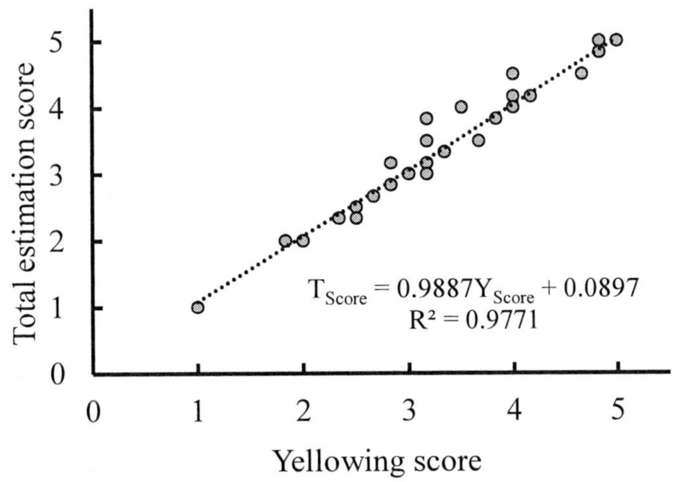

Figure 5 Relationship between yellowing score (Y_{Score}) and total estiamtion score (T_{Score})

From this analysis, we attempted to construct a mathematical model to predict changes in the commercial quality of broccoli based on the occurrence of yellowing. When we analyzed the relationship between the cumulative temperature (storage temperature [°C] × duration [h]) and the yellowing score (Y_{Score}), it could be expressed by the following equation.

$$Y_{Score} = (0.0024 - 0.0018 \times T) \times h + 4.8501 \quad (2)$$

Thus, Equation (3) holds by integrating Equations (1) and (2).

$$T_{Score} = 0.9887 \times \{(0.0024 - 0.0018 \times T) \times h + 4.8501\} + 0.0897 \quad (3)$$

Using Eqution (3), we can estimate the threshold of cumulative temperature during transportation, storage, and sale to keep the commercial quality of broccoli transported by the thermal insulation box.

4. CONCLUSIONS

We proposed a mathematical model to predict changes in the commercial quality of broccoli using a storage test under several temperature conditions. This model would contribute to the estimation of the cumulative temperature required to maintain the commercial quality of broccoli during transport using a thermal insulation box. Our future research will focus on designing and improving the thermal insulation box by considering actual transportation conditions, such as duration and temperature. Moreover, we will explore the effect of harvest season on the changes in quality of broccoli.

5. ACKNOWLEDGEMENT

Part of this study was funded by the Asahi Kasei Corporation, Japan, under the joint research program "Construction of Smart Food Supply Chain".

6. REFERENCES

[1] Lu F, Ishikawa Y, Kitazawa H. and Satake T. (2010). Effects of sampling intervals on truck transport vibration levels. Journal of the Society of Agricultural Structures, Japan, 41(3), pp. 103-110.

[2] Gao J, Si Y, Zhu Y, Luo F, Yan S. (2018). Temperature abuse timing affects the rate of quality deterioration of postharvest broccoli during different pre-storage stages, 227, pp. 207-212, DOI: 10.1016/j.scienta.2017.09.034

[3] Gillies SL and Toivonen PMA. (1995). Cooling method influences the postharvest quality of broccoli. HortScience, 30(2), pp. 313-315. DOI: 10.21273/HORTSCI.30.2.313

[4] Dirapan P, Boonyakiat D. and Poonlarp P. (2021). Improving shelf life, maintaining quality, and delaying microbial growth of broccoli in supply chain using commercial vacuum cooling and package icing. Horticulturae, 7(11), 506, DOI: 10.3390/horticulturae7110506

[5] Sugino N, Watanabe T, Ohishi M, Sato F. and Kitazawa H. Effect of harvest season variation on mass loss and yellowing of broccoli (*Brassica oleracea* L. *italica*) during storage. Journal of Packaging Science and Technology, Japan, Submitted.

FP-O01

Application of Biodegradable Film as Modified Atmosphere Packaging for Red Chili (Capsicum annuum cv. Jinda)

Dane Archibald G. Balanon[a], Amporn Sane[a], Piyawanee Jariyasakoolroj[a], Pattarin Leelaphiwat[a,b]*

[a]*Department of Packaging and Materials Technology, Faculty of Agro-Industry, Kasetsart University, Bangkok, Thailand*
[b]*Center of Advanced Studies for Agriculture and Food, Kasetsart University, Bangkok, Thailand*

***Correspondence to:** pattarin.le@ku.ac.th*

ABSTRACT: Biodegradable blend films containing thermoplastic starch (TPS, polylactic acid (PLA) and poly(butylene adipate-co-terephthalate) (PBAT) were assessed for red chili storage under modified atmosphere packaging (MAP) at 10°C. Performances of biodegradable blend films were compared to polypropylene (PP) film and the control (unpacked). Gas permeabilities and permeability ratios were determined to measure the ability of the films to create an optimal atmosphere for red chili storage. PP film attained values close to the recommended MAP gas concentrations and maintained desirable chili quality parameters up to the 15th day. Biodegradable films did not achieve the recommended MAP gas concentrations but high vapor permeability prevented moisture loss and maintained firmness and overall quality. No discoloration was recorded in all MAP treatments, while biochemical parameters showed a decreasing trend, with measure levels significantly higher in MAP samples than unpacked samples. MAP for red chili was achieved using a packaging system, with similar gas permeabilities to PP film but significantly higher water vapor permeability.

Keyword: biodegradable blend film, red chili, modified atmosphere packaging, permeability, permeability ratio

CONFREENCE PROCEEDINGS :

PACKAGING FOR HAZARDOUS & DANGEROUS GOODS

• ORAL PRESENTATION

• PEER REVIEWED STREAM

HP-O01 Development of an analysis and testing concept for the evaluation of impact targets in the mechanical safety testing of dangerous goods packagings

HP-O02 Flow properties of powdery or granular filling substances of dangerous goods packagings - Comparison of the measurement of the angle of repose and the determination of the Hausner ratio

HP-O01

Development of an analysis and testing concept for the evaluation of impact targets in the mechanical safety testing of dangerous goods packagings

Nikolaos Lengas[1], Karsten Müller[1], Eva Schlick-Hasper[1], Marcel Neitsch[1]
Sergej Johann[1], Manfred Zehn[2]

[1]BAM Federal Institute for Materials Research and Testing, Berlin, Germany
[2]Institute of Mechanics, TU Berlin, Berlin, Germany

***Correspondence to:** Nikolaos Lengas, BAM Federal Institute for Materials Research and Testing, Berlin, Germany*
Email: nikolaos.lengas@bam.

ABSTRACT: The mechanical and geometrical properties of impact targets greatly influence the outcome of a drop test. The International Agreement concerning the International Carriage of Dangerous Goods by Road (ADR) as well as ISO 2248 describe the characteristics of impact targets for drop tests of dangerous goods packagings. According to these regulations, the impact target's surface needs to be unyielding, under testing conditions non-deformable, flat and integral with a mass at least 50 times that of the heaviest packaging to be tested. The problem is that many production facilities, especially manufacturers of corrugated fibre board boxes, do not have their own testing device with the required 50 times mass ratio of the impact target for a regulation compliant drop test during series production. Furthermore, at UN level, it is considered necessary to revise these requirements. In the present paper, the impact target requirements are examined in detail and compared with those in other technical areas (e.g., impact target for container for the transport of radioactive materials). Are search method is being developed to investigate the dependency between the mass ratio of the packaging and the target as well as the damage resistance of a drop tested package in relation to specific design characteristics. The results are of high relevance for industry purposes and intended to ensure a uniform level of safety assessment for the mechanical testing of dangerous goods packagings.

Keyword: corrugated fibre board boxes, dangerous goods packagings, drop test, impact target, structural dynamics

HP-O02

Flow properties of powdery or granular filling substances of dangerous goods packagings - Comparison of the measurement of the angle of repose and the determination of the Hausner ratio

Eva Schlick-Hasper[1], John Bethke[1], Nico Vogler[1], Thomas Goedecke[1]

[1]*Federal Institute for Materials Research and Testing (BAM), Berlin, Germany*

ABSTRACT: To ensure safety when transporting dangerous goods, it is important to specify the flow properties of the respective solid filling substance of the packagings. For this purpose, the angle of repose is currently used for the UN approvals in Germany. Measurements were carried out on 12 powdery or granular substances applying the angle of repose measuring methods customary in the test centres. The results of the methods differ significantly from each other. In addition, some of the techniques cannot be applied for very cohesive or coarse-grained materials. The results for the angle of repose show a strong scatter for some constellations (coefficient of variation more than 20 %). As a consequence, an alternative parameter to characterize the flow properties of bulk materials should be considered, such as the Hausner ratio. This approach leads to more precise test results for the substances examined (maximum coefficient of variation 2.8 %). It also has advantages in terms of applicability and occupational safety. Since dangerous goods are repeatedly released during transport from packagings that are actually intact, and flowability is an important influencing factor for this, the testing practice should be improved and standardized.

Keyword: sift-proofness; powders; flow properties; flowability; angle of repose; Hausner ratio

CONFREENCE PROCEEDINGS :
PACKAGING MATERIAL

• ORAL PRESENTATION

• GENERAL STREAM

MT-O01

Determination and Pre-treatment of Hazardous Substances in Food Can

Rao Fu, Shuai Wu, Ruijia Liu, Jiacheng Li, Zhe Chen, Zhihang Deng,
Ruomei Wu*

College of Packaging and Materials Engineering, Hunan University of Technology, Zhuzhou, China

ABSTRACT: The components of harmful substances in food contact materials (FCMs) are complex and unknown. The key of food safety evaluation is to detect the migration of harmful substances in food packaging. The variety of food substances is complex, so the food simulants are used in test. Such as distilled water, acetic acid(3g/100mL), ethyl alcohol, and edible oil are used to simulate aqueous food, acid food, alcoholic food, and fatty food, respectively. Then the liquid-liquid extraction, solid-phase extraction and auxiliary extraction are used to extract substances form food simulants. Finally, the harmful substances are quantitatively analyzed by high-performance liquid chromatography (HPLC) or gas chromatography-mass spectrometry (GC-MS), so as to evaluate the safety of food packaging. In view of the problems of safety research in food packaging materials, such as unclear migration mechanism and imperfect detection methods of harmful substances, it is the future research direction to explore efficient pretreatment methods combined with rapid detection methods, and improve the migration mechanism of harmful substances and establish migration models.

Keyword: Food Can; Hazardous Substances migration; Pre-treatment; Detection

1. INTRODUCTION

FCMs refers to substances that comes into direct contact with food and may migrate its components to the food in use. The harmful substances in FCMs are divided into intentionally added substances and unintentionally added substances. Intentionally added substances are artificially added substances to improve the performance of food packaging, and its using range and content have corresponding national regulations. Unintentionally added substances refer to the harmful substances formed due to the degradation of food packaging and its additives[1,2]. As a common food packaging, metal packaging has the advantages of excellent barrier performance, high mechanical strength, recycling and so on. However, the chemical stability of the metal is poor. In order to prevent corrosion of packaging materials caused by metal packaging in contact with food, coating on the inner wall is required to protect the tank wall[3]. The inner wall coating of metal food packaging is in direct contact with the food itself, and some harmful substances contained in the coating, such as free formaldehyde, bisphenol A and its derivatives, will migrate to the food and be ingested [4,5]. Long-term ingestion of such harmful substances will harm to human health. Therefore, it is of great significance to accurately monitor the migration of harmful substances in food packaging and study its migration rules. Food packaging generally goes through the testing process shown in Figure 1. In this paper, the harmful substances in metal food packaging are elaborated, the pre-treatment methods and common detection methods of food packaging safety testing are studied, and the detection methods are prospected.

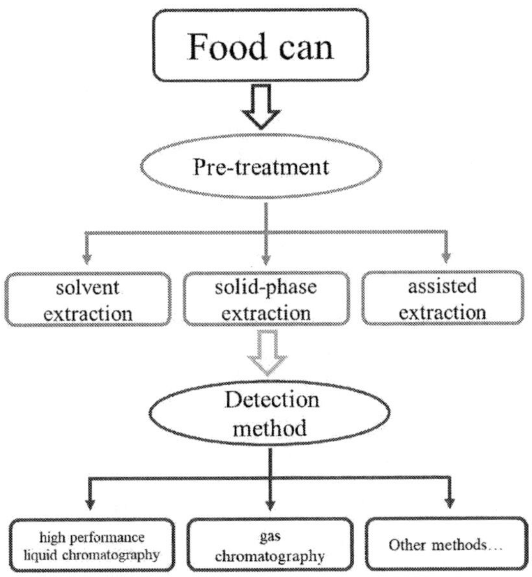

FIG. 1 Food packaging safety testing process

2. The Hazardous Substances of Food can coating

The safety problem of metal food packaging is mainly the migration of harmful substances in the coating to food. The harmful substances in the inner coating film of food packaging are generally exogenous endocrine disruptors, also known as environmental hormones. When chemicals enter the human body,they do not directly harm the body as a toxic substance,but interfere with the normal secretion of hormones in the body, play a similar role to estrogen in the human body, will cause harm to the reproductive system, immune system and nervous system and so on[6].

2.1 Bisphenol substance

Bisphenol substances generally refer to substances with two hydroxyphenyl structures, including bisphenol A (BPA),bisphenol S (BPS) and so on[7]. Bisphenol A is currently the most widely used, which is an important raw material for the production of polycarbonate, epoxy resin, and is widely used in food packaging. Long-term intake of BISphenol A can cause endocrine disorders, affecting the reproductive, nervous and immune systems. China's Ministry of Health has banned the sale of baby food containers containing BPA, as the safety of bisphenol A has been questioned. Therefore, bisphenol S and other substitutes are used in infant food packaging, but some studies have shown that bisphenol S intake can interferes with metabolism and lead to obesity[8].

According to China's national standard GB 9685-2016 (Standard for Additives used in Food Contact Materials and Products), the specific migration amount of Bisphenol A in food packaging materials and inner wall coatings is 0.6mg/L, and bisphenol A shall not be used in infant food packaging materials[9]. Bisphenol F can be added to the coating according to the production needs, but the migration amount should not exceed 0.5mg/L.

2.2 Novolac Glycidyl Ether (NOGE)

NOGE can be obtained by reaction of phenolic aldehyde with epichlorohydrin, which is the general term of many novolac multivariate glycerol ether mixtures with different molecular weights. The gastrointestinal tract can absorb compounds with molecular weights less than 1000Da, so generally only 3 to 6 rings NOGE is studied. NOGE has strong fat-soluble and is easy to migrate to some fatty foods to form a series of derivatives. The European Union limits its content to less than 1 mg/kg and prohibits the use of materials containing NOGE after January 2005[10,11].

2.3 Other hazardous substances

Phthalates (PAEs), as a common plasticizer, are used in metal packaging sealing and other aspects. It can enter the human body through respiratory tract, skin, etc., and has been proved to be reproductive toxicity, which may affect fetal development. According to relevant state regulations, only diallyl phthalate(DAP) dibutyl phthalate (DBP), di(2-ethylhexyl) phthalate(DEHP), diisononyl phthalate (DINP) can be used in food packaging[12]. Formaldehyde is colorless and has a irritating smell, has an irritating effect on the eyes and nose, long-term ingestion will damage the nervous system, liver and kidney, is listed as a level 1 carcinogen. Food packaging used in adhesives and inner coating films will introduce formaldehyde. If the formaldehyde reaction is incomplete during the production process, the residual formaldehyde will migrate into the food after long-term contact with packaging materials or high-temperature cooking [13]. According to GB4806.10-2016 "National Standard for Food Safety", free formaldehyde in food packaging shall not be higher than 0.1mg/L[14]. The structure and properties of common harmful substances are shown in Table 1.

Table 1 structure and properties of common harmful substances

Name of hazardous substance	Structure	Nature
BPA	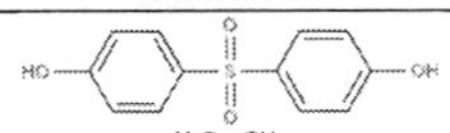	Fat-soluble
BPS	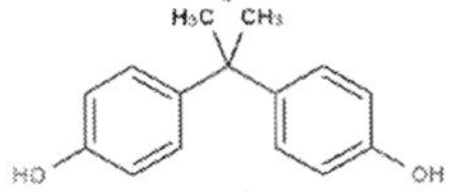	Fat-soluble
PAEs	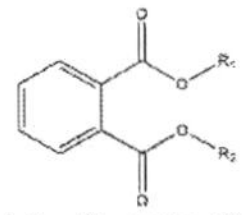	Fat-soluble
Formaldehyde	$H_2C=O$	Fat-soluble and water-soluble

3. The Pre-treatment of Food can

Before testing harmful substances in food contact materials, samples need to be pre-treated. Most of the experiments were conducted using food simulants.Then solvent extraction, solid phase extraction, auxiliary extraction and other methods are used to complete the enrichment of the target substances, in order to meet the requirements of subsequent detection.

3.1 The food simulants

At present, most migration research experiments are carried out through food simulants, and the content of harmful substances in the migration from packaging materials to food simulants is detected by changing simulated species, contact time and contact temperature. And the types of food simulants are specified, as shown in Table 2.

Table 2 food simulants list

Types of food	Food simulants standards
Non-acidic food	Water / 10% alcohol
Acidity food	4% acetic acid
Greasy food	N-hexane/edible oil
10%< Ethanol content≤20%	20% alcohol
20%< Ethanol content≤50%	50% alcohol
Ethanol content>50%	95% alcohol

Although the migration of harmful substances can be accurately determined, the migration amount of the foodt mimemet is greater than that in the food.Therefore, in order to accurately detect, it is better to use real food in the migration experiment.

3.2 Pre-treatment of tested sample

3.2.1 solvent extraction

Solvent extraction method, also known as liquid-liquid extraction method, is the use of a selected solvent to extract a certain component of mixed solution. Generally, acetonitrile, n-ethylene, methanol, etc. are used as extractants, and because of their versatility,they play an important role in the extraction of food hazardous substances. In the extraction process, in order to ensure that harmful substances can be completely extracted, it is necessary to carry out multiple extraction, so more solvents are used and time consuming. Garcia-ibarra et al. [15] used acetonitrile as the extractant to extract diethyl phthalate, dibutyl diphthalate, phthalate, tributyl acetyl citrate and benzophenone for detection. Wang Dongxu et al. [16] used acetonitrile as an extractant to extract seven bisphenol A, bisphenol F and other bisphenol substances in coffee cans, and determined them by liquid chromatography-mass spectrometry. The results showed that the linear correlation coefficients of seven bisphenol compounds were greater than 0.998 in the range of 2.0-200.0 μg /kg, and the recoveries were in the range of 80.2%-97.4%. The equipment used in the above method is simple and easy to operate.

Dispersive phase liquid-liquid microextraction is to inject the dispersant and extractant into the sample solution of analyte to make the trace extractant contact with the sample solution. After centrifugation, analyte is transferred to the extractant to achieve enrichment of the analyte. Yan Meng Meng et al. [17] pretreated food contact materials with 1-butyl-3-methyl-imidazolium hexafluorophosphate and 1-hexyl-3-methyl-imidazolium hexafluorophosphate solutions as dispersants and acetonitrile as extraction agent, and detected perfluorooctane acid (PFOA) and perfluorooctane sulfonic acid (PFOS) in food packaging contact materials by liquid chromatography-tandem mass spectrometry. The average recoveries of water, ethanol and acetic acid were 86.4%-115.3% and 89.8%-116.9%, and the relative standard deviations were 4.3%-12.2% and 4.6%-14.4%. This method uses less solvent and has high enrichment ratio. However, the dispersed phase liquid-liquid microextraction can not directly extract compounds in solid phase materials. The compounds to be analyzed need to be transferred to an appropriate solvent before extraction.

3.2.2 solid-phase extraction

Solid phase extraction (SPE) is a traditional extraction method, which uses solid phase adsorbent to adsorb the target in the liquid phase, and then elution with eluent for detection, also known as liquid-solid extraction. The key of SPE is to choose suitable filler according to different target materials. The commonly used fillers are polymer fillers and inorganic fillers. Wu Ziyang [18] used SPE extraction to extract phenols and optimized extraction conditions. Good extraction results were achieved when the pH of the sample was 2, the pH of the methanol elution solution was 4 and the dosage was 6 mL. SPE consumes less solvent and has a high recovery rate, but the fillers are easily blocked and the extraction time is long. In recent years, carbon nanotubes (CNTS) as a new type of filler have attracted extensive attention. Jiang et al [19] used the prepared doped magnetic carbon nanotubes to detect bisphenol A for canned drinks. The adsorption of bisphenol A was completed in only 10min with 5.76mg adsorbent. Carbon nanotubes have a high specific surface area, with the advantages of strong adsorption capacity and high adsorption capacity, which can improve detection sensitivity and reduce detection costs when applied to food testing.

The process of solid phase microextraction is simple, and the extraction head can complete the extraction and concentration of harmful substances, which greatly reduces the amount of solvent. Lin et al. [20] used fiber-solid phase microextraction technology to enrich mercury metal and analyzed it in combination with gas chromatography-dielectric barrier discharge-optical emission spectrometry. The detection limits of mercury, methyl mercury and ethyl mercury were 0.5, 0.75, 1.0 μg/kg, and the spiked recoveries were 90%-105%. At present, the commercial extraction head has a single type, poor adsorption capacity and high price, which limits its application scope. Lv Fangying [21] extracted polychlorinated biphenyls (PCBs) from inner wall coating by introducing metal-organic framework materials (MOF) with high specific surface area and covalent organic framework materials (COF), and detected PCBs in samples by gas chromatography-mass spectrometry. The adsorption capacity of the prepared extractor was more than 4 times that of the commercial extractor.

3.2.3 assisted extraction

Ultrasound-assisted extraction is to use the strong vibration generated by ultrasonic waves to promote the migration of harmful substances to solvents and shorten the extraction time. Lin et al. [22] extracted plasticizers and bisphenol from inner wall coatings of metal cans such as aluminum cans with tert-butylmethyl ether as extraction agent, and completed extraction after 20 minutes with ultrasonic assistance. Microwave-assisted extraction is a method that utilizes microwave assisted extraction to provide extraction efficiency. Zhang Xian-chen et al. [23] extracted 48 kinds of harmful substances in food contact materials with methanol as the extraction solution assisted by microwave, and the extraction efficiency reached the highest within 10min. Liquid chromatography-mass spectrometry was used for detection. The results showed that the limit of detection for harmful substances was between 0.1-1.0 μg /kg. The average recoveries were 71.2%-108.8% with relative standard deviations (RSDS) of 2.2%-11.8%. Assisted extraction has a great prospect in the detection of food contact materials because of its rapid and high efficiency.

4 The Detection method of Hazardous Substances

4.1 High performance liquid chromatography

High performance liquid chromatography, also known as high pressure liquid chromatography, uses a high-pressure system to pump the mobile phase into a column containing a stationary phase. Samples are analyzed according to the time at which each group is shunt out of the column. Ye Minli et al. [24] used liquid chromatography to detect the contents of benzene and toluene migrating from the outer layer of cans to the inner wall. The results showed that benzene and toluene were not detected in the inner wall of cans without stacking.But after stacking for two months, benzene and toluene were detected in the inner wall of cans. In the concentration range of 1-20mg/L, the correlation coefficient of benzene was 0.9992, and the detection limit was0.04 mg/L. The correlation coefficient of toluene was 0.9993, and the detection limit was 0.06mg/L. It is concluded that the effects of stacking temperature, stacking pressure and stacking relative humidity on benzene and toluene migration decrease successively. HPLC is suitable for the analysis of compounds with strong polarity and poor thermal stability.

HPLC can be matched with fluorescence detectors and mass spectrometers for detection. Sun Xilan et al. [25] used high performance liquid chromatography-fluorescence detection method to detect the content of bisphenol A in the inner wall coating of aluminum cans. The results showed that the concentration of bisphenol A was in the range of 0.1-20 μg /mL, the correlation coefficient was 0.9985, the detection limit was 0.05μg /mL, and the detection recovery rate was 95%-100.9%. Li Quanzhong et al. [26] used liquid chromatography-mass spectrometry to measure the migration of

28 primary aromatic amines (PAAs) in the inner coating film of aluminum cans. The limits of detection and quantification of 28 kinds of PAAs were 0.02-0.3μg/ Kg and 0.1-1.0μg/Kg, and the correlation coefficients were greater than 0.9915 in the concentration range of 1.0-1000μg/L. Liang Xizhen et al. [27] simultaneously detected bisphenol A, bisphenol F and bisphenol S in food contact materials by liquid chromatography-triple quadrupole mass spectrometry, and the linear relationship between the three substances in the concentration of 1-50μg/L was good. The lowest detection limits of bisphenol A and F were 0.3 μg /L, and the lowest detection limits of bisphenol S were 0.05 μg /L.

4.2 gas chromatography

Gas chromatography takes the gas as the mobile phase, the sample into the chromatographic column, the different components of the sample and the force of the solid phase are different, and the time through the column is also different, so as to achieve the separation effect of different components.Darowska et al. [28] used gas chromatography to detect formaldehyde and acetaldehyde contents in food contact materials, and showed that the contents of formaldehyde and acetaldehyde migrated into food were related to the contents of formaldehyde and acetaldehyde in bottle materials. Zhang Jie et al. [29] detected diallyl phthalate by gas chromatography-mass spectrometry using cyclohexane as extraction solution with ultrasonic assisted extraction, and obtained that the detection limit of diallyl phthalate was 0.15 μg /kg, and the recovery rate was 90%-98%. Nanyang et al. [30] used gas chromatography-mass spectrometry to detect 32 volatile organic compounds in food contact materials, and found that in the range of 0.01-0.17mg/m^2, the correlation coefficient was greater than 0.99, the spiked recovery rates were 90.3%-109.9%, and the relative standard deviation was less than 9%. Gas chromatography-mass spectrometry (GC-MS) is suitable for the detection of volatile small molecules and can directly detect non-polar and highly volatile substances, with high speed and sensitivity. Difficult volatile substances can not be directly detected, and need to be converted into the corresponding volatile derivatives for detection.

4.3 Other methods

Buculei et al. [31] studied the migration of Cd, Pb, Cu, Fe, Zn and Sn in canned meat products by flame and graphite furnace atomic absorption spectrometry. Qiao Zhaohua et al. [32] measured the migration amount of aluminum in aluminum-plastic packaging materials by inductively coupled plasma-mass spectrometry. After determination,the average recoveries of acetic acid and distilled water simulants were 92.2%-103.4% and 94.5%-105.5%, with relative deviations between 1.20%-5.52% and 1.27%-7.78%, respectively. Limits of detection and quantitation were 0.17 μg /L and 0.56 μg /L, respectively. Lu[33] et al used immunoassay to detect bisphenol A in canned vegetables and canned drinks, with detection limits of 0.5μg/L and 100 μg/L, respectively. Currently, there are also nuclear magnetic resonance spectroscopy, anodic stripping voltammetry and other methods, which cannot be widely used in the detection of food contact materials due to relatively high experimental conditions [34].

5. CONCLUSIONS

In the pretreatment of food packaging safety testing, solvent extraction steps are cumbersome and consume more solvents. Compared with solvent extraction, solid-phase extraction has a small amount of extraction phase and is easier to elute solutes. It can be combined with microwave-assisted extraction to accelerate extraction speed. The key of solid phase microextraction is the extraction head. Currently, the commercially available extraction head has poor adsorption capacity and single adsorption substance, so it is necessary to develop the extraction head that can absorb a variety of substances in the future. Gas chromatography-mass spectrometry (GC-MS) and liquid chromatography-mass spectrometry (LC-MS) are widely used in migration experiments. GC-MS has the advantages of high sensitivity and small sample quantity, but it requires strict detection conditions and long time consuming,so it is necessary to explore new detection methods. For degradable food packaging, its own non-toxic, but in the process of degradation will produce harmful substances, for degradable food packaging needs to develop new detection methods for detection. For plastic packaging materials, there are mature digital models to predict the migration of harmful substances, but the migration model of metal coating is complicated and there is no mature mathematical model. Therefore, it is a low-cost way to establish data models through mathematics and computers.

There have been a lot of reports about metal food packaging, but there are still some problems to be further studied. The main research directions in the future are: to develop a simple and fast pre-treatment method, combined with high sensitivity, can detect a variety of harmful substances at the same time, rapid detection of harmful substances in food packaging. The toxicological principle of harmful substances and migration mechanism of metal coatings were clarified to provide theoretical support for the establishment of migration model of harmful substances in metal packaging.

6. REFERENCES

[1] González-Castro MI, Olea-Serrano M F, Rivas-Velasco A M, et al. Phthalates and Bisphenols Migration in Mexican Food Cans and Plastic Food Containers[J].Bulletin of Environmental Contamination and Toxicology,2011,86.

[2] Huang Xiang-lu, Li Li, Cao Jin, et al. Research progress on safety testing of food contact materials in China [J]. Chin Pharm Aff, 2012, 26(5):513-516.

[3] He M, Yang C, Geng R, et al. Monitoring of phthalates in foodstuffs usinggas purge microsyringe extraction coupled with GC-MS [J]. Anal ChimicaActa, 2015, 879: 63-68.

[4] Wang T, Ma M, Shao M, et al. Simultaneous determination of migration ofbisphenol s, bisphenol a and 4, 4'-dichlorodiphenyl sulfone in plastic foodcontact materials by HPLC [J]. Phys Test Chem Anal Part B: Chem Anal,2017, 53(12): 1398–1402.

[5] Zhou LY, Luo HQ. The domestic and foreign regulations and detectionmethods for bisphenol A in the food contact materials [J]. Chin J FoodHyg, 2016, 28(6): 787–790.

[6] Liu K, Du J, Wu J, et al. Superhydrophobic gecko feet with high adhesive forces towards water and their bio-inspired materials[J]. Nanoscale, 2012, 4(3): 768-772.

[7] B. Lokahita, M. Aziz, K. Yoshikawa, et al. Energy and resource recovery from Tetra Pak waste using hydrothermal treatment [J]. Applied Energy, 2017, 207: 107-113.

[8] C. I. K. Diop, J. M. Lavoie. Isolation of Nanocrystalline Cellulose: A Technological Route for Valorizing Recycled Tetra Pak Aseptic Multilayered Food Packaging Wastes [J]. Waste & Biomass Valorization, 2016, 8(1): 1-16.

[9] GB 9685-2016 Food safety national standards-Use of additives in foodcontact materials and products [S].

[10] GB 4806.1-2016 Food safety national standards-General rules for foodcontact materials and products [S].

[11] GB 31604.1—2015, General Rules for MigrationTesting of Food Contact Materials and Products underNational Food Safety Standards[S].

[12] Shiheng Yin,Dongxiao Wu,Ji Yang,Shumei Lei,Tong chun Kuang,Bin Zhu. FabricationandSurface characterization of biomimic superhydrophobic copper surface bysolution immersion and self-assembly[J]. Applied Surface Science ,2011 (257) 8481–8485.

[13] DegangXie,Wen Lia,.A Novel simple approach to preparation of superhydrophobic surfaces of aluminum alloys[J]. Applied Surface Science, 2011 (258) 1004–1007.

[14] GB 4806.10—2016, National Food Safety Standard Coatings and Coatings for Food Contact[S].

[15] Garcia-Ibarra V, Rodriguez-Bernaldo DQA, Paseiro-Losada P, et al.Identification of intentionally and non-intentionally added substances inplastic packaging materials and their migration into food products [J].Anal Bioanal Chem, 2018, 410(16): 3789-3803.

[16] Wang Dong-xu,Wang Xin-cai,Hu Qi-jie,Wang Feng-li,Chen Chu-jian,Li Fen.Determination of 7 bisphenols in canned coffee drinks by salting-out assisted liquidliquid extraction coupled with ultra-performance liquid chromatography-tandem mass spectrometry[J]. Chinese Journal of Analysis Laboratory,2020,39(08):986-992.

[17] Yan Meng-meng,Chen Meng,Meng Xians-huang,Guo Xiang-yu,Bai Hua,Ma Qiang,Wang Penglong,Lei Haimin. Determination of migration of perfluorooctanoic acid and perfluorooctane sulfonic acid from food contact materials by ionic liquid-based dispersive liquid-liquid microextraction and ultra-performance liquidchromatography-tandem mass spectrometry[J]. Chinese Journal of Chromatography,2018,36(08):738-744.

[18] Wu Zi-yang,He Fei,Li He-he. Determination of phenolic active compounds in qingxiang liquor by solid phase extraction combined with high performance liquid chromatography[J/OL]. Chinese Journal of Analytical Chemistry,2020,7(4): 1-14.

[19] Jiang H L, Lin Y L, Li N, et al. Application of magnetic N—doped carbon nanotubes in solid—phase extraction of trace bisphenols from fruit juices [J].Food chemistry,2018,269:413—418.

[20] Lin Y, Yang Y, Li Y. Ultrasensitive speciation analysis of mercury in rice by headspace solid phase microextraction using porous carbons and gas chromatography-dielectric barrier discharge optical emissionspectrometry [J]. Environ Sci Technol, 2016, 50(5): 2468–2476.

[21] Lv Fang-ying. Research of solid phase micro-extraction method and pretreatment coupled with GC-MS for determination of persistent organic pollutants in environment and food[D].Ning Bo: Ningbo University.

[22] Lin QB, Cai LF, Wu SJ, *et al.* Determination of four types of hazardouschemicals in food contact materials by UHPLC-MS/MS [J]. PackagTechnol Sci, 2015, 28(5): 461-474.

[23] Zhang Xian-chen,Zhang Peng-jie,Sshi Chen-gyu,Hua Hong-bo,Rong Yu-tang,Lu Jun-wen,Yang Fang. Rapid determination of 48 contaminant residues in food contactplastic products by ultra performance liquid chromatographyquadrupole/electrostatic field orbitrap high resolution massspectrometry coupled with microwave-assisted extraction[J]. Chinese Journal of Chromatography,2018,36(07):634-642.

[24] Ye Min-li，Xiang Hong，Li Da-wei，Fan Xiao-ping，Sun Kui-kui. Influencing Factors of Migration of Benzene and Toluene from External Coatings to Internal Coatings of Metal Cans[J]. PACKAGING ENGINEERING,2015,36(23):43-47+57.

[25] Sun Xilan，Zhu Zhengli，Shan Yingying. Determinationof Bisphenol A in Coatings of Metal Food Cans by HPLCFluorescenceDetectionMethod[J]. Packaging Engineering，2013，34(13)：27-30.

[26] Li Quan-zhong,Zhou Ming-hui.Liquid chromatography-tandem mass spectrometry was used to determine the migration of 28 primary aromatic amines in food contact materials[J]. Chinese Journal of Analytical Chemistry,2013,41(01):25-30.

[27] LiangXi-zhen1,Sui Hai-xia，Li Dan,Zhong Huai-ning. Determination on Migration of Bisphenol A,Bisphenol F and Bisphenol S in FoodContact Materials by Liquid Chromatography—Tandem Mass Spectrometry[J]. Journal of Instrumental Analysis, 2018,37(01):87-91.

[28] Darowska A, Borcz A, Nawrocki J, et al. Aldehyde contamination of mineral water stored in PET bottles [J]. Food Addit Contam A, 2003,20(12): 1170-1177.

[29] Zhang Jie,Zhou Ming-yan,Xia Jie,Li Baoyuan1,Luo R ong,Xiong Bo. Determination of diallyl phthalate in plastic lunch boxes by high performance gas chromatography — mass spectrometry[J]. Yunnan Chemical Technology,2020,47(08):106-108.

[30] Nan Yang, Du Yue, Li Chang-xing. Determination of VOCs in Food Packaging Materials by GC-MS[J]. PACKAGING ENGINEERING,2018,39(01):58-63.

[31] Buculei A, Amariei S, Oroian M, et al. Metals migration between product and metallic package in canned meat [J]. LWT-Food Sci Technol, 2014,58(2): 364-374.

[32] Qiao Zhao-hua,Shan Li-jun,Lin Qin-bao,Song Huan,GUO Jie,Wang Rong-zhen. Aluminum migration from aluminum-plastic composite packaging to food simulants[J].Science and Techonlogy of Food Industry,2016,37(13):250-253.

[33] LU Y, PETERSON J R, GOODING J J, et al. Development of sensitive direct and indirect enzyme-linked immunosorbent assays (ELISAs) for monitoring bisphenol-A in canned foods and beverages[J]. Analytical and Bioanalytical Chemistry, 2012,403(6): 1607-1618.

[34] Jakmunee J, Junsomboon J. Determination of cadmium, lead, copper and zinc in the acetic acid extract of glazed ceramic surfaces by anodic stripping voltammetric method [J]. Talanta, 2008, 77(1): 172-175.

MT-O02

Determination of stretch film properties with hysteresis for freight transport applications

Elora Leguebe*[1,2,3], Baptiste Boutrige[1,2,3], Victor Huart[1], Nicolas Krajka[1], Jérôme Pellot[1], Marie Le Baillif[2,3], Damien Erre[2,3]

[1]R&D, Metropack,Reims, 51100,France
[2]ITheMM, Université de Reims Champagne-Ardenne, Reims, 51100, France
[3]ESIREIMS, Reims,51100, France

***Correspondence to:** R&D, Metropack, 30-32 Rue du capitaine Georges Madon, Reims, 51100, France. Elora.leguebe@metropack.fr*

ABSTRACT: Nowadays, product protection during transportation has become essential as import-export is increasing drastically. As a result, wrapping is used to stabilize and protect the goods and must be performed properly. The most commonly used wrapping material is the linear low-density polyethylene (LLDPE), either manually or through a mechanical equipment. This study will focus on the adequation of stretch film standard to wrapping and transport applications. Both manual and mechanical films were analyzed. The anisotropy of the materials was investigated. Characterization of the material was focused on tensile properties, elastic recovery, permanent deformation, and stress retention in order to correspond to pallet wrapping applications. It was found that both materials showed anisotropic behaviors. Standard testing conditions were found to not be described properly transport wrapping applications. Relaxation time and extension speed were found to significantly affect hysteresis properties.

Keyword: stretch film; wrapping film; stretch film behavior; hysteresis; transport

1. INTRODUCTION

The stretch film is widely used in the packaging industry. The main application is tertiary packaging (for transport or storage) [1]. Pallet wrapping is the most popular technique for stabilizing a pallet. A packaging system presents many constraints which can damage it, such as shocks, vibrations during transport as well as various manipulations. All these operations can be responsible for a lot of damage. The purpose of laboratory testing of shipping packaging is to simulate or replicate these hazards [2]. The choice of the wrapping solution (material, stretching, technique, etc.) is therefore essential to optimize load stability [3].

Nowadays, the most used standard for the characterization of packaging film is ASTM D4649 [4]. This standard refers to other standard methods for laboratory film characterization, such as testing [5] and hysteresis tests [6]. It also describes some "field-test" methods which are then further detailed in ASTM D8314 [7]. If these laboratory characterizations provide interesting information on the mechanical properties of the film materials, these results do not describe wrap film behaviors after wrapping [8].

Prior transportation, wrap film are stretched at very high speed up to 100% to 300%, with a given pattern (number of layers, orientation). Then, the load is stored for some time (1 hours to several days or months) until the actual transport. Relaxation will occur during this storage time. Testing parameters described in ASTM D5459 standard do not allow to reproduce such conditions. ASTM D8314 standard describes several characterization methods of laboratory wrap films. However, the obtained results are always dependent of the wrapping parameters such as wrapping pattern and wrapping speed. Moreover, the described containment force testing methods are destructive and can't be compared one to each other.

The present study was focused on developing a hysteresis test method based on the ASTM D5459 standard, that would better simulate transport wrapping applications. The effect of testing parameters such as the speed, the relaxation time and recovery time on the measured film properties were investigated.

2. MATERIALS AND METHODS

The investigation was performed using cast co-extruded LLDPE stretch films (Resinex 581 grade machine wrapping and LC2 grade manual wrapping) supplied by Embal'Vit company (Reims, France). Specimens of LLDPE were prepared according to IS 14995 standards [9]. The polymer films were set between two sheets of paper prior cutting to avoid catching by the cutting machine blade. Specimens were cut according to standards in transverse (TD) and longitudinal directions (LD). Each specimen was taped on a cardboard frame (see Figure 1) and conditioned at room T°23°C (73.4°F) prior testing.

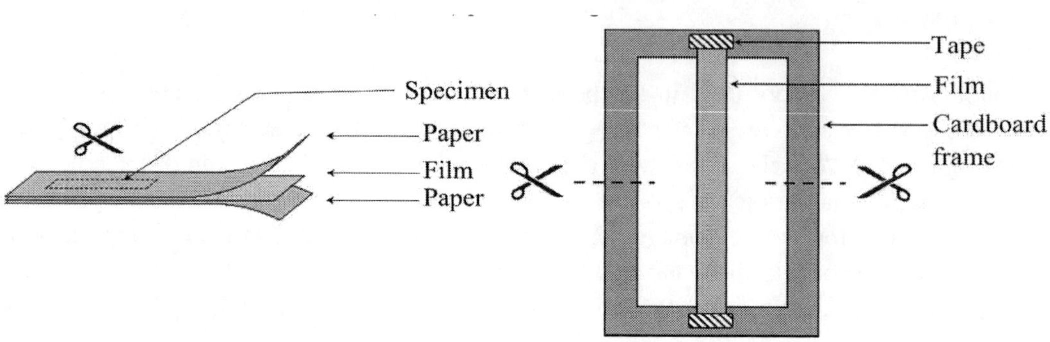

Figure 1 Assembly and cutting of test specimens

Mechanical characterization was performed through hysteresis tests on an MTS tensile testing machine. The accuracy of the testing machine was verified according E4 Practices [10]. The presented results were calculated out of 5 parallel measurements. Both machine and manual films were tested in longitudinal and transverse directions to analyze the anisotropy of the properties. The cardboard frame holding the specimen was set into the testing equipment to avoid any pre-stretching or stressing of the specimen. The frame was cut after the proper tightening of the grips prior starting the measurement.

Figure 2a shows the extension profile of the hysteresis test (where v is the chosen extension speed, ε_{max} the chosen extension level, t_{relax} the relaxation time applied at ε_{max} and $t_{recovery}$ the recovery time applied) and Figure 2b a typical resulting stress/strain curve. Elastic recovery (ER), permanent deformation (PD) and stress retention (SR) were calculated according to equations 1 to 3 (where GE also correspond to the residual stress σ_{∞}). Since elastic recovery and permanent deformation are complementary, it has been chosen to discuss on the elastic recovery only in the results.

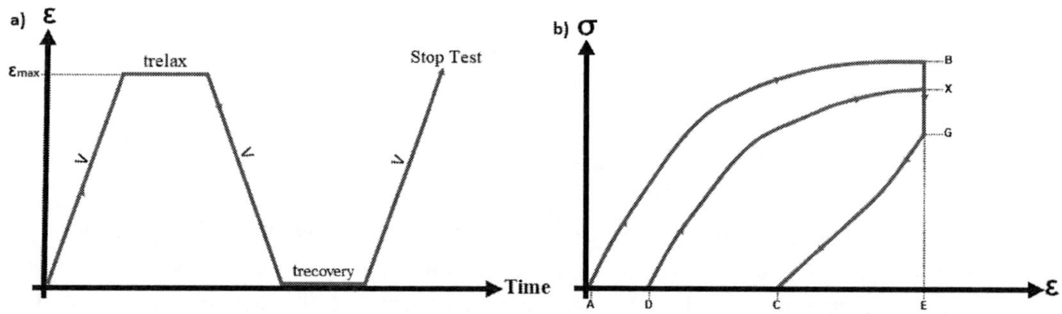

Figure 2 Hysteresis test: (a) extension profile, (b) typical stress-strain curve (A, B, G, C, D, X)

$$PD_{\%} = (AD/AE) \times 100 \quad \text{(Eq.1)}$$

$$ER_{\%} = (DE/AE) \times 100 \quad \text{(Eq.2)}$$

$$SR_{\%} = (GE/BE) \times 100 \quad \text{(Eq.3)}$$

The mechanical characterization was composed of hysteresis tests all based on the ASTM D5459 standard with an extension level of 200%. The parameters of each series were chosen to fit to either the ASTM D5459 standard or wrapping applications. Earlier studies have shown that residual stress was reached after 15 min of relaxation [11]. This value was therefore chosen for the high relaxation time. Wrapping applications show no recovery time and very high wrapping speed. Recovery time was therefore set to 0 min and high speed was chosen as the maximum speed allowed by the testing equipment (450 mm/min). This resulted in several series of hysteresis tests, which are described in Table 2. One may note that serie 5 corresponds to hysteresis test described in ASTM D5459 standard.

Table 2 Parameters for different tests.

Parameters \ N°	**1**	**2**	**3**	**4**	**5**	**6**	**7**	**8**	**9**	**10**	**11**
V	127	450	127	450	127	450	127	450	31.75	288.5	381
T_{relax}	1	1	15	15	1	1	15	15	15	15	15
$T_{recovery}$	0	0	0	0	3	3	3	3	0	0	0

3. Results and Discussions

3.1 Traction, relaxation, and hysteresis

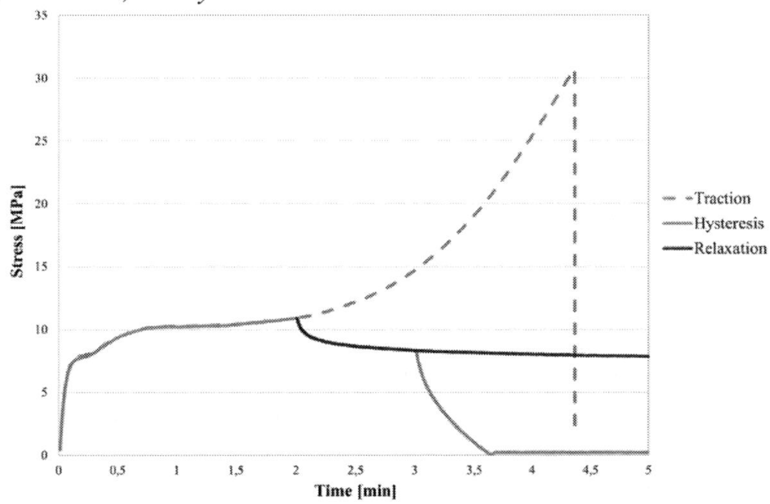

Figure 3 Traction, relaxation, and hysteresis of manual MD

Figure 3 presents the hysteresis compared to the tensile test of ASTM D822 [12] with a tensile speed of 127mm/min and to the relaxation test of ISO 2285[13] with a stress level of 200%. These tests were carried out previously [14] and are calculated on 10 parallel measurements for each film and in both directions. The curves overlap perfectly. Thus, the values obtained by the previous tests can be applied to the hysteresis. The test was also carried out at 450 mm/min and the results are identical.

3.2 Standard Test

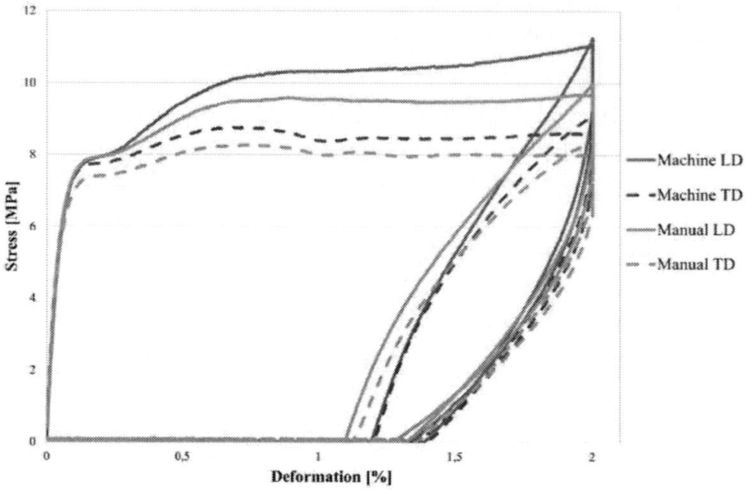

Figure 4 Hysteresis of machine and manual film in longitudinal (LD) and transverse (TD) orientations

Figure 4 presents hysteresis curves of the LLDPE films in longitudinal (LD) and transverse (TD) directions. It is clearly visible that both machine and manual films had anisotropic properties. Longitudinal stress was higher than transverse stress for both films, and machine film showed higher stresses than manual film.

Table 3 Properties of standard hysteresis

Materials		$ER_{\%}$	$PD_{\%}$	$SR_{\%}$
Manual	LD	44.66 ± 0.12	55.33 ± 0.12	74.41 ± 0.17
	TD	42.56 ± 0.48	57.43 ± 0.48	74.93 ± 0.20
Machine	LD	40.35 ± 0.41	59.64 ± 0.41	74.93 ± 0.43
	TD	39.04 ± 0.14	60.95 ± 0.14	74.74 ± 0.27

Hysteresis properties are described in Table 3 with the mean values and the standard errors. Elastic recovery was found to higher for manual films than machine films, and higher in longitudinal direction than transverse direction for both films. The stress retention was found to not depend on the type of film nor the testing direction. The differences between manual and machine films properties were most probably due to their manufacturing. They may have different pre-stretching levels and different average molecular weight. Further comparison between the type of films would be of little use for the present study. The higher elastic recovery in longitudinal direction than transverse direction shows the anisotropy of the films. It is known that LLDPE is a linear polymer. The cast process tends to align the polymer chain in the longitudinal direction [15]. It was therefore expected to measure higher elastic recovery in longitudinal direction.

3.3 Effect of extension speed

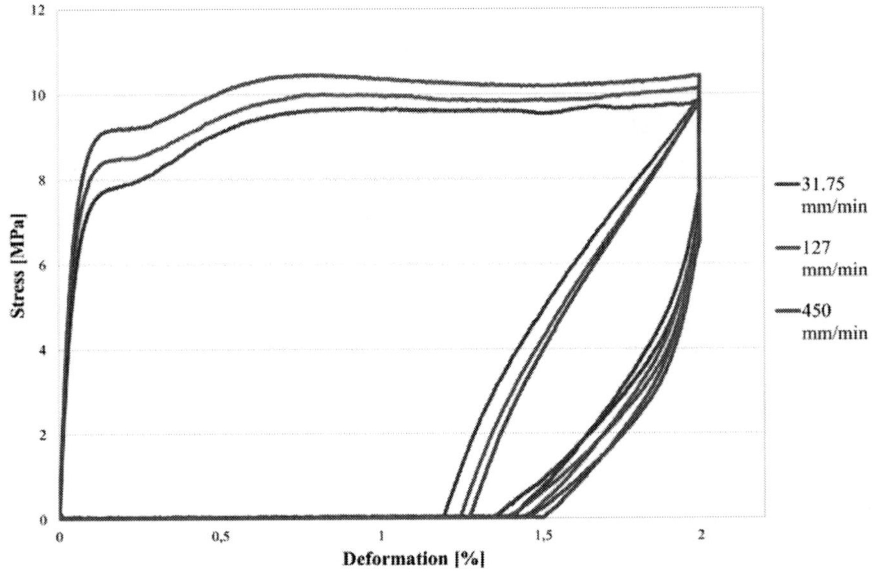

Figure 5 Different speed in curves of hysteresis at 15min of relaxation and 0 min of recovery

Figure 5 presents the effect of extension speed on the manual longitudinal direction film hysteresis curves, Table 4 the corresponding hysteresis properties and Figure 6 presents the residual stress of manual longitudinal direction film at different extension speed with 15 min of relaxation.

The increased measured stress with increasing speed was an expected effect of the viscoelastic behavior of a polymer. It can be observed that increasing the extension speed increased the extension the stress during the first extension. On the contrary, it decreased both stress retention and residual stress after relaxation and caused a slight decrease on elastic recovery.

However, it has been reported that low speed extension could lead to morphological changes in the polymer such as crystallinity changes, with larger and more uniform crystals [16]. These changes would result in higher stress measurement for low-speed straining. In the present study, morphology changes may not be sufficient to overcome the effects of viscoelastic responses during the straining. However, higher crystallinity would reduce the ability of the polymer to relax, explaining the effect of extension speed on residual stress and stress retention.

Table 4 Properties of hysteresis with different speed

Speed [mm/min]	$ER_{\%}$	$SR_{\%}$	$\sigma\infty$
31.75	40.00 ± 0.12	72.36 ± 0.23	7.11 ± 0.04
127	37.40 ± 0.14	67.71 ± 0.33	6.94± 0.07
288.5	36.77 ± 0.11	63.44 ± 0.60	6.55 ± 0.08
381	36.10 ± 0.23	62.50 ± 0.32	6.53 ± 0.04
450	35.28 ± 0.26	62.40 ± 0.29	6.47 ± 0.05

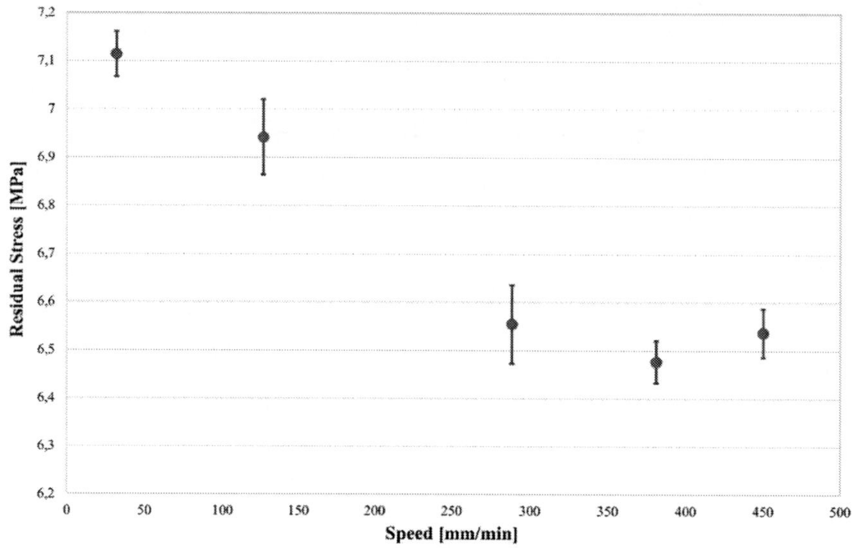

Figure 6 Effect of extension speed in residual stress of hysteresis

Similar trends were measured with shorter relaxation time, with specimen cut in the transverse direction and also with machine wrapping film. Further investigation of the effect of extension speed on the film's morphology would be necessary to confirm this effect.

3.4 Time of recovery

Table 5 presents the effect of recovery time on elastic recovery. Slight changes of elastic recovery were observed with recovery time. However, the obtained standard errors do not allow any significant tend. Similar results were obtained for all testing parameters, for both longitudinal direction and transverse direction and for both manual and machine films.

Table 5 Properties of hysteresis for machine film in LD

N° of series	Parameters	$ER_{\%}$	$PD_{\%}$
1	127 mm/min – 1 min – 0 min	42.52 ± 0.44	57.48 ± 0.44
5	127 mm/min – 1 min – 3 min	44.66 ± 0.12	55.33 ± 0.12
4	450 mm/min – 15 min - 0 min	35.28 ± 0.26	64.72 ± 0.26
8	450 mm/min – 15 min - 3 min	37.68 ± 0.29	62.31 ± 0.29

3.5 Effect of relaxation time

Figure 7 presents elastic recovery (Figure 7a), stress retention (Figure 7b) and residual stress (Figure 7c) of manual longitudinal direction films tested with short and long relaxation times. The relaxation is known to be an energy dissipation phenomenon through molecular rearrangement which leads to a decreased stress with time and permanent deformation [17,18,19]. The measured elastic recovery, stress retention and residual stress results were therefore expected followings of the longer relaxation time. Similar results were obtained for manual film.

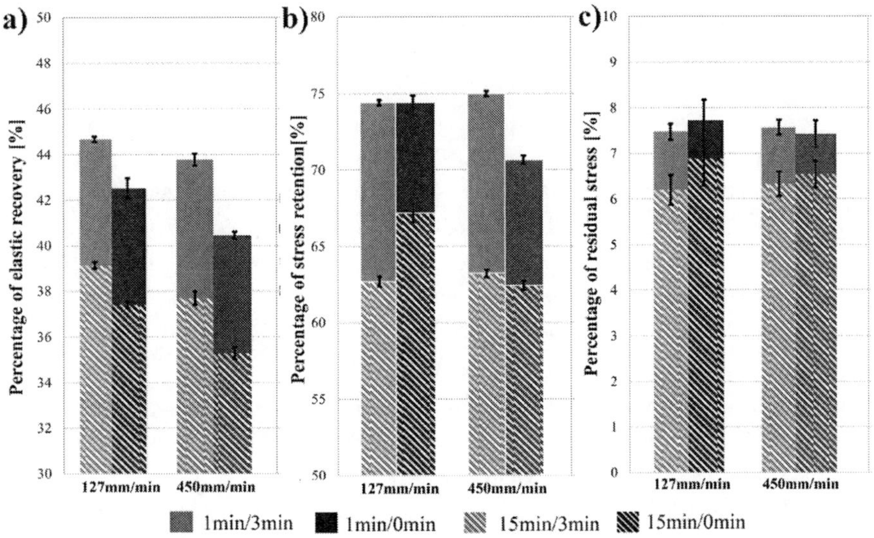

Figure 7 ER%, SR% and σ∞ with different hysteresis of machine LD

4. CONCLUSIONS

There tests have proved the importance of testing parameters, however, one must be cautious with any extrapolation with the extension speed. Wrapping being usually performed at speed of 8000 to 16000 mm/min [20]. These values are outside the experimental domain.

Mechanical properties of machine and manual wrapping films were investigated and both films showed similar and anisotropic properties. The study has shown the importance of testing parameters on the film mechanical properties. Extension speed and relaxation time were found to have significant effect on residual stress. The use of ASTM standard for optimizing load stabilization would require to use testing parameters more related to real wrapping conditions.

Choosing the adequate wrapping in view of load stabilization would require the knowledge of actual residual stress. The study has shown possible changes in film morphology due to testing parameters that could affect it. Further investigation on films morphology in real wrapping condition should be performed. The study of film responses to dynamical stress after relaxation would probably be of great interest for predicting proper load stabilization system.

The goal of wrapping is the stabilization of the load, against transport constraints. The intensity of these constraints will depend on the nature of the transportation but also the nature of the load. The stabilization would result from a combination of wrapping pattern and film ability to maintain the load. In that regard, the film property of interest would be the residual stress. The calculation of elastic recovery, stress retention and permanent deformation as proposed in the ASTM standard would not be sufficient since they are only relative values. The study of wrap films responses to dynamic load after relaxation would be of interest for predicting suitable stabilization.

5. ACKNOWLEDGMENTS

This research was largely supported by the METROPACK company (www.metropack.fr) with the financial support of a PhD.

6. REFERENCES

[1] J. Fauville – Emballage des produits industriels – Protection mécanique – AG6202, Techniques de l'ingénieur, 2002. https://doi.org/10.51257/a-v1-ag6202

[2] H. Marcel – Essais et programmes d'essais sur les emballages de transport – AG6290, Techniques de l'ingénieur, 2002. https://doi.org/10.51257/a-v1-ag6290

[3] D. Finnemore – Stretch wrap film and pallet load stability – CSC Publishing, powderbulk.com, 2017.

[4] ASTM D4649-20: Standard Guide for Use of Stretch Films and Wrapping Application.

[5] ASTM D882-18: Standard Test Method for Tensile Properties of Thin Plastic Sheeting.

[6] ASTM D5459-12: Standard Test Method for Machine Direction Elastic Recovery and Permanent Deformation and Stress Retention of Stretch Wrap Film.

[7] ASTM D8314-20: Standard Guide for Performance Testing of Applied Stretch Films and Stretch Wrapping.

[8] J. Bisha – The effect of load stabilizer selection on load shift within unit loads – Virginia Polytechnic Institute and State University, 2008. http://hdl.handle.net/10919/33035

[9] IS 14995 (2001): Stretch Cling Films [PCD 12: Plastics]

[10] E4 Practices for Force Verification of Testing Machines

[11] E. Leguebe - Study and characterization of palletizing films used in the field of freight transport - IAPRI Mexico 2020.

[12] ASTM D882-18: Standard Test Method for Tensile Properties of Thin Plastic Sheeting.

[13] NF ISO 2285-19: Caoutchouc vulcanisé ou thermoplastique – Détermination de la déformation rémanente sous allongement constant et de la déformation rémanent, de l'allongement et du fluage sous charge constante de traction.

[14] E. Leguebe – Determination of stretch film properties for freight transport applications - IAPRI 2021.

[15] G. McNally – The effect of polymer properties on the mechanical behavior and morphological characteristics of cast polyethylene film for stretch and cling film applications – Journal of Plastic film & sheeting, Vol 21, 2005. https://hal.archives-ouvertes.fr/hal-00572056

[16] J. V. Bisha - Correlation of the Elastic Properties of Stretch Film on Unit Load Containment - Ph.D. thesis, Virginia Polytechnic Institute and State University, 2012. http://hdl.handle.net/10919/27992

[17] Stress relaxation and creep – Brush Wellman Inc. Materion, Issue N°12, 2009.

[18] H. Ornaghi – Stress relaxation, creep, and recovery of carbon fiber non-crimp fabric composites – Composites science and technology, Elsevier, 2020. https://doi.org/10.1016/j.jcomc.2020.100051

[19] M. Sreekala - Stress-relaxation behaviour in composites based on short oil-palm fibres and phenol formaldehyde resin – Composites science and technology, Elsevier, 2001. https://doi.org/10.1016/S0266-3538(00)00214-1

[20] https://www.cenpac.fr/machines-systemes-emballage/banderoleuse/banderoleuse-plateau-tournant-modele-rotoplat-standard/p20144.

MT-O03

Evaluating the Relationship Between the Quality of 3D Printing and Rheological Properties of PLA and PLA\PHA

Hadis Torabi[1], Hayden Mcgreal[2], Ehsan Behzadfar*[1,2]

[1]Chemical Engineering Department, Ryerson University, Toronto, ON, Canada
[2]Sustainable Packaging Lab, School of Graphic Communications Management, Ryerson University, Toronto, ON, Canada M5B 2K3

***Correspondence to:** Dr. Ehsan Behzadfar, Ryerson University, Toronto, Canada,*
E-mail: behzadfar@ryerson.ca

ABSTRACT: There has been a growing interest in using Additive Manufacturing of biodegradable polymers to fabricate complex shapes of package prototypes and finished parts. While the quality of printed products has been well studied for conventional polymers, there is limited data in the literature that addresses the processing challenges of biodegradable polymers such as polylactic acid (PLA) and Polyhydroxy alkenoates (PHA) in the Fused Filament Fabrication (FFF) process. In this study, processing parameters such as nozzle temperature, bed temperature, printing speed, and material parameters such as rheological properties were investigated to obtain a quality product of PLA and PLA/PHA filaments. Our finding showed that extensional viscosity plays a significant role in determining the adhesion strength between the layers in the FFF process. In addition to the type of polymer, nozzle temperature, bed temperature, and print speed are effective on the quality of the printed products. Experimental results indicate that while the PLA/PHA samples showed the highest adhesion strength, toughened PLA required less peeling force compared to virgin PLA. Also, increasing the nozzle temperature has the most significant effect on the adhesion strength. On the other hand, the lowest print speed at 60 mm/s resulted in the highest adhesion strength between the layers.

Keyword: PLA, PHA, 3D printing, additive manufacturing, rheology

1. INTRRODUCTION

Fused Filament Fabrication (FFF) is considered one of the most common 3D printing methods and is still gaining attention. FFF is popular because of its simple design, generating non-significant amounts of waste, nontoxicity, environmental safety, and affordable cost of a 3D printing system [1]–[3]. Conventional plastics have created rising environmental concerns because of the constant rise in their carbon footprint during production and disposal [4], [5]. As a result, biodegradable polymers are viable options for reducing municipal solid waste generated [6]. There has been a lot of effort in replacing fossil fuel-based products with environmentally friendly bio-based/ biodegradable materials in recent years. Combining 3D printing technology and a renewable source of filaments is an almost environmentally friendly and creative way to produce biodegradable packaging [7]–[9]. PLA is one of the most commonly used thermoplastics in the FFF method in commercial applications since it comes from renewable resources and has good mechanical characteristics. PLA, like other thermoplastics, can be using traditional thermoplastic processing methods, including extrusion or injection molding. However, some issues with PLA, such as brittleness, low heat deflection temperature, poor processability, and low melt strength, need to be tackled [5], [10]. In order to overcome these problems and get the most benefit from PLA, many researchers and companies are working in different ways, such as copolymerization and blending PLA with flexible polymers. Also, it is common to use some additives like impact modifiers to improve the toughness and mechanical properties[4], [10]–[12]. PHAs are bio-based polymers that can be produced and degraded by microorganisms and broken down into non-toxic, water-soluble oligomers and monomers. However, its high melting point, difficult processing, and unstable molten state are PHAs' weaknesses for 3D printing applications. Nevertheless, blending PHAs with other polyesters significantly improves their mechanical properties and makes materials less brittle. Recently, many companies have tried to make filament for 3D printing out of PHAs and PLAs to overcome both materials' limits [13]–[15].

Due to variations in viscosity and melt strength, processing parameters need to be chosen carefully for biopolymers. Also, these parameters could affect the final properties of printed parts. Layer adhesion of a 3D printed part could be affected by FFF process parameters as well. Interlayer adhesion is responsible for the mechanical strength of printed objects, and interdiffusion between layers is directly related to the viscosity and rheological properties of the polymer[16]. In this study, we looked for interlayer adhesion in different materials with different process parameters and its relation with extensional viscosity.

2. METHODS AND METERIALS

Commercially available tough and virgin PLA and PLA/PHA filaments with a diameter of 1.75 mm were investigated in this study. The materials used have been listed in *Table 1.*

Table 1. Material information

Material	Grade	Supplier
PLA virgin	3D 850	NatureWorks
PLA tough	3D 870	NatureWorks
PLA/PHA	Natural filaments	colorFabb

The samples were manufactured using a Creality CR-10 printer, with each sample consisting of four layers at 100% infill percentage. To investigate the effect of processing parameters on the final properties of the printed samples, three different sets of values have been used for nozzle temperature, bed temperature, and printing speed*(Table 2).*

Table 2. Processing parameters for the three unique filaments PLA3D850 and PHA/PLA.

	Bed Temperature (°C)	Nozzle Temperature (°C)	Print Speed (mm/s)
Baseline	60	200	100
Level 1	55	185	60
Level 2	65	215	80

3 Results and Discussion

PLA/PHA interlayer adhesion was too strong to be peeled with our testing device. Peeled samples after peeling are shown in Figure 1. The highest peeling force belonged to virgin PLA at a nozzle temperature of 215°C, which was around 3 N. All other samples needed less than 1 N to be peeled. Both PLAs were extremely sensitive to nozzle temperature. The lowest nozzle temperature (185°C) indicated the smallest value of peel strength, confirming that the adhesion between every printed layer is weaker compared to the samples printed at a higher temperature, such as 215 °C. Also, bed temperature showed a significant effect on the peeling test. The result showed that in higher bed temperature, higher peeling force is needed. It seemed that better diffusion between PLA layers could occur due to higher nozzle and bed temperature, resulting in stronger layer adhesion. In terms of printing speed, the strongest layer adhesion was seen in virgin PLA films at a printing speed of 60 mm/s, which was the slowest printing speed. Besides, higher printing speed demonstrated the lower needed forces for peeling. At slower speeds, layers have sufficient time to properly diffuse to each other, increase entanglement, and create stronger bonding.

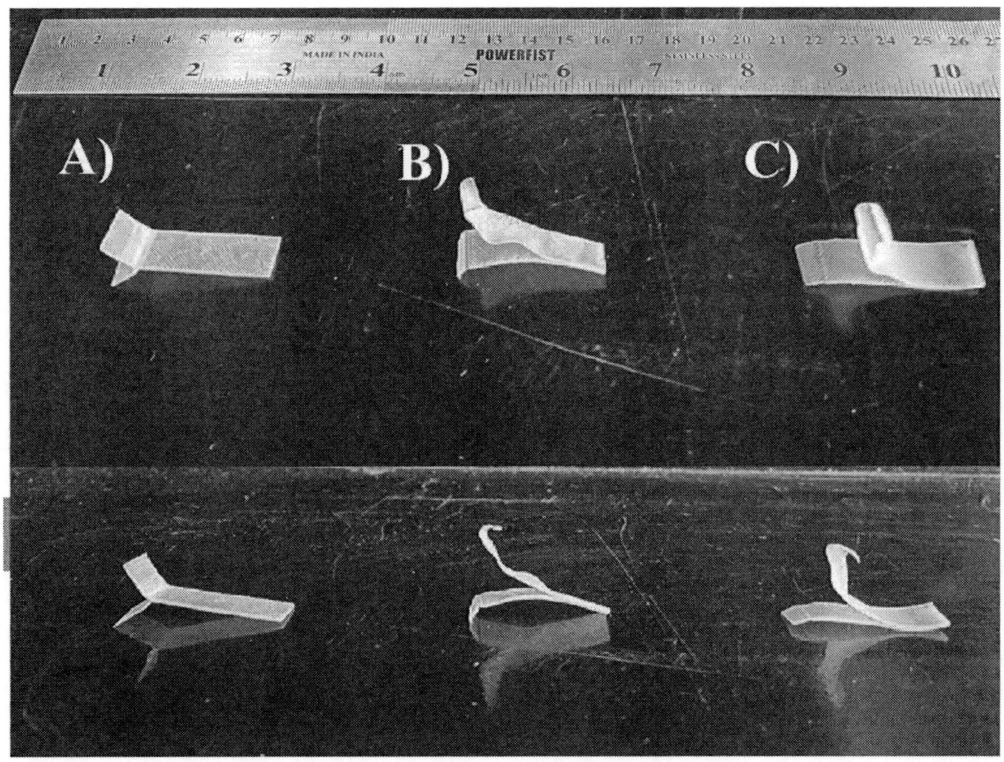

Figure 1. Peeled A) PLA/PHA, B) Virgin PLA, and C) Tough PLA Samples.

As shown in Figure 1, virgin PLA showed higher interlayer strength since it needs a higher force to be peeled compared to toughed PLA.

The extensional viscosity results are illustrated in Figure 2. The collected data revealed the lower extensional viscosity for PLA/PHA and virgin PLA compared to tough PLA. Therefore, as the PLA/PHA and the virgin PLA samples showed higher interlayer strength, it can be considered that higher extensional viscosity results in lower interlayer diffusion and strength.

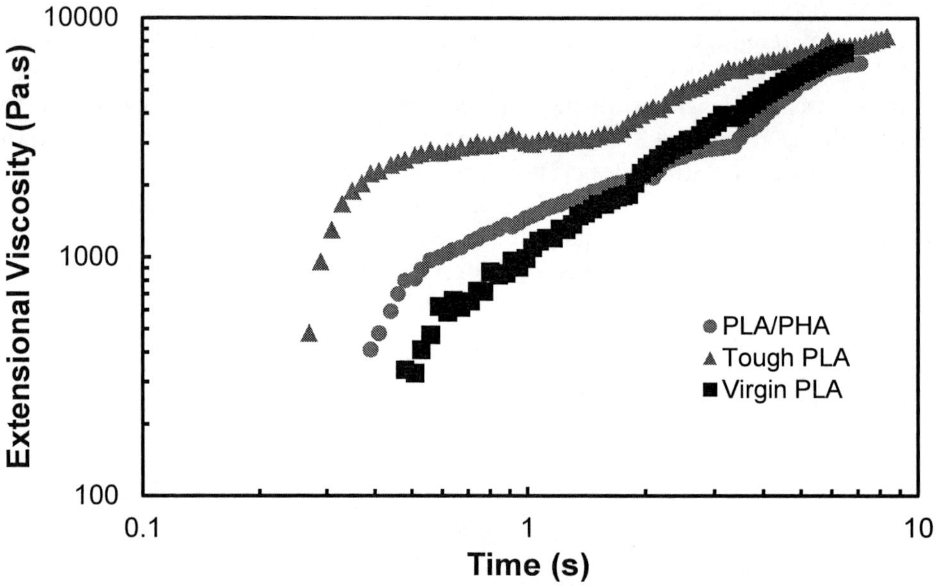

Figure 2. Extensional viscosity of virgin PLA, tough PLA, and PLA/PHA at the shear rate of 0.1 s^{-1} at 200°C.

As shown in Figure 1, virgin PLA showed higher interlayer strength since it needs a higher force to be peeled compared to toughed PLA.

The extensional viscosity results are illustrated in Figure 2. The collected data revealed the lower extensional viscosity for PLA/PHA and virgin PLA compared to tough PLA. Therefore, as the PLA/PHA and the virgin PLA samples showed higher interlayer strength, it can be considered that higher extensional viscosity results in lower interlayer diffusion and strength.

4. CONCLUSION

PLA/PHA and virgin PLA showed higher interlayer adhesion strength and lower extensional viscosity than tough PLA. Both virgin PLA and tough PLA were extremely sensitive to nozzle temperature. Also, bed temperature showed almost the same result in the peeling test. In contrast, stronger layer bonding and higher peel strength values resulted from the lower printing speed.

5. REFERENCES

[1] A. Dey, I. N. R. Eagle, and N. Yodo, "A review on filament materials for fused filament fabrication," *Journal of Manufacturing and Materials Processing*, vol. 5, no. 3. MDPI AG, Sep. 01, 2021. doi: 10.3390/jmmp5030069.

[2] Krishnanand, S. Soni, and M. Taufik, "Design and assembly of fused filament fabrication (FFF) 3D printers," in *Materials Today: Proceedings*, 2020, vol. 46, pp. 5233–5241. doi: 10.1016/j.matpr.2020.08.627.

[3] Y. Tao *et al.*, "A review on voids of 3D printed parts by fused filament fabrication," *J. Mater. Res. Technol.*, vol. 15, pp. 4860–4879, Nov. 2021, doi: 10.1016/j.jmrt.2021.10.108.

[4] N. Tripathi, M. Misra, and A. K. Mohanty, "Durable Polylactic Acid (PLA)-Based Sustainable Engineered Blends and Biocomposites: Recent Developments, Challenges, and Opportunities," *ACS Eng. Au*, vol. 1, no. 1, pp. 7–38, Oct. 2021, doi: 10.1021/acsengineeringau.1c00011.

[5] C. O'Mahony *et al.*, "Determination of thermal and thermomechanical properties of biodegradable PLA blends: for additive manufacturing process," *J. Therm. Anal. Calorim.*, vol. 142, no. 2, pp. 715–722, Oct. 2020, doi: 10.1007/s10973-020-09859-6.

[6] S. T. Sam, M. A. Nuradibah, H. Ismail, N. Z. Noriman, and S. Ragunathan, "Recent Advances in Polyolefins/Natural Polymer Blends Used for Packaging Application," *Polymer - Plastics Technology and Engineering*, vol. 53, no. 6. Taylor and Francis Inc., pp. 631–644, 2014. doi: 10.1080/03602559.2013.866247.

[7] J. Gonzalez Ausejo *et al.*, "A comparative study of three-dimensional printing directions: The degradation and toxicological profile of a PLA/PHA blend," *Polym. Degrad. Stab.*, vol. 152, pp. 191–207, Jun. 2018, doi: 10.1016/J.POLYMDEGRADSTAB.2018.04.024.

[8] Y. Zhu, C. Romain, and C. K. Williams, "Sustainable polymers from renewable resources," *Nat. 2016 5407633*, vol. 540, no. 7633, pp. 354–362, Dec. 2016, doi: 10.1038/nature21001.

[9] Y. J. Chen, "Bioplastics and their role in achieving global sustainability," *J. Chem. Pharm. Res.*, vol. 6, no. 1, pp. 226–231, 2014.

[10] R. M. Taib, Z. A. Ghaleb, and Z. A. M. Ishak, "Thermal, Mechanical, and Morphological Properties of Polylactic Acid Toughened with an Impact Modifier," *J Appl Polym Sci*, vol. 123, pp. 2715–2725, 2012, doi: 10.1002/app.34884.

[11] N. Petchwattana, P. Naknaen, and B. Narupai, "Combination effects of reinforcing filler and impact modifier on the crystallization and toughening performances of poly(Lactic acid)," *Express Polym. Lett.*, vol. 14, no. 9, pp. 848–859, 2020, doi: 10.3144/expresspolymlett.2020.70.

[12] W. Li *et al.*, "The Effect of Core–Shell Ratio of Acrylic Impact Modifier on Toughening PLA," *Adv. Polym. Technol.*, vol. 36, no. 4, pp. 491–501, Dec. 2017, doi: 10.1002/ADV.21632.

[13] A. Kovalcik, "Recent Advances in 3D Printing of Polyhydroxyalkanoates: A Review," *EuroBiotech J.*, vol. 5, no. 1, pp. 48–55, Jan. 2021, doi: 10.2478/ebtj-2021-0008.

[14] F. I. Butt, N. Muhammad, A. Hamid, M. Moniruzzaman, and F. Sharif, "Recent progress in the utilization of biosynthesized polyhydroxyalkanoates for biomedical applications – Review," *Int. J. Biol. Macromol.*, vol. 120, pp. 1294–1305, Dec. 2018, doi: 10.1016/J.IJBIOMAC.2018.09.002.

[15] C. H. Lee, S. M. Sapuan, R. A. Ilyas, S. H. Lee, and A. Khalina, "Development and Processing of PLA, PHA, and Other Biopolymers," *Adv. Process. Prop. Appl. Starch Other Bio-Based Polym.*, pp. 47–63, Jan. 2020, doi: 10.1016/B978-0-12-819661-8.00005-6.

[16] B. B. Shahriar, C. France, N. Valerie, C. Arthur, and G. Christian, "Toward improvement of the properties of parts manufactured by FFF (fused filament fabrication) through understanding the influence of temperature and rheological behaviour on the coalescence phenomenon," *AIP Conf. Proc.*, vol. 1896, no. 1, p. 040008, Oct. 2017, doi: 10.1063/1.5008034.

MT-O04

Compatibilization of polyethylene-polypropylene blends via ionic liquid to enhance performance for packaging applications

Zhijie Yan[1], Sarvesh Kinariwala[1], Maria Portie[1,2], Ken Medlin[2], Hunaid Nulwala[2], Chau Tang[2], Carlos A. Diaz[1]*

[1]*Packaging Science, Rochester Institute of Technology, Rochester, NY 14623, USA*
[2]*Roco Global, Pittsburgh, PA 15205, USA*

***Correspondence to:** Packaging Science, Rochester Institute of Technology, Rochester, NY 14623, USA. cdamet@rit.edu*

ABSTRACT: Plastics are essential to modern life. Over 8000 million metric tons of plastic have been produced, of which >50% has been lost to the environment in the last 50 years. Plastic reuse and recycling are a long-term environmental challenge. The main challenge is that when non-miscible plastics are mixed, they lose their properties. Current compatibilizers in the market are based on block copolymers and require a high loading (i.e., 20 wt.%) to achieve properties like those of virgin plastic. This study uses ionic liquid-based compatibilizer technology to improve mixed plastics' miscibility and the quality of post-consumer recycled polyethylene mixed with polypropylene. Furthermore, this study demonstrates that the proposed technology is more cost-effective than the traditional block copolymers approach. Melt compounding was used to prepare samples according to a design of experiments, and the tensile properties were measured. A similar mechanical performance compared to a benchmark compatibilizer was obtained using ILs at low concentrations. DSC and rheology characterization confirmed the miscibility of blends using ILs.

Keyword: Recyclability, Polypropylene, Polyethylene, Ionic liquids, Mechanical properties

1. INTRODUCTION

Plastics are essential to modern life and find applications in every industry due to their high barrier, toughness, and elasticity. Additionally, plastic is low cost compared to glass and metal [1]. Due to high demand, plastic applications have significantly increased in the past decades. The application of plastic such as polyethylene (PE), polypropylene (PP), poly(ethylene terephthalate) (PET), and other kinds of polymer in different industries is quite common, especially PE and PP take the most significant proportion in the plastic production industry [2].

The most prevalent use of plastic is packaging such as flexible pouches, bottles, and containers. Packaging applications (like containers and bags) consume over one-third of plastics produced globally [1], [3]. Petroleum and natural gas are the most common feedstocks to make plastics. About 4% of fossil fuels are converted to plastic products annually. Due to the high durability of plastic, it takes a long time for the plastic to deteriorate in the natural environment, which leads to its accumulation and plastic pollution [4].

From 1950 to 2015, the world generated 6300 metric tons of plastic waste [5]. About 79% of the plastic waste was dumped into landfills or discarded into natural surroundings, around 12% was incinerated, and only 9% of the plastic waste has been recycled or reused. If mankind continues to discard plastic waste per this trend, by 2050, about 12,000 metric tons of plastic waste will be generated and disposed of in landfills or released into the environment [3]. Moreover, plastic pollution in seas and oceans has caught several authoritative organizations' attention, emphasized the importance and challenge of plastic recycling [6].

The disposal of plastic packaging waste is a long-term environmental challenge. To decrease plastic pollution, plastic recycling is the most important and effective action to reduce the impact of plastic waste and disposal. It is estimated that only 10% of total energy is used in remaking products from recycled materials, which is significant. It also means that 0% of new fossil is extracted when a polymer is recycled. The amount of petroleum consumption and plastic waste could decrease through exemplary activities such as reducing plastics consumption, developing a better recycling strategy, and increasing material reuse and repurposing. There are four categories of plastic recycling: primary (mechanical reprocessing into a product with equivalent properties), secondary (mechanical reprocessing into products requiring lower properties), tertiary (recovery of chemical constituents), and quaternary (recovery of energy) [4]. Primary recycling is the most effective and economical way because of material reuse. However, the current recycling strategies have poor efficiency because of the challenge of mixing immiscible plastics such as PE and PP.

Mixed plastics lose their properties compared to virgin material after several processing cycles. Due to the variation in molecular weight, crystal structure, and molecular weight distribution, the blends of PE and PP in the absence of a compatibilizer exhibit severe phase separation and reduced mechanical properties, even though these polyolefins have similar chemical structures [6]. Due to the positive Gibbs free energy, these mixtures produce an interface and reduce their interfacial adhesion. This phase separation results in the loss of properties. As a result, non-miscible polymers tend to have significantly lower properties and cannot be used in higher-value items. [7], [8]. Generally, tensile strength and the percent elongation are reduced due to contaminants in the recycled materials [9]. This deficiency in mechanical properties is usually overcome by adding compatibilizers into the recycled PE and PP mixtures.

Compatibilizers improve the adhesion between separate phases; hence, these materials improve the final product's physical properties [10]. Current compatibilizers in the market are based on block copolymers and require a high loading (i.e., up to 10-20 wt.%) to achieve properties like virgin plastics. The research of mixing PET bottles and PE bags with poly(ethylene-co-methacrylic acid) copolymer has been studied [11]. It was found that three wt.% of the poly(ethylene-co-methacrylic acid) copolymer is sufficient to make the 3:1 mass ratio of PET bottle and PE bags compatible. Furthermore, poly(ethylene adipate) and poly(tetramethylene ether glycol) were used as compatibilizers in the biodegradable polymer blend of poly(L-lactic acid)/poly (butylene succinate-*co*-L-lactate)[12] Another research to develop a high-performance wood plastic composite consisting of wood flour, maleated PE as a compatibilizer, and recycled PET reveals that the optimal content of compatibilizer is 10 (Parts per Hundred Resins (PHR) [13]. The compatibilization of 164:18:18 PET/PE/PP blends requires 10 g ethylene-glycidyl methacrylate copolymer to yield the best performance [14]. Karaagac et al. [15] investigated using an olefin block copolymer to improve tensile elongation and tensile impact strength of high-density polyethylene (HDPE) contaminated with PP. HDPE-PP blend with 15 wt.% PP requires eight wt.% of the olefin block copolymer to recover the impact strength to a level similar to that of the uncontaminated HDPE [22]. Compatilibizers based on block copolymer technology have to be designed for specific systems and require high loadings. There is a need for versatile compatibilizers that can work at low loadings. Ionic liquids technology is an alternative to address this need.

Ionic liquids (ILs) are salts comprising cations and ions in a liquid form [16]-[18]. ILs are versatile substances that show excellent thermal and chemical stability, have wide electrochemical windows, are an excellent alternative to traditional solvents, and are unique catalysts [19], [20]. ILs as compatibilizers have been demonstrated in the literature [21], [22]. Recently, studies conducted by Leroy et al. on improving the plasticization of starch-zein blends have discovered that adding ionic liquid also has a compatibilizing effect, improving phase adhesion between starch and zein [21]. In another study, Yousfi et al. [22] showed a small amount of ionic liquid's compatibilization effect when added to a thermoplastic blend of PP and nylon 6.

This study aims to assess the effects of adding ionic liquid as compatibilizers in PE/PP blends to enhance the properties of recycled plastics; hence, it would increase the recycling rates. Plastic recycling has become an increasingly crucial topic in the modern world and needs to be emphasized. This study provided a new perspective and novel solution for plastic recycling.

2. MATERIALS AND METHODS

2.1 Materials

HDPE (melt flow index of 2.2 to 2.4 g/10 min (190 °C, 2.16 kg)) and PP (melt flow index of 3 to 5 g/10 min (230 °C, 2.16 kg) were purchased from Sigma-Aldrich. INFUSE™ 9504 from DOW Chemical Company was purchased from Entec Polymers (Florida, US) and was used as the benchmark in this study.

ILs were mixed in through a masterbatch process separately by Roco Global. IL-59 is a reactive phosphonium based ionic liquid made by Roco Global. IL-59 was mixed with a 50:50 HDPE/PP blend. Masterbatches were created on a twin-screw Brabender (40-mL cup) at 165 □ for 10 minutes. Benchmarks were made by adding 4, 6, and 8 wt.% INFUSE™ 9504. Benchmark producing process is based on the study of the compatibilization with olefin block copolymer (OBC) to or its commercial name INFUSE [22].

2.2 Sample Preparation

All the masterbatches, benchmarks, and samples were made through an internal shear mixer, CWB Brabender. The masterbatches with IL-59 ionic liquid as the additive were mixed with a 50:50 PE/PP blend in the twin-screw Brabender at 165 °C for 10 min. The benchmarks with different amounts of INFUSE and samples with different ILs were mixed with varying compositions of PE: PP blends, as shown in Table 1.

Table 1 Formulation of PE: PP blends and additives.

PE: PP Blend Composition (wt.%)	Benchmark INFUSE (wt.%)	Ionic Liquid IL-59 (wt.%)
95:5	4	0.5, 1, 2
90:10	6	0.5, 1, 2
85:15	8	0.5, 1, 2

Figure 1 shows the process flow for all the samples. Before the compounding process, PP, PE, and additives were weighed separately, and then these ingredients were mixed in the bowl. A Brabender mixer compounded the mixture. The blending process was maintained at 175 °C, the twin-screw was running at 60 rpm speed, and the running process was kept for 10 minutes with a nitrogen purge. After the compounding process, the blended mixture was pressed in the Carver press machine for the compression molding. The post compounding press recipe includes two segments. Firstly, the sample was compressed under 175°C and 1000lbs pressure for 1min; then, the model was compressed under 175 °C and 1500lbs pressure for 2min. The mixture was pressed between 2 pieces of 12 x 12.5-inch Kapton film and two stainless steel plates.

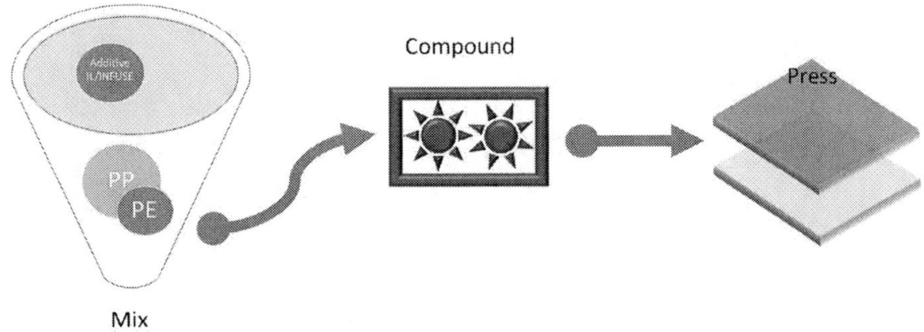

Figure 1 Process flow

2.3 DSC analysis

Differential scanning calorimetry (DSC) was used to characterize the thermal properties of the sample. The test instrument is TA Q-1000 DSC and has a heating/cooling rate of 10 °C/min. The weight of each piece was approximately 10mg (ranging from 6mg-12mg). Samples were sealed in aluminum pans and were put in the testing chamber before the machine's measurements. The testing chamber was first equilibrated at 30 °C, heated to 200 °C with ten °C/min, and then called down with ten °C/min to -30 °C ending the first cycle. The same steps were repeated for a total of 3 cycles. The whole process was performed under a nitrogen atmosphere with a 50 mL/min flow rate.

2.4 Tensile tests

The tensile tests were conducted on PE/PP polymer blends containing compatibilizers at varying concentrations to observe the physical properties of the resulting polymeric material. The specimens were punched out from compression-molded sheets using ASTM D638-14. Each sample produced five identical Type-V specimens of narrow section length (7.62 mm) and width (3.18 mm). The tensile test was performed on the Instron Universal Testing Machine at 18 °C using a 5 KN load cell with a 10 mm/min extension rate. Averaged tensile properties were calculated from five measurements.

2.5 Hardness Shore D Test

The hardness of various samples was tested using a Shore D durometer as per ASTM 2240. Measurements were made at three different locations for each of the formulations mentioned above, and the average value was reported.

2.6 Rheological analysis

The rheological properties of the samples were investigated using the parallel-plate mode of the Anton-Paar MCR 302 rheometer equipped with a heating chamber under nitrogen. A 25-mm diameter parallel plate was used to run tests with a 1 mm gap size. Temperature sweeps from 190 °C to 60 °C, constant angular frequency of 1 rad/s and constant amplitude of 1%. Frequency sweep with ranges as seen on graphs, at constant temperature 190 °C and constant amplitude 1%. Amplitude sweeps from the range seen on graphs, at a constant frequency 1Hz and a constant temperature of 190 °C.

3. RESULTS AND DISCUSSION

3.1 Miscibility analysis

Miscibility of PE/PP blends with varying amounts of IL-59 ionic liquid was investigated via DSC for up to 15 wt.% of PP content. Figure 2 shows how the addition of ionic liquid at low amounts decreased the enthalpy of crystallization of the blend. The decreasing crystallization enthalpy indicates the increasing miscibility of blends. An increase in miscibility was observed using IL in PE/PP blends containing up to 15 wt.% PP. In combination with 15 wt.% PP, 2% IL was required to outperform the benchmark. On the other hand, incorporating 0.5 and 1% IL into PE/PP blend with five wt.% PP significantly reduces the crystallization enthalpy below the benchmark.

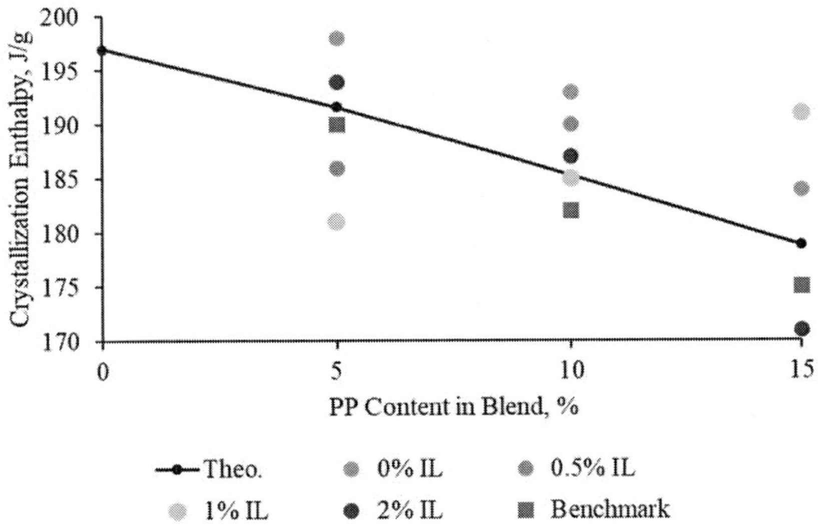

Figure 2 Crystallization enthalpy of PE-HD with varying amounts of wt.% of PP as a contaminant.

3.2 Tensile tests

Figure 3 shows the tensile test results. It was observed that elongation at break drops steeply as the contamination level of PP increases (Figure 3b), while the modulus of elasticity and tensile strength remained less affected. The elongation at break correlates with other important performance properties such as impact strength.

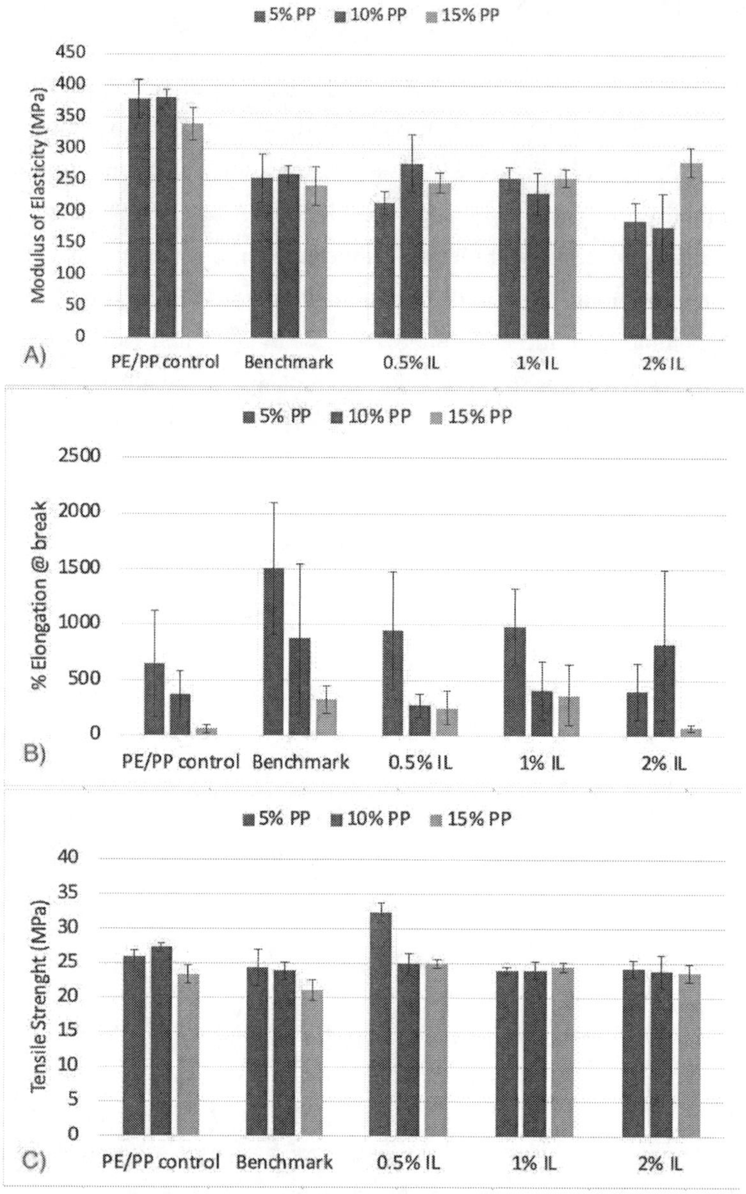

Figure 3 Tensile tests of PE contaminated at different wt.% of PP compatibilized using varying amounts of IL-59.

When the INFUSE 9504 copolymer was used as the benchmark, it reduced the modulus of elasticity of all the samples (Figure 3a) and increased the elongation at break (Figure 3b). Although affected by the amount of PP and compatibilizers, the tensile strength remained at around 25 MPa for all the samples. Similar levels of elongation at break to those of the benchmark were achieved using ILs. At 15 wt.% PP in the blend, the addition of 1% IL-59 showed the same improvement in elongation at break as the benchmark and 5-fold higher than the blend that does not contain any compatibilizer. This indicates that IL-59 (when added in small concentrations) improves the elongation at break when contamination levels are high compared to the benchmark.

3.3 Shore D Test

Figure 4 shows the effect of ILs on the hardness of the different blends. In general, samples showed a slight decrease in hardness irrespective of the compatibilizer, benchmark, or ILs. Interestingly, at 1% IL, the hardness increased with increasing PP content. This trend seems to be opposite to the samples without a compatibilizer—the 2% IL samples correlated with the benchmark.

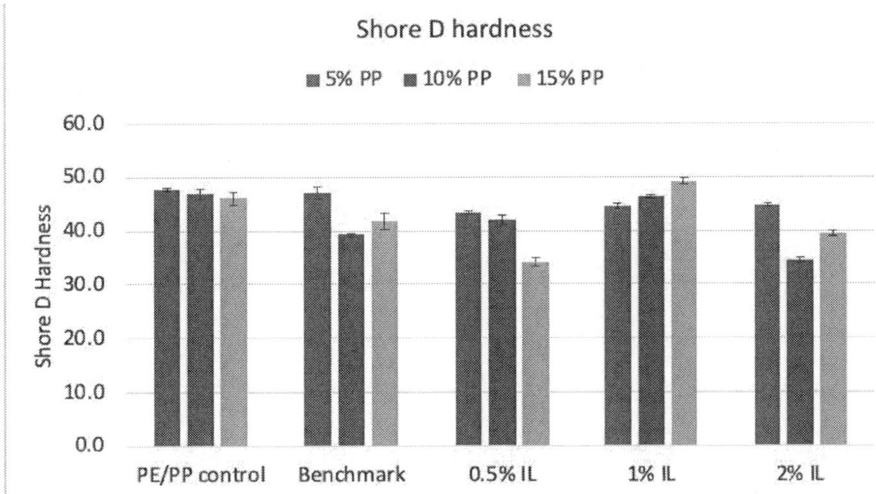

Figure 4 Shore D hardness of PE/PP Blends with IL-59.

3.4 Rheological analysis

Figures 5 and 6 show the rheological properties of the benchmark with 8% INFUSE 9504 and the PP/PE blends with 0.5 and 1 wt.% IL addition. Figure 5 shows the crossover point of the storage and loss modulus, which is related to the relaxation time. A shift to the left implies a longer relaxation time attributed to an increase in the degree of molecular chain entanglement [22]. The result proves that IL-59 is an effective compatibilizer to improve interfacial adhesion. According to Figure 6, the addition of 8 wt.% INFUSE 9504, 0.5 wt.% IL or 1 wt.% IL has a similar effect reducing the complex viscosity. Moreover, the performance of 0.5 and 1 wt.% IL is better than the addition of 8 wt.% INFUSE 9504.

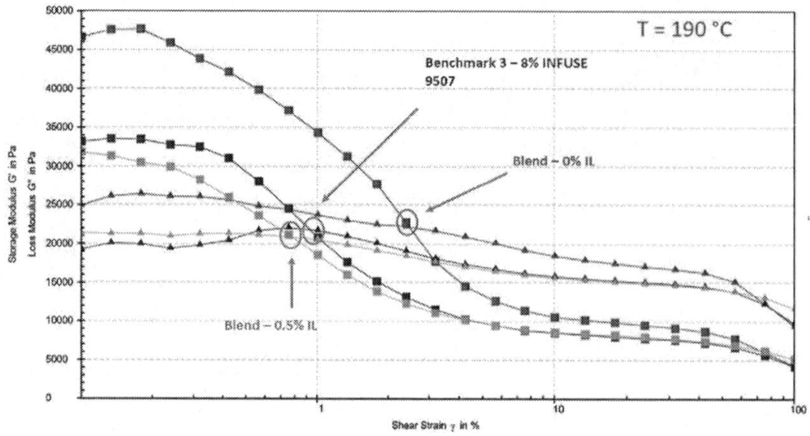

Figure 5 Storge modulus of 85:15 PE/PP blend with 8 wt.% INFUSE (benchmark), 0% IL and 0.5 wt.% IL.

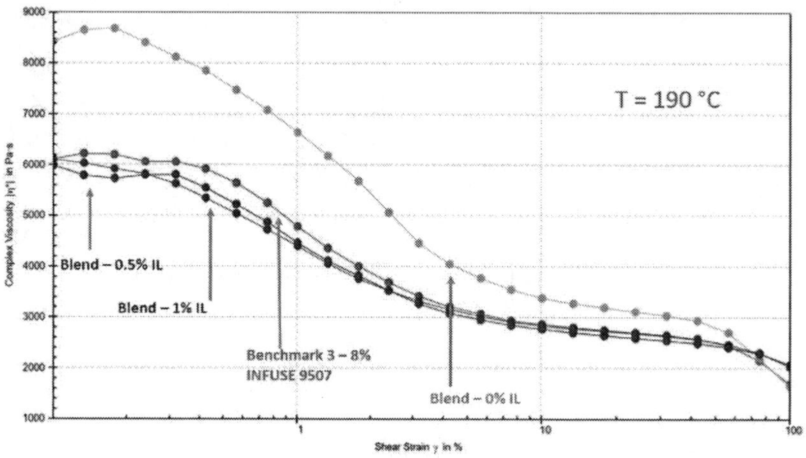

Figure 6 Complex viscosity of 85:15 PE/PP blend with 8 wt.% INFUSE (benchmark), 0% IL and 0.5 wt.% IL.

4. CONCLUSIONS

This study demonstrated the compatibilization effect of ionic liquid (i.e., IL-59) in the PE-rich blends with PP. PP concentrations of 5, 10, and 15 wt.% were analyzed. IL-59 increased the miscibility of the blends, as evidenced by DSC. Reduction in the enthalpy of crystallization exceeded that of the benchmark, and the optimum concentration of IL depended on the amount of PP in the blend. The rheological analysis showed that the 0.5 and 1 wt.% addition of IL-59 had a positive compatibilizing effect on improving interfacial adhesion and increasing the degree of molecular chain entanglement. IL-59 had a minor effect on the hardness of the plastic. This study reveals that IL-59 effectively improves the mechanical, thermal, and rheological performances of PP/PE blends. Moreover, a similar mechanical performance compared to a benchmark compatibilizer was obtained using IL-59 at a low concentration. Similar to that of the benchmark, property enhancement was achieved using IL-59 at lower concentrations.

5. ACKNOWLEDGEMENT

This work was possible through the support of the DOE Award Number: DE-SC0021793.

6. REFERENCES

[1] A. L. Andrady and M. A. Neal, "Applications and societal benefits of plastics," *Philosophical transactions. Biological sciences*, vol. 364, no. 1526, pp. 1977–1984, 2009, doi: 10.1098/rstb.2008.0304.

[2] J. M. Eagan *et al.*, "Combining polyethylene and polypropylene: Enhanced performance with PE/iPP multiblock polymers," *Science (1979)*, vol. 355, no. 6327, pp. 814–816, Feb. 2017, doi: 10.1126/SCIENCE.AAH5744.

[3] G. Roland, J. J. R, and L. K. Lavender, "Production, use, and fate of all plastics ever made," *Science Advances*, vol. 3, no. 7, p. e1700782, Feb. 2022, doi: 10.1126/sciadv.1700782.

[4] J. Hopewell, R. Dvorak, and E. Kosior, "Plastics recycling: challenges and opportunities," *Philos Trans R Soc Lond B Biol Sci*, vol. 364, no. 1526, pp. 2115–2126, Jul. 2009, doi: 10.1098/rstb.2008.0311.

[5] C. J. RHODES, "Plastic pollution and potential solutions," *Science Progress (1933-)*, vol. 101, no. 3, pp. 207–260, 2018, [Online]. Available: https://www.jstor.org/stable/26508614

[6] J. Qu *et al.*, "Industrial-Scale Polypropylene–Polyethylene Physical Alloying Toward Recycling," *Engineering*, Aug. 2021, doi: 10.1016/J.ENG.2021.02.021.

[7] F. S. Bates, "Polymer-Polymer Phase Behavior," 1991.

[8] F. S. Bates and G. H. Fredrickson, "Conformational Asymmetry and Polymer-Polymer Thermodynamics," *Macromolecules*, vol. 27, no. 4, pp. 1065–1067, Jul. 2002, doi: 10.1021/MA00082A030.

[9] C. Meran, O. Ozturk, and M. Yuksel, "Examination of the possibility of recycling and utilizing recycled polyethylene and polypropylene," *Materials and Design*, vol. 29, no. 3, pp. 701–705, 2008, doi: 10.1016/J.MATDES.2007.02.007.

[10] E. Karaagac, T. Koch, and V. M. Archodoulaki, "The effect of PP contamination in recycled high-density polyethylene (rPE-HD) from post-consumer bottle waste and their compatibilization with olefin block copolymer (OBC)," *Waste Management*, vol. 119, pp. 285–294, Jan. 2021, doi: 10.1016/J.WASMAN.2020.10.011.

[11] D. Rajasekaran and P. K. Maji, "Recycling of plastic wastes with poly (ethylene-co-methacrylic acid) copolymer as compatibilizer and their conversion into high-end product," *Waste Management*, vol. 74, pp. 135–143, 2018, doi: https://doi.org/10.1016/j.wasman.2018.01.018.

[12] P. M. Chou, M. Mariatti, A. Zulkifli, and M. Todo, "Effect of secondary forces in the compatibility of two incompatible biodegradable polymers," *Polymer Bulletin*, vol. 69, no. 4, pp. 455–469, 2012, doi: 10.1007/s00289-012-0748-4.

[13] Z. Chen, H. Shahrajabian, S. A. Bagherzadeh, A. Maleki, and Q. V. Bach, "The impact energy analysis by genetic algorithm and response surface methods to study the plastic composite, compatibilizer, and recycled poly effects," *Journal of Thermal Analysis and Calorimetry*, vol. 141, no. 1, pp. 421–433, Jul. 2020, doi: 10.1007/S10973-019-09188-3/TABLES/6.

[14] A. S. Adekunle, A. A. Adeleke, C. v. Sam Obu, P. P. Ikubanni, S. E. Ibitoye, and T. M. Azeez, "Recycling of plastics with compatibilizer as raw materials for the production of automobile bumper," *Cogent Engineering*, vol. 7, no. 1, Jan. 2020, doi: 10.1080/23311916.2020.1801247.

[15] E. Karaagac, T. Koch, and V. M. Archodoulaki, "The effect of PP contamination in recycled high-density polyethylene (rPE-HD) from post-consumer bottle waste and their compatibilization with olefin block copolymer (OBC)," *Waste Management*, vol. 119, pp. 285–294, Jan. 2021, doi: 10.1016/J.WASMAN.2020.10.011.

[16] G. Ondrey, "Ionic liquids," *Chemical Engineering*, vol. 112, no. 1, p. 18, Jan. 2005, [Online]. Available: https://ezproxy.rit.edu/login?url=https://www.proquest.com/trade-journals/ionic-liquids/docview/194439063/se-2?accountid=108

[17] S. v Dzyuba and R. A. Bartsch, "Synthesis, properties, and applications of ionic liquids," Texas Tech University, Ann Arbor, 2002. [Online]. Available: https://ezproxy.rit.edu/login?url=https://www.proquest.com/dissertations-theses/synthesis-properties-applications-ionic-liquids/docview/305511162/se-2?accountid=108

[18] J. Yuan and M. Antonietti, "Poly(ionic liquid)s: Polymers expanding classical property profiles," *Polymer (Guildf)*, vol. 52, no. 7, pp. 1469–1482, Mar. 2011, doi: 10.1016/J.POLYMER.2011.01.043.

[19] M. Reslan and V. Kayser, "Ionic liquids as biocompatible stabilizers of proteins," *Biophysical Reviews*, vol. 10, no. 3, pp. 781–793, Jun. 2018, doi: 10.1007/S12551-018-0407-6/TABLES/2.

[20] D. W. Fang, M. C. Xia, J. tao Zuo, Z. kai Zhou, J. Li, and J. zhen Yang, "Physicochemical properties of ionic liquids [Cnmim][SbF6] (n = 4,5,6)," *The Journal of Chemical Thermodynamics*, vol. 131, pp. 360–368, Apr. 2019, doi: 10.1016/J.JCT.2018.07.018.

[21] E. Leroy, P. Jacquet, G. Coativy, A. L. Reguerre, and D. Lourdin, "Compatibilization of starch–zein melt processed blends by an ionic liquid used as plasticizer," *Carbohydrate Polymers*, vol. 89, no. 3, pp. 955–963, Jul. 2012, doi: 10.1016/J.CARBPOL.2012.04.044.

[22] M. Yousfi, S. Livi, and J. Duchet-Rumeau, "Ionic liquids: A new way for the compatibilization of thermoplastic blends," *Chemical Engineering Journal*, vol. 255, pp. 513–524, Nov. 2014, doi: 10.1016/J.CEJ.2014.06.080.

[23] L. Qi, L. Wu, R. He, H. Cheng, B. Liu, and X. He, "Synergistic toughening of polypropylene with ultra-high molecular weight polyethylene and elastomer-olefin block copolymers," *RSC Advances*, vol. 9, no. 41, pp. 23994–24002, Jul. 2019, doi: 10.1039/C9RA01073D.

MT-O05

The Effect of Alcoholic Solutions on the Thermomechanical Properties of Poly (lactic acid)

Dian Xu[1], Rafael Auras[1]*, Loong-Tak Lim[2], Maria Rubino[1], Muhamad Rabnawaz[1]

[1]School of Packaging, Michigan State University, East Lansing, 48824, USA
[2]Department of Food Science, University of Guelph, Guelph, N1G 2W1, Canada

***Correspondence to:** School of Packaging, Michigan State University, 448 Wilson Rd, East Lansing, 48824, USA. aurasraf@msu.edu*

ABSTRACT: Organic substances such as aroma compounds can interact with food and medical packaging polymers such as poly(lactic acid) (PLA) and change their properties and performance. So, it is crucial to accurately measure the change in their thermo-mechanical properties and sorption capacity as they are in contact with organic substances to fully develop PLA applications. The effects of aqueous and deuterium oxide solutions of straight-chain alcohols with one to four carbon atoms on the thermomechanical properties of poly(lactic acid) films were assessed by dynamic mechanical analysis (DMA). The glass transition temperature (Tg) of PLA films immersed in 25%, 50%, 75%, and 100% (v/v) alcoholic solutions were compared with the Tg's of dry and immersed films in 100% HPLC water and deuterium oxide. The Tg's of PLA films immersed in methanol and ethanol (one to two carbon atoms) aqueous solutions decreased as the concentration of the alcohol solutions increased. For methanol and ethanol, the Tg's decreased by 8.6 °C and 7.6 °C per every 25% (v/v) increase of alcohol, respectively. Tg's of PLA films immersed in 1-propanol and 2-propanol (three carbon atoms) decreased by 8.6 °C and 8.9 °C for every 25% increase of propanol, respectively, and the Tg's reduced until 75% (v/v). The concentration of 1-butanol (four carbon atoms) at 100% (v/v) was assessed since 1-butanol is not miscible with water. Aqueous and deuterium oxide solutions had similar effects on Tg. The results of this work help develop PLA films needed to be in contact with aqueous-alcoholic solutions.

Keyword: PLA, alcoholic solutions, DMA, glass transition temperature

1. INTRODUCTION

Understanding the interaction of organic compounds with polymers is crucial for applications' development in medicine, food packaging, and environmental remediation. Organic substances such as aroma compounds can interact with polymers such as polyolefins[1–3], polycarbonate[4], poly(ethylene terephthalate) - PET[5,6], as well as poly(lactic acid) – PLA[7–9], and change their valuable properties and performance. The interactions of organic substances in solutions with aliphatic and aromatic polyesters such as PLA and PET can cause swelling or plasticization of the polymer matrix during their useful life, thereby affecting the polymers' performance[10,11].

PLA has gained attention as a new polymer for the circular economy since it can be disposed of in any waste management scenario. Still, it is mainly targeted for in industrial composts where it is converted to CO_2 and water by the action of microorganisms in the environment. PLA has one of the lowest environmental footprints per kilogram of resin (0.62 kg CO_2-eq for Ingeo™ from NatureWorks LLC and 0.5 kg CO_2-eq for the Total Corbion PLA from Total Corbion – both values reported including credit for biogenic carbon)[12,13]. Therefore, its use has increased in food packaging, biomedicine, and agriculture[14]. However, hydrolysis affects PLA, and its properties are easily affected by environmental conditions such as temperature and humidity. To expand PLA's applications, many studies on PLA have been focused on improving its properties by processing[15], blending[16], and compounding[17–19]. However, limited studies have addressed PLA interaction with organic compounds and their aqueous solutions[11,20–22]. Thus, the objective of this work was to address the effects of aqueous and deuterium oxide solutions of straight-chain alcohols with one to four carbon atoms on the thermo-mechanical properties of PLA films when submerged in these alcoholic solutions.

2. MATERIALS AND METHODS

PLA Ingeo™ 2003D resin was procured by NatureWorks LLC (Minnetonka, MN, USA). PLA cast films were produced using Randcastle Extrusion Systems, Inc. (Cedar Grove, NJ, USA) with a temperature from 193 to 215 °C and 49 rpm with a volume of 34 cc and a screw length and diameter L/D ratio of 1.5875 cm and 24:1, respectively. PLA pellets were dried in a vacuum oven at 60 °C for 24 h before extrusion. The thickness of the produced PLA film varied from 15 to 25 µm. PLA films were stored in a refrigerator (-10 °C) and conditioned at 23 °C (50% RH) for 24 h before use to minimize physical aging.

Alcohols used as solvents in the immersion tests were straight-chain alcohols with the number of carbon atoms C1–C4 in the main chain (i.e., methanol, ethanol, 1-propanol, 1-butanol) and branched-chain alcohol (2-propanol). All the alcohols were purchased from Sigma-Aldrich (St. Louis, MO, USA). HPLC grade water (VWR, Radnor, PA, USA) and deuterium oxide (D_2O, 99.9 atom % D, Sigma-Aldrich) were used to produce solvent mixtures with the alcohols for the immersion test. All materials and chemicals were used as received.

To ensure that the PLA films were fully amorphous, a hot-pressing process was performed using a manual hydraulic compression press (PHI, City of Industry, CA, USA). The heating temperature was set as 160 °C, which is higher than the melting temperature (T_m) of the PLA samples (150 °C), and the heating time was 10 min. Two layers of Teflon® (polytetrafluoroethylene—PTFE) were used to sandwich the original PLA film. After hot-pressing, the film was cooled immediately by clamping the PLA films to cold platens at -9.7 °C overnight preventing crystallization, as confirmed by dynamical scanning calorimetry.

To measure the thermomechanical properties of the PLA films, an RSA-G2 Solids Analyzer (TA Instruments, New Castle, DE, USA) DMA instrument was used. The films were tested before and during the immersion process. The sample size of PLA films was 10 mm × 50 mm. Parameters of the DMA test were set as follows: preload force 100 g, strain 0.2%, loading gap

15 mm, max gap changes up 5–10 mm, max gap changes down 1 mm, frequency 1 Hz, temperature ramp rate 5 °C/min. For the dry film, the start temperature was 25 °C, and for the alcoholic solution immersions, the start range of the temperature ramp was 0 to 10 °C. The ramp ended at around 90 °C for dry film and 60 °C for alcoholic solution immersions. Before starting each immersion test, the alcohol solvents were precooled to 5 °C. Liquid nitrogen was used to control the temperature.

The thickness of each film was measured in five different positions using a digital micrometer (Testing Machines Inc., Ronkonkoma, NY, USA) with 120 V, 60 Hz, and the readability of 0.001 mm. Three replicates were tested for each film. The immersion condition tests were conducted in alcohol-aqueous solution at 0, 25%, 50%, 75% and 100% (v/v) - volume ratio of alcohol to water (or D_2O) in the solution.

The thermal properties of the PLA films before testing were measured by a differential scanning calorimeter DSC Q100 (TA Instruments) calibrated with indium standards. The weight of each sample was between 5 and 10 mg, and samples were sealed in the hermetic aluminum pans.

There was one heating cycle from 0 °C to 210 °C with 70 mL/min flow rate. The temperature ramp rate was 10 °C/min, and each sample was tested on three replicates. The T_g and the crystallinity (X_C) of PLA films measured by DSC were obtained using the TA Instruments Universal Analysis 2000 version 4.5A software.

The statistical significance was determined by one-way analysis of variance (ANOVA), and the mean differences were determined by Tukey's honestly significant difference (Tukey's HSD) test ($\alpha = 0.05$) using SAS Analysis Software University Edition (SAS Institute Inc., Cary, NC).

3. RESULTS AND DISCUSSION

3.1 Thermal properties of pre-immersion PLA film

The T_g, T_m, the cold crystallization temperature (T_c), and the X_C of the original PLA films without hot-pressing were initially measured by DSC. The results were $T_g = 58.0 \pm 1.2$ °C, $T_m = 149.9 \pm 0.5$ °C, $T_c = 123.7 \pm 3.3$ °C and $X_C = 1.3 \pm 0.2\%$. A hot-pressing process was used before sorption and DMA testing to ensure that the PLA films were totally amorphous. The PLA films' thermal properties after the hot-pressing process were, $T_g = 58.5 \pm 0.6$ °C, $T_m = 150.6 \pm 0.6$ °C, $T_c = 125.9 \pm 0.9$ °C, and $X_C = 0.4 \pm 0.2\%$. X_C was close to 0%, so the hot-pressed PLA films can be fully amorphous.

3.2 Thermal properties of immersed PLA film in pure alcohol

Figure 1 and Table 1 summarize the T_g's of PLA films immersed in different solvents and the reduction of T_g's (compared with the no immersion sample measured by DMA). There was no significant difference between the T_g's of PLA films immersed in water and D_2O. Still, there were substantial differences between the T_g's of no immersion PLA film and PLA films immersed in pure methanol, ethanol, 1-propanol, 1-butanol, and water. However, there was no significant difference between the T_g's of PLA films immersed in 1-propanol and 2-propanol.

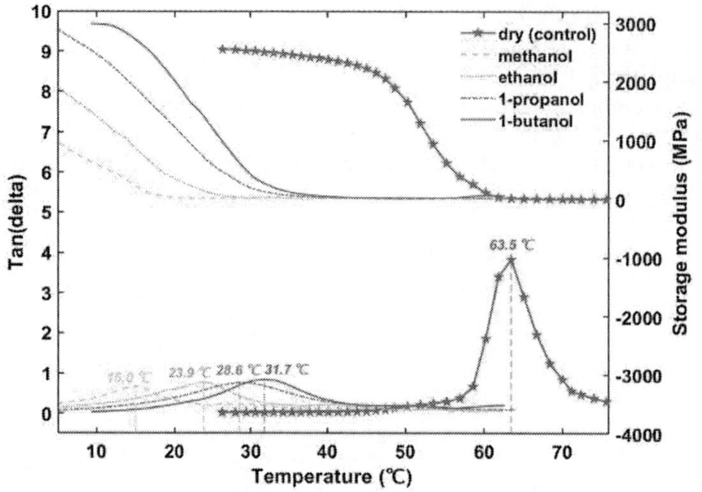

Figure 1 Tan δ, the storage modulus of PLA films immersed in different alcohols as a function of temperature

Table 1 T_g of PLA films immersed in different solvents as tested by DMA

Solvent	#C	T_g, °C	T_g reduction, %
none (no immersion)	-	62.8±0.4 ^{a}	-
methanol	1	15.2±0.1 ^{b}	75.8
ethanol	2	24.3±0.4 ^{c}	61.3
1-propanol	3	27.7±0.8 ^{d}	55.9
2-propanol	3	27.9±0.5 ^{d}	55.6
1-butanol	4	31.8±0.1 ^{e}	49.4
water	-	51.6±0.3 ^{f}	17.8
deuterium oxide	-	52.7±0.3 ^{f}	16.1

$^{a-f}$Lowercase superscript letters indicate different comparisons based on Tukey's HSD tests at a significance level of $\alpha = 0.05$. There are no significant differences between values with the same letters. #C is the number of carbon atoms in the main chains of alcohols

The T_g of the PLA films immersed in water and D_2O decreased but not as much as when compared with the T_g of the PLA films immersed in alcohols (Table 1). Hydrogen bonding strength could impact the depression of T_g since the PLA chains can be restricted by the secondary bonds between the carbonyl group and D_2O and H_2O. The OH bond length in H_2O is ~3% longer than the OD bond length in D_2O[23]. However, the difference in the length of the OH/OD bond and the difference in the length of the hydrogen bond between water and D_2O to PLA seems not to affect the depression of T_g of PLA film differently.

The T_g of the PLA films immersed in methanol decreased up to 76% compared with no immersed films' T_g (Table 1). The drop of T_g became smaller as the number of carbon atoms in the main chain of alcohols increased. T_g's of the PLA films immersed in small molecular weight alcohols were significantly reduced, indicating that these alcohols may plasticize the PLA chains, increasing mobility and the free volume of the system, thereby causing the PLA chains to move and leading to the decreased T_g. The PLA films immersed in alcohols with more than two carbon atoms in the main chain have a smaller reduction in T_g, indicating that these alcohols may be sorbed at lower concentrations and have a less plasticizing effect on the PLA matrix. So, alcohols with a low number of carbons are quickly diffused through the PLA matrix.

Pure 1-propanol and 2-propanol have a similar effect on the PLA films, indicating that the different carbon chain structures did not impact the T_g reduction of PLA films whether the situation in the alcohol-aqueous solution is the same as in the pure alcohol was described in the following section.

3.3 Thermal properties of immersed PLA film in alcohol-aqueous solutions

Commonly PLA films will not be in contact with pure alcohol solutions but with a range of concentrations of water and alcohol solutions, such as in medical and food packaging. Hence, the *in-situ* DMA test was used to determine the effect of alcohol solution concentrations on Tg of PLA. Figure 2a and Figure 2b show the T_g of PLA films immersed in 0%, 25%, 50%, 75%, and 100% (v/v) alcohol and aqueous and deuterium oxide solutions.

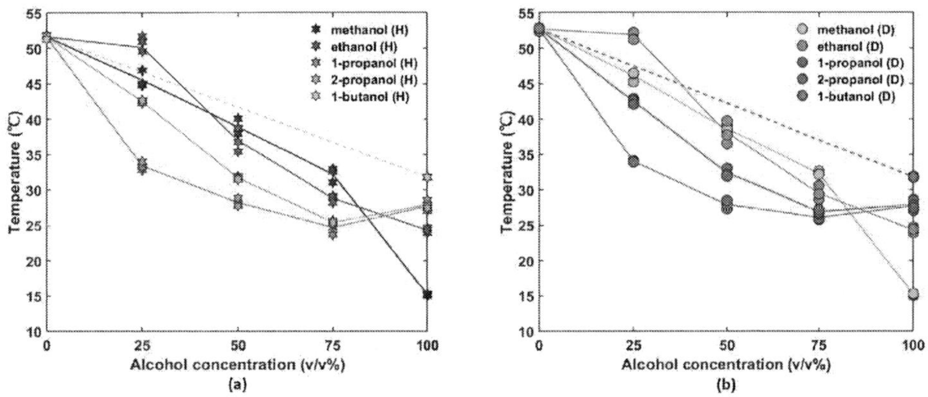

Figure 2a T_g of PLA immersed in alcohol-aqueous solutions as a function of alcohol concentration. (H) denotes respective alcohol (i.e., methanol, ethanol, 1-propanol, 2-propanol, and 1-butanol) in water solutions; Figure 3b T_g of PLA immersed in alcohol-D_2O solutions as a function of alcohol concentration. (D) denotes alcohol in deuterium oxide

The interaction between alcohol molecules and PLA leads to swelling and plasticization of the PLA matrix, resulting in the movement of the PLA segments, thereby reducing the T_g[24]. The extent of the swelling and T_g reduction is not yet fully theoretically predicted.

The T_g was decreased by 8.6 °C per every 25% (v/v) increase of methanol (alcohol with 1 carbon atom) and 7.6 °C per every 25% increase of ethanol (2 carbon atoms). When the PLA films were immersed in alcohol solutions with 3 to 4 carbon atoms, the T_g's decreased until 75% (v/v). The reduction in T_g of 1-propanol at a concentration of 25% was more significant than that of 2-propanol at the same concentration. As the concentration of alcohol (v/v) increased, the difference between straight-chain alcohols (1-propanol) and alcohols containing branched chains (2-propanol) on the decrease in T_g of PLA became smaller. Although there are few studies discussing the effect of branching alcohol in PLA, similar studies about other polymers can provide some insights. Chandra et al.[25] studied the equilibrium sorption and kinetics of lower alcohols in PET films. They showed that the different positions of the -OH group in 1-propanol and 2-propanol, which determine the shapes and dimensions of the molecules, lead to an order of magnitude decrease in the diffusion coefficient. This may explain why the T_g-reducing effect of 2-propanol was smaller than that of 1-propanol in our results (Figure 2a, Figure 2b). Figure 2a also shows that the T_g of PLA immersed in 1-butanol is reduced by about 20 °C compared with PLA immersed in water.

Figure 2a and Figure 2b indicate that the aqueous and D_2O solutions have similar effects on the changes in T_g's of PLA, which means that the plasticization effect of alcohol in D_2O solutions is identical to water. Therefore, D_2O could be used instead of water to configure the solution in a system that is hard to quantify water. Using D_2O can also avoid the influence of water in the environment when the absorption of alcohol is measured.

4. CONCLUSIONS

This study aimed to determine the thermomechanical properties of PLA film immersed in alcohol-aqueous solutions. The changes in T_g of PLA films immersed in alcohol-aqueous/D2O solvents were determined using *in-situ* immersion DMA. Immersion of PLA in water and alcohol affected the T_g of PLA. Due to the plasticization of the PLA matrix by the solvents, the mobility of the polymer system increased so that the PLA's T_g is decreased.

Specifically, compared with the PLA film not immersed in the alcohol solution, the T_g of the film immersed in pure methanol decreased by 76%. Compared to the T_g of dry PLA film, the T_g of PLA film immersed in pure ethanol, 1-propanol, 2-propanol and 1-butanol decreased 61%, 56%, 56%, and 49%, respectively. The PLA films' T_g's were also affected by the concentration of the alcohol solutions. Aqueous and D_2O solutions have similar effects on T_g changes. In addition, the sorption of the alcoholic solution by the film also changed with the immersion time.

5. REFERENCES

[1] Randová A, Bartovská L, Friess K, Hovorka Š, Izák P. Fundamental study of sorption of pure liquids and liquid mixtures into polymeric membrane. *European Polymer Journal*. 2014;61:64-71. 10.1016/j.eurpolymj.2014.09.021

[2] van Willige RWG. *Effects of Flavour Absorption on Foods and Their Packaging Materials*.

[3] Peychès-Bach A, Moutounet M, Peyron S, Chalier P. Factors determining the transport coefficients of aroma compounds through polyethylene films. *Journal of Food Engineering*. 2009;95(1):45-53. 10.1016/j.jfoodeng.2009.04.012

[4] Wang YC, Teng MY, Lee KR, Lai JY. Comparison between the pervaporation and vapor permeation performances of polycarbonate membranes. *European Polymer Journal*. 2005;41(7):1667-1673. 10.1016/j.eurpolymj.2005.01.011

[5] Lin H, Freeman BD. Gas solubility, diffusivity and permeability in poly(ethylene oxide). *Journal of Membrane Science*. 2004;239(1):105-117. 10.1016/j.memsci.2003.08.031

[6] Chandra P, Koros WJ. Sorption and transport of methanol in poly(ethylene terephthalate). *Polymer (Guildf)*. 2009;50(1):236-244. 10.1016/j.polymer.2008.10.031

[7] Widiastuti I, Sbarski I, Masood SH. Sorption characteristic of organic liquid and its effect on the mechanical performance of a PLA-based plastic. *Journal of Applied Polymer Science*. 2016;133(20). 10.1002/app.43250

[8] Iñiguez-Franco F, Auras R, Rubino M, Dolan K, Soto-Valdez H, Selke S. Effect of nanoparticles on the hydrolytic degradation of PLA-nanocomposites by water-ethanol solutions. *Polymer Degradation and Stability*. 2017;146:287-297. 10.1016/j.polymdegradstab.2017.11.004

[9] Byun Y, Whiteside S, Thomas R, Dharman M, Hughes J, Carolina S. The Effect of Solvent Mixture on the Properties of Solvent Cast Polylactic Acid (PLA) Film. 2011:13-18. 10.1002/app

[10] Cho HH. *Swelling Properties of Poly(Ethylene Terephthalate) Fiber in Mixed Solvent(II)-Equilibrium Swelling and Shrinkage.*; 2016. https://www.researchgate.net/publication/264029628.

[11] Salazar R, Domenek S, Courgneau C, Ducruet V. Plasticization of poly(lactide) by sorption of volatile organic compounds at low concentration. In: *Polymer Degradation and Stability*. Vol 97. ; 2012:1871-1880. 10.1016/j.polymdegradstab.2012.03.047

[12] Vink ETH, Davies S. Life Cycle Inventory and Impact Assessment Data for 2014 Ingeo® Polylactide Production. *Industrial Biotechnology*. 2015;11(3):167-180. 10.1089/ind.2015.0003

[13] Morão A, de Bie F. Life Cycle Impact Assessment of Polylactic Acid (PLA) Produced from Sugarcane in Thailand. *Journal of Polymers and the Environment*. 2019;27(11):2523-2539. 10.1007/s10924-019-01525-9

[14] Peng T. Energy Modelling for FDM 3D Printing from a Life Cycle Perspective. *International Journal of Manufacturing Research*. 2017;12(1):83-98. 10.1504/IJMR.2017.10003722

[15] Lim LT, Auras R, Rubino M. Processing technologies for poly(lactic acid). *Progress in Polymer Science (Oxford)*. 2008;33(8):820-852. 10.1016/j.progpolymsci.2008.05.004

[16] Lee W, Lee J, Chung JW, Kwak SY. Enhancement of tensile toughness of poly(lactic acid) (PLA) through blending of a polydecalactone-grafted cellulose copolymer: The effect of mesophase transition on mechanical properties. *International Journal of Biological Macromolecules*. 2021;193:1103-1113. 10.1016/j.ijbiomac.2021.09.205

[17] Shahdan D, Flaifel MH, Ahmad SH, Chen RS, Razak JA. Enhanced magnetic nanoparticles dispersion effect on the behaviour of ultrasonication-assisted compounding processing of PLA/LNR/NiZn nanocomposites. *Journal of Materials Research and Technology*. 2021;15:5988-6000. 10.1016/j.jmrt.2021.11.046

[18] Yu B, Meng L, Fu S, et al. Morphology and internal structure control over PLA microspheres by compounding PLLA and PDLA and effects on drug release behavior. *Colloids and Surfaces B: Biointerfaces*. 2018;172:105-112. 10.1016/j.colsurfb.2018.08.037

[19] Loyo C, Moreno-Serna V, Fuentes J, et al. PLA/CaO nanocomposites with antimicrobial and photodegradation properties. *Polymer Degradation and Stability*. 2022;197:109865. 10.1016/j.polymdegradstab.2022.109865

[20] Sato S, Gondo D, Wada T, Kanehashi S, Nagai K. Effects of various liquid organic solvents on solvent-induced crystallization of amorphous poly(lactic acid) film. *Journal of Applied Polymer Science*. 2013;129(3):1607-1617. 10.1002/app.38833

[21] Sonchaeng U, Auras R, Selke S, Rubino M, Lim LT. In-situ changes of thermo-mechanical properties of poly(lactic acid) film immersed in alcohol solutions. *Polymer Testing*. 2020;82(October 2019):106320. 10.1016/j.polymertesting.2019.106320

[22] Thakur KAM, Kean RT, Hall ES, et al. Concurrent solvent induced crystallization and hydrolytic degradation of PLA by water-ethanol solutions. *Polymer (Guildf)*. 2012;99(4):315-323. 10.1016/j.polymer.2016.07.018

[23] Soper AK, Benmore CJ. Quantum differences between heavy and light water. *Physical Review Letters*. 2008;101(6). 10.1103/PhysRevLett.101.065502

[24] Meyers M, ed. *Mechanical Behavior of Materials*. second. Cambridge university press; 1999.

[25] Chandra P, Koros WJ. Sorption of lower alcohols in poly(ethylene terephthalate). *Polymer (Guildf)*. 2009;50(17):4241-4249. 10.1016/j.polymer.2009.06.066

MT-O06

Barrier Performance of recyclable thermoforming sheets for sous-vide potatoes

Marit Kvalvåg Pettersen[1]*, Magnhild S. Grøvlen[1], Simon Saxegaard[2], Thomas Eie[3]

[1]*Nofima AS, Norwegian Institute of Food, Fisheries and Aquaculture Research, Aas, 1430, Norway*
[2]*NORSUS, Norwegian Institute for Sustainability Research, Kråkerøy, 1671, Norway*
[3]*Bama Industri AS, Tranby, 3408, Norway*

***Correspondence to:** 1Nofima AS, Norwegian Institute of Food, Fisheries and Aquaculture Research, Aas, 1430, Norway, marit.kvalvag.pettersen@nofima.no.*

ABSTRACT: The increasing focus on circular economy enables a drive towards development of food packaging concepts using recyclable materials. The objective of this study was to evaluate the effect of using recyclable materials on the quality of sous-vide potatoes. The performance of the materials was evaluated in two steps with presently available recyclable materials. In the first experiment, two materials consisting of PP/EVOH/PE (150μm and 170μm bottom webs) were compared to the reference (PA/EVOH/PE). OTR of flat films and thermoformed packages were similar for all materials at 0 % internal RH at 4 °C/90 % RH and 23 °C/50 % RH, while 3 times higher OTR for test materials with 100% internal humidity at both temperatures. After 5 weeks of storage the flavour of the potatoes stored in the test materials was regarded as not acceptable for sale. In the second experiment PE/EVOH (170μm) and PP/EVOH (200μm) were compared with the reference. These materials showed similar OTR at dry internal condition and 3-4 times higher OTR at 100 % internal RH compared to the reference. Minor variation in flavour scores were measured storage although no significant differences between the samples were found. After 9 weeks of storage, flavour scores were not acceptable. Although the light exposure did not affect the odour and flavour of the potatoes, the effect was clearly visible through a significant decrease in yellowness for potatoes in test materials while only no significant reduction in the reference material was observed. The environmental assessment of the packages indicates a generally lower climate change profile associated to the direct life cycle of the recyclable alternatives compared to the reference. Additionally, the environmental impact was lower for the recyclable lighter materials. This study demonstrated that thermoforming, process and storage condition affect the properties and packaging performance in terms of food quality.

Keyword: Recyclable, Thermoforming, barrier properties, sous vide potatoes, Food quality

1. INTRODUCTION

EU's Circular Economy Action Plan (CEAP) push product design towards circular economy. A linear design with energy efficiency optimization is no longer sufficient to achieve sustainable products [1]. This imposes pressure on stakeholders for development and implementation of food packaging concepts that not only allow the substitution of non-recyclable materials with recyclable food packaging materials but also permit improvement of systems for sorting, recycling, and use of recycled materials. At the same time, efforts are put in the endeavour to preserve food and obtain optimal shelf life, it is important not to blindly optimize toward a circular economy at the cost of increased food waste [2].

For many food products optimal packaging material is a crucial part, to fulfil required barrier and mechanical properties, processability, and sealability [3]. In some cases, to it is also necessary to withstand processes with high temperatures, pressure and/or humidity (such as sous-vide) while maintaining physical and chemical integrity [4,5]. For decades, materials have been designed by combining different types of polymers into multi-material constructions. High oxygen barrier materials are often required to obtain relatively long shelf life. Ethylene vinyl alcohol (EVOH) is currently predominantly used in food packaging due to its excellent oxygen barrier properties in dry conditions [6]. However, the barrier properties of EVOH are significantly reduced in humid conditions, due to its hydrophilic properties [7]. Consequently, EVOH is often sandwiched between materials with high water vapor barrier in multi-layered materials, such as polyolefins (PE and PP) [8]. PA has been widely used for fresh food products and for in-pack processes such as autoclave, due to its excellent thermoforming, strength, and barrier properties. Although multi-material constructions are optimal to protect and preserve the food, such complexed materials face a challenge in terms of recycling [9,10].

Thermoforming is an efficient, cost-effective manufacturing process which is widely used for food products. The thermoforming process involves heating of the material to its softening point and forming, in terms of stretching into a mould by use of different techniques [11,12]. Consequently, properties such as thickness of the material are altered, often leading to a non-uniform thickness in walls, bottom, and corners. The type of polymer, and the processing parameters can affect the wall thickness uniformity [11]. Thermoforming may also have impact on mechanical and gas barrier properties [11].

With CEAP as founding principles, it is important to identify alternative recyclable materials with sufficient barrier or protective properties and to explicitly evaluate if there is an environmental gain. Today many materials are optimized towards high barrier and protective properties and low weight. A low material weight is often associated with low environmental burdens. Also, there must be appropriate recycling sorting systems that ensure that materials designed for recycling are actually recycled.

The aim of this study was to investigate the suitability of recyclable materials for thermoforming and in-pack processing for sous-vide potatoes and the effect on food quality to replace the non-recyclable PA/EVOH/PE laminate. The performance of the materials with respect to food quality was evaluated in two steps with presently available recyclable materials, named first- and second-generation recyclable materials. Material thickness and material composition were investigated, and three strategies were addressed, material change, material design for recycling and reducing overpacking. Furthermore, light exposure was investigated in the second experiment.

2. MATERIALS AND METHODS

2.1 Packaging materials, packaging process and product

The potatoes were vacuum packaged in thermoformed packages with base web and top web as described in Table 1. using commercial thermoforming machines at Bama Industri AS (Moss, Norway) in two separate experiments. The size of the packaging/cavity was 180x125mm and the drawing depth was 25mm. The potatoes were autoclaved using standard industrial parameters shortly after packaging. The packages were cooled to 4 °C directly after autoclaving and transported to Nofima the next day for storage and evaluation. The potatoes were stored for up to 9 weeks at 4 °C in the dark. In the second experiment, some samples were exposed to light (LED light, 2000 lux) for 1 week after 5 weeks of dark storage – simulating realistic storage conditions at retail.

2.2 Material Thickness

The thickness of the packages was measured using a Model 543 Film Thickness Gauge (Qualitest, Japan). The thickness was measured across the packaging from edge/wall every 1cm (Figure 1). Number of replicates were three and the results are given in μm. In addition, the thickness was measured in the corners of the packages. Three replicates were taken for each material, at three spots in each corner.

Table 1 Packaging materials; product name, thickness and composition for bottom and top webs included in Experiment 1 and 2.

		Material bottom web		Material top web	
	Sample ID	**Product name and thickness**	**Material composition**	**Product name and thickness**	**Material composition**
EXP1: First generation recyclable	$PE/EVOH/PP_{150}$	Nice Eco XX 8 (150μm) [a]	PE/EVOH/PP/PE	Biaxop XX 90 (90μm) [a]	BOPP/PE/EVOH/PE
	$PE/EVOH/PP_{170}$	Nice Eco XX 12 (170μm) [a]	PE//EVOH/PP/PE	Biaxop XX 90 (90μm) [a]	BOPP/PE/VOH/PE/EVOH
EXP2: Second generation recyclable	$PP/EVOH_{200}$	Nice Eco13 XX PP (200μm) [a]	PP/EVOH/PP	Biaxop ECO XX 70 XPP (70μm) [a]	BOPP/EVOH/PP
	$PE/EVOH_{170}$	NFO BTM Eco 170xx (170μm) [a]	PE/EVOH/PE/EVOH/PE	NFO Top Eco 100xx (100μm) [a]	PE/EVOH/PE/EVOH/PE
Reference material	REF_{150}	PA + PE G EVOH B (150μ) [b]	PA/PE/EVOH	RB (100μ) [b]	PA/PE/EVOH

[a] Wipak Nastola, Finland; [b] SP Group, Cordoba, Spain

2.3 The oxygen transmission rate (OTR)

The oxygen transmission rate (OTR) of the bottom webs were measured before and after thermoforming for the first-generation recyclable materials (EXP1) at 23°C/50% Relative Humidity (RH) using Dualperm 8001 (Industrial Physics, USA). The OTR of the thermoformed packages (including top web) in EXP1 was measured at 23°C/50% RH and at 4°C/90% RH with both dry (0% RH) and 100% internal humidity (by adding 10ml water), and in EXP2 the OTR was measured at 4°C/90% RH with 0% and 100% internal humidity by using the ambient oxygen ingress rate (AOIR) method [13].

2.4 Colour

The colour of the potato was measured through the top web by using a Minolta Chromameter CR-400 (Minolta Konica Sensing Inc., Osaka, Japan) with an 8 mm viewing port and illuminant D_{65}. The used colour model is decided by the International Commission on Illumination (CIE), CIE L*a*b* (lightness (L*), redness (a*), yellowness (b*). Colour was measured in the same ten potato packages per material throughout the time of storage and two potatoes in each package were measured. All food quality analyses were performed after 3 and 5 weeks in Experiment 1, and after 2, 4, 6, 8 and 9 weeks of storage in Experiment 2.

2.5 Odour and Flavour

Evaluation of odour and flavour were performed by a lab panel of three to four assessors. The samples were ranked on the 1-3 scale used by Bama Industri as quality check; where 1 indicates good quality, 2 indicates alteration, but still acceptable for sale, and 3 indicates unacceptable product quality. Odour and flavour were performed on pieces taken from the middle of the potatoes, containing both the internal and external parts. Appearance/colour was also evaluated. Same potatoes were used for all characterization. Three replicates of each variant were randomized and given a three-digit number and evaluated at each sampling times.

2.6 Analyses of environmental assessment for the packaging materials- Methods and scope

In this study's environmental evaluation, life cycle assessment (LCA) was used with inventory from: 1) industry regarding material compositions and inbound transport distances, and 2) database data (ecoinvent 3.8) for background and foreground activities. This LCA study presents results that are approximations to the packages in Table 1 and cannot be used for marketing purposes or external commercial communication. The LCI was assembled in SimaPro 9.3.0.3 and the LCIA method used is Climate change (IPCC2013 GWP 100a v1.03). The scope of the LCA was cradle to grave: raw material production, extrusion and thermoforming, recycling and/or incineration, and transport between each activity. Each packaging is attributed all burdens until end-of-waste state is achieved, including the recycling and energy recovery activities (Directive 2008/98/EC). Potential material (100% fossil PE) and heat and electricity (Norwegian and German grid mixes) substitution is included. Material losses during manufacturing 10.7 (bottom) and 10.9% (top) and post-use need sorting to recycling (74.5%, GPN 2022) are included. Losses during use is neglected.

The functional unit is: *One pk primary food packaging used for X g of sous-vide potatoes sold in Norwegian supermarkets.*

3. RESULTS AND DISCUSSION

3.1 Thickness

In the first experiment, the thickness distribution was measured in $PE/EVOH/PP_{170}$ and REF_{150} after thermoforming. Similar profiles in thickness were observed in both materials. As expected, lowest thickness was measured in the edge/corners and highest thickness on the sides close to the sealing section (Figure 1). The thickness in the corners was reduced to 74% and 68% reduction of thickness after thermoforming in $PE/EVOH/PP_{170}$ and REF_{150}, respectively. In EXP2, the thickness distribution after thermoforming in both recyclable materials $PE/EVOH_{170}$ and $PP/EVOH_{200}$ was compared with the REF_{150} (Figure 1). Analogous to the first experiment, lowest thickness in the corners and highest thickness on the sides close to the sealing was observed. The thinning was somewhat higher in the edge (position 5/6 and 13) in

REF_{150}, resulting in 59% reduction compared to 41 and 50% reduction in PE/EVOH$_{170}$ and PP/EVOH$_{200}$, respectively. These results, with highest reduction in thickness closer to the bottom correlates with earlier findings [12,14].

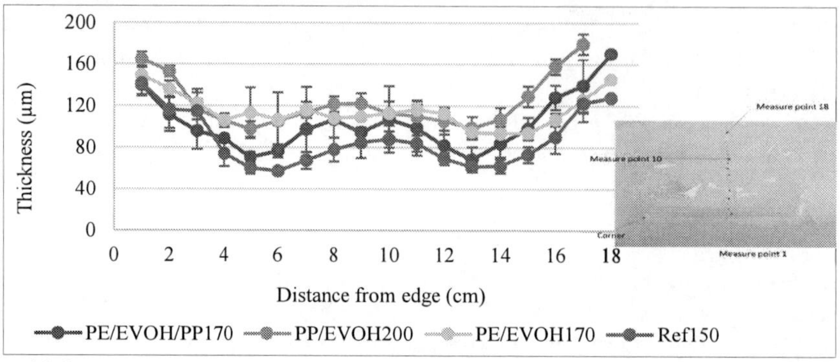

Figure 1Thickness measured in PE/EVOH/PP$_{170}$ (EXP1), PP/EVOH$_{200}$, and PE/EVOH$_{170}$ (EXP2) and references (REF_{150}) (EXP 1 and 2). Measured positions are shown in the picture.

3.2 Oxygen transmission rate

The oxygen transmission rate (OTR) of flat films before thermoforming were < 1 ml/m²/24h for the materials in EXP1; 0.29 ± 0.01 ml/m²d and 0.55 ± 0.02 ml/m²d for PE/EVOH/PP$_{150}$ and PE/EVOH/PP$_{170}$, respectively (measured at 23°C/50%RH). After thermoforming the OTR in the bottom web increased to 1.56 ± 0.01 ml/m²d and 1.58 ± 0.0 ml/m²d, due to increased surface area and thinning of the materials [11].

All packages in EXP1 had similar OTR at 0% internal RH at both conditions and a slight increase in OTR at 23°C/50%RH compared to 4°C/90% RH (Table 2). Conversely to REF_{150}, the humidity clearly affected the OTR for the first-generation materials with a 2-3-fold increase in the latter materials at 4°C/90% RH, and a 6-fold increase at 23°C/50% RH. At 100% internal RH, the OTR was almost 3 times higher in PE/EVOH/PP$_{150}$ and PE/EVOH/PP$_{170}$ compared to REF_{150} at 23°C/50% RH. The reference material was more affected by the temperature than the internal humidity, similar to a previous studies [12,15]. Maes et al. (2018) [15] stated that, despite of the lower moisture barrier for PET and PA compared to PP and PE, EVOH layer was well protected from moisture, resulting in lower oxygen permeability.

Table 2 Oxygen transmission rate in packages (ml/package d), Experiment 1 and 2

Experiment	Packages-Sample ID	4°C 90% Relative Humidity (RH)		23°C 50% Relative Humidity (RH)	
		0% internal RH	100% internal RH	0% internal RH	100% internal RH
EXP 1:	PE/EVOH/PP$_{150}$	0.05 ± 0.01	0.11 ± 0.02	0.09 ± 0.02	0.65 ± 0.16
	PE/EVOH/PP$_{170}$	0.08 ± 0.04	0.23 ± 0.15	0.11 ± 0.06	0.76 ± 0.17
	REF_{170}	0.07 ± 0.02	0.04 ± 0.03	0.17 ± 0.07	0.18 ± 0.01
EXP 2:	PP/EVOH$_{200}$	0.12 ± 0.04	0.54 ± 0.04		
	PE/EVOH$_{170}$	0.41 ± 0.19	1.56 ± 0.42		
	REF_{150} *)	0.07 ± 0.02	0.04 ± 0.03		

*) Results from Exp1

In EXP2, the OTR of PE/$EVOH_{170}$ was significantly higher (3-4 times) compared to PP/$EVOH_{200}$ in dry condition (Table 2) which could be due to the fact that OTR of LDPE is higher than for PP [15]. The thickness of the different layers may also play an important role on the increased OTR. The effect of internal humidity was similar in the two recyclable materials, resulting in a 3-4-fold increase in the OTR. The differences in OTR between these samples were almost the same as in dry conditions.

3.3 Colour

The colour and appearance of food is an important parameter for consumer acceptance at purchase. In EXP1, the lightness (L*) increased during 5 weeks dark storage in all packaging materials; which was 67.8 ± 1.9 to 75.0 ± 1.1, 67.2 ± 2.1 to 75.1 ± 1.3 and 65.5 ± 1.9 to 73.6 ± 1.2 for PE/EVOH/PP_{150}, PE/EVOH/PP_{170} and REF_{150}, respectively. No clear differences were observed between potatoes stored in recyclable or non-recyclable materials after 3 and 5 weeks of storage. According to Catauro et al. (2012) [16] darkening has been reported as colour change for starch products. Increase in lightness, redness and yellowness due to microwave-assisted pasteurization systems have also been reported for mashed potatoes [17].

In EXP2, as well as in EXP1 the preponderant differences in colour were related to yellowness (b* value), which deceased during 9 weeks of dark storage; 14.0 ± 2.3 to 9.9 ± 3.8 for PP/$EVOH_{200}$, 12.2 ± 1.9 to 9.1 ± 2.2 for PE/$EVOH_{170}$, and 17.3 ± 2.8; to 10.7 ± 2.5) for REF_{150}. Although some changes during storage was observed, no significant differences between the potatoes stored in non-recyclable material (REF_{150}) and the recyclable materials were observed during 9 weeks storage despite minor variation.

3.4 Odour and Flavour

In EXP1, no clear difference in odour between the samples were detected after 3 weeks of storage, while after 5 weeks of storage the odour of potatoes stored in the recyclable materials were regarded as less fresh and were given score higher than the reference (1.8 ± 0.3, 2.3 ± 0.3, and 1.5 ± 0.3 for PE/EVOH/PP_{150}, PE/EVOH/PP_{170} and REF_{150}, respectively).

After 5 weeks of storage, potatoes stored in the reference were regarded as acceptable in flavour (score 1.6 ± 0.4) while potatoes stored in the recyclable materials were regarded as not acceptable with score of 2.3 ± 0.5 and 2.2 ± 0.3 for PE/EVOH/PP_{150}, and PE/EVOH/PP_{170}, respectively. Based on these results, and with max score 3 (not acceptable) given by 3 of the assessors, the product quality was regarded as not acceptable (also by the product specialist from BAMA) and the experiment was closed.

In EXP2 the odour score was below score 2 (acceptable for sale) for all samples during 9 weeks of storage and no difference in odour between the recyclable materials (PP/$EVOH_{200}$ and PE/$EVOH_{170}$) and the non-recyclable material (REF_{150}) could be detected.

Regarding the flavour of the potatoes apparently some variation in scores were given. However, there were no differences between potatoes stored in the second-generation recyclable materials compared to the reference at any sampling time. After 9 weeks of storage, flavour scores were 2.3 ± 0.6 for PP/$EVOH_{200}$ and 2.1 ± 0.5 for both PE/$EVOH_{170}$ and REF_{150}, indicating not acceptable flavour for potatoes stored in recyclable materials, as well as in the reference.

3.4 Effect of light exposure

The effect of light exposure was clearly observable and resulted in a large and significant decrease in the b*-value for potatoes stored in the recyclable materials, in contrast to storage in REF_{150} (no significant reduction) (Figure 2). All materials in EXP2 had relatively low OTR in dry internal conditions. The difference was more pronounced at 100% internal RH. This could explain the difference in colour and appearance. However, the difference was not detected for dark storage, thus, the combination of presence of oxygen and light exposure can elicit the discoloration of the potatoes. The lack of difference in discoloration in $PE/EVOH_{170}$ and $PP/EVOH_{200}$ although different OTR at humid condition could indicate that the required OTR at high humidity to avoid discoloration is below 0.5 ml/package. However, it should be noted that conversely to the colour and appearance, none of the samples exposed to light were regarded as unacceptable in terms of odour and flavour.

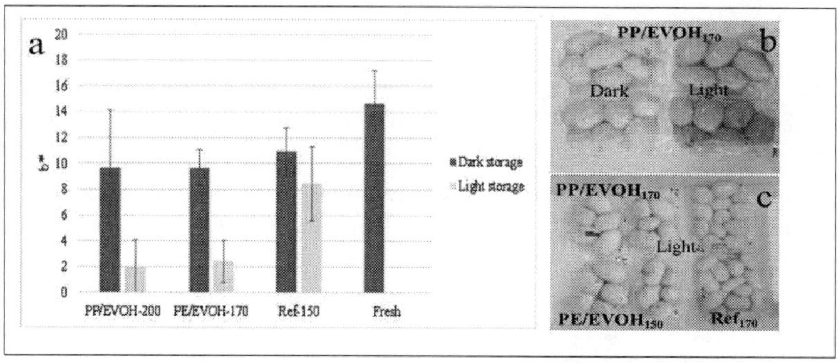

Figure 2 Colour (b* value) (Fig. a) and appearance (Fig b and c) of potatoes stored 6 weeks in dark and 5 weeks in dark + 1 week in light, $PP/EVOH_{200}$, $PE/EVOH_{170}$ and REF_{150}.

3.5 LCA - recyclability

The LCA-results indicate that the REF_{150} packaging cause higher CO_2-eq impact than the EXP1 and EXP2 materials (Figure 3). The higher CO_2-eq. impact of REF_{150} is mainly attributed to the use of PA which has six times the CO_2-impact compared to polyolefins and longer inbound transport. There is relatively low impact variation between the EXP1 and EXP2 materials. Recycling influences net CO2-eq less than the change in material from PA to polyolefins.

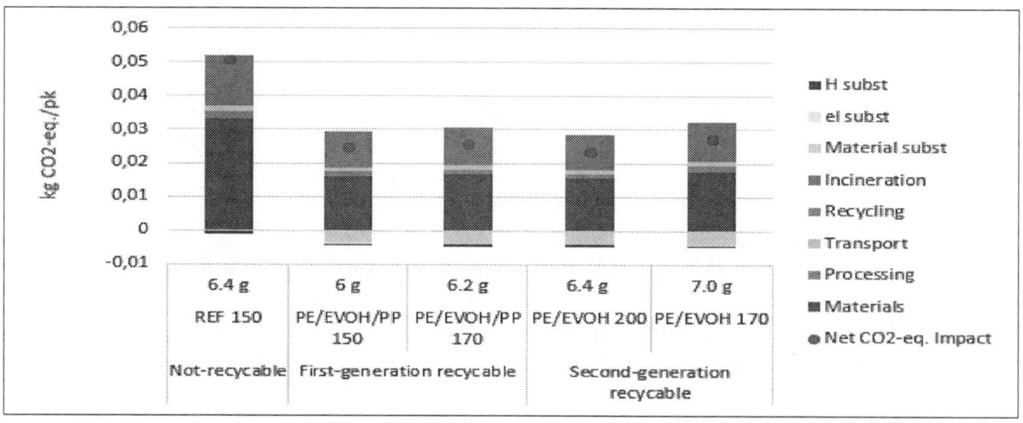

Figure 3 Caused, avoided and net CO_2-eq. / pk for materials in EXP1 and EPX2.

In the experiment it was found that the EXP1 and EXP2 materials did not satisfy all requirements after respectively 5 weeks and 9 weeks. Therefore, it is important to evaluate functionality when designing recyclable primary food packaging to identify such potential tradeoffs, which is in line with previous research [18-19]. This study shows that by increasing material thickness and thus weight, EXP2 materials provided similar protection as REF_{150}, although not sufficient shelf life, whilst causing less CO_2-eq. than REF_{150}. However, the tradeoff between shelf-life and recyclability must be evaluated further in terms of acceptability and its potential risk of increasing food waste.

4. CONCLUSIONS

This study clearly demonstrated that even if the materials have similar barrier properties before thermoforming; the application, process and storage conditions affect the properties and thus packaging performance in terms of food quality. The new recyclable materials in EXP2 did not provide 9weeks of storage with optimal quality. However, the flavour was not considered different from the reference. Thus, the recyclable material was regarded as a promising candidate, but some improvements are needed. Our study revealed challenges in obtaining required shelf life, e.g., related to oxidation and discoloration. Furthermore, it is evident that light exposure accelerates the rancidity development and discoloration. However, polyolefins, which are recyclable, cause less CO_2-eq compared to PA. Therefore, special care must be taken when selecting packaging material in line with the CEAP goals without compromising on required packaging properties and potentially causing increased food waste.

5. ACKNOWLEDGEMENT

This research was conducted within the ReducePack project (Grant No. 296335) funded by the Research Council of Norway, and in the strategic program FutureFoodControl (Project No. 314743) funded by the Norwegian Levy on Agricultural Products. The authors would like to thank Merete R. Jensen and Anette W. Åsli (Nofima), Kjetil Pettersen and Eskild Kellmer (Wipak Norway), and Therese A. Ellis and Kosta Hansen (Bama Industri AS, Moss) for their valuable contribution in the studies.

6. REFERENCES

[1] COMMISSION, E. 2021. *Waste Framework Directive* [Online]. Environment. Available: https://ec.europa.eu/environment/topics/waste-and-recycling/waste-framework-directive_en#ecl-inpage-632 [Accessed].

[2] Wikström F, Verghese K, Auras R, Olsson A, Williams H, Wever R, Grönman K, Pettersen MK, Møller H and Soukka R (2019). "Packaging Strategies That Save Food: A Research Agenda for 2030." Journal of Industrial Ecology 23(3): pp. 532-540, DOI: 10.1111/jiec.12769

[3] Massey LK. (2003). Introduction. In Massey LK (Eds) (Second Edition), Permeability Properties of Plastics and Elastomers. William Andrew Publishing: Norwich, NY, pp.1-56.

[4] Siracusa V, Blanco I, Romani S, Tylewicz U, and Dalla Rosa M. (2012) Gas Permeability and Thermal Behavior of Polypropylene Films Used for Packaging Minimally Processed Fresh-Cut Potatoes: A Case Study. Journal of Food Science 77 (10), pp E264-E272 DOI: 10.1111/j.1750-3841.2012.02905.x

[5] Galotto MJ, Ulloa PA, Hernandez D, Fernandez-Martın F, Gavara R and Guarda A. (2008). Mechanical and thermal behaviour of flexible food packaging polymeric films materials under high pressure/temperature treatments. Packaging Technology and Science, 21, pp297–308. DOI:10.1002/pts.807

[6] Zhang Z, Britt, IJ and Tung MA. (2001). Permeation of oxygen and water vapor through EVOH films as influenced by relative humidity. J. of Applied Polymer Science 82, pp.1866–1872. DOI:10.1002/app.2030

[7] Hodge RM, Bastow TJ, Edward GH, Simon GP and Hill AJ. (1996). Free volume and mechanism of plasticization in water swollen polyvinyl alcohol. Macromolecules, 29(25), pp. 8137–8143. DOI 10.1021/ma951073j

[8] Mokwena KK, Tang J, Dunne CP, Yang TCS and Chow E. (2009). Oxygen transmission of multilayer EVOH films after microwave sterilization. J. of Food Eng. 92 (3), pp291–296, DOI: 10.1016/j.jfoodeng.2008.11.011

[9] Ragaert K, Delva L and Van Geem K. (2017). Mechanical and chemical recycling of solid plastic waste. Waste Management, 69, pp. 24-58, DOI: 10.1016/j.wasman.2017.07.044

[10] Pauer E, Tacker M, Gabriel V and Krauter V. (2020) Sustainability of flexible multilayer packaging: Environmental impacts and recyclability of packaging for bacon in block. Cleaner Environmental Systems. Vol1 (Dec (2020), 100001, pp. 1-13, DOI:10.1016/j.cesys.2020.100001

[11] Buntinx M, Willems G, Knockaert G, Adons D, Yperman J, Carleer R and Peeters R. (2014). Evaluation of the Thickness and Oxygen Transmission Rate before and after Thermoforming Mono- and Multi-layer Sheets into Trays with Variable Depth. Polymers, 6 (12), pp. 3019-3043; DOI:10.3390/polym6123019

[12] Pettersen MK, Gällstedt M and EieT. (2004). Oxygen Barrier Properties of Thermoformed Trays Manufactured with Different Drawing Methods and Drawing Depths Packaging. Technology and Science 17, pp. 43–52 DOI:10.1002/pts.

[13] Larsen H, Kohler A, Magnus EM. (2000), Ambient oxygen ingress rate method – an alternative method to Ox-Tran for measuring oxygen transmission rate of whole packages. Packaging Technology and Science,13, pp.233–241. DOI: 10.1002/pts.519
[14] Piergiovanni L, Rho V, Bernig W. (1996) Thinning of Barrier Layers in Multiwall thermoformed Packages. Packaging Technology and Science, 9, pp.313–326. DOI: 10.1002/(SICI)1099-1522(199612)9:6<313::AID-PTS374>3.0.CO;2-9
[15] Maes C, Luytena W, Herremansa G, Peetersb R, Carleerc R, and Buntinx M. (2018) Recent Updates on the Barrier Properties of Ethylene Vinyl Alcohol Copolymer (EVOH): A Review. POLYMER REVIEWS, 58 (2), pp. 209–246, DOI 10.1080/15583724.2017.1394323
[16] Catauro, P. M., & Perchonok, M. H. (2012). Assessment of the long-term stability of retort pouch foods to support extended duration spaceflight. Journal of Food Science, 77(1), pp. S29–39. DOI: 10.1111/j.1750-3841.2011.02445.x.
[17] Sonara CR, Parhia A, Liua F, Patela J, Rascob B, Tanga J and Sablania SS. (2020). Investigating thermal and storage stability for vitamins in pasteurized mashed potatoes packed in barrier packaging films. Food Packaging and Shelf Life, 24, 100486, pp. 1-9, DOI: 10.1016/j.fpsl.2020.100486
[18] Svanes E, Vold M, Møller H, Pettersen MK, Larsen H and Hanssen OJ. (2010). Sustainable packaging design: A holistic methodology for packaging design. Packaging Technology and Science, 23, pp. 161-175. DOI: 10.1002/pts.887
[19] Molina-Besch K, Wikstöm F and Williams H. (2019). The environmental impact of packaging in food supply chains—does life cycle assessment of food provide the full picture? International Journal of Life Cycle Assessment, 24, pp. 37-50. DOI: 10.1007/s11367-018-1500-6

MT-O07

Evaluation of Melt-Processed Polyhydroxyalkanoates and Zinc Oxide Nanocomposite Films as Flexible Packaging Materials

Chris Vanheusden[1]*, Natalia Rosa Andrade Lopes[1], Pieter Samyn[2], Roos Peeters[1], Mieke Buntinx[1]

[1]*Materials and Packaging Research & Services, Institute for Materials Research (IMO-IMOMEC), Hasselt University, Wetenschapspark 27, 3590 Diepenbeek, Belgium*
[2]*SIRRIS – Smart Coatings Lab, Wetenschapspark 3, 3590 Diepenbeek, Belgium*

***Correspondence to:** chris.vanheusden@uhasselt.be*

ABSTRACT: Polyhydroxyalkanoates (PHAs) show great potential as renewable packaging materials due to their biobased and biodegradable nature along with high versatility. The copolymers poly(3-hydroxybutyrate-co-3-hydroxyvalerate) (PHBV) and poly(3-hydroxy-butyrate-co-3-hydroxyhexanoate) (PHBHHx) are the most promising because of increased flexibility and intermediate barrier properties, depending on the copolymer content. However, the relation between melt processing parameters and mechanical performance remains to be elucidated in order to optimize processing conditions. In addition, the incorporation of ZnO nanoparticles could further improve the required thermal stability and barrier properties. Therefore, the aim of this study is the incorporation of ZnO into a suitable PHA matrix, targeting sufficient flexibility while minimizing embrittlement by crystallization. In this study, process parameters such as extrusion and mold temperature, screw speed and cooling time were systematically varied in a full factorial design of experiments. The injection molded PHBV samples show increased tensile strength and Young's modulus under a critical set of mold temperatures. The improved mechanical properties are in parallel with increased crystallinity caused by lower cooling rates at high mold temperatures. However, the stretchability of PHBV with low valerate content (<3%) remains insufficient compared to PHBHHx, limiting its applicability to high stiffness and strength applications (e.g. cutlery). Therefore, the more flexible PHBHHx copolymer was selected as a matrix for the preparation of 1-10 wt.% ZnO/PHBHHx nanocomposite films via a combination of twin-screw extrusion and compression molding. As such, the feasibility of ZnO/PHBHHx nanocomposite films

Keyword: poly(3-hydroxybutyrate-co-3-hydroxyhexanoate), PHBHHx, poly(3-hydroxy-butyrate-co-3-hydroxyvalerate), PHBV, zinc oxide, ZnO, nanocomposites, extrusion

1. INTRODUCTION

Bioplastics show great promise to substitute fossil based polymers as part of a circular economy [1]. Among these bioplastics, polyhydroxyalkanoates (PHAs) are a group of biobased and bacterially synthesized polyesters that show great promise for (food) packaging applications [2]. The exceptional stereo-chemical regularity of the mainly investigated poly(3-hydroxy-butyrate) (PHB) leads to a highly crystalline polymer, being stiff and brittle. An approach to eliminate or reduce the brittleness and thermal instability of PHB above its melting point is the production of co-polyesters containing PHB units with other 3-hydroxyalkanoate units, such as e.g. 3-hydroxyvalerate (3HV) and 3-hydroxyhexanoate (3HHx). The copolymers poly(3-hydroxybutyrate-co-3-hydroxyvalerate) (PHBV) and poly(3-hydroxybutyrate-co-3-hydroxy-hexanoate) (PHBHHx) show lower crystallinity, lower melting temperature, decreased stiffness (decreased Young's Modulus), increased elongation at break and increased impact strength compared to PHB. However, the thermomechanical properties of PHBV can widely vary depending on the 3HV content [3], being stiff and brittle at low 3HV content. Although PHBV with low 3HV content is mainly available, little research has been performed on the effect of melt processing conditions on its mechanical properties (strength and ductility). Our previous study showed that the mechanical properties of PHBHHx can be tuned by specific processing conditions [4]. Hence, in order to verify whether the mechanical properties of PHBV can also be tuned, the first aim of this study is the optimization of the melt processing conditions of PHBV.

In addition, having medium oxygen barrier, slow crystallization rate, relatively low strength and poor heat resistance, the PHBV or PHBHHx copolymers need to be functionalized to broaden their application in packaging areas. The incorporation of ZnO nanoparticles via melt processing could improve the heat stability and barrier properties of PHAs, but the nanoparticles may simultaneously influence the mechanical properties through modified crystallization [5]-[9]. Accordingly, the second aim of this study is the melt incorporation of ZnO into a suitable PHA matrix, targeting sufficient flexibility while minimizing embrittlement by crystallization.

2. MATERIALS AND METHODS

2.1 Materials

PHBV granulate (trade name PHI 002) with a 3-hydroxyvalerate content <3% was purchased from Natureplast (Ifs, France). PHBHHx granulate (trade name Aonilex X151A) with a 3-hydroxyhexanoate (3HHx) content of 10.5 mol% [10] was kindly provided by Kaneka Corporation (Westerlo-Oevel, Belgium). Zinc oxide nanoparticles (ZnO NPs) with triethoxycaprylylsilane surface modification (trade name Zano 20 Plus) were kindly provided by EverZinc (Loncin, Belgium). The materials were dried for at least 3 days at 65°C prior to melt processing to remove moisture.

2.2. PHBV processing methodology

A lab-scale co-rotating twin-screw extruder (Process 11, Thermo Scientific, Karlsruhe) and piston injection molding apparatus (Haake MiniJet II, Thermo Scientific, Karlsruhe) were used to produce dumbbell shaped samples. The PHBV samples were conditioned at 23°C and 50% RH (relative humidity) for at least 3 days prior to characterization.

A 2^k full factorial design of experiments (FFD) with four factors at two levels and one center point was developed to investigate the main extrusion and injection molding parameters that influence the mechanical properties of PHBV. The four factors include: (i) mold temperature **M**, (ii) extrusion screw speed **S** (iii) processing temperature **P** and (iv) cooling time **C**. The injection molding parameters, such as injection pressure, injection time, post pressure, and post pressure time were held constant. The parameter settings (Table 1) resulted from a pre-screening to define all suitable combinations of upper (+) and lower limit settings (□) that led to reproducible and testable samples for analysis.

Table 1. Factors of the FFD with low (-), middle (0) and high (+) values.

Variables	Symbol	Low (-)	Middle (0)	High (+)
Processing temperature (°C)	**P**	175	182.5	190
Mold temperature (°C)	**M**	45	67.5	90
Screw speed (rpm)	**S**	50	100	150
Cooling time (s)	**C**	60	180	300

2.3. Tensile testing

Tensile testing of injection molded PHBV dumbbell samples (ISO 527-2 type 5A) was performed with a with a 10-M systems tensile tester (MTS) at a speed of 20 mm/min (n = 5). Tensile testing of PHBHHx and PHBHHx/ZnO compression molded films was executed with a Tinius Olsen 5ST at a speed of 1 mm/min (n = 4). Tensile testing was performed at 23°C and 50% RH.

2.4. PHBHHx/ZnO nanocomposite film production

PHBHHx pellets were grinded to powder using a basic analytical mill (IKA A11) with liquid nitrogen and were dry mixed with ZnO NPs using a SpeedMixer (Synergy Devices Ltd., High Wycombe, UK) in ratios of 1, 3, 5 and 10 wt.% ZnO. The PHBHHx/ZnO dry mix was further melt-mixed with a lab-scale co-rotating twin-screw extruder (Process 11, Thermo Scientific, Karlsruhe) at a processing temperature of 145°C and screw speed of 50 rpm [4]. The extruded pellets were compression molded into 10x10 cm films using a hot press (PCH-600DG, Henan Chuanghe Laboratory Equipment Co. Ltd, Zhengzhou, China) and a 150 μm thick stainless-steel mold.

2.5. Differential scanning calorimetry (DSC)

The melting and crystallization behavior were evaluated under inert atmosphere (50 ml/min nitrogen) using a Q200 instrument (TA Instruments). For PHBV, temperature was varied between □30°C and 200°C. For PHBHHx and PHBHHx/ZnO, temperature was varied between □30°C and 160°C. The heating/cooling rate was 20°C/min. Crystallinity calculations were based on the first heating scan, taking into account melting enthalpies for 100% crystalline material of 146 J/g [11], [12] and 115 J/g [13] for PHBV and PHBHHx respectively.

2.6. Opacity and colorimetric analysis

Colorimetric and opacity analysis was performed using a Check 3 spectrophotometer (Datacolor, Luzern, Switzerland). The opacity Y (in %) can be calculated as the ratio between the opacity of a sample on a black standard Y_b and the opacity on a white standard Y_w. Colorimetric results are presented in CIELAB color coordinates (L^*, a^*, and b^*). L^*, a^*, and b^* represent degree of lightness, green/red, and blue/yellow, respectively.

3. RESULTS AND DISCUSSION

3.1. PHBV processing optimization

Full Factorial Design (FFD) analysis

The FFD matrix with settings of melt processing parameters **P, M, S** and **C** (Table 1) and corresponding results are shown in Table 2. The tensile strength ranges between 38.0 and 54.7 MPa. The highest tensile strength is obtained with a factor combination of **P** = 175°C, **M** = 90°C, **S** = 150 rpm and **C** = 60 s, while the lowest tensile strength is obtained with a factor combination of **P** = 190°C, **M** = 45°C, **S** = 50 rpm and **C** = 300 s. The Young's modulus ranges between 5656 and 4180 MPa. The highest Young's modulus is obtained with a factor combination of **P** = 190°C, **M** = 45°C, **S** = 150 rpm and **C** = 300 s, while the lowest Young's modulus is obtained with a combination of **P** = 190°C, **M** = 90°C, **S** = 150 rpm and **C** = 60 s.

The elongation at break (ε_b) is not included in the FFD because PHBV remains very brittle with ε_b values under 3%. The obtained mechanical properties are in line with supplier specifications and in range with literature values of PHB and low valerate PHBV [2], [3], [14].

Table 2. FFD with factors (**P, M, S** and **C**), levels, and response data for tensile strength and Young's modulus (n = 5).

	P (°C)	M (°C)	S (rpm)	C (s)	TS (MPa)	E (MPa)
1	-	-	-	-	45.0 ± 0.4	5008 ± 173
2	+	+	-	-	49.2 ± 0.9	5004 ± 91
3	+	+	+	+	50.1 ± 1.3	5127 ± 105
4	+	-	+	+	46.2 ± 1.0	5656 ± 123
5	+	+	-	+	52.6 ± 0.4	5365 ± 81
6	-	+	+	-	54.7 ± 2.0	5530 ± 265
7	-	-	-	+	49.0 ± 1.6	5575 ± 128
8	-	-	+	+	39.8 ± 0.7	4549 ± 53
9	-	+	+	+	44.7 ± 0.4	4539 ± 37
10	-	-	+	-	48.7 ± 1.4	5321 ± 107
11	+	-	-	-	38.8 ± 0.8	4653 ± 19
12	+	-	+	-	39.0 ± 0.4	4534 ± 117
13	+	+	+	-	42.6 ± 0.2	4180 ± 32
14	+	-	-	+	38.0 ± 1.1	4431 ± 46
15	-	+	-	+	45.6 ± 0.4	4650 ± 77
16	-	+	-	-	45.5 ± 2.7	4622 ± 92
17	0	0	0	0	43.3 ± 0.6	4449 ± 44

Trends of mechanical properties with changing processing parameters are shown in the main effect plots for Young's modulus (Figure 1a) and tensile strength (Figure 1b). Main effect plots show the mean response value at high and low factor level (+, −) and are a relative measure of factor significance on the system response [15]. If the line from low to high factor levels is steep, factors have a high effect on the response. Main effects are calculated with the following equation:

$$E_{f,X} = \bar{F}_{(+1)} - \bar{F}_{(-1)} \quad (1)$$

with $E_{f,X}$ the main effect of of parameter X on the response, $\bar{F}_{(+1)}$ the average response at high-level factor setting and $\bar{F}_{(-1)}$ the average response at low-level factor setting.

Figure 1a shows that cooling time **C** has the highest impact on Young's modulus, followed by processing temperature **P**. Higher cooling times result in higher average Young's modulus values (positive main effect), while higher processing temperatures result in lower average Young's modulus values (negative main effect). Mold temperature **M** has the highest effect on tensile strength (Figure 1b). Higher mold temperatures result in higher average tensile strength values (positive main effect).

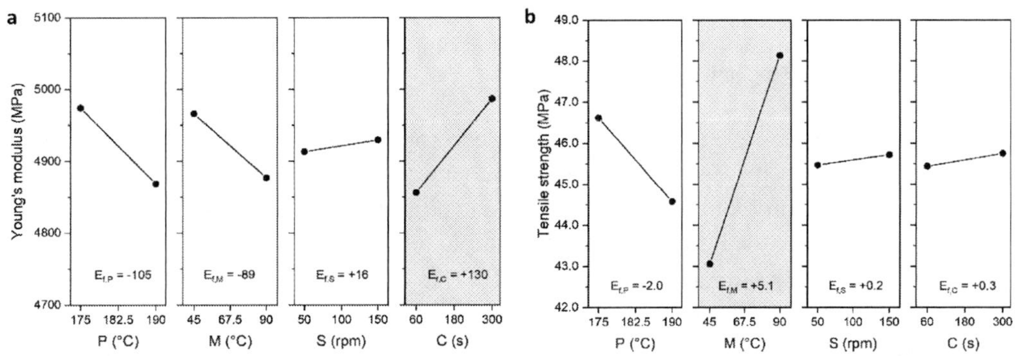

Figure 1. Main effect plot for a) Young's modulus and b) tensile strength showing average values across the FFD matrix for the parameters **P, M, S,** and **C**. Highest main effects are indicated in green.

3.2. Comparison of PHBV and PHBHHx matrices

PHBV and PHBHHx processing window

Additional processing runs were performed to further investigate the PHBV processing window. Tensile strength of PHBV in function of processing temperature **P** and mold temperature **M** is shown in Figure 2a as a (smoothed) average contour plot. It is clear that tensile strength increases at higher mold temperatures close to **M** = 90°C. Tensile strength decreases towards **M** = 45°C and especially in combination with processing temperatures close to **P** = 190°C. Despite improvements in tensile strength, PHBV shows very stiff and brittle behavior, limiting its use to rigid packaging applications like e.g. cutlery and straws. On the other hand, PHBHHx has shown more potential for use in flexible packaging applications [4]: the mechanical properties of PHBHHx can be tuned under specific processing conditions with tensile strength ranging between 20-22 MPa (Figure 2b), elongation at break ranging between 19-342% and Young's modulus ranging between 883-1205 MPa.

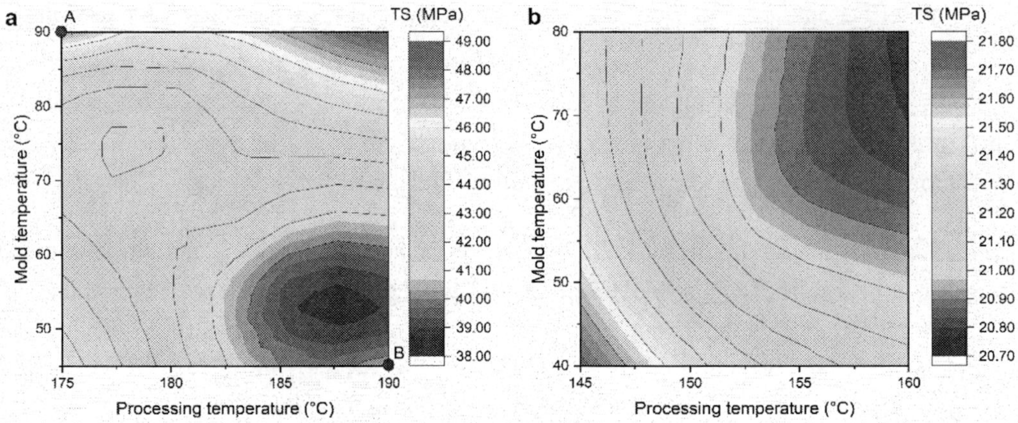

Figure 2. PHBV (a) and PHBHHx (b) processing window as (smoothed) average contour plot for tensile strength (TS) in function of processing temperature **P** and mold temperature **M**. This plot includes average FFD values.

Crystallinity and melting behavior

The crystallinity of specific PHBV samples was investigated by DSC analysis to elucidate the relation between processing, morphology and mechanical properties. Table 3 shows the melting properties and crystallinity for different processing conditions A and B, as depicted in the contour plot of Figure 2a. The crystallinity increases with mold temperature from 59% at **M** = 45°C to 67% at **M** = 90°C, resulting in stronger, but stiffer PHBV. No clear shift in peak melting temperature ($T_{m,p}$) is apparent. These findings confirm evidence from previous observations that higher crystallization temperatures (slower cooling at higher mold temperatures) can result in higher crystallinity, larger spherulites and increased lamellar thickness [11], [16], [17]. The increased crystallinity can cause increased TS, higher stiffness (higher E) and brittle failure [18], [19]. On the contrary, injection molded PHBHHx shows lower crystallinity of about 38-39%, independent of mold temperature [4].

Table 3. Melting enthalpy, melting temperature and degree of crystallinity of samples with different processing conditions (A and B).

	P (°C)	**M (°C)**	**ΔH_m (J/g)**	**$T_{m,p}$ (°C)**	**X_c (%)**
A	175	90	98.1 ± 2.2	175.1 ± 0.2	67.2 ± 1.5
B	190	45	86.3 ± 0.6	175.9 ± 0.3	59.1 ± 0.4

Due to the stiff and brittle behavior of PHBV with high crystallinity, PHBHHx is selected as best suited matrix for incorporation of ZnO NPs (as further shown in this paper) because of its superior flexibility, sufficient strength and good processability.

3.3. Production and characterization of PHBHHx/ZnO nanocomposites

Mechanical properties

The mechanical properties of PHBHHx/ZnO films with a thickness of 160-170 μm and a ZnO loading of 1-10% are shown in Table 4. There is no significant change in both TS and E with ZnO addition. In contrast, [5] found a gradual increase of PHBHHx/ZnO properties with the highest increase of E and TS at 5 wt.% ZnO of 41% and 27% respectively. An increase of TS and E with ZnO loading was also found for PHBV/ZnO [7] and PHB/ZnO nanocomposites [8]. The addition of ZnO in PHBHHx confirms that high strength and flexible mechanical properties can be maintained for flexible packaging materials.

Table 4. Mechanical properties tensile strength (TS) and Young's modulus (E) of PHBHHx/ZnO nanocomposites for different ZnO loading (n=4).

wt.%	Tensile strength (MPa)	Young's modulus (MPa)
0	20.8 ± 0.4	940 ± 17
1	20.8 ± 2.6	892 ± 131
3	19.4 ± 1.0	933 ± 38
5	19.7 ± 0.4	955 ± 15
10	20.2 ± 1.5	993 ± 89

Thermal properties

The melting and crystallization properties of the PHBHHx/ZnO nanocomposites are shown in Table 5 and 6. The crystallinity of the nanocomposite films remains nearly constant with addition of ZnO. There is no significant change in crystallization, suggested by a similar crystallization onset temperature ($T_{c,o}$), peak temperature ($T_{c,p}$) and enthalpy (ΔH_c). A very little increase in the width at half peak (ΔW_{hp}) during cooling for higher ZnO loading suggests a slightly slower crystallization. These results agree with the relatively small variations in mechanical properties that were observed.

Table 5. DSC first heating and cooling cycle details of PHBHHx/ZnO nanocomposites with varying wt.% ZnO (n = 3).

wt.%	First heating	Cooling			
	X_c (%)	$T_{c,o}$ (°C)	$T_{c,p}$ (°C)	ΔH_c (J/g)[a]	ΔW_{hp} (°C)
0	39.5 ± 0.6	77.2 ± 0.1	68.5 ± 0.1	34.0 ± 0.2	9.76 ± 0.02
1	38.9 ± 0.3	77.7 ± 0.0	69.1 ± 0.1	34.2 ± 0.2	9.89 ± 0.05
3	38.7 ± 0.3	77.7 ± 0.2	69.1 ± 0.3	34.2 ± 0.4	9.99 ± 0.04
5	38.7 ± 0.4	77.6 ± 0.1	68.9 ± 0.2	33.9 ± 0.5	10.03 ± 0.03
10	39.1 ± 0.3	77.8 ± 0.1	68.8 ± 0.2	34.0 ± 0.2	10.16 ± 0.05

[a] Enthalpy values are calculated with respect to polymer weight fraction.

Table 6. DSC second heating cycle details of PHBHHx/ZnO nanocomposites with varying wt.% ZnO (n = 3).

wt.%	Second heating			
	T_g (°C)	$T_{m,1}$ (°C)	$T_{m,2}$ (°C)	ΔH_m (J/g)[a]
0	-0.89 ± 0.22	114.8 ± 0.0	129.1 ± 0.0	36.6 ± 0.9
1	-0.66 ± 0.25	114.9 ± 0.1	129.7 ± 0.0	37.2 ± 0.3
3	-0.72 ± 0.14	114.9 ± 0.2	129.6 ± 0.2	37.3 ± 0.2
5	-0.58 ± 0.43	114.8 ± 0.2	129.4 ± 0.2	37.2 ± 0.4
10	-0.79 ± 0.14	114.9 ± 0.1	129.4 ± 0.2	37.4 ± 0.2

[a] Enthalpy values are calculated with respect to polymer weight fraction.

Optical properties

The opacity and colorimetric properties of PHBHHx/ZnO nanocomposites are shown in Table 7. The opacity of the films increases with ZnO loading due to increased light scattering. This can be associated with presence of materials with different refractive index (polymer, ZnO and/or aggregates) [9]. A yellowing effect is apparent with increasing ZnO loading, as indicated by an increased b* value. Material yellowing can be induced by thermal degradation during processing and/or the intrinsic white/yellowish color of the ZnO NPs. However, thermal degradation would also possibly affect the mechanical properties and melting behavior, which has not been found. Therefore, it can be concluded that gradual haze formation without macroscopic aggregates suggests a homogeneous incorporation of ZnO nanoparticles with increasing concentration.

Table 7. Opacity and colorimetric properties of PHBHHx/ZnO nanocomposites for different ZnO wt.%.

wt.%	Opacity Y (%)	L*	a*	b*
0	12.9 ± 0.1	94.17 ± 0.13	-0.78 ± 0.05	4.62 ± 0.21
1	19.1 ± 0.5	94.57 ± 0.03	-0.92 ± 0.02	4.56 ± 0.08
3	32.4 ± 0.3	94.83 ± 0.05	-0.99 ± 0.01	4.95 ± 0.05
5	39.9 ± 0.4	94.84 ± 0.09	-1.08 ± 0.01	5.44 ± 0.11
10	52.4 ± 0.6	95.08 ± 0.11	-1.23 ± 0.02	6.22 ± 0.26

4. CONCLUSIONS

This paper underlines the significant effect of melt processing conditions on the mechanical properties of PHBV and PHBHHx. The results suggest that tensile strength and Young's modulus of PHBV can be increased under a critical set of processing conditions in parallel with higher crystallinity. PHBV shows great potential for use in rigid (packaging) applications where strength and stiffness are required like cutlery, straws or plates. However, PHBHHx shows more potential for applications with sufficient flexibility due to its lower Young's modulus, higher elongation at break, but sufficient tensile strength. Therefore, PHBHHx was selected as matrix for the development of ZnO nanocomposites. PHBHHx/ZnO nanocomposite films were developed via twin-screw extrusion and compression molding. No significant effect of ZnO was found on the crystallization and mechanical properties while maintaining the flexible packaging properties. Addition of ZnO gradually increased haze formation without macroscopically visual aggregates, suggesting homogeneous ZnO incorporation. Future work will focus on different ZnO melt dispersion techniques and interaction with the PHBHHx matrix, in relation to the thermal, mechanical, gas barrier and migration properties.

5. ACKNOWLEDGEMENT

The authors acknowledge Prof. Wouter Marchal (Analytical & Circular Chemistry, IMO-IMOMEC, UHasselt) for access to DSC and Prof. Louis Pitet (Advanced Functional Polymers, IMO-IMOMEC, UHasselt) for use of the grinding and compression molding equipment. This research was funded by the Special Research Fund (BOF) of Hasselt University, grant number BOF20DOC06.

6. REFERENCES

[1] J.-G. Rosenboom, R. Langer, and G. Traverso, "Bioplastics for a circular economy," *Nat Rev Mater*, vol. 7, no. 2, pp. 117–137, Feb. 2022, doi: 10.1038/s41578-021-00407-8.

[2] P. Ragaert *et al.*, "Polyhydroxyalkanoates for Food Packaging Applications," in *Reference Module in Food Science*, Amsterdam: Elsevier, 2019. doi: 10.1016/B978-0-08-100596-5.22502-X.

[3] A. Anjum, M. Zuber, K. M. Zia, A. Noreen, M. N. Anjum, and S. Tabasum, "Microbial production of polyhydroxyalkanoates (PHAs) and its copolymers: A review of recent advancements," *International Journal of Biological Macromolecules*, vol. 89, pp. 161–174, Aug. 2016, doi: 10.1016/j.ijbiomac.2016.04.069.

[4] C. Vanheusden *et al.*, "Extrusion and Injection Molding of Poly(3-Hydroxybutyrate-co-3-Hydroxyhexanoate) (PHBHHx): Influence of Processing Conditions on Mechanical Properties and Microstructure," *Polymers*, vol. 13, no. 22, p. 4012, Nov. 2021, doi: 10.3390/polym13224012.

[5] A. M. Díez□Pascual, "Poly(3-hydroxybutyrate-co-3-hydroxyhexanoate) with zinc oxide nanoparticles for food packaging," *J Food Process Eng*, Jul. 2021, doi: https://doi.org/10.1111/jfpe.13814.

[6] M. Abbas, M. Buntinx, W. Deferme, and R. Peeters, "(Bio)polymer/ZnO Nanocomposites for Packaging Applications: A Review of Gas Barrier and Mechanical Properties," *Nanomaterials*, vol. 9, no. 10, p. 1494, Oct. 2019.

[7] A. M. Díez-Pascual and A. L. Díez-Vicente, "ZnO-Reinforced Poly(3-hydroxybutyrate-co-3-hydroxyvalerate) Bionanocomposites with Antimicrobial Function for Food Packaging," *ACS Appl. Mater. Interfaces*, vol. 6, no. 12, pp. 9822–9834, Jun. 2014.

[8] A. Díez-Pascual and A. Díez-Vicente, "Poly(3-hydroxybutyrate)/ZnO Bionanocomposites with Improved Mechanical, Barrier and Antibacterial Properties," *IJMS*, vol. 15, no. 6, pp. 10950–10973, Jun. 2014.

[9] J. L. Castro-Mayorga, M. J. Fabra, A. M. Pourrahimi, R. T. Olsson, and J. M. Lagaron, "The impact of zinc oxide particle morphology as an antimicrobial and when incorporated in poly(3-hydroxybutyrate-co-3-hydroxyvalerate) films for food packaging and food contact surfaces applications," *Food and Bioproducts Processing*, vol. 101, pp. 32–44, Jan. 2017.

[10] J. Vandewijngaarden *et al.*, "Poly(3-hydroxybutyrate-co-3-hydroxyhexanoate)/Organomodified Montmorillonite Nanocomposites for Potential Food Packaging Applications," *J Polym Environ*, vol. 24, no. 2, pp. 104–118, 2016, doi: 10.1007/s10924-016-0751-1.

[11] P. J. Barham, A. Keller, E. L. Otun, and P. A. Holmes, "Crystallization and morphology of a bacterial thermoplastic: poly-3-hydroxybutyrate," *J Mater Sci*, vol. 19, no. 9, pp. 2781–2794, Sep. 1984, doi: 10.1007/BF01026954.

[12] V. Jost and H.-C. Langowski, "Effect of different plasticisers on the mechanical and barrier properties of extruded cast PHBV films," *Eur. Polym. J.*, vol. 68, pp. 302–312, Jul. 2015, doi: 10.1016/j.eurpolymj.2015.04.012.

[13] Y. Xie, I. Noda, and Y. A. Akpalu, "Influence of cooling rate on the thermal behavior and solid-state morphologies of polyhydroxyalkanoates," *J. Appl. Polym. Sci.*, vol. 109, no. 4, pp. 2259–2268, Aug. 2008, doi: 10.1002/app.28278.

[14] Y.-M. Corre, S. Bruzaud, J.-L. Audic, and Y. Grohens, "Morphology and functional properties of commercial polyhydroxyalkanoates: A comprehensive and comparative study," *Polymer Testing*, vol. 31, no. 2, pp. 226–235, Apr. 2012, doi: 10.1016/j.polymertesting.2011.11.002.

[15] J. Antony, *Design of experiments for engineers and scientists*, 2nd edition. London: Elsevier, 2014.

[16] A. El-Hadi, R. Schnabel, E. Straube, G. Müller, and S. Henning, "Correlation between degree of crystallinity, morphology, glass temperature, mechanical properties and biodegradation of poly (3-hydroxyalkanoate) PHAs and their blends," *Polym. Test.*, vol. 21, no. 6, pp. 665–674, Jan. 2002, doi: 10.1016/S0142-9418(01)00142-8.

[17] L. M. W. K. Gunaratne and R. A. Shanks, "Melting and thermal history of poly(hydroxybutyrate-co-hydroxyvalerate) using step-scan DSC," *Thermochimica Acta*, vol. 430, no. 1–2, pp. 183–190, Jun. 2005, doi: 10.1016/j.tca.2005.01.060.

[18] C. Fischer and D. Drummer, "Crystallization and Mechanical Properties of Polypropylene under Processing-Relevant Cooling Conditions with respect to Isothermal Holding Time," *International Journal of Polymer Science*, vol. 2016, pp. 1–11, 2016, doi: 10.1155/2016/5450708.

[19] S. J. A. Rizvi, "Effect of injection molding parameters on crystallinity and mechanical properties of isotactic polypropylene," *Int J Plast Technol*, vol. 21, no. 2, pp. 404–426, Dec. 2017, doi: 10.1007/s12588-017-9194-3.

MT-O08

Microencapsulation and Thermal Enhancement of Fish Oil Based Phase Change Material Through Epoxidation Process

Yeyen Laorenza[1], Bambang Riyanto[2], Nathdanai Harnkarnsujarit[1]*

[1]*Department of Packaging and Materials Technology, Faculty of Agro-Industry, Kasetsart University, Bangkok, 10900, Thailand*
[2]*Department of Aquatic Product Technology, Faculty of Fisheries and Marine Science, Bogor Agricultural University, Bogor, 16680, Indonesia*

***Correspondence to:** 50 Ngam Wong Wan Rd, Latyao, Chatuchak, Bangkok, 10900, Thailand.*
E-mail: nathdanai.h@ku.ac.th

ABSTRACT: Fatty acid based organic phase change material (PCM) has been widely used as low thermal energy storage integrated with packaging and other cooling systems. However, oil based PCM has low melting point (25-60 °C), restricting the application as thermal energy storage. This research investigated the effect of epoxidation process using hydrogen peroxide and acetic acid on improving thermal properties of fish oil based PCM. The microstructure of microencapsulated PCM was determined for stability. The result showed that epoxidation process extremely accelerated oxidation of the fish oil, increasing peroxide value (11.58-1594.15 meq/kg) and free fatty acid (2.77-32.74%), while reducing iodine value (116.99-43.11 g I_2/100g) resulting in bleached fish oil due to oxidation and bleaching reaction by H_2O_2. The epoxidation process reduced numbers of double bond in mono and poly unsaturated fatty acid chains which effectively increased melting point from 26.72 °C to 48.42°C due to saturation. The oil based PCM was microencapsulated with gum Arabic and maltodextrin with globular shape of microcapsule (size of 8.00-13.06 μm). Findings suggested that the epoxidation process effectively improved the thermal properties of fish oil based PCM and enhanced the function as thermal energy storage likely to be integrated with thermal insulation in packaging system.

Keyword: Epoxidation, Fish oil, Microencapsulation, PCM

1. INTRODUCTION

Phase change material (PCM) has been widely used as thermal energy storage. Different level of latent heat storage capacity depends on types and ability of material to absorb and release the latent heat during its phase change. PCM can be applied as thermal insulation in building [1, 2, 3] or packaging [4,5] to maintain the temperature. The sources of PCM for thermal insulators are divided into organic, inorganic, and eutectic PCM [6].

Remarkably, fatty acid is widely investigated in recent years as one of the organic PCM that is suitable for thermal energy storage under 20°C which is equal with their melting point. The promising fatty acid-based PCM were reported such as myristic acid and palmitic acid with the excellent melting point of 49-51°C and 60°C, respectively [7]. Fish oil is one of the promising marine resources as fatty acid-based PCM. Otherwise, the oil with various fatty acid content such as coconut oil is reported to have a low melting point between 25-60°C, limiting its application as a thermal insulator [6]. Meanwhile, the PCM can stabilize the temperature at the melting point range [8]. The utilization of fish oil waste as PCM could be investigated to optimize the clean industry by increasing their melting point.

Epoxidation is a catalyst to double bond that forms the oxirane ring marked by the presence of proxy acid [9]. In this study, the epoxidation process aims to convert the double bond of fish oil to increase the melting point since the double bond has a low melting point. Moreover, PCM needs to be micro-capsulated to prevent leakage of melted PCM during phase transition, prevent reaction between PCM and environment or other composite material, and increase heat transfer rate [10]. Microencapsulation is defined as coating a single droplet of the liquid phase of material with a continuous core to produce the micro size of capsules.

This research investigated the effect of the epoxidation process on the chemical and thermal properties of fish oil-based PCM. Moreover, the PCM has been microencapsulated using gum Arabic and maltodextrin to observe the structure. It was hypothesized that the epoxidation process can be increased the thermal properties of fish oil and be suitable with the core material.

2. MATERIAL AND METHOD

Fish oil Epoxidation

Sardinella fish oil (Maya Food Industries Ltd., Pekalongan, Indonesia) was centrifuged at a speed of 11200 g at 4°C for 30 min. Crude fish oil (50g) was mixed with 6.5 mL CH_3COOH 99% and 0.4 mL H_2SO_4 98% in the three-neck flask followed by stirring with a speed of 700 rpm and stable heated at 58-60°C. The 20.2 mL solution of 50% H_2O_2 was added gradually for 30 min. The solution was stirred for 4 h after the mixing process in the closed system. The phase change material was stored at the closed inert bottle at 4°C.

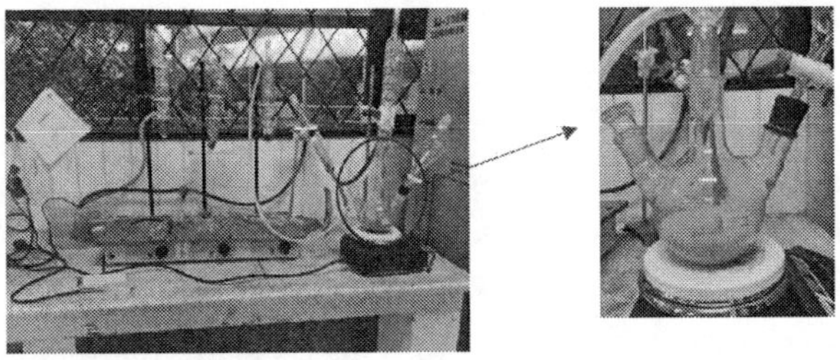

Fig 1 Epoxidation process of fish oil using hydrogen peroxide and acetic acid

Microencapsulation of PCM

The PCM was microencapsulated using coacervation-phase separation method [11]. Maltodextrin and Arabic gum were mixed at the ratio of 2:3 and added by 15% water followed by heating at 60°C and cooled down to 45°C. The solution was homogenized using high shear homogenizer AE300L-H with the speed of 13000 rpm for 15 min. The epoxidized fish oil or PCM was added to the solution and homogenized. The solution was dried using mini spray dryer Buchi 190 at T_{inlet} and T_{outlet} of 160°C and 95°C. The microencapsulated PCM was stored at the sealed aluminum bag at 4°C.

Peroxide Value

The oxidation level of fish oil was determined by peroxide value analysis [12]. Fish oil (5 g) was added by 30 mL of CH_3COOH 99% solution and chloroform (3:2). Saturated potassium iodide (0.5 mL) was added into the solution followed by adding 30 mL of distilled water and starch indicator 1%. The dark blue solution was titrated with $Na_2S_2O_3$ (0.01 N) was finished by the color changed to yellow at the first time and calculated for peroxide value as follows.

$$\text{Peroxude value (meq/kg)} = \frac{\text{V x M x 1000}}{\text{g}} \qquad (1)$$

Where V is volume of $Na_2S_2O_3$ (mL); M is concentration of $Na_2S_2O_3$; and g is sample weight (g).

Iodine Value

Iodine value of fish oil was tested using wijs-titrimetri method [13]. The 0.5 g of fish oil was added by chloroform and 25 mL potassium iodide 10% and shaken. The solution was titrated with 0.1 N $Na_2S_2O_3$ solution. The color of the solution was changed to light yellow and was titrated by 5 drops of amylum indicator 1% until the blue color disappear and the iodine value was calculated as follow.

$$\text{Iodine value} = \frac{\text{(mL blangko} - \text{mL sample)x N } Na_2S_2O_3 \text{ x 12.69}}{\text{sample weight}} \qquad (2)$$

Free Fatty Acid

Fish oil (2 g) was added by 25 mL alcohol and heated. The solution was added by 2 mL of phenolphthalein indicator and titrated by 0.1 N KOH until the color was changed to pink color and the free fatty acid was calculated as follow.

$$\text{Acid value (\%)} = \frac{\text{A x N x M}}{\text{10 x G}} \qquad (3)$$

Where A is volume KOH; N is normality KOH; G is sample weight; and M is dominant fatty acid molecular weight (Palmitic acid = 256.42 g/mol).

Fatty Acid Identification

Fatty acid of fish oil was identified by AOAC Official Method 969.33 2005 [14]. The 20-30 mg of fish oil was added by 1 mL of NaOH 0.5 N and heated at 55 °C for 20 min followed by centrifugation. The sample 1 µL was injected into the gas chromatography HP 5890A Series II MH-61 with the flowing rate of N2 was 30 mL/min with the temperature of injector and detector were set at 220°C and 240°C. Helium was used as a carrier with the flow rate was set at 30 mL/min. The flow rate of the air was set at 400 mL/min. Retention time and peak for each component were recorded and compared to the standard of FAME 17 and calculated as follows.

$$\frac{\frac{Ax}{As} \times C_{standard} \times \frac{Vsample}{100}}{G} \times 100\% \qquad (4)$$

Where Ax is sample area; As is standard area; $C_{standard}$ is concentration of FAME 17; V sample is volume of sample; and G is sample weight (g).

***Differential Scanning Calorimetry* (DSC)**

The melting profile of the foam was determined by *Differential Scanning Calorimetry* (DSC) 7 Perkin Elmer. The 10 g sample was subjected to crucible and heated at the temperature range of 0-140°C with the heating rate was set at 10°C/min. The result namely melting point, area, and latent heat were recorded.

3. RESULT AND DISCUSSION

Visual PCM

The visual of fish oil is shown in Figure 2. Fish oil waste was brighter with yellow color after centrifugation. This color was affected by the presence of carotenoid and tocopherol. Fish oil changed to creamy and turbid after epoxidation using H_2O_2 for 4 h with the catalyst of H_2SO_4 due to the oxidation reaction of fish oil during the epoxidation process. Hydrogen peroxide is a bleaching agent causing discoloration through an oxidation reaction.

Figure 2 Appearance of fish oil namely a) before centrifugation, b) after centrifugation, and c) after epoxidation

Chemical Analysis

The peroxide value increased after the epoxidation process due to oxidation and double bond breakage. The reaction between acetic acid and hydrogen peroxide created a peroxy acid involving the oxygen in the chain. Moreover, the heat process during the epoxidation process increased the free fatty acid (FFA) due to the reaction between double bond and heat [15]. This result was supported by reducing iodine value due to the conversion of a double bond into epoxide as the main product and hydroxyl group as a by-product while oxirane ring was open and exposed by water and acid (Table 1). In this research, the iodine value can represent the conversion of a double bond into epoxide value for 64%. [16] reported that the iodine value of linseed oil was decreased after the epoxidation process with a conversion of 0.81%. The long-chain fatty acid has a low water solubility that causes a low conversion level.

The epoxidation also affects the fatty acid profile, giving decreased total saturated fatty acid, monounsaturated fatty acid, and polyunsaturated fatty acid of fish oil. Oleic acid, linoleic acid, and linolenic acid decreased 79.73%, 85.43%, and 97.24% respectively after the epoxidation process. The composition of the fatty acid chain had a significant effect on the epoxidation rate. The double bond in oleic acid and linoleic acid has a similar reactivity level otherwise double bond in linolenic acid has twice as high [16]. The reducing double bond after the epoxidation process explained the conversion of the double bond to epoxide value resulted in fish oil being dominated by saturated fatty acid. This result suggested that the epoxidation process reduced the double bond of fish oil which would improve the thermal stability.

Table 1 Chemical analysis of fish oil namely peroxide value, free fatty acid, iodine value, and fatty acid.

Analysis	Parameters	Fish Oil	
		Crude fish oil	Epoxidized fish oil
Chemical	Peroxide value (meq/kg)	11.5789±0.81	1594.15±1.78
	Free fatty acid (%)	2.773±0.2073	32.74±0.0706
	Iodine value (g I2/100g)	116.99 ± 0.09	43.11 ± 0.57
	Saturated Fatty Acid (SAFA)	19.5	14.3
	Monounsaturated Fatty Acid (MUFA)	15.56	2.37
	Polyunsaturated Fatty Acid (PUFA)	28.73	0.52

Differential Scanning Calorimetry

Epoxidation increased the melting point of fish oil due to the conversion of the unsaturated fatty acid to epoxide value resulting in saturation. Saturated fatty acid has a higher melting point compared to unsaturated fatty acid [16]. The increasing melting point and latent heat are shown in Figure 3. In this research, the epoxidized fish oil is dominated by saturated fatty acid namely palmitic, myristic, and stearic acid which has the highest melting point at the range of 50-70°C compared to other fatty acids [17]. The shorter intermolecular in the chain of saturated fatty acid results in a higher melting point [16, 18]. The higher hydrocarbon molecule results in increasing reaction between the area causing the attraction force in the intermolecular chain effect on increasing viscosity, melting point, and boiling point. This result suggested that epoxidized showed higher thermal stability and was applied as PCM.

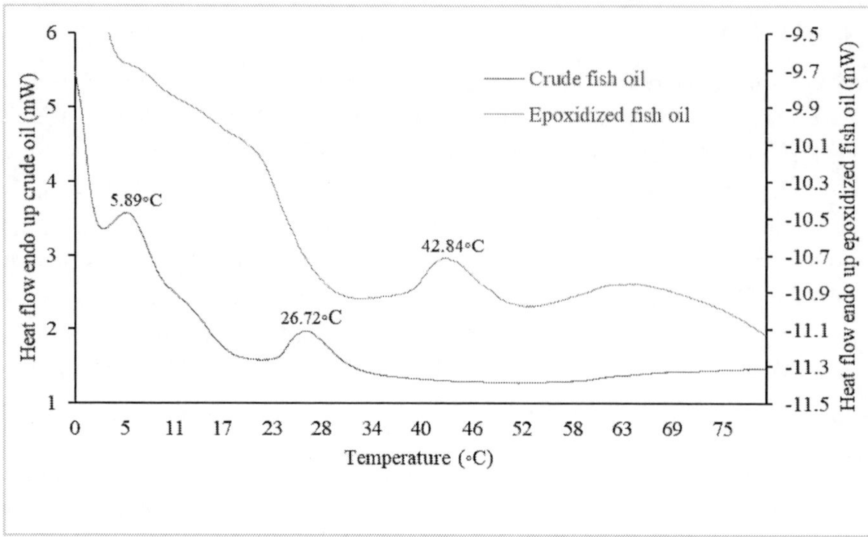

Figure 3 Differential scanning calorimetry curve of crude and epoxidized fish oil

Stability of Oil Droplet

Droplet of fish oil was dispersed uniformly in the liquid continuous phase at 15, 30, 45, and 60 minutes before drying had the average size of 43.29±1.60 µm, 49.237±0.81 µm, 48.0153±2.57 µm, 53.7526±3.36 µm, respectively (Figure 4). Size of the droplets increased with time, suggesting reduction of droplet stability due to oil's tendency to coalesce before drying. However, the globular stability of fish oil was maintained due to the hydrophobicity differences between the fish oil and the liquid core. The microcapsule was formed during coacervation process due to phase separation caused by the inability of fish oil to intermixed with water and polymer. The result suggested that the oil was formed dispersed-continuous phase before drying resulting in the stability of microencapsulation.

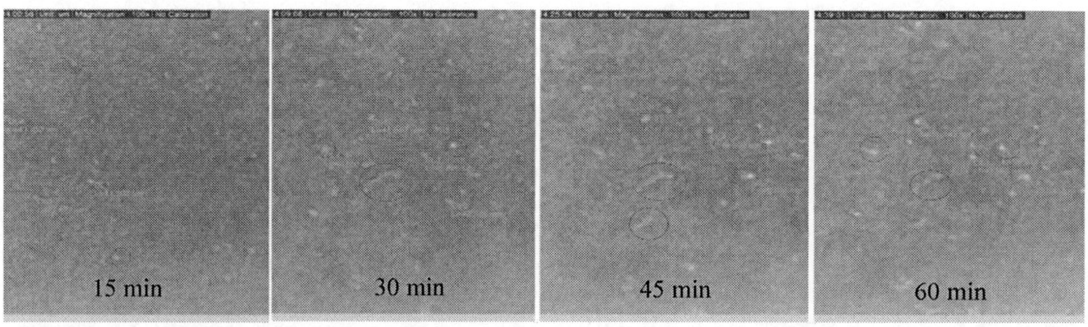

Figure 4 Globular stability of microencapsulated epoxidized fish oil before drying

Microstructure of Microencapsulated PCM

The dried microencapsulated PCM had a round shape with a smooth surface and curvature (Figure 5). High stability with no cracks that would further prevent the oxidation and leakage of the PCM during phase change. The diameter of microencapsulated PCM was 8.005-13.06 µm.

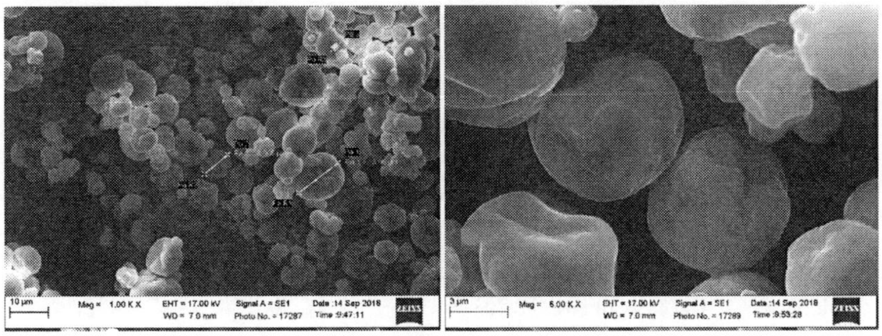

Figure 5 Microstructure of microencapsulated epoxidized fish oil at magnification of a) 1000x and b) 5000x

4. CONCLUSION

The epoxidation process was successfully increased the thermal properties of fish oil via the double bond of fatty acid chain destruction by hydrogen peroxide and acetic acid. The double bond was converted into epoxide value resulting in mono and polyunsaturated fatty acid reduction. Epoxidation process was also affected on severe oxidation consequence increasing peroxide value and free fatty acid, meanwhile reducing iodine value. The PCM was also successfully microencapsulated using gum Arabic and maltodextrin as a core with a globular shape and little hollow.

5. ACKNOWLEDGEMENT

This work was partially supported by Kasetsart University Research and Development Institute (KURDI), FF(KU) 1.65.

6. REFERENCES

[1] Ismail KA, Castro JN, and Lino FA. (2015). Thermal insulation of walls and roofs by PCM: modeling and experimental validation. International Journal of Engineering and Applied Sciences, 2(9), pp.83-93. 257820.

[2] Diaconu, B and Cruceru, M. (2010). Phase Change Material (PCM) composite insulating panel with high thermal efficiency. In International conference on energy and environment technologies and equipment, PP.105-110. April 20-22, 2010, Bucharest, Romania.

[3] Wang, D, Zhang, W. and Han, B. (2020). New generation of cement-based composites for civil engineering. In New Materials in Civil Engineering. Oxford: Butterworth-Heinemann, 777-795.

[4] Madhan, S., Santo, C. E., Andrade, L. P., Silva, P. D., & Gaspar, P. D. (2020). Active and intelligent packaging with phase change materials to promote the shelf-life extension of food products. *KnE Engineering*, 232-241. doi.org/10.18502/keg.v5i6.7037.

[5] Vennapusa JR, Konala A, Dixit P, and Chattopadhyay S. (2020). Caprylic acid based PCM composite with potential for thermal buffering and packaging applications. *Materials Chemistry and Physics*, *253*, 123453. doi.org/10.1016/j.matchemphys.2020.123453

[6] Velmurugan K, Kumarasamy S, Wongwuttanasatian T and Seithtanabutara V. (2021). Review of PCM types and suggestions for an applicable cascaded PCM for passive PV module cooling under tropical climate conditions. *Journal of Cleaner Production*, 126065. doi.org/10.1016/j.jclepro.2021.126065

[7] Jebasingh BE. (2016). Preparation of organic based ternary eutectic fatty acid mixture as phase change material (PCM), optimizing their thermal properties by enriched solar treated exfoliated graphite for energy storage. *Materials Today: Proceedings*, *3*(6), 1592-1598. doi.org/10.1016/j.matpr.2016.04.047

[8] Mjallal I, Feghali E, Hammoud M, Habchi C and Lemenand T. (2021). Exploring the colligative properties of Arachidic acid for potential use as PCM. *Solar Energy*, *214*, 19-25. doi.org/10.1016/j.solener.2020.11.020.

[9] Bashiri S, Ghobadian B, Soufi MD, and Gorjian, S. (2021). Chemical modification of sunflower waste cooking oil for biolubricant production through epoxidation reaction. *Materials Science for Energy Technologies*, *4*, 119-127. doi.org/10.1016/j.mset.2021.03.001

[10] Jamekhorshid A, Sadrameli SM, and Farid M. (2014). A review of microencapsulation methods of phase change materials (PCMs) as a thermal energy storage (TES) medium. *Renewable and Sustainable Energy Reviews*, *31*, 531-542. doi.org/10.1016/j.rser.2013.12.033

[11] Huo X, Li W, Wang Y, Han N, Wang J, Wang N, and Zhang X. (2018). Chitosan composite microencapsulated comb-like polymeric phase change material via coacervation microencapsulation. *Carbohydrate polymers*, *200*, 602-610. doi.org/10.1016/j.carbpol.2018.08.003

[12] AOAC Official Method 969.33. 2005. 1: Peroxide value of acids in oils and fats. In Official Method of Analysis of the Association of Official Analytical Chemists (18th ed.). Virginia: Accoc. Off. Anal. Chem.

[13] AOAC Official Method 993.20. 2005. 1: Iodine value of fats and oils. In Official Method of Analysis of the Association of Official Analytical Chemists (18th ed.). Virginia: Accoc. Off. Anal. Chem.

[14] AOAC Official Method 969.33. 2005. 1: Fatty acids in oils and fats. In Official Method of Analysis of the Association of Official Analytical Chemists (18th ed.). Virginia: Accoc. Off. Anal. Chem.

[15] Nawar WW. (1984). Chemical changes in lipids produced by thermal processing. Journal of Chemical Education. 16(4). 299-302.

[16] Kim, J. R., & Sharma, S. (2012). The development and comparison of bio-thermoset plastics from epoxidized plant oils. *Industrial Crops and Products*, *36*(1), 485-499. doi.org/10.1016/j.indcrop.2011.10.036

[17] Yuan, Y., Zhang, N., Tao, W., Cao, X., & He, Y. (2014). Fatty acids as phase change materials: a review. *Renewable and Sustainable Energy Reviews*, *29*, 482-498. doi.org/10.1016/j.rser.2013.08.107

[18] Markley, K.S. (1947). Fatty Acids. Their Chemistry and Physical Properties. New York: Interscience Publ., Inc.

MT-O09

Factors Influencing Damage on Tinplate Coatings

Mieke van den Berg[1]*, Hein van Jaarsveld[2]

[1]*Department of Design, Production and Management, Faculty of Engineering Technology, University of Twente, Drienerlolaan 5, Enschede, 7522 NB, the Netherlands*
[2]*Trivium Packaging, Zweedsestraat 7, Deventer, 7418 BG, the Netherlands*

***Correspondence to:** Mieke van den Berg, Department of Design, Production and Management, University of Twente, the Netherlands.*
Email: m.j.vandenberg@utwente.nl

ABSTRACT: This research deals with issues in processing polymer coated tinplate. The coating is damaged in some, but not all, cases in a process where tinplate is deformed and locally stretched to about 50% of the initial length. This multiple-step process occurs in less than one second and as a final step the outside of the material is coated electrolytically and cured for two minutes at 200°C. Until recently it was unclear where the damage initiates, so this research explores the cause of the issue with material and production parameters that affect it. This is done by explorative research in a production facility to compare various coatings, production temperatures and tooling settings. At the same time some steps are simulated in a lab environment, such as bending coated sheets to increase stresses and a simulation of the final curing step by using a lab oven at various temperatures. The results are evaluated by using a porosity measurement system, which uses a conduction set-up to determine the amount of bare-laying tinplate, i.e., the degree of damaged coating. The various available coatings show varying levels of damage before the final curing step. There is a clear increase of coating damage after the curing step for all systems. The damage increases by increasing curing temperature, which was tested from ~50°C to 225°C. It was also found that the tooling that determines the shape during the deformation and stretching steps has a relevant effect on the initial damage. The coating damage therefore seems to occur due to a combination of production, material and tooling factors, caused by mechanical stresses and temperature related processes.

Keyword: metal packaging, tinplate, polymer coating, thermal stress relaxation, crosslinking

1. INTRODUCTION

The metal packaging sector has implemented changes on material level in the last years. It seems that this is causing some issues that were previously not experienced in production. The material changes are largely driven by EU regulations that steer production companies to implement different coatings. Coatings based on epoxy are phased out due to regulations regarding BPA. Instead, BPA-NIA (non-intentional added) coatings are more prevalent. These types of systems are often polyester based [1]. The most important function of coatings is to protect the metal from its surroundings. The inside of the packaging naturally comes into contact with foodstuff, so the requirements for the coating vary based on the properties of the contents. The coating system provides further protection against migration and corrosion. Depending on the type of coating and requirements, the coating system has a thickness of 5-15 µm. The production process often slightly damages the coating, so the finished product needs to meet requirements measured as porosity (mA) for damage. This meaning that the coating needs to be flexible enough for the deformation steps and deal with the temperature fluctuations.

Other than the polymer coating, metal packaging is comprised of several other layers to provide a suitable protection, visualised in Figure 1. The steel base is required to be hard, but not brittle, which is why low carbon steel is used. Roughness is inherent to the production process of the thin steel base (of 0.18-0.24 mm) and the Ra is between 0.25-0.45 µm for a fine stone finish [2]. A thin tin layer is electrolytically applied and remolten. This tin layer is often between 0.2 and 1.6 µm thick. The tin reflow process results in a thin layer of an iron-tin alloy, which is around 0.15 µm thick [3]. This alloy occurs in peaks or 'needles' and is individually an inert layer. However, the space in between the needles is sometimes not covered with tin which could affect the corrosion behaviour [3]. The passivation layer of chromium is even thinner, in nanometre range. The result of the various layers in coated metal packaging is many interfaces and a complex system.

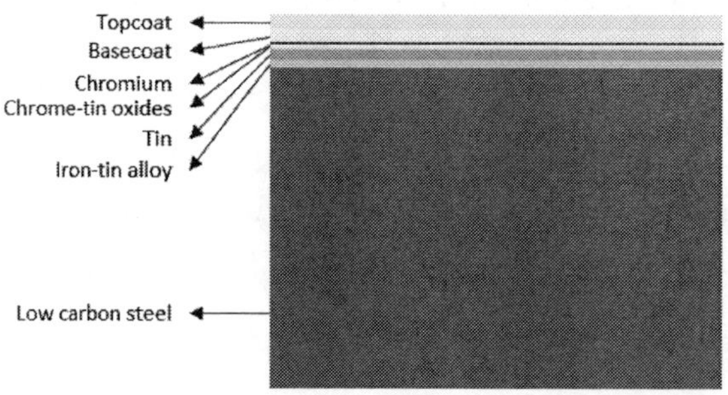

Figure 1: Layers of coated tinplate for metal packaging

The production process of metal packaging is improved over the years, directed not only by the changes in material but also by optimisation steps. This means that the tooling for shaping steps were adjusted, temperatures were changed and the relaxation time after application and curing was extended. Some process steps in can making include large deformation or stretching. This stretching step can change the material properties in a way that is sometimes difficult to predict. An important final step in the process is that of curing a very thin coating layer to repair damages on the outside of the metal packaging part. The bare laying steel is electrolytically coated and cured in about two minutes at 200°C.

As mentioned before, the cause of increased coating damage is unclear. Relaxation and crosslinking are seen as two possible causes of this issue. It is not previously determined at which temperature the damage starts to occur and this may give an indication of the underlaying cause. Explorative research is done in this short paper to see if simple tests can provide an answer to what the cause of the damage after an oven treatment is.

2. MATERIALS AND METHODS

The oven plays an important role in the damage of coatings. The contact with tooling during deformation seems to determine the initial damage, which was inspected with SEM at University of Twente. SEM was also used to inspect the damage after the oven treatment. SEM images of a section of comparable coated tinplate were made to see how the damage increases. The same zoom is used in both images as is visible in Figure 2. The damage is larger in the right picture of the figure, which is the damage after an oven treatment of 200°C.

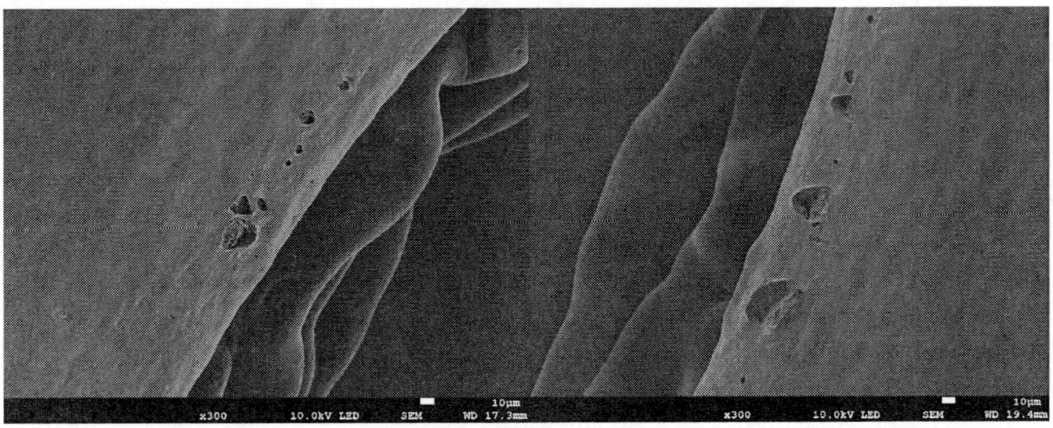

Figure 2: SEM images of damaged lacquer, before (left) and after (right) 200°C oven pass

Now knowing that the initial damage gains in size by expanding, more tests can be done to explore this phenomenon and find the cause. Most likely, this has to do with the material properties and its behaviour in higher temperatures. However, the initial damage occurs still, most likely due to tooling contact.

2.1 Oven simulation

The initiation of increased damage compared to samples before the oven treatment needs to be determined to understand the material better. The effect of lower treatment temperatures therefore could provide insight. The temperatures range from 50°C up to 225°C, since tin reflow starts at 231°C which should be avoided. The final oven treatment is at 200°C so it is included in this range. To get an idea of the effects of temperature another oven treatment test is done. While temperature remains below the crosslink temperature the effect of time on porosity can show the effect of the relaxation process. This temperature is chosen based on the temperature range test. The regular oven treatment temperature of 200°C is chosen to see the possible effect of the crosslinking.

Stretching introduces residual tensile stresses. Stress relaxation can occur because of the residual stresses and temperature. Stress relaxation is relevant in deformation of polymers [4]. The severity of this in polymer coatings on metal packaging is not widely reported. A comparison is made between the temperature treatment below curing temperature and at regular curing temperature. Time is important in this case, so several treatment times are used to see the effect on porosity. The expectation is that a long treatment time at a low temperature will represent the effect of relaxation on porosity, while the higher temperature with any time will represent the effect of crosslinking on porosity. Perhaps a combination of relaxation and crosslinking effects is found at higher temperature.

2.2 Porosity measurements

One of the important quality control systems that is in place for the integrity of coated metal packaging is the porosity test. This measures the amount of electricity that is conducted over a certain surface within a limited time. Electricity is conducted through an electrode on one side via the steel, where the coating is scratched off, through a fluid that is in contact with the damaged coating side and the other electrode. The amperage is measured and used as a measure of bare laying steel. This is done in a slight vacuum. With the lab set-up it is possible to measure on a specific area of interest in two seconds.

2.3 Materials

Previous work regarding coating issues on metal packaging has focused on one specific coating system [5]. It will be interesting to see how different coatings compare regarding the issues. The coating systems are as follows. The coating system with the polyester base coating and organosol top coating (system A) is used as a reference, an aluminium pigmented polyester coating with the organosol coating on top (system B) and a system with the same aluminium pigmented polyester coating with a transparent polyester coating on top (system C). These three coating systems are tested on the sensitivity of the post-curing oven treatment. System A is also tested on ECCS. Moreover, the polyester-organosol coating system is tested for various process temperatures in combination with the 200°C oven treatment as explained in section 2.1.

The polyester coating is crosslinked with the help of a phenolic crosslink agent, this type of coating is often dissolved in styrene [6]. It is a system that is classified as a BPANIA – a bisphenol-A non-intentional added, i.e., if there are any BPA traces found it was not added intentionally. This coating starts to crosslink from temperatures around 190°C and evaporation of solvents occurs at slightly lower temperatures. Pigments such as zinc oxide or aluminium can be added to obtain different properties [7].

The organosol coating is a system that uses small PVC particles in a dispersed state, using plasticisers and dilutants in a dispersant with phenolic crosslinker like the polyester coating. The PVC chains gain mobility, unfold and fuse due to the temperature treatment, starting at around 185°C. This coating provides resistance against the most aggressive foodstuffs, such as those with sulfuric acid [8]. It also remains flexible due to the limited crosslinking [7]. However, the PVC coating is not suitable for direct application on tinplate, due to poor adhesion [1]. This requires a basecoat, such as in this case the polyester coating.

A relevant part of a durable can is the use of a suitable coating. Even though the protection against aggressive foodstuff is provided by this coating, still some molecules could migrate and colour the tinplate. This is especially the case in systems with a low plating weight, since there are more iron-tin alloyed peaks on the surface of the tinplate. Iron is already at or very close to the surface at lower plating weights such as 1.4 g/m^2 or 2.8 g/m^2, while for higher plating weights such as 5.6 g/m^2 or 8.4 g/m^2 the tin is detected from a deeper level [3]. For a suitable protection of the steel base, also a suitable amount of tin is required.

2.4 Wedge bend

The wedge bend test is used to test the flexibility of coated metal sheets. It is bent over a 3 mm radius axle with a weight dropped from about 50 centimetres. With the metal sheet placed on a slanted block to ensure a decrease in radius in the bent sheet from 3 mm to 0.1 mm. Any coated sheet can be used for this, bent along or perpendicular to the rolling direction. The smaller the radius, the higher the deformation. After this is done, the samples are placed in a copper sulphate solution for about 30 seconds. The copper from the copper sulphate solution deposits on the bare iron (if there is any).

The radius at which a change between damaged and undamaged coating is observed can be measured and used as an indication of the flexibility and cohesion of the coating system. It is important to only include the damage on top of the radius, since the weight dropped on the sides possibly damages the coating. The samples get an oven treatment of 200°C after the full wedge bend is done, as to evaluate how this test correlates to the production process. The wedge bend test is done on 5.6 g/m^2 material with system A, B and C.

3. RESULTS AND DISCUSSION

The coating system tested for various temperatures is system A. The temperatures are on a range of elevated temperatures as described in section 2.1, to see the starting point of the increased porosity. The average porosity values are plotted in Figure 3 and an S-curve becomes visible. Oven temperatures higher than 210°C would come close to the tin-reflow, which is around 231°C. The effect of the oven temperature starts between ~80°C and 100°C. The largest increase is between 100°C and 175°C. After 175°C the increase stagnates. The crosslink temperature is from ~185°C.

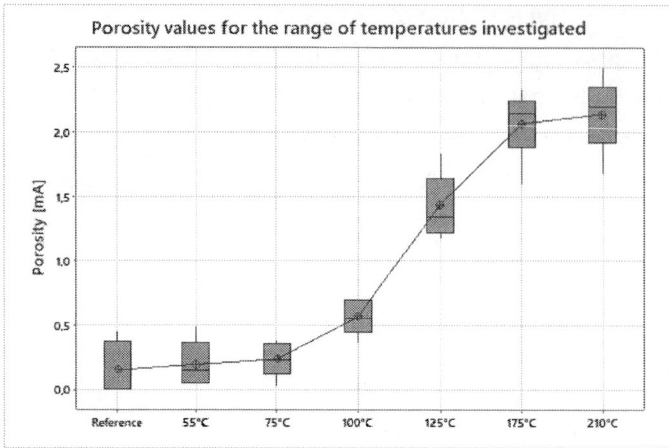

Figure 3: Change of porosity due to oven treatment for a range of oven temperatures

As a continuation of this test, the variable time is introduced. As described in section 2.1, a relatively low temperature is chosen to compare the resulting porosity to that of an oven treatment at crosslinking temperature. In addition, the effect of time is measured, with five, ten and 60 minutes for the lower temperature treatment and five and ten minutes for the higher temperature treatment. 100°C was chosen for the lower temperature oven treatment based on the results in Figure 3.

A different set of material than the temperature range test is used to test the effect of time at low and high temperatures. Treatment time has an interesting effect on porosity in the lower temperature treatment test, as in Figure 4 it can be roughly seen that porosity decreases with increasing time. However, there is a widespread in nearly each sample. Overlap between the samples is a result and a there is only a significant difference between the five- and 60-minute samples.

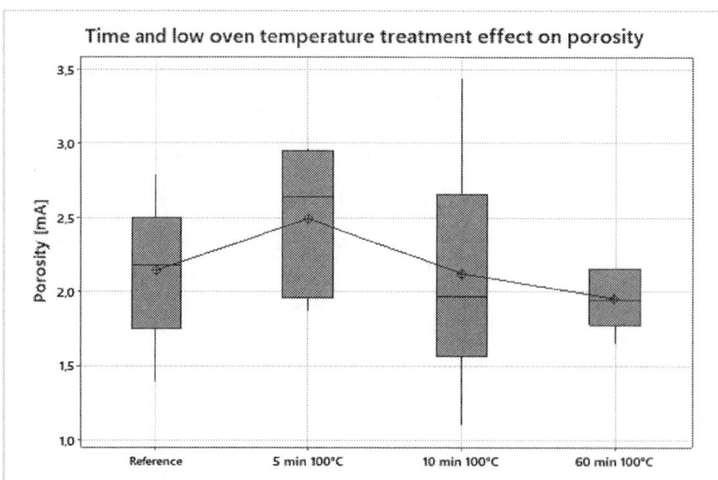

Figure 4: Porosity because of time at 100°C oven treatment compared to non-treated material

The effect of time on the higher temperature oven treatment is in Figure 5. There is a slight increase with increasing time compared to the reference samples. The highest average porosity comes from the samples of 5 minutes at 100°C, with 10 minutes at 200°C the second highest porosity values. A longer oven treatment time at 100°C decreases the initial damage. However, this does not occur in regular production. It looks like a longer time at 100°C might heal the coating in a way, since the level of porosity first increases and then decreases over time. However, it must be noted that the porosity values largely overlap and are rather different from earlier tests.

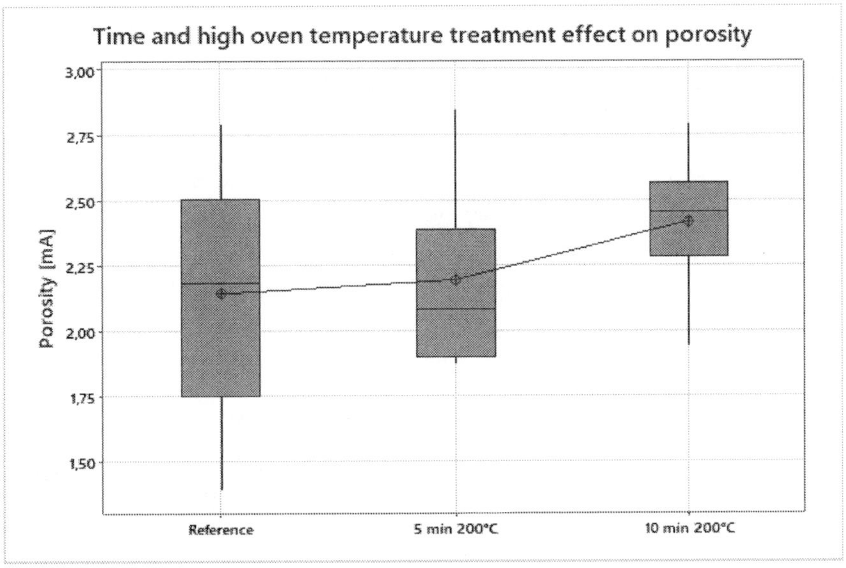

Figure 5: Porosity because of time at 200°C oven treatment compared to non-treated material

The porosity values of systems A, B and C are measured to establish and quantify the effect of the oven treatment, as in Figure 2. This is done by treating a set of samples of each coating system in the oven and not doing so for another set of each coating system. Afterwards these samples were tested on the porosity. In Figure 6 is the comparison of the coating systems. The ECCS substrate with coating system A was also added to see the difference, since it was already shown that ECCS provides better adhesion [9]. The batch that the samples in this test were taken from is different than those used in Figure 5. Perhaps due to the age of the samples or other differences, the results are quite different. This is most clear from the steep increase in porosity normally seen, as in Figure 6

However, looking at the samples of the same coating system in Figure 6, the increase of porosity is much less steep. However, the averages after the oven treatment are similar.

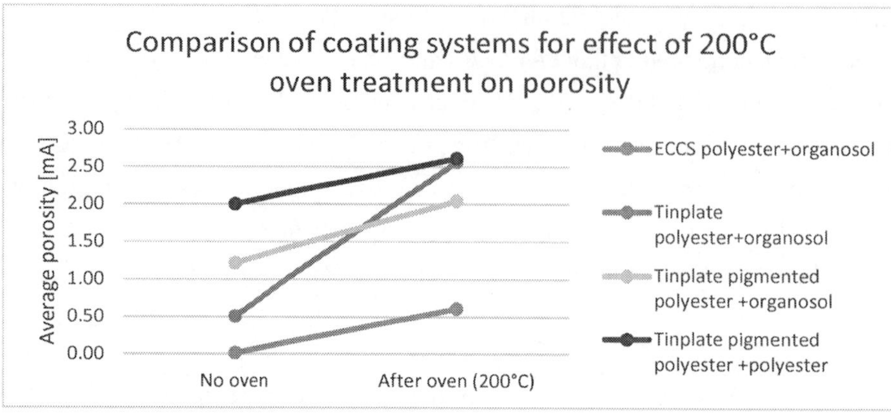

Figure 6: The effect of the 200°C oven treatment on the porosity of various coating systems.

The results of the wedge bend test for coating systems A, B and C on tinplate are in Table 1. The effect of the oven was tested, like the samples as in Figure 6. The trend is that there is an increase after the oven treatment of 200°C. However, the severity of the increase is not like the samples in Figure 6. Although system A has the largest increase, system C barely increases. This may be due to the type of coating, since systems B and C both have a pigmented basecoat.

Table 1: Wedge bend results of no oven and post-oven

Sample	*No oven, radius pores [mm]*	*With oven, radius pores [mm]*
System A	0.87	1.13
System B	0.75	0.82
System C	0.70	0.72

4. CONCLUSION

The results from initial tests show that there is an increase of porosity of polymer coated metal packaging after an oven treatment at 200°C. Several experimental tests were done to get an understanding of the cause of this increase. The temperature range test shows that there is an onset of increased porosity, starting at around 100°C. This gives the indication that damage already increases below the crosslinking temperature. The more detailed test shows that time also has an effect. However, the variation is still large and overlaps for nearly each sample.

The organosol coating could have a large contribution to the issues found and not enough focus was placed on the specific coating properties. The wedge bend test was used to see if it is a suitable test to simulate the oven treatment in an intermediate step of the process. The result is not in line with the measurements from production. A steep increase is often found, while the wedge bend test only shows a very slight increase after the oven.

5. REFERENCES

[1] J. S. LaKind, "Can coatings for foods and beverages: issues and options," *Int. J. Technology, Policy and Management,* vol. 13, 1, pp. 80-95, 2013.

[2] *Cold reduced tinmill products - electrolytic tinplate and electrolytic chromium/chromium oxide coated steel*, B. Standards, 2001.

[3] F. Cova Caiazzo, L. Brambilla, A. Montanari, and S. Mischler, "Chemical and morphological characterization of commercial tinplate for food packaging," *Surface and Interface Analysis,* vol. 50, no. 4, pp. 430-440, 2018, doi: https://doi.org/10.1002/sia.6386.

[4] Crow. "Stress Relaxation." https://polymerdatabase.com/polymer%20physics/Relaxation.html (accessed 2022).

[5] M. v. d. Berg, "Research after lacquer failure in metal packaging," presented at the IAPRI conference, online, 2021.

[6] FibermaxComposites. "Types of resin families." https://www.fibermaxcomposites.com/shop/index_files/resinsystems.html (accessed 2022).

[7] TemaNord, *Lacquers in cans. Technology, legislation, migration and toxicology.* Copenhagen: Nordic Council of Ministers, 1998.

[8] what-when-how. "Organosol coatings." http://what-when-how.com/materialsparts-and-finishes/organosol-coatings/ (accessed 2022).

[9] A. Allman, E. Jewell, A. d. Vooys, and R. Hayes, "Inter-layer Adhesion Performance of Steel Packaging Materials for Food Cans Under Retort Conditions," *Journal of Packaging Technology and Research,* vol. 2, 2018.

CONFREENCE PROCEEDINGS :
PACKAGING SUSTAINABILITY

• ORAL PRESENTATION

• GENERAL STREAM

• PEER REVIEWED STREAM

• POSTER PRESENTATION

ST-O01

Investigating the Environmental Benefits of Increasing the Pallet Top Deck Stiffness in Unit Load System

Saewhan Kim[1], Laszlo Horvath[1]*, Jonghun Park[2], Jennifer D. Russell[1]

[1]*Department of Sustainable Biomaterials, Virginia Polytechnic Institute and State University, Blacksburg, 24061, USA*
[2]*School of Graphic Communications Management, Ryerson University, Toronto, M5B 2K3, Canada*

***Correspondence to:** Department of Sustainable Biomaterials, Virginia Polytechnic Institute and State University, 1650 Research Center Drive, Blacksburg, 24061, USA. lhorvat@vt.edu*

ABSTRACT: Trends focusing on packaging sustainability are emerging, not only in consumer packaging, but also in distribution packaging transported in unit load form. Wooden pallets transporting corrugated boxes are the most common unit load type in the United States. Recently, it has been observed that modifying pallet top deck stiffness can significantly affect the strength of corrugated boxes depending on various unit load design factors. It was also recently discovered that this phenomenon can be used as a tool for packaging cost optimization. However, there is still a knowledge gap regarding how this unit load optimization method could affect the environment. This study was conducted to investigate the environmental effect of optimizing a unit load by increasing the stiffness of its pallet's top deck and reducing the board grade of the corrugated boxes using life cycle analysis. Numerous unit load designs were investigated using different wood species, pallet top deck thicknesses, corrugated boxes sizes, corrugated flutes, and board grades. The initial and optimized unit load scenarios were created to ensure that the unit loads offered the same performance. The scope of this study was cradle-to-grave, including raw material production, packaging manufacture, use, and the end-of-life phases of wood pallets and corrugated boxes. The results showed that optimizing the unit load offsets its environmental impacts by up to 23%. The environmental benefits in most of the impact categories were primarily due to the reduction of the corrugated material used (except for ozone depletion). Ozone depletion was mainly governed by a change in the amount of pallet materials. This study showed that the unit load optimization method involving increasing the stiffness of the top decks and reducing the corrugated board grade could be an effective way of reducing the overall environmental footprints of unit loads.

Keyword: Packaging sustainability, distribution packaging, unit load, pallet, corrugated box.

1. INTRODUCTION

Nowadays, packaging is widely used to contain, protect, preserve, and transport goods. Besides these essential functions, end of life scenarios for packaging is also a major concern for our current society. As sustainability issues become the most important debates in many areas, the sustainability of packages has also drawn tremendous attention due to the potentially high volume of waste generation. In 2018, about 28% (82 million tons) of municipal solid waste generated in the United States was reported as packaging related materials [1].

Despite the increasing focus on improving the sustainability of packaging, distribution packaging is often neglected as many don't realize that it is also an essential factor in the supply chain. Distribution packaging plays a crucial role behind the scenes in transporting goods, in the primary packaging, safety, efficiency, and cost-effectiveness in getting goods to wherever consumers can easily access them. Eighty percent (80%) of distribution packaging in the United States is in unit load form which consists of packaging, pallets, and the material handling system [2–3]. Pallets are ubiquitous in the supply chain with 2.6 billion pallets circulating in the United States annually [4]. Out of all pallet materials, wood dominates the industry with 94% of the market share [5]. Approximately 804 million wooden pallets were newly produced or recycled in 2016 from various industries in the United States [6]. In addition, the predominant type of packaging is the corrugated box, which accounts for 72% of the packaging materials used to build unit loads [7].

As the unit load accounts for a large share of the distribution packaging system, it is crucial to understand and evaluate the environmental impact of the unit load. Investigating the environmental impact of unit load components has been the focus of numerous research papers. Several studies compared the environmental impact of wooden pallets to the impact of pallets made from different materials, manufacturing systems, or alternative distribution packaging systems using LCA [8–11]. Other studies investigated environmental hotspots and developed a life cycle inventory for wooden pallets to help decide on designing or choosing an environmentally friendlier pallet [12–15]. Many previous researchers from various fields have assessed the environmental impact of corrugated boxes when they are used as part of a whole packaging system and as compared with reusable alternative packaging systems [16–23].

Meanwhile, the environmental impacts of unit load components have been studied focusing on the impact of pallets and corrugated boxes separately. Up until now, there have been no attempts to holistically optimize the environmental impact of the unit load using the interactions between these two components. Understanding the interactions between unit load components is essential to optimizing them to be safer, more cost-effective, and environmentally friendly. In recent years, the effect of the stiffness level of the wooden pallet top deck boards on the performance of corrugated boxes was broadly studied by researchers [24–27]. Quesenberry et al. [26] found that the stiffness of wooden pallets' top deck boards affects the strength of corrugated boxes up to 37% when corrugated boxes are asymmetrically supported. He also found that this phenomenon can be used as a cost optimization method for unit loads by increasing the stiffness of the wooden pallet top deck boards and decreasing the board grade of corrugated boxes [26]. Kim et al. [27] further investigated this unit load cost optimization method, and they found that it can be affected by various unit load design factors like pallet wood species, top deck thickness of initial unit load scenario, corrugated box size, and board grade. Both studies expressed that this optimization method could, not only be used for cost optimization, but could also possibly be a method of improving the environmental performance of current unit load designs at the same time.

Therefore, this unique study investigates the environmental effects of the unit load optimization method previously proposed by Quesenberry et al. [26] which allows for the decrease of the board grade of corrugated boxes by increasing the thickness of the pallets' top deck boards. This work (1) adds to our knowledge about how unit loads can be holistically environmentally optimized using the interactions between components, and (2) demonstrates hotspots of environmental impact when optimizing unit load designs.

2. MATERIALS AND METHODS

This study mainly employed the life cycle assessment (LCA) method to compare the environmental impact of both the initial and the optimized unit load designs, which comply with international standard LCA guidelines ISO 14040:2006 and ISO 14044:2006 [28], [29].

2.1 Goal and Scope Definition

The main aim of this study was to investigate the environmental effects of optimizing a unit load by increasing the stiffness of the pallets' top deck boards and reducing the board grade of its corrugated boxes using LCAs. The LCAs compared the environmental performance of multiple initial and optimized unit load scenarios with a cradle to grave perspective. This study included raw material production, packaging manufacturing, distribution, and end-of-life options (EOL). However, this study excluded the life cycle of packaged items from the system boundary due to their high variability. Figure 1 presents a drawing of the system boundaries considered in this study. Also, the geographical scope was limited to the southeastern United States due to the variability of local wood species supplied for wood pallet manufacturing.

The optimization method employed in this study required different amounts of pallet top deck material increase. It also resulted in different amounts of corrugated box material decrease depending on various unit load design factors [27]. Hence, this study compared a wide range of paired initial and optimized unit load scenarios to investigate at what point they cross the line to show measurable environmental benefits through this unit load optimization method. Therefore, this study was not able to select specific load capacities for the functional unit. The functional unit for this study was defined as a double-stacked unit load with the same maximum safe load capacity on floor stacking conditions.

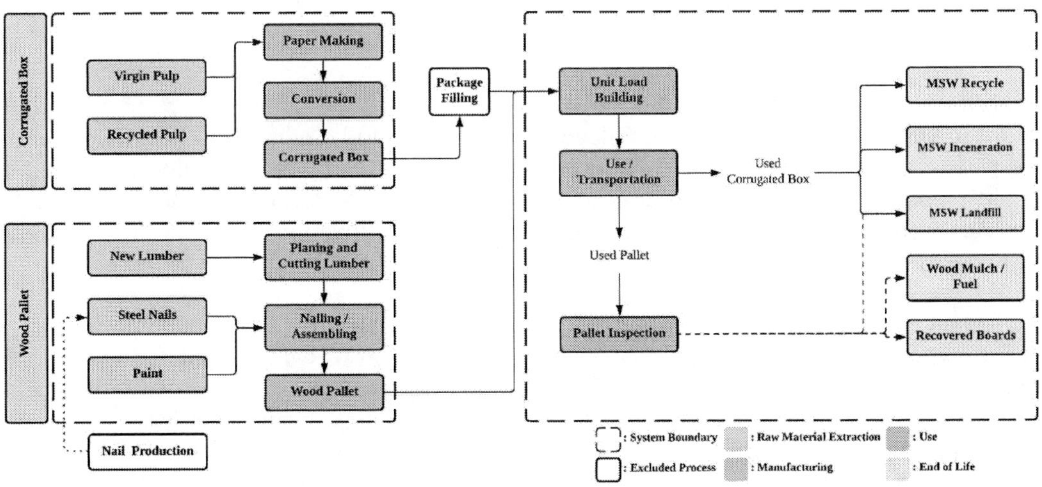

Figure 1 The schematization of the unit load system boundary.

2.2 Unit Load Scenario Employment

In this study, the environmental performance of multiple pairs of initial and optimized unit load scenarios, composed of wooden pallets and corrugated boxes, were compared to each other. A total of 108 pairs of unit load scenarios were employed from the authors' previous study Kim et al. [27]. Kim et al. [27] designed an extensive list of unit load scenarios with varying factors such as initial top deck board thickness, pallet wood species, box size, and board grade covering a wide range of common unit load designs. The unit load scenarios were classified into three grades according to the amount that the top deck board thickness increased. Among these unit load scenarios, only scenarios with which it was found to be feasible to apply the unit load optimization method from the manufacturing standpoint, grade 1 (less than 12.7 mm top deck thickness increase) and grade 2 (12.7 mm to 25.4 mm top deck thickness increase), were used for this study. The optimized unit load scenarios listed in that research paper were created from the initial unit load scenarios by increasing the stiffness (thickness) of the pallet top deck boards and reducing the board grade of corrugated boxes while maintaining identical box performance.

2.3 Ranked Scenario Analysis

In order to effectively evaluate a wide range of unit load scenarios, four different pallet wood species groups were analyzed separately and ranked using a unit load optimization ratio (UOR) developed by the authors.
UOR is a ratio of the decreased weight of corrugated materials to the increased weight of pallet materials; both of which change during the unit load design optimization process. The UOR was calculated by following equation 1:

$$UOR = \frac{C_i - C_o}{P_o - P_i} \qquad (1)$$

where:
UOR = Unit load optimization ratio
C_i = Weights of corrugated boxes from initial unit load design
C_o = Weights of corrugated boxes from optimized unit load design
P_i = Weights of wooden pallets from initial unit load design
P_o = Weights of wooden pallets from optimized unit load design

Higher ratios reflect a smaller amount of weight increase of pallet materials to decrease in the weight of corrugated materials, which also can be assumed to be closer to a best-case scenario. The worst-case solutions have a lower ratio, which means there was a greater increase in the weight of pallet materials required to decrease the weight of corrugated materials. Unit load scenarios were ranked and listed in the order of material utilization according to the UOR of each scenario.

The weight of wooden pallets and corrugated boxes from each pair of initial and optimized unit load scenarios were determined to calculate UOR. The weights of wooden pallets and corrugated boxes were computed by the industry-accepted pallet design software Pallet Design System™ (PDS™) v. 6.2 (National Wooden Pallet & Container Association, Alexandria, VA, USA) and the unit load design software Best Load™ v.4.0 (White & Company LLC, Blacksburg, VA, USA), respectively. Specifications of wooden pallets and corrugated boxes obtained from the previous study were entered into the corresponding software.

Once the unit load scenarios were ranked by UOR, six pairs of initial and optimized unit load scenarios from HD HW group, LD HW group, and GSYP group were selected for life cycle analysis at a regular intervals. Exceptionally, all available unit load scenarios (only five pairs) were selected from the KD SYP group due to the limited number of optimizable scenarios. Among these scenarios, the unit load scenario with the lowest UOR from each wood species group was defined as the worst case scenario. And, the unit load scenario with the highest UOR was defined as the best case scenario in terms of proportional material utilization.

2.4 Environmental Performance Difference Analysis

The difference in environmental performance of the initial unit load designs and the optimized unit load designs from each of the scenarios selected at regular intervals for each wood species group were studied to investigate whether the unit load optimization method had environmental benefits or not. Positive environmental impact difference (+%) indicates that optimizing these unit load designs lower environmental burdens, and on the contrary, negative environmental impact difference (-%) reflects the fact that optimizing these unit loads generate environmental burdens instead.

2.4.1 Life Cycle Inventory Analysis

Life cycles of unit load scenarios were modeled through commonly used LCA software SimaPro 9.0 (PRe Consultants, Amersfoort, The Netherlands). The secondary inventory data regarding pallet weights were obtained from PDS™ v. 6.2. Unit load design software Best Load™ v.4.0 was utilized to obtain the secondary inventory data of corrugated box weights. The US LCI database, which traditionally represents U.S. regions, was mainly utilized for LCA modeling in SimaPro 9.0. The Ecoinvent v.3 database was also employed for LCA modeling to fill the existing inventory gaps within the US LCI database.

Unit Load Raw Material Production

Inventory data of wooden pallet raw materials production was modified from the most up-to-date life cycle inventory of wooden pallets in the United States developed by Alanya-Rosenbaum et al. [14]. Values were adjusted according to the density of the four different wood species used in pallet manufacturing: green high-density hardwood, green low-density hardwood, green southern yellow pine, and kiln-dried southern yellow pine. The density of each wood species was calculated from PDS™ by dividing the weight (kg) of the pallets made from different wood species by each pallets' volume (m3). Inventory data was also broken down to the level of 'per 1 kg of wooden pallet raw material production' in order to be able to universally apply it to the various unit load scenarios.

The inventory data for corrugated box raw material production was adopted from the comprehensive life cycle assessment study done for average corrugated products in the U.S. by National Council for Air and Stream Improvement [30].

Unit Load Manufacture

For modeling the pallet assembly and corrugated box manufacturing, inventory data developed from the previous studies were also employed without modification since the input and output of data for unit load component manufacturing does not dramatically change due to unit load design factors. The average input and output data for the pallet manufacturing process was collected from Alanya-Rosenbaum et al. [14]. Inventory data for the corrugated box production process was obtained from the National Council for Air and Stream Improvement [30].

Transportation

Assumed transportation modes and travel distances for wooden pallets and corrugated boxes are presented in Table 1. Road transportation was mainly considered for wooden pallets due to this study's geographical scope, which also showed that wooden pallets are supplied and consumed locally. Only single use pallet scenarios were considered. Both road and rail distribution were considered for corrugated boxes since they could be supplied from anywhere in the United States. However, road transportation was regarded as the primary mode of distribution for corrugated boxes in order to follow the geographical scope of this study.

Table 1 Transportation data: distribution modes and distances considered for unit load components.

Product	Item	Truck		Rail	
		Weighting (%)	Distance (km)	Weighting (%)	Distance (km)
Wooden pallet	Raw material to manufacturer	100	148	-	-
	Pallet manufacturer to product manufacturer	100	148	-	-
	Unit load transportation (use)	100	1207	-	-
	EOL transportation	100	148	-	-
Corrugated box	Wood logs to pulp and paper mils	98.4	159	1.6	1577
	Wood chips to pulp and paper mils	94.5	299	5.5	1674
	Recovered fiber to pulp and paper mils	85.4	241	14.6	505
	Pulp to pulp and paper mils	80.1	262	19.8	1511
	Chemicals	72	217	28	1333
	Containerboard to converting facility	80.1	262	19.9	1511
	Corrugated sheets to product manufacturer	80.1	262	19.9	1511
	Product to use	95.7	283	4.3	2446
	Unit load transportation (use)	100	1207	-	-
	EOL transportation	87.4	241	12.6	505

End of Life of Unit Load

Unit load components' end of life (EOL) phases were modeled based on the United States national data and details from the literature. At the end of a wooden pallet's life cycle, unbroken boards were recovered for reuse or repair of the other pallets, and the rest were landfilled, used for boiler fuel, or used for mulch and animal bedding [14]. Corrugated boxes end up being recycled in many cases, combusted for energy, or landfilled in fewer cases [31]. Table 2 presents the distribution of the EOL scenarios for each unit load component.
Table 2 End of life stage of unit load.

	Wooden pallets	Corrugated boxes
Recovered boards	37.3 %	N/A
Fuel	17.3 %	N/A
Mulch and animal bedding	40.4 %	N/A
Landfill	5 %	15.4 %
Combustion	N/A	3.7 %
Recycle	N/A	80.9 %

2.4.2 Life Cycle Impact Analysis

A midpoint-oriented life cycle impact analysis method, Tool for the Reduction and Assessment of Chemical and other Environmental Impacts 2.1 (TRACI 2.1), was utilized to calculate the environmental impact generated by the inputs and outputs of the unit load life cycle. This evaluation method was selected for calculations due to its matching geographical scope with this study. TRACI 2.1 was developed to reflect the environmental situation of U.S locations by U.S. Environmental Protection Agency (EPA). A total of ten impact categories were calculated and presented including ozone depletion (kg CFC-11 eq), global warming (kg CO2 eq), smog (kg O3 eq), acidification (kg SO2 eq), eutrophication (kg N eq), carcinogenics (CTUh), non carcinogenics (CTUh), respiratory effects (kg PM2.5 eq), ecotoxicity (CTUe), and fossil fuel depletion (MJ surplus).

3. RESULTS AND DISCUSSION

3.1 Ranked Scenario Analysis Results

Figure 2 plotted the adopted unit load scenarios in ascending order of ranking based on the unit load optimization ratio (UOR) calculated for each wood species group. Columns colored red in Figure 2 indicate the unit load scenarios chosen at regular intervals to be utilized for the life cycle analysis. The best case scenarios were found to be unit load scenarios with UOR of 3.200, 3.733, 4.480, and 1.918 for HD HW, LD HW, GSYP, and KD SYP, respectively. The worst case scenarios from HD HW, LD HW, GSYP, and KD SYP had UOR of 0.035, 0.021, 0.023, and 0.16, respectively.

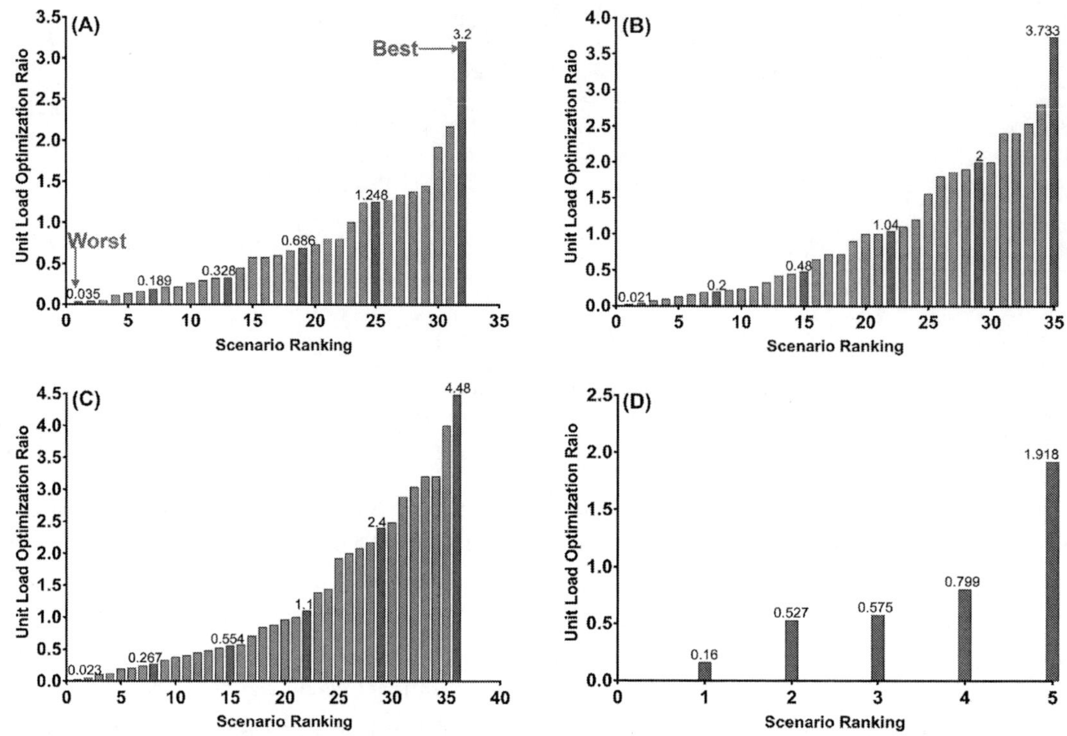

Figure 2 Rankings of investigated unit load scenarios from various wood species groups based on unit load optimization ratio. (A) displays results of HD HW group, (B) displays results of LD HW group, (C) displays results of GSYP group, and (D) displays results of KD SYP group.
* Note: Red columns within plots indicate unit load scenarios selected for environmental analysis.

3.2 Environmental Performance Difference Analysis Results

Researchers compared and plotted the differences in environmental impacts between the initial unit load designs and the optimized unit load designs from each wood species group in order to investigate whether the unit load optimization method generated environmental benefit or environmental burden.

Figure 3 reports the environmental impact differences between initial unit load designs and optimized unit load designs within the HD HW group. It was observed that the unit load optimization method does not always create environmental benefits in the HD HW group. Optimized unit load design scenarios with a UOR of 0.035 generated up to 22.57% more negative environmental impacts than the initial unit load design in all impact categories except acidification. Although the unit load optimization method was not environmentally beneficial for the low UOR scenario, it started generating environmental benefits as the UOR increased. More than half of the impact categories (seven impact categories - excluding ozone depletion, eutrophication, and ecotoxicity) for the unit load scenario with UOR of 0.189 showed environmental benefits by optimizing unit load design. Furthermore, it was discovered that optimizing unit load scenarios with UOR of 1.248 or higher improved environmental performance up to 22.93% in all investigated impact categories. Additionally, the unit load scenario with 1.248 UOR was employed as the base scenario in the minimum required condition analysis. This was the first scenario that crossed the environmental performance difference break-even line during regular interval analysis.

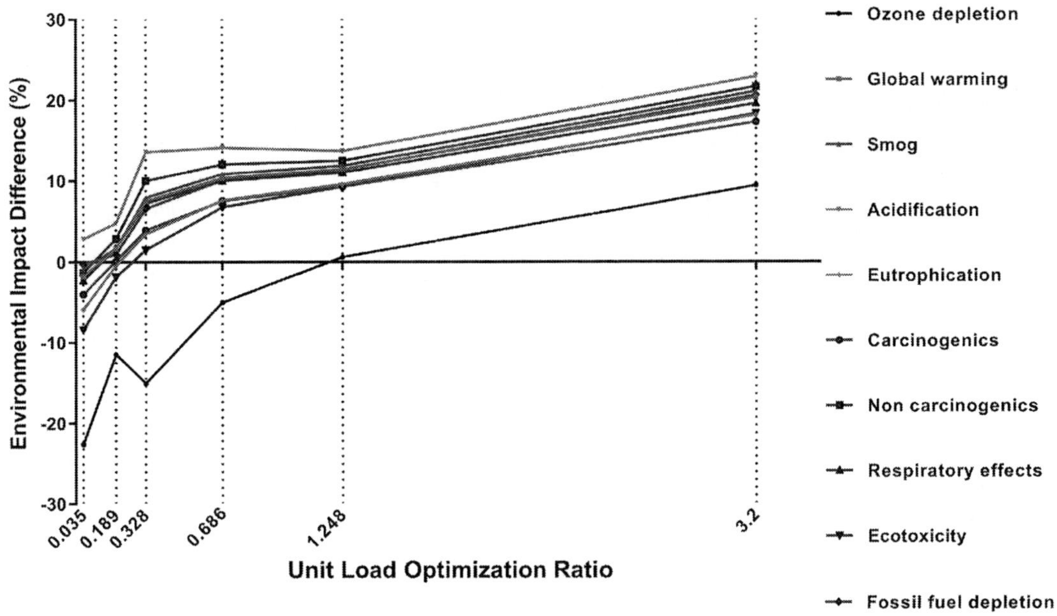

Figure 3 The environmental impact difference of initial and optimized unit load designs within the HD HW group.

Figure 4 presents the environmental impact difference of the LD HW group's initial unit load designs and optimized unit load designs. Like the HD HW group, optimizing the first investigated unit load scenario with a UOR of 0.021 decreased environmental performance up to 35% in most impact categories except acidification. As the UOR increased, more and more impact categories started to show environmental benefits from the unit load optimization method. The unit load scenario with a UOR of 0.2 reported improvements in environmental performance from six impact categories, excluding ozone depletion, eutrophication, carcinogenics, and ecotoxicity. Moreover, all impact categories indicated that optimizing the unit load could generate environmental benefits as much as 22.85% within LD HW group. The base scenario for minimum required condition analysis was set to 2 UOR in this case.

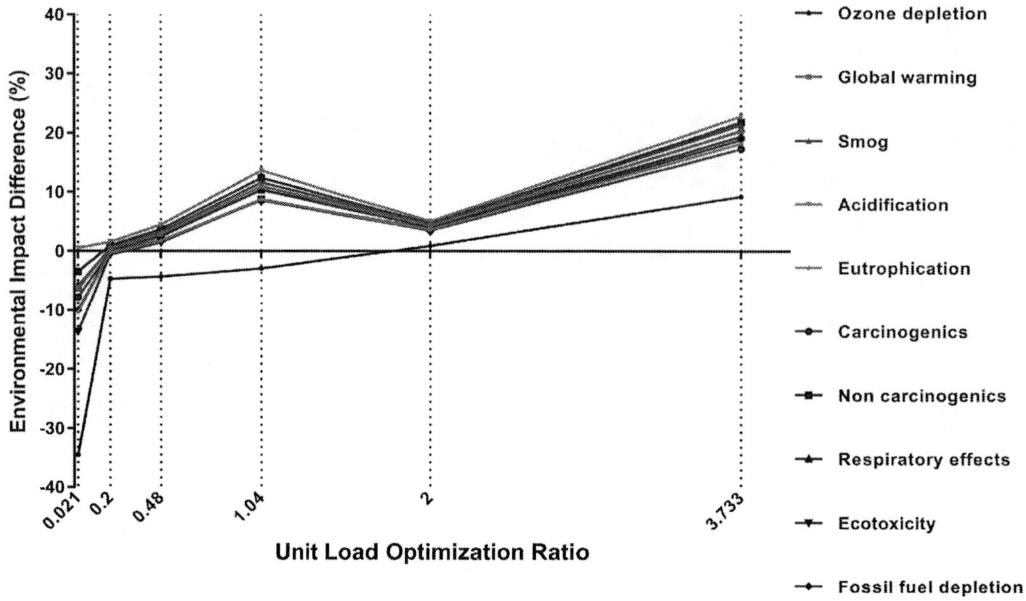

Figure 4 The environmental impact difference of initial and optimized unit load designs within the LD HW group.

Figure 5 displays the difference in the environmental performance of initial unit load designs versus the optimized unit load designs from the GSYP group. The trend of optimizing unit load scenarios with lower UORs causing diminishing environmental performance was also observed in the GSYP group. Optimizing the worst case scenario (0.023 UOR) decreased the overall environmental performance by as much as 35.23%, and an environmental benefit was only observed for acidification. GSYP group results also followed the trend discovered in the HD HW and LD HW groups where the number of impact categories showing environmental benefits from unit load optimization escalates as UOR increases. Consequently, the best case scenario in the GSYP group (4.48 UOR) showed as much as 20.48% increase in environmental benefits in all impact categories when applying the unit load optimization method. There were environmental benefits in all impact categories when the unit load was designed with a UOR of 2.4 or higher in the GSYP group.

Figure 6 shows the environmental impact difference of the initial and optimized unit load designs belonging to the KD SYP group. As the UOR escalates, improvement in the environmental performance through the optimization of the unit loads was also observed in many impact categories in the KD SYP group. The best case scenario improved the environmental impact of unit load up to 13.16%. However, the KD SYP group reported a slightly different trend from the other wood species groups. Previous results from the HD HW group, LD HW group, and GSYP group showed that optimizing the unit load can improve the environmental performance of unit load in all impact categories when the UOR is higher than specific points. In contrast, the KD SYP group could not enhance the environmental performance of unit load in terms of ozone depletion even with the maximum possible UOR. In other words, KD SYP unit load scenarios cannot expect full environmental benefits from this unit load optimization method.

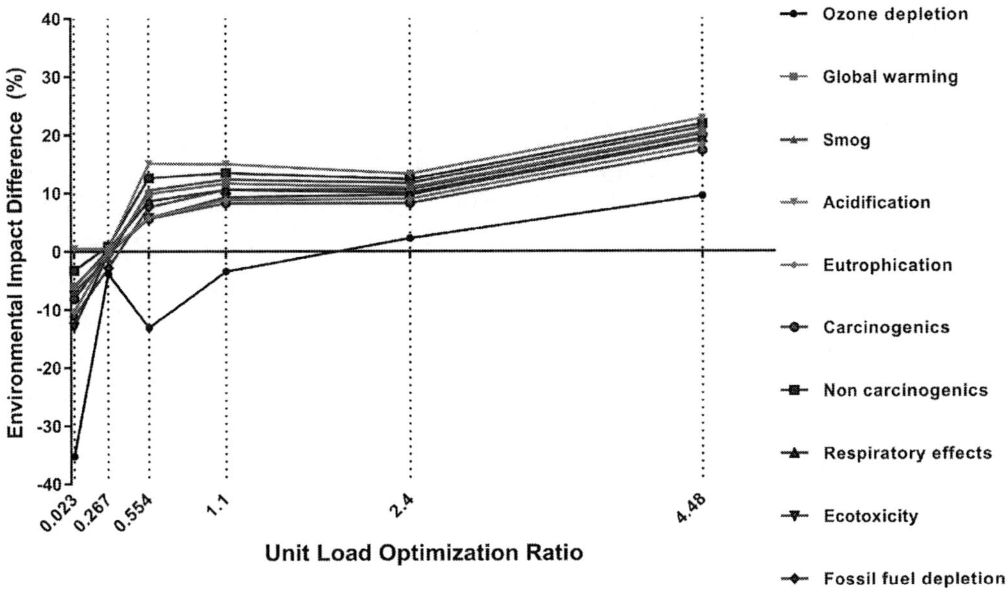

Figure 5. The environmental impact difference of initial and optimized unit load designs within the GSYP group.

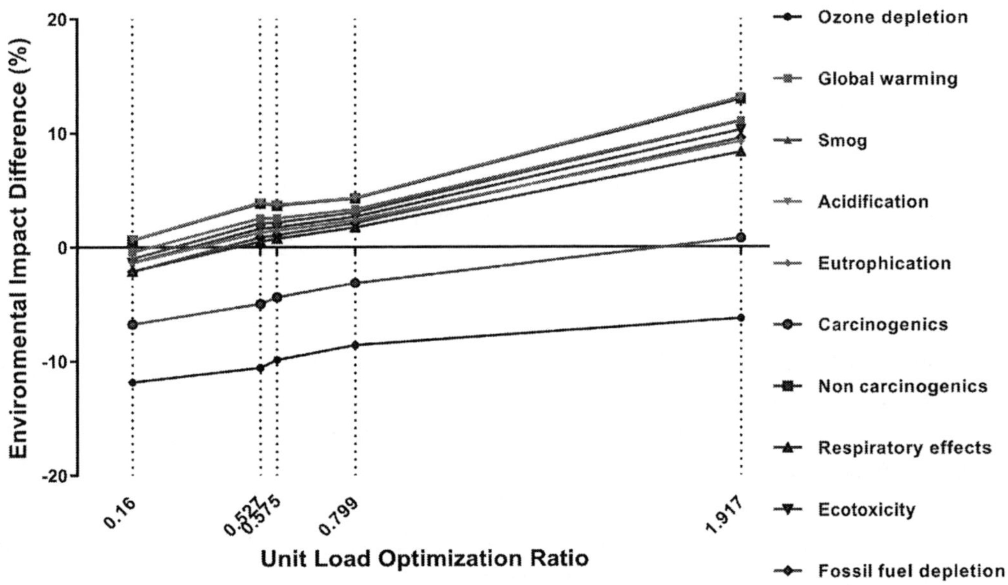

Figure 6 The environmental impact difference of initial and optimized unit load designs within the KD SYP group.

Overall, it was found that the investigated unit load optimization method could improve the environmental performance of unit loads when the unit load scenario has a specific unit load optimization ratio or higher. The general trend of environmental benefits observed from more and more impact categories as the UOR increases remains unchanged within all investigated wood species groups.

Furthermore, it was discovered that unit load scenarios with a higher UOR do not necessarily mean they'll show higher environmental performance improvement. Up and down in the level of environmental impact differences were consistently observed regardless of the amount of UOR increase. This could be due to the fact that UORs were only based on the proportion of material utilization, and do not account for the absolute amount of pallet and corrugated material change. In other words, unit load scenarios with lower UORs may have higher amounts of pallet and corrugated material changes during optimization than unit load scenarios starting with higher UORs. Since the UOR is only based on material efficiency, unit load scenarios with higher than or equal UORs to unit load scenarios that have crossed the break-even line for the first time can always expect environmental benefits by applying the unit load optimization method, but to different degrees. Therefore, the UOR that first crosses the environmental performance break-even line suggests the minimum required conditions to improve the environmental performance of the initial unit load design. These specific minimum required conditions for each wood species group were further investigated in the next section. It also means unit load designers need to run their own environmental performance difference analysis in order to figure the exact degree of environmental benefit received by applying the unit load optimization method to their specific unit load designs.

Figure 7 presents contribution analyses on six environmental impact categories: global warming potential, smog creation, fossil fuel depletion, eutrophication, acidification, and ozone depletion for each wood species groups' best and worst case scenarios. The most significant contributors to global warming potential and acidification was the corrugated box raw materials production phases. Two leading contributors to smog creation and fossil fuel depletion were corrugated box raw materials production and transportation phases. Eutrophication was predominantly affected by corrugated raw materials production and box manufacturing processes. Ozone depletion was heavily affected by the pallet raw materials production phase, followed by the corrugated box raw materials production phase.

Regarding the trend of ozone depletion, this trend was exceptionally difficult to generate a positive impact through the unit load optimization method; especially when compared to all other impact categories. It required a relatively high UOR before observing any environmental benefit in terms of ozone depletion, and unit load scenarios in the KD SYP group were not able to create the high UOR. This was mainly because ozone depletion is governed by pallet associated factors, while corrugated box related processes controlled the other impact categories. This study's adopted unit load optimization method basically adds chemically unprocessed wood materials to the pallets top deck in order to decrease the chemically processed materials in the corrugated boxes. Consequently, environmental benefits in terms of ozone depletion, which is mainly governed by the pallet raw materials production phase, cannot easily be achieved until there is sufficient reduction of corrugated box raw materials phase to mitigate the effects of the pallet raw materials production phase.

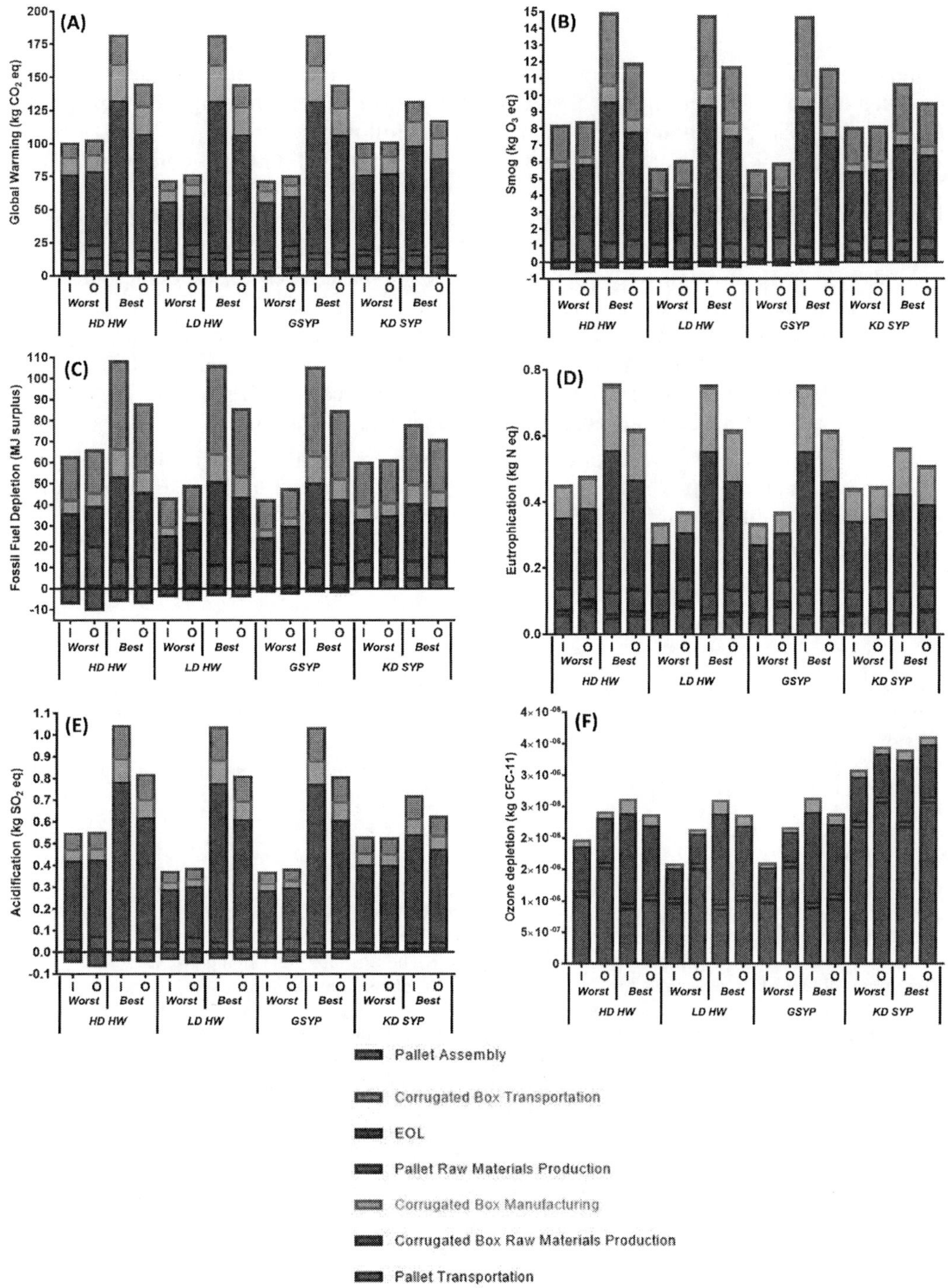

Figure 7 Contribution analyses for the best and worst case scenarios from the HD HW, LD HW, GSYP, and KD SYP groups. (A) shows global warming potential, (B) shows smog creation, (C) shows fossil fuel depletion, (D) shows eutrophication, (E) shows acidification, and (F) shows ozone depletion. I indicates initial case and O indicates optimized case.

4. CONCLUSIONS

This study was mainly conducted to investigate the environmental performance of the unit load optimization method of increasing top deck board thicknesses and decreasing the board grade of corrugated boxes. The key findings of this study were as follows:

- Environmental performance of unit load was improved through optimization using physical interaction between components in many cases as much as 23%.
- Out of all impact categories, ozone depletion consistently showed a lower performance increase in all wood species groups. This was mainly due to not enough corrugated material decrease compared to pallet material increase since ozone depletion is more heavily affected by the pallet associated factors.
- There are chances that unit load scenarios with relatively low absolute amount of pallet and corrugated box materials to have higher UOR than unit load scenarios with the relatively high absolute amount of materials used because UOR is only based on the ratio of changes in pallet and corrugated box material utilization.
- Therefore, it is highly recommended that designers conduct unit load design-specific environmental analysis in order to understand the actual degree of environmental advantages gained by optimizing a particular unit load scenario.

5. ACKNOWLEDGEMENT

The research was financially supported by the Industrial Affiliate Program of the Center for Packaging and Unit Load Design at Virginia Tech. The PDS™ software package used for data acquisition was provided by the National Wooden Pallet and Container Association. The Best Load™ software package used for data acquisition was provided by White and Company LLC.

6. REFERENCES

[1] Environmental Protection Agency. (2020). Advancing Sustainable Materials Management: 2018 Fact Sheet Assessing Trends in Materials Generation and Management in the United States.

[2] White, M. S., & Hamner, P. (2005). Pallets MoveThe World The Case for Developing The Case for Developing System-Based Designs for Unit Loads. Forest Products Journal, 55(3). https://www.researchgate.net/publication/237554331

[3] Raballand, G., & Aldaz-Carroll, E. (2007). How Do Differing Standards Increase Trade Costs? The Case of Pallets. World Economy, 30(4), 685–702. https://doi.org/10.1111/J.1467-9701.2007.01009.X

[4] Freedonia Group. (2015). Pallets - Market Size, Market Share, Market Leaders, Demand Forecast, Sales, Company Profiles, Market Research, Industry Trends and Companies. https://www.freedoniagroup.com/industry-study/pallets-3314.htm

[5] Modern Materials Handling. (2021). Annual Pallet Report 2021: Short Supply Meets High Demand – Modern Materials Handling. https://www.mmh.com/article/annual_pallet_report_2021_short_supply_meets_high_demand

[6] Gerber, N., Horvath, L., Araman, P., & Gething, B. (2020). Investigation of new and recovered wood shipping platforms in the United States. BioResources, 15(2), 2818–2838. https://doi.org/10.15376/biores.15.2.2818-2838

[7] McCrea, B. (2020, October 12). Annual Pallet Report: 2020's market evaluation. https://www.mmh.com/article/annual_pallet_report_2020s_market_evaluation

[8] Lee, S. G., & Xu, X. (2004). A Simplified Life Cycle Assessment of Re-usable and Single-use Bulk Transit Packaging. Packaging Technology and Science, 17(2), 67–83. https://doi.org/10.1002/pts.643

[9] Bengtsson, J., & Logie, J. (2015). Life cycle assessment of one-way and pooled pallet alternatives. Procedia CIRP, 29(December), 414–419. https://doi.org/10.1016/j.procir.2015.02.045

[10] Kočí, V. (2019). Comparisons of environmental impacts between wood and plastic transport pallets. Science of the Total Environment, 686, 514–528. https://doi.org/10.1016/j.scitotenv.2019.05.472

[11] Anil, S. K., Ma, J., Kremer, G. E., Ray, C. D., & Shahidi, S. M. (2020). Life cycle assessment comparison of wooden and plastic pallets in the grocery industry. Journal of Industrial Ecology, 24(4), 871–886. https://doi.org/10.1111/jiec.12974

[12] García-Durañona, L., Farreny, R., Navarro, P., & Boschmonart-Rives, J. (2016). Life Cycle Assessment of a coniferous wood supply chain for pallet production in Catalonia, Spain. Journal of Cleaner Production, 137, 178–188. https://doi.org/10.1016/J.JCLEPRO.2016.07.032

[13] Carrano, A. L., Thorn, B. K., & Woltag, H. (2014). Characterizing the carbon footprint of wood pallet logistics. Forest Products Journal, 64(7–8), 232–241. https://doi.org/10.13073/FPJ-D-14-00011

[14] Alanya-Rosenbaum, S., Bergman, R. D., & Gething, B. (2021). Assessing the life-cycle environmental impacts of the wood pallet sector in the United States. Journal of Cleaner Production, 320, 128726. https://doi.org/10.1016/j.jclepro.2021.128726

[15] Park, J., Horvath, L., & Bush, R. J. (2018). Life Cycle Inventory Analysis of the Wood Pallet Repair Process in the United States. Journal of Industrial Ecology, 22(5), 1117–1126. https://doi.org/10.1111/jiec.12652

[16] Berlin, J. (2002). Environmental Life Cycle Assessment (LCA) of Swedish Semi-hard Cheese. International Dairy Journal, 12, 939–953.

[17] Dimyati, A. F., & Singgih, M. L. (2019). Environmental Impact Evaluation Using Green Value Stream Mapping (Green-VSM) and Life Cycle Assessment (LCA). JURNAL TEKNIK ITS, 8(2).

[18] Hospido, A., Moreira, M. T., & Feijoo, G. (2003). Simplified life cycle assessment of galician milk production. International Dairy Journal, 13(10), 783–796. https://doi.org/10.1016/S0958-6946(03)00100-6

[19] Munasinghe, M., Deraniyagala, Y., Dassanayake, N., & Karunarathna, H. (2017). Economic, social and environmental impacts and overall sustainability of the tea sector in Sri Lanka. Sustainable Production and Consumption, 12(July), 155–169. https://doi.org/10.1016/j.spc.2017.07.003

[20] Silvenius, F., Grönman, K., Katajajuuri, J.-M., Soukka, R., Koivupuro, H.-K., & Virtanen, Y. (2014). The Role of Household Food Waste in Comparing Environmental Impacts of Packaging Alternatives. Packaging and Technology and Science, 27, 277–292. https://doi.org/10.1002/pts.2032

[21] Levi, M., Cortesi, S., Vezzoli, C., & Salvia, G. (2011). A Comparative Life Cycle Assessment of Disposable and Reusable Packaging for the Distribution of Italian Fruit and Vegetables. Packaging and Technology and Science, 24, 387–400. https://doi.org/10.1002/pts.946

[22] Koskela, S., Dahlbo, H., Judl, J., Korhonen, M. R., & Niininen, M. (2014). Reusable plastic crate or recyclable cardboard box? A comparison of two delivery systems. Journal of Cleaner Production, 69, 83–90. https://doi.org/10.1016/j.jclepro.2014.01.045

[23] Thorbecke, M., Pike, A., Dettling, J., & Eggers, D. (2019). Life Cycle Assessment of Corrugated Containers and Reusable Plastic Containers for Produce Transport and Display. In Quantis, Corrugated Packaging Alliance. https://26mvtbfbbnv3ruuzp1625r59-wpengine.netdna-ssl.com/wp-content/uploads/2019/06/CPA_Comparative_LCA_Quantis.pdf

[24] Baker, M. (2016). Effect of Pallet Deckboard Stiffness on Corrugated Box Compression Strength. Packaging Technology and Science, 29(4–5). https://doi.org/10.1002/pts.2201

[25] Phanthanousy, S. (2017). The Effect of the Stiffness of Unit Load Components on Pallet Deflection and Box Compression Strength. Virginia Polytechnic Institute and State University, Blacksburg, VA, USA.

[26] Quesenberry, C., Horvath, L., Bouldin, J., & White, M. S. (2020). The effect of pallet top deck stiffness on the compression strength of asymmetrically supported corrugated boxes. Packaging Technology and Science, 33(12). https://doi.org/10.1002/pts.2533

[27] Kim, S., Horvath, L., Russell, J. D., & Park, J. (2021). Investigation of the effect of pallet top-deck stiffness on corrugated box compression strength as a function of multiple unit load design variables. Materials, 14(21). https://doi.org/10.3390/ma14216613

[28] ISO. (2006a). ISO 14040:2006 Environmental Management - Life Cycle Assessment - Principles and Framework.

[29] ISO. (2006b). ISO 14044: 2006 Environmental Management - Life Cycle Assessment - Requirements and Guidelines.

[30] National Council for Air and Stream Improvement. (2017). 2014 Life Cycle Assessment of U.S. Average Corrugated Product. http://www.corrugated.org/wp-content/uploads/PDFs/LCA/NCASI_2014_LCA_Final.pdf

[31] U.S. Environmental Protection Agency. (2020). Advancing Sustainable Materials Management: 2018 Fact Sheet Assessing Trends in Materials Generation and Management in the United States.

ST-O02

Physical Properties Improvement of In Situ Modified Poly(lactide) Film for Fresh Produce Packaging

Piyawanee Jariyasakoolroj[1,2*], Pattarin Leelaphiwat[1,2], Amporn Sane[1,2]

[1]*Department of Packaging and Materials Technology, Faculty of Agro-Industry, Kasetsart University, Bangkok, 10900, Thailand*
[2]*Center for Advanced Studies for Agricultural and Food (CASAF), Kasetsart University, Bangkok, 10900, Thailand*

***Correspondence to:** Department of Packaging and Materials Technology, Faculty of Agro-Industry, Kasetsart University, Bangkok, 10900 Thailand.*
E-mail: fagipnj@ku.ac.th

ABSTRACT: This study proposes an in situ reactive plasticization of poly(lactide) (PLA) with poly(ethylene glycol) (PEG) using poly(ethylene glycol) diglycidyl ether (PEGDE) as coupling agent, and N,N'-ethylene bis-stearamide (EBS) as lubricant. The miscibility improvement was obtained through PEGDE reacting with PLA and PEG chains, and consequently promoted the mobility of PLA chains and nucleation of PLA crystallization. The content of EBS in the PLA/PEG/PEGDE (PLA/mPEG) blend was systematically varied. It was found that glass transition temperatures of PLA in the PLA/mPEG/EBS blends gradually decreased, according to increasing of EBS contents. It correlated to flexibility enhancement. Moreover, the presence of EBS influenced on the increment of PLA degree of crystallinity, and the transition of PLA crystals in perfect form (α-phase). This provided the effective improvement of tensile strength and Young's modulus of the PLA/mPEG/EBS films. Furthermore, water vapor barrier can be achieved by incorporating EBS in the blend system. However, the addition of EBS with high contents obstructed the ordered alignment of PLA chains, the determined oxygen transmission values of the PLA/mPEG/EBS films are higher than those of PLA film.

Keyword: Biodegradable film, Poly(lactide), Lid film, Reactive extrusion, Food packaging

1. INTRODUCTION

Nowadays, it has been already recognized that biodegradable plastics show low greenhouse gas emission index which fit in the circular economy model. Several biodegradable plastics are commercially available, for example, poly(lactide) (PLA), Poly(butylene succinate) (PBS), poly(butylene-adipate-co-terephthalate) (PBAT), and polyhydroxylkanoates (PHAs) [1]. Among these, PLA shows the attractive properties of high transparency and high tensile strength. However, the brittleness and low thermal resistance limit for single-use packaging [2]. It is caused by high glass transition and crystallization temperatures, and required helical structure formation for crystallization [3].

Plasticization is a promising approach to improve flexibility of PLA with maintained high transparency. Lauryl alcohol-based compounds, citrate esters, fatty acids, di-2-etylhexyladipate, polymeric adipates, and poly(ethylene glycol)s (PEGs) are widely used as plasticizers for PLA [4]. PEGs with various types were revealed the effective improvement of PLA ductility with maintained high transparency. The molecular weight of PEG affected on crystallization behaviors and mechanical properties of PLA [5-7]. However, lose mechanical strength and PEG migration limit PLA/PEG blends to utilize, especially at high relative humidity [8, 9].

In addition, low molecular weight aliphatic amides (i.e., *N,N'*-ethylene bis-stearamide (EBS) and *N,N'*-ethylene bis(12-hydroxystearamide) (EBHS)) also showed the efficiency of PLA chain mobility and nucleation acceleration. These amide compounds are organic additives with good environmentally friendly and act as lubricant or releasing agent to improve processability of polymers [10-12]. The crystallization of PLA can be improved by adding these amide compounds because the hydroxyl group (-OH) and amide (-NH) of these compounds can react with carboxylic group at end chain of PLA, and they can form intermolecular interaction with carbonyl group on PLA chains [10].

The fact that the problem of plasticizer leaching or migration towards PLA surface is always found because of ease of mobility for low molecular weight plasticizer with low glass transition temperature, and difference in recrystallization rates between PLA and plasticizer [9, 13]. In order to prevent the migration of plasticizer from PLA matrix, *in situ* reactive blending has been widely applied. This technique is continuously developed for industrial-scale production due to cost-effectiveness [9, 13-14]. Although the plasticizers consisting of reactive functional groups have been widely reported, the additional initiator or inhibitor was required. Moreover, the complicated steps of production were also concerned.

Furthermore, the reactive modified PLA with PEG is limited to use as flexible packaging by relatively hydrophilic PEG phase. This contributes to increase water vapor transmission, which is hard to reserve moisture in fresh produce. Storing at high relative humidity is a key factor to prolong shelf-life and directly impacts on loss of firmness and weight loss [15].

In this study, the reactive blending of PLA and high molecular weight PEG with presence of poly(ethylene glycol) diglycidyl ether (PEGDE) as cross-linking agent is focused to produce a PLA/mPEG blend. EBS with systematically varied contents was incorporated in the PLA/mPEG blend to investigate the crystallization behaviors and thermal properties. Moreover, the effect of adding EBS on oxygen and water vapor permeabilities, and tensile properties was clarified.

2. MATERIALS AND METHODS

2.1 Materials

Commercial PLA (Ingeo™ Biopolymer 4043D) resin was purchased from NatureWorks LLC, Minnetonka, MN, U.S.A. PEG flask (number average molecular weight (M_n) of 35,000 Da), poly(ethylene glycol) diglycidyl ether (PEGDE) with M_n of 500 Da, and *N,N'*-ethylenebis(stearamide) (EBS) were purchased from Sigma-Aldrich Corp., Saint Louis, MI, U.S.A. All chemicals were used without further purification.

2.2 Preparation of sample films

PLA resin was blended with PEG and PEGDE at a weight ratio of 95/4/1 to produce PLA/mPEG using a Labtech Engineering LTE 20–40 twin-screw extruder (Labtech Engineering, Samut Prakan, Thailand). The temperature profile was set in the range of 150 °C – 170 °C with a screw speed of 100 rpm. The obtained PLA/mPEG blend was cooled down in a water bath at 20 °C, and pelletized with 2.5 mm length and dried overnight at 50 °C in a hot air oven.

In addition, an EBS additive was also added into the PLA/mPEG blend with various draw ratios (i.e., EBS 0.5, 0.75, and 1.0 phr). All compounds (i.e., PLA/mPEG and PLA/mPEG/EBS blends) were further blown as films by a Labtech Engineering LE-25-30C single-screw extruder equipped with a Labtech Engineering LF-400 blown film unit. The barrel temperature was set in the range of 160 °C – 175 °C with a screw speed of 50 – 60 rpm and a nip roll speed of 3 m/min. In similar, the extruded PLA and the blended PLA with PEG (M_n 35,000 Da) at a weight ratio of 95/5 (PLA/PEG 95/5) films were prepared as blank samples.

2.3 Characterization

Crystallization behavior and thermal properties were investigated using a Mettler Toledo DSC1 STAR^e differential scanning calorimeter (Mettler-Toledo (Schweiz) GmbH, Greifensee, Switzerland). The sample was heated from 0 °C to 200 °C (heat-cool-heat condition) with a heating rate of 10 °C/min, and cooling rate of 20 °C/min under a nitrogen flow of 50 mL/min. The degree of crystallinity (X_c) of PLA was calculated using Eq. (1):

$$X_{c,PLA} = [(\Delta H_{m,PLA} - \Delta H_{cc,PLA})/(93.1\ J/g \times W_{PLA})] \times 100\% \quad (1)$$

where $\Delta H_{m,PLA}$ and $\Delta H_{cc,PLA}$ are enthalpy changes of PLA fusion and cold crystallization, respectively. 93.1 J/g is the enthalpy change of perfectly crystalline PLA. W_{PLA} is PLA weight fraction.

Oxygen transmission rate (OTR) of the sample was determined by a Systech Illinois 8500 oxygen permeation analyzer (Systech Illinois, Thame, UK). The OTR experiment was according to ASTM D3985-81, and the obtained OTR value was further converted to oxygen permeability coefficients (OP) by using Eq. 2:

$$OP = (OTR \times T) \times (\Delta p)^{-1} \quad (2)$$

where OTR and T are the overall oxygen volume passing through the sample in time unit (*cc/m^2.day*), and thickness of sample film in *mm*, respectively. Δp is the difference in oxygen partial pressure across the film sample in *atm*.

Water vapor transmission rate was estimated using a Mocon PERMTRAN-W® model 398 Water Vapor Transmission Rate (WVTR) tester (MOCON, Inc., Brooklyn Park, MN, U.S.A.) under a nitrogen gas flow of 500 mL/min and 90% RH at 37.8 °C. The samples were conditioned at 25 °C over a saturated calcium chloride solution with 32% RH.

$$\text{WVP} = (\text{WVTR} \times \text{T}) \times (\Delta p)^{-1} \quad (3)$$

where WVTR is the amount of water vapor passing through a certain area film sample over time (*g/m²*.*day*). T is the film thickness in *mm* and Δp is the difference in the partial pressure of water vapor (*atm*) across the film sample.

Tensile properties were characterized by an Instron universal testing machine (Instron, Norwood, MA, U.S.A.) with a cross-head speed of 12.5 mm/min and tensile load of 1 kN, according to ASTM D882-12. The samples were cut into 2.54 cm width × 10 cm length, and stored at 25 °C and relative humidity (%RH) of 52% for 24 hours before test. The tensile test was performed in both machine (MD) and cross (CD) directions of the film samples, with repeating 10 times for each sample.

3. RESULTS AND DISCUSSION

3.1 Crystallization behavior and thermal properties

Typically, PLA is classified as a low crystallization rate material with degree of crystallinity (X_c) of 2-3%. It shows high glass transition temperature (T_g) in the range of 50-60 °C, and high cold crystallization temperature (T_{cc}) at 112.3 °C, as shown in Figure 1a. Additionally, the melting temperatures (T_m) is mainly observed at 147.7 °C, overlapping with smaller endotherm at 152.9 °C. The double T_ms indicated the different crystalline patterns of PLA, and less perfect crystalline phase in α' of δ-form predominantly presented, as compared to the stable perfect crystal of α-form. These double melting peaks of PLA were also found and carefully clarified in other research [16-17].

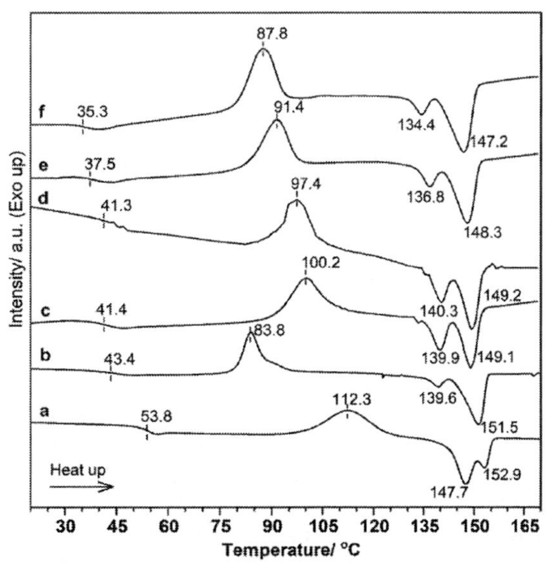

Figure 1 Second-heating-scan DSC thermograms: (a) PLA, (b) PLA/PEG, (c) PLA/mPEG, (d) PLA/mPEG/EBS 0.5, (e) PLA/mPEG/EBS 0.75, and (f) PLA/mPEG/EBS 1.0.

After blending with 5%wt PEG M_n of 35,000 Da, T_g and T_{cc} of PLA/PEG 95/5 significantly decrease to 43.4 °C and 83.8 °C, respectively (Figure 1b). Table 1 shows the X_c of PLA increased from 2.33% to 7.49%. This pointed out the plasticization function of PEG which can promote the mobility of PLA chains and the nucleation of PLA crystallization as widely reported in previous [4-7]. Notably, change in melting endotherm is observed, the melting peak is broadened and T_m peak mainly presents at 151.5 °C. This result suggests that PEG can accelerate the crystallization of PLA and promote the transition of δ- to α-form crystals.

Table 1 Degree of crystallinity of PLA in PLA/mPEG and PLA/mPEG/EBS blends

Sample	Degree of crystallinity (X_c, %)
PLA	2.33
PLA/PEG 95/5	7.49
PLA/mPEG	10.80
PLA/mPEG/EBS 0.5	9.83
PLA/mPEG/EBS 0.75	12.42
PLA/mPEG/EBS 1.0	14.99

In the case of reactive blends (Figure 1c), T_g of PLA in the PLA/mPEG continuously decreased to 41.4 °C, whereas T_{cc} shifted to higher at 100.2 °C and double T_ms peaks strongly present. It implied the success of reaction between oxirane rings in PEGDE and PLA or PEG chains, and PEGDE effectively promoted the PLA crystallization. T_{cc} shifting is similar to the study using poly(ethylene glycol) diacrylate monomer to reduce migration of plasticizer [9]. Moreover, Figure 1d – 1f combined with the determined X_c suggest that adding EBS can gradually improve the mobility of PLA chains and completion of PLA crystallization in α-form, according to increased EBS content.

3.2 Gas permeabilities

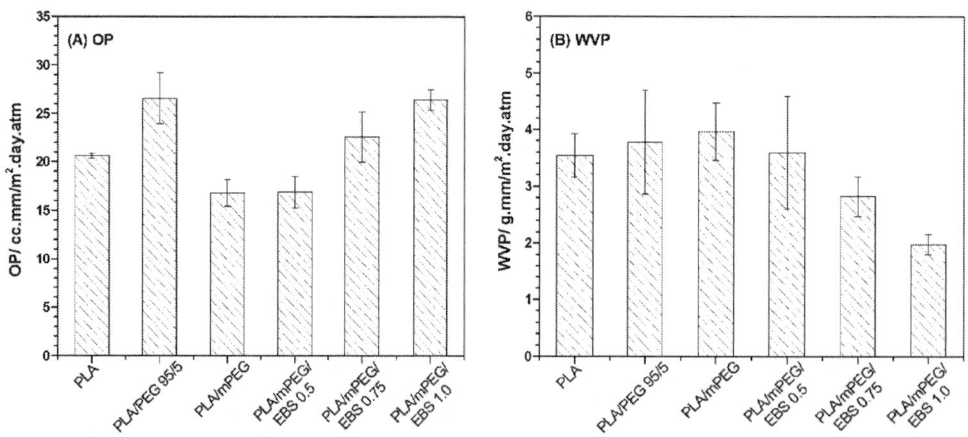

Figure 2 (a) OP and (b) WVP of PLA, PLA/mPEG and PLA/mPEG/EBS films

Figure 2a show the OP value of PLA which increases after blending with PEG M_n of 35,000 Da. It is referred to phase separation due to difference hydrophilicity and crystallization rate between PLA and PEG, resulting in the formation of microvoid in PLA matrix, and consequently high oxygen transmission. In case of PLA/mPEG, the OP value was decreased to 15 cc.mm/m^2.day.atm which indicated the achievable miscibility improvement by PEGDE. However, EBS at high contents can disturb the packing structure of PLA chains in ordered state, the free amorphous phase was expanded. This brings to increase of OP values of PLA/mPEG/EBS 0.75 and 1.0 films. Although the X_c of PLA was enhanced by incorporating PLA with mPEG and EBS, the crystalline phase cannot increase torturous path for oxygen passing through.

For WVP, it is well known that PEG and mPEG are relatively high hydrophilic because of polar molecule, the WVP values of the PLA/PEG and PLA/mPEG films are higher than those of hydrophobic PLA film. But, when EBS was added into the PLA/mPEG blend, the water vapor barrier was successfully improved. It suggests that the long hydrocarbon chains on EBS repel the local diffusion of water molecules across the PLA/mPEG/EBS films.

3.3 Tensile properties

Figure 3 presents tensile properties of pure PLA film with high Young's modulus of 2.5 GPa, high tensile strength 60 MPa, but low elongation at break of 3%. It pointed out that PLA is rigid material. When plasticized PLA with PEG M_n of 35,000 Da, Young's modulus drastically reduces to 1.4 GPa with slightly decreased tensile strength of 45 MPa and increased elongation at break to 7%. It suggests that phase separation between PLA and PEG phases, the propagation of cracking can easily occur, resulting in loss of mechanical strength.

After the miscibility of PLA and PEG molecules by coupling with PEGDE molecules, the mechanical strength of the PLA/mPEG film was improved significantly, as evident by increased Young's modulus (2.75 GPa), tensile strength (57 MPa) and elongation at break (9.5%) (see Figure 3). It was also found in previous studies by Fang et al. [9], Kfoury et al. [13], and Choi et al. [14]. Regarding to EBS addition, Young's modulus slightly decreased to 2.2 GPa, while tensile strength gradually decreased to 42 MPa with maintained elongation at break of approximately 12%. The results correspond to the increment of X_c and more perfect crystalline phase which affected on mechanical performances.

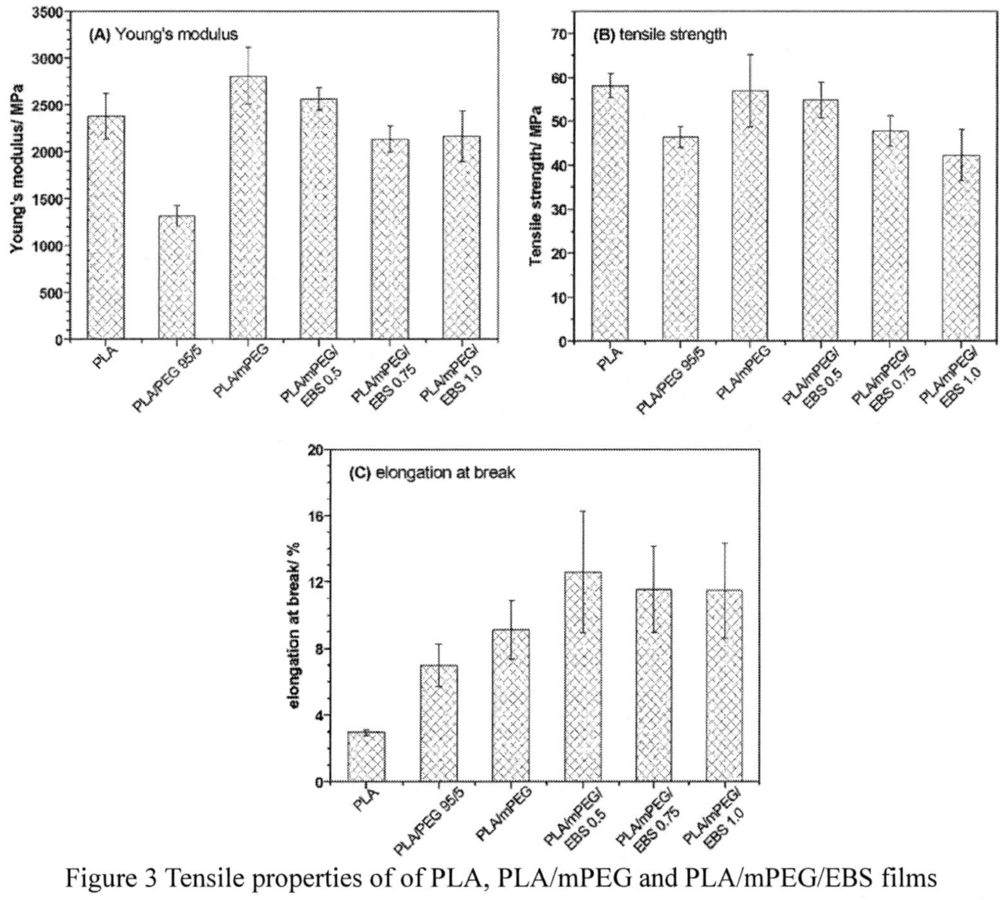

Figure 3 Tensile properties of of PLA, PLA/mPEG and PLA/mPEG/EBS films

4. CONCLUSIONS

PEGDE was used to couple with PLA chains and high molecular weight PEG *via in situ* reactive melt blending, in order to improve miscibility and prevent plasticizer leaching. In addition, EBS was incorporated in the PLA/mPEG blend with various contents to reduce water absorption of PEG molecule. The miscible blend of PLA/mPEG provided the improvement of PLA toughness, as evident by decreased T_g, increased X_c, and enhanced mechanical strength with increased elongation. In the case of PLA/mPEG/EBS blends, T_g and T_{cc} of PLA shift towards lower temperature, the materials become soften due to EBS function as lubricant. The mobility of PLA chains and nucleation of PLA crystallization were effectively promoted when the content of EBS was increased. However, the free amorphous phase was simultaneously enhanced by EBS, resulting in less ordered structures of PLA crystals. It contributed to high oxygen permeability. Whereas the water vapor barrier of the PLA/mPEG/EBS films was improved due to long hydrocarbon chains of EBS. Moreover, the homogeneous PLA/mPEG/EBS films presented the toughness improvement which has further potential use as packaging of fresh-cut produce.

5. ACKNOWLEDGEMENTS

This project is funded by National Research Council of Thailand with research program Grant No. N25A640017 (subproject code: 110898).

6. REFERENCES

[1] Haider TP, Völker C, Kramm J, Landfester K. and Wurm FR. (2019). Angewandte Chemie International Edition, 58(1), pp. 50-62, https://doi.org/10.1002/anie.201805766.

[2] Ahmed J. and Varshney SK. (2011). Polylactides—Chemistry, properties and green packaging technology: A review. International Journal of Food Properties, 14(1), pp. 37-58, DOI: 10.1080/10942910903125284

[3] Hébert JS, Wood-Adams P, Heuzey MC, Dubois C. and Brisson J. (2013). Morphology of polylactic acid crystallized during annealing after uniaxial deformation. Journal of Polymer Science: Part B. Polymer Physics, 51(6), pp. 430-440, https://doi.org/10.1002/polb.23231

[4] Saeidlou S, Huneault MA, Li H. and Park CB. (2012). Poly(lactic acid) crystallization. Progress in Polymer Science, 37(12), pp.1657-1677, https://doi.org/10.1016/j.progpolymsci.2012.07.005.

[5] Hu Y, Hu YS, Topolkaraev V, Hiltner A. and Baer E. (2003). Crystallization and phase separation in blends of high stereoregular poly(lactide) with poly(ethylene glycol). Polymer, 44(19), pp. 5681–5689, DOI:10.1016/s0032-3861(03)00609-8

[6] Hu Y, Rogunova M, Topolkaraev V, Hiltner A. and Baer E. (2003). Aging of poly(lactide)/poly(ethylene glycol) blends. Part 1. Poly(lactide) with low stereoregularity. Polymer, 44(19), pp. 5701–5710, DOI:10.1016/s0032-3861(03)00614-1.

[7] Sungsanit K, Kao N. and Bhattacharya S. (2012), Properties of linear poly(lactic acid)/polyethylene glycol blends. Polymer Engineering and Science, 52(1), pp. 108-116. https://doi.org/10.1002/pen.22052

[8] Lai WC, Liau WB. and Lin TT. (2004), The effect of end groups of PEG on the crystallization behaviors of binary crystalline polymer blends PEG/PLLA. Polymer, 45(9), pp. 3073-3080, https://doi.org/10.1016/j.polymer.2004.03.003.

[9] Fang H, Jiang F, Wu Q, Ding Y. and Wang Z. (2014). Supertough polylactide materials prepared through in situ reactive blending with PEG-based diacrylate monomer. ACS Applied Materials & Interfaces, 6(16), pp. 13552-13563. https://doi.org/10.1021/am502735q

[10] Xing Q, Zhang X, Dong X, Liu G and Wang D. (2012). Low-molecular weight aliphatic amides as nucleating agents for poly (L-lactic acid): Conformation variation induced crystallization enhancement. Polymer, 53(11), pp. 2306-2314, https://doi.org/10.1016/j.polymer.2012.03.034.

[11] Xing Q, Li R, Zhang X, Dong X, Wang D. and Zhand L. (2015). Tailoring crystallization behavior of poly (L-lactide) with a low molecular weight aliphatic amide, Colloid and Polymer Science 293, pp. 3573–3583 https://doi.org/10.1007/s00396-015-3730-5

[12] Harris A.M. and Lee E.C. (2008). Improving mechanical performance of injection molded PLA by controlling crystallinity. Journal of Applied Polymer Science, 107(4), pp. 2246-2255. https://doi.org/10.1002/app.27261

[13] Kfoury G, Hassouna F, Raquez JM, Toniazzo V, Ruch D. and Dubois P. (2014).Tunable and durable toughening of polylactide materials via reactive extrusion. Macromolecular Materials and Engineering, 299: 583-595. https://doi.org/10.1002/mame.201300265

[14] Choi KM, Choi MC, Han DH, Park TS. and Ha CS. (2013). Plasticization of poly(lactic acid) (PLA) through chemical grafting of poly(ethylene glycol) (PEG) via in situ reactive blending. European Polymer Journal, 49(8), pp. 2356-2364, https://doi.org/10.1016/j.eurpolymj.2013.05.027.

[15] Volpe S, Mahajan P, Rux G, Cavella S. and Torrieri E. (2017). Condensation and moisture regulation in packaged fresh-cut iceberg lettuce. Journal of Food Engineering, 216, pp.132-137, DOI:10.1016/j.jfoodeng.2017.08.015.

[16] Pan P, Kai W, Zhu B, Dong T. and Inoue Y. (2007). Polymorphous crystallization and multiple melting behavior of poly(L-lactide) molecular weight dependence. Macromolecules, 40(19), pp. 6898–6905. https://doi.org/10.1021/ma071258d

[17] Zhang L, Zhao G. and Wang G. (2021). Investigation on the α/δ crystal transition of poly(L-lactic acid) with different molecular weights. Polymers, 13, pp. 3280. https://doi.org/10.3390/polym13193280

ST-O03

Sorting plastic packages with 3D relief codes

Roland ten Klooster[1]*, Ulphard Thoden van Velzen[2]

[1]*Cluster Information Integrated Manufacturing & Packaging, University of Twente, Drienerlolaan 5, Enschede, 7522NB, Netherlands*
[2]*Wageningen Food & Biobased Research, University of Wageningen, Wageningen, Bornse Weilanden 9, 6708WG, Netherlands*

***Correspondence to:** Cluster Information Integrated Manufacturing & Packaging, University of Twente, Drienerlolaan 5, Enschede, 7522NB, Netherlands. r.tenklooster@utwente.nl*

ABSTRACT: The European Circular Economy Action Plan strives to use more recycled plastics in Fast Moving Consumer Goods and hence contribute to the more sustainable use of plastics. Mechanical recycling of well-designed, food packages can yield highly pure recycled plastics that in principle could be used to produce new food packages, but unfortunately EFSA rulings demand that the recycling feedstock is composed of guaranteed high levels of food packaging only. As this guarantee cannot be given for sorted products from EPR schemes, the circular economy stifles. It is therefore crucial to develop sorting techniques that can discriminate between previous food-use and non-food-use of wasted plastic packages. A new sorting technology is proposed that is based on a 2D print technique used for anti-counterfeiting, with curved patterns of dots, not visible by the naked eye. The patterns of dots can be recognized by reflection of light with RGB cameras and specific developed software. The technique is translated to 3D with dots in relief in the packaging material. The aim is to find dots of such a size that they are hardly visible to the naked eye, but flawlessly readable for cameras. At first, patterns with dots with a diameter of 0.5 mm and a depth of 100 micron were studied. The patterns of dots are taken up in moulds of injection moulded and thermoformed packaging of in total 6 PP and 3 PET items like cups, trays, lids, for food packaging with different characteristics like transparent, coloured, and different surface qualities, produced on various machines. Additionally, a dynamic test was executed with a tailor-made sorting machine with different conveyor belt speeds and lighting strengths. The sorting yield was over 90% with no false positive showing that 3D relief dots can be used for this purpose. Thermoforming requires more control of the production process to obtain well-readable relief codes than injection moulding. A procedure is developed to take up a 3D relief code in packaging items. In next steps sorting will be tested based on realistic waste processing scenarios.

Keyword: Sorting of plastic packaging, markers, recycling, 3D relief code

1. INTRODUCTION

The European Circular Economy Action Plan [1] strives to reuse and recycle more plastic packaging. The Commission will propose mandatory requirements for recycled content and waste reduction measures for Fast Moving Consumer Goods such as packaging.

The use of recycled plastics in food contact materials is strictly regulated in European directives; EC 282/2008, under the umbrella legislation EC 1935/2004. Recycling companies can file a request for a positive opinion of their process-feedstock combination at the European Food Safety Authority (EFSA). To limit potential hazards with incidental contamination, EFSA set threshold values for the share of non-food-packages allowed to be present in recycling feedstock, which is at least 95% for PET bottles and 99% for HDPE bottles. Most positive opinions were given to companies recycling PET beverage bottles and one to a company recycling HDPE milk bottles in a closed collection system. No recycling company has received a formal approval, yet, from the European Commission. EC 282/2008 is currently under revision because the risk assessment EFSA executes is based on a triple worst case approach and too conservative effectively hindering the progress towards a more circular economy [2].

Within the current legal context, recycling companies have to prove and guarantee that their feedstock is almost exclusively composed of food packages. Producing high quality recycled plastics also requires that the recycling company can discriminate between different qualities (grades) of plastics from the same polymer class, with different melt flow indexes. Current sorting techniques are sorting on the level of main polymers like PE, PP, PET, etc. but cannot distinguish between various types or grades of a polymer class, for instance for PE, between HDPE, MDPE, LDPE and LLDPE. Near-infrared (NIR) is the current standard technology in plastic sorting. It uses the frequency range of 600 to 2500 nm of common light sources [3,4]. The light is shone onto the plastic items, from which the polymers absorb specific frequencies and reflect that modulated radiation back into spectrophotometer, which can accurately identify the main type of polymer within microseconds. The technique is used for sorting packages on fast moving conveyor belts. Software determines a contour of an item and its position on the conveyor belt and will determine if this item needs to be blown out by the air nozzles. Within sorting facilities that use a cascade of NIR sorting machines only the most common polymer types are sorted; PE, PP, PET and PS.
Sorting packaging waste with NIR machines does not give detailed information about the grade of plastic that is sorted, nor about its previous use. NIR is also used to sort flakes of plastic. There are more sorting techniques of material available like elutriation, wind-sifting, optical colour recognition, static electro charge, X-ray fluorescence and magnetic density separation [4], to mention some, but they all sort on base of simple characteristics of the materials and are not able to make a distinction between food and non-food used materials yet.

Companies are searching for ways to be able to recycle the material of the packaging they use themselves back again to a new packaging item and are investigating possibilities to be able to circularly recycle the packaging material they used previously. From plastics used in Europe 39.6% of 50.7 million tons is used in 2020 for packaging, in total 20 million tons [5]. Food accounts for about 2/3 of this amount, but there are hardly clear figures to underline this. In the Netherlands 520.000 tons of plastic is used for packaging yearly [6].
Sorting on the level of packaging can be a solution to overcome uncertainties about grades, additives and previous use. Several techniques are promising to recognize packaging from waste streams: sorting on base of digital marking and by using camera's with recognition software with Artificial Intelligence (AI). In the HolyGrail project [7] the use of tracers and digital marking are used. By adding elemental tracers to the polymer, the material can be recognized with X-ray fluorescence equipment. With digital marking patterns are taken up in

the print (2D) or in relief in the material (3D) as tested with two different techniques: Digimarc and CurvCode. The patterns can be recognized by reflection of light of the dots and the use of conventional cameras and specific software. In the HolyGrail project it was concluded that digital marking techniques are more promising than the use of tracers [8]. For using digital marking technique the packages have to be changed. The patterns have to be taken up in the print, or in the material in relief. For plastic packaging or items without print or large labels or sleeves, the patterns have to be taken up in the mould to reproduce the code.
Digimarc is using unique codes per SKU by which it is possible to add specific information to the sorted item. The Digimarc codes have a dimension of ca. 2.2 x 2.2 cm and are taken up all over the packaging to ensure they can be scanned even if they are crushed and dirty. The Digimarc technique can be used in every stage of the lifecycle of the packaging, like scanning the packs on the paying desk of a supermarket, for consumer engagement, and also for waste sorting. Every packaging has u unique code which means that for waste sorting a big database of codes is needed and this will set requirements to the sorting equipment and process. The CurvCode is using codes per type of material. The codes are only used for sorting packaging items for recycling. The number of codes in the database is limited. For every type of material a different code for food as for non-food is used.

A technique in which the packaging does not have to be changed and that possibly also can be used for sorting is based on recognition of images scanned with RGB camera's with use of Artificial Intelligence. For face recognition patterns can be used as a matrix because every face has the same structure. With packaging a common pattern cannot be used. Software must be trained to recognize different types by all kind of characteristics, including pressed packs and covered with dirt. In the coming decade, these AI technologies are likely to be used for sorting plastic waste.

This paper explores a digital marking technique in more detail; the CurvCode, a code with 11 dots in the shape of a curve with a width of 20 mm, between the outer dots in a straight line. The aim is to explore the possibility of sorting plastic waste with simple codes that are applied in relief in the packaging material. Furthermore, it will also address the design of this code and application method. This exploratory study was limited to thermoforming and injection moulding as production processes.

The research will focus on a sorting technique with which it is able to sort between food and non-food packaging items, that uses a limited amount of codes, that are only used in the stage of sorting of waste, with the focus on 3D codes. Two important issues need to be considered:.

- Speed of current sorting lines determines the minimum required speed for recognizing a package on a conveyor belt.
- The way of applying 3D relief codes in plastic packages made by thermoforming and injection moulding.

At first the CurvCode will be described in more detail. Then attention will be spend on the mentioned issues.

2. THE CURVCODE

The CurvCode is formed by dots (2D) or rolls (3D) positioned at a Bezier curve, a parametric curve with discrete control points. The dots are equal in size, positions of the dots are determined by dividing them equally across the curve. The distance between visible and invisible positions is equal. The location of the dots across the curve is symmetric, meaning the code can be read from left to right and right to left.

2D codes can be used for packaging that are covered with a label, (wrap around) sleeve, in-mould-label (IML), sticker etc., and 3D codes can be used for packaging or packaging items with only small labels or without labels. Also a combination of 2D and 3D is possible.
The 2D technique is an existing technique which has proven to function. It is used among others for anti-counterfeiting of printed products. For 2D codes the angle of the position of the cameras is irrelevant. By use of light a code that is made with small dots of different colour in artwork, is recognized by software and converted in a computer code.
The code is built up with 27 fixed positions. Eleven dots are placed on the curve. Six dots determine the material code. The middle dot is used as control; reading a code from left to right has to end at the same position as reading from right to left. On both sides the code starts with two dots at fixed positions. 27 minus the dot in the middle, makes 26 dots, so 13 positions for a code, minus four fixed positions makes 9 dots. Using 9 positions, and including two angle values of the lines determined by the two starting dots, make 2^{10} number of combinations possible. The system cannot read code 0, leaving 1024-1, in total 1023 possible codes. This is taken as enough for sorting materials.

For 3D codes the intention is to make the code invisible or almost invisible, meaning as small as possible. At first tests are executed in a static situation with codes of dimensions 15 mm, heart to heart from the first dot to the last dot in a straight line. First samples were made by making small holes in metal blocks because metal is used for moulds to make plastic packages. The idea of the size of the dots was taken from the codes as taken up in the heel of glass bottles and also used for sorting. Tests showed that the codes taken up in metal were readable very well. Discussing the options to put the code in a plastic packages with injection blow moulders made clear that a code of that size would be hard to create. The blow moulders preferred a code with width of 30 mm but for marketing reasons and not to influence the outer appearance of the packaging too much, a length of 19.5 mm was chosen for next tests, heart to heart of first and last dot at a straight line. A code that is larger also needs a larger surface of a crushed packaging that has to be scanned. From that perspective smaller codes are to be preferred. The diameter of the dots or balls is chosen at 0.5 mm, so from the outer diameter of the first dot to the outer diameter of the last dot the length of the line is 20.0 mm.
A dot or ball cannot be bigger than 0.7 mm, because the space between dots is 0.5 mm. Larger dots imply smaller spaces which makes the code unreadable. A code starts with two dots on both sides at fixed positions. Light is shone at the packaging surface and reflected. The reflected image is analyzed in reversed black and white. The software scans the processed image to find two black dots beside each other, being the start of the curve. If two dots are found, the software scans in a circle of 20 mm around the dots two find two other dots. If these are found, two lines are made, one from the first two dots and one from the other two dots. The angle in which the lines cross has to be one of a pattern that is in the register. If not, the software stops with the image and continues searching. If the angle is according one of the registered standards, the other dots are analyzed and the pattern is compared with the register. In this way the pattern of dots can get a registered number. The number can activate a blowing unit of the sorting machine to positively sort out the packaging.

Figure 1 – Example of a CurvCode

By coding plastic packaging they can be sorted on base of use as food or non-food. This can be done by only using codes in food packaging, or by using codes in both, food as well as non-food. The format that has been chosen is to determine the value of the angles between the two lines that are defined by the first two dots on each side of the curve. The code for a material that is food grade is a different angle than the code for non-food.

The next possible examples of codes to be able to recycle the plastic back to food use for PP and PET can be considered. For PP, transparent, white opaque, black opaque, carbon black, other colours opaque and for PET transparent, white opaque, black opaque, carbon black, other colours, opaque. This example makes ten codes for the materials and twenty in total if for food grade different codes are used as for non-food plastic packaging. It will depend on the volumes used in the market what a feasible choice for codes will be in practice. The research will not focus on economic feasibility.

3. THE FUNCTIONING OF 3D CURVCODE AND DOT SIZES

If light is shone on the surface of a plastic package with relief, the relief will reflect a part of the light. Cameras with recognition software can detect patterns of light reflection in the same way as used for 2D recognition of patterns. For reflection of light by 3D shapes, the light should shine under a certain angle, which is different than for 2D codes.
Questions that arise are about the dimensions and locations of the codes on an item.

- How big should the 3D codes be to be readable in the same way as the 2D codes?
- How many codes are needed on the package to be readable?
- Does orientation of the codes influence recognition results?
- What to do with flat items that are not transparent, like coloured lids of trays, jars or buckets? Does a code has to be taken up on both sides?
- Does transparency influence the readability of the code?
- Does the surface appearance like gloss or matt surface influence the readability of the code?

3.1 Relief of a dot

The higher a relief, the better the reflection corner will be, the better readable the code will be. Figure 1 shows a low relief with reflection in a smaller angle than the reflection shown in figure 2. The green lines represent light shining at the dots, the red lines represent reflected light. The relief in figure 1 can only be read very close to the cameras, not in a wider field. This will limit the speed of the conveyor and the readability of packaging in random positions.

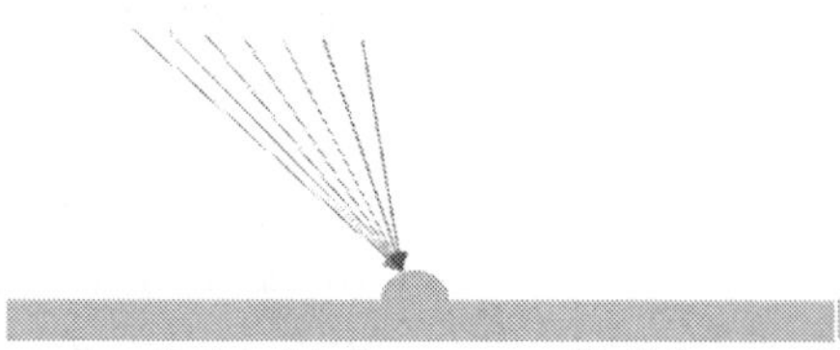

Figure 1 – light shining on a relief dot

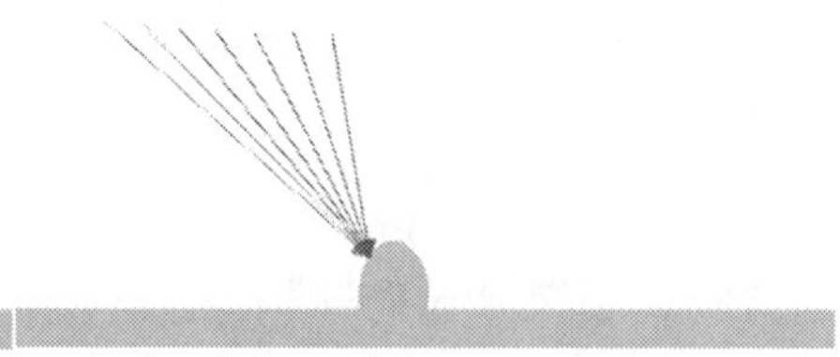

Figure 2 – light shining on a higher relief dot

It is also possible to use debossing besides embossing. In figure 3 an example is shown. The light reflects in a different way creating a pattern.

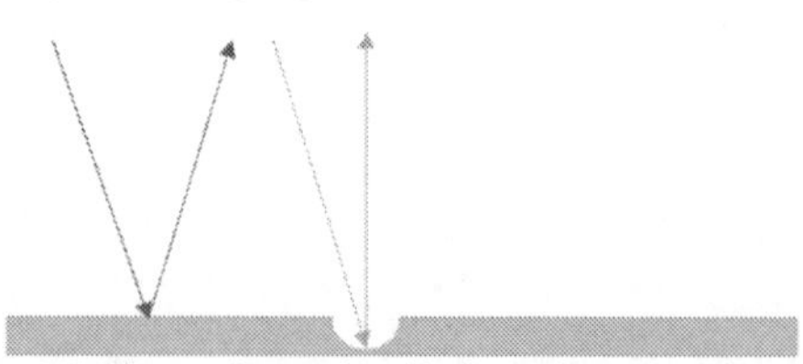

Figure 3 – Hollow shaped relief, embossing in a material

3.2 The light source

In trial tests with an injection moulded lid, white coloured, with codes on both sides, several light sources are tested.

The better the light source the more light will be reflected. The next light sources were used in sequence of order:

- First tests were executed with indirect led light, 6.450 lux., recognized 72% from 132 lids, 76% from 138 lids
- The next test with 4Pico LED, direct lightning, 25K lux., recognized 84% from 119 lids
- Led light, direct, 26K+ lux, provisional mounted (direct), 100% from 100 samples
- Led light, direct, 26K+ lux, mounted on a frame (direct), 100% from 100 samples and 96% from 100 samples

Final dynamic testing has been executed with the latest tested light conversion: Led light, direct, 26K+ lux, mounted on a frame (direct).

3.3 Dimensions of a relief dot

First tests were executed with a 3D code with a diameter of 0.5 mm and a depth of 100 micron. To ensure the readability of the code, first a static test was done. Indeed these codes proved to be well-readable.

To make testing more practical, an app for a smartphone was developed. The app uses the camera of the smartphone and this helped to greatly simplify the static testing of Curvcodes. If one code put in the mould of a package can be recognized by the app on the smartphone, the code is readable. Then more codes can be taken up in the mould with the first as template. In this way a first selection on base of colour, surface and transparency is executed.

3.4 The amount of needed codes on the package

Multiple codes are needed on a package because wasted packages can be distorted, crushed and fouled, which can limit the readability of the code. But putting too many codes on the surface, could produce unrecognizable patterns. Changing an existing mould cost money. The less changes, the lower the costs. A balance has to be found, also to avoid that the appearance of a package is visibly affected by the amount of codes.

On vertical surfaces of cups or trays with high side walls it can be difficult to get a relief code in the package. This accounts especially for thermoformed packaging. A dot made by a relief of 0.5 mm diameter with a depth of 100 micron in a mould used for injection moulding will get the dimensions of the relief. For injection moulding an issue can be the draft angles of the article. The dots should not frustrate relieving of the package from the mould, and still have to be readable. In thermoforming it is harder to get a dot of 0.5 mm in the mould. Every relief should have a vacuum channel which makes the mould rather expensive. By making the diameter of the cavity in the mould larger, up to 0.7 mm, the dots that came out were close to 0.5 mm and could be recognized by the cameras. If the dots do not have a vacuum channel, forming of the dot was hard and reading often did not work out well. Also the operational temperature and used pressure/vacuum influences the forming of the dots. The softer the material, so the closer to the glass transition temperature, the better the forming process will be, but the more energy and time are needed to form the article.

3.5 Orientation on the package

Articles with codes in the same direction were tested. Sometimes the articles were not recognized. The positioning of the article on the conveyor is of importance. It turned out that positioning of CurvCode in more directions functions better, making the positioning of the article independent from the light sources and reflection.

3.6 Type of material

Different types of materials with different colours and surface have different light reflecting characteristics resulting in different contrast ratios between positive and negative elements of the CurvCode. The software needs to use different configurations to be able to read transparent PET and white PP for example.

4. DYNAMIC TEST

To perform a dynamic test a machine has been constructed with the following functions:

- A head band to transport packaging to a conveyor on which they are dropped
- A conveyor which can be set at different speeds
- A beam with light sources and cameras with an overlap in the view of 12 mm
- An electronic connection with a blowing unit for sorting

The packaging are put one by one on the head band to transport them to the conveyor on which they pass the light beams and the cameras. If a CurvCode is recognized they are sorted out with the blowing unit. At the end the percentage sorted packages can be calculated.

Several brand and private label owners were willing to cooperate in the test and they adjusted their packaging. All the packages in which CurvCodes have been applied, were first tested in a static situation. Different surface characteristics like gloss and matt surface and different colours are chosen, including black. All the packaging used for the test are packaging that are currently used in the market, so the results will be representative for practice.

Figure 4 – Picture of the test machine for dynamic testing

Description of the picture: At the back the head band can be seen, from the middle running to the top at the right. At the right the packages fall on the conveyor behind the grey screen and are transported from right to left. The software is taken up in the grey coloured computer. The beam with light and cameras are positioned in the middle, the mushroom shaped blue part above the CurvCode logo. At the left the blowing unit is taken up with a cover with the Filigrade logo to make sure that the packages are not blown away out of the area of collecting.

The packages listed in Table 1were fitted with CurvCodes in relief and were used in the dynamic test. The packaging were put on the headband so they got randomly positioned on the conveyor. The tests were repeated 4 to 17 times, until the number of 100 sorted packages is reached each time, depending on the amounts that were available. No false positive sorting was found, proving the reliability of the technique.

Table 1 – Packaging used in the dynamic test with the test results in percentage

Packaging	**Dimensions (ca.)**	**Colour**	**Appearance**	**Position of codes**	**Dot size**	**Percentage recognized at 1.25 m/s**
PP injection moulding, mono layer material						
1. Lid for a 10 liter bucket	Diameter 230 mm Height 15 mm	Un-coloured, hazy	With IML on top	Codes on both sides	Ø 0.5 mm	100%
2. Lid for 1 liter bucket	Diameter 130 mm Height 9 mm	White	Without print	Codes on both sides	Ø 0.5 mm	98%
3. Tray	Rectangular with rounded corners 114x128 mm, height 25 mm	Un-coloured	Hazy, matt look	Codes on both sides	Ø 0.5 mm	80%
4. Tray	Diameter 170 mm Height 40 mm	Green	IML on vertical wall and a round IML diameter 7.6 cm at the bottom.	Outside surface is matt. Codes on both sides. Inside surface is not gloss, but also not very matt	Ø 0.5 mm inside and outside	82%
5. Tray	Diameter 170 mm Height 40 mm	Green	Matt surface on vertical wall and a round IML diameter 7.6 cm at the bottom, inside more glossy then other green tray	Outside surface is matt. Codes on both sides. Inside surface is more glossy than other green tray	Ø 0.5 mm outside and Ø 0.56 mm inside	87%
6. Tray	Diameter 170 mm Height 40 mm	Black	A high gloss circle in the bottom (for putting a label on) 7.3 cm diameter	Outside surface is matt. Codes on both sides	Ø 0.5 mm outside and Ø 0.56 mm inside	82%
7. Cap	Diameter 82 mm	Brown	Normal PP gloss	Codes only on the outside	Ø 0.5 mm	88%
PET thermoforming, monolayer						
8. Tray	Rectangular 140x180 mm, Height 35 mm	Transparant	No labels	Codes only on the outside	Ø 0.6 mm	98%
9. Tray	Rectangular 170x100 mm Height 10 mm	Transparant	No labels	Code only on outside, can be seen on inside	Ø 0.6 mm	99%
10. Tray	Rectangular 170x100 mm Height 10 mm	Black	No labels	Code only on outside, can be seen on inside	Ø 0.6 mm	98%

The results prove that the cameras can scan the codes rather well and recognize the packaging. This is the first dynamic test of scanning packaging for codes that is executed with the test machine with three main parameters that can be varied independandly; the type of lightening, the position of the cameras and the conveyor belt speed. The tests were used to establish a procedure to apply Curvcodes in packages in the best possible manner.

5. PROCEDURE FOR TAKING UP A CURVCODE IN A PACKAGING

The insights have been translated to a procedure in which steps are taken up for putting a code in a packaging. The procedure is called: **the Embedding Guidelines for 3D CurvCode**. The procedure starts from the perspective of a potential user that has no previous experience with Curvcode and ends by providing the user with the coded package, ready to be sorted. The guidelines contain 10 points.

1. Explanation of CurvCode
   The use of the code is explained as follows: The CurvCode is a technique to recognize packaging on base of reflection of light. It can be used on many materials and as far as known for all kind of plastics, transparent, coloured, matt surface and high gloss. The codes that are used are related to materials like PE, PP, PET etc. The codes can be used to make a distinction between food and non-food packaging.
2. The appearance of the CurvCode
   The code can be taken up as 2D in artwork or 3D in relief. This guideline is specifically developed for 3D relief codes. The code is made by 12 dots that are put in relief in the plastic by use of the mould. The relief can be positive or negative: embossing or debossing.
3. Dimensions of the code
   The code has the shape of a Bezier Curve with length from the first to the last dot of 19.5 mm, heart to heart, 20 mm on the outline. The diameter of a dot is 0.5 mm. Pictures of codes as they appear in plastic packaging are available to be shown to potential users.
4. Depth of the code
   The relief of a dot has to be around 100 microns after leaving the mould. For injection moulding this can be used as the depth for a dot taken up in the mould. For thermoforming best it to use vacuum channels under the dots and making the width 0.7 mm with a curve that is deeper than 100 microns. If 150 microns is used, the dots are very well readable. If dots are taken up in surfaces without a vacuum channel, the diameter of 0.7 mm and depth of 150 microns are minimum dimensions. The result is mostly readable, but has to be tested. The forming temperature and vacuum pressure is also of importance for forming the dots.
5. Compensating shrinkage or expanding of the plastic after leaving the mould
   The code has to be 19.5 mm heart to heart after leaving the mould and cooling down. Possibly the code in the mould is larger or smaller, depending on shrinkage or expanding of the package after leaving the mould. Scaling of dimensions has to be considered.
6. The amount of codes on a packaging
   The more codes on a package the better, but the codes may not overlap. A distance of 1 cm between the codes at least is needed to overcome false recognition of patterns. Also a distance of minimum 2 mm between other relief codes like a recycle logo is needed. It is advised to take up at least 6 codes per packaging, more or less one on every side of a package; 3 dimensions with two sides each.
7. Integration of the code in the design
   Best is to take up codes on places that are expected to be visible after crushing the package. Trays with sides that go up, cups with high sides, will be flattened during waste collection and bottoms will most likely remain flat and readable, so at least codes have to be taken up in the bottom. Codes in the side can help.
8. Putting the codes on a 3D curved surface
   Best is to design the code as seen rectangular to the 3D curved surface as a projection on the surface. In software the code can be scaled and projected on the surface. In this way the distortion is smallest seen from several directions.
9. Orientation of the code
   For reading the code it does not matter what the axis of the code is. The best is to put codes in different directions on the surface so the light shines on the curves in different directions.

10. Orientation on side of the packaging
    A code taken up in a thermoformed article can be read from both sides, if the dots are deep enough the light is reflected even if the code is taken up on one surface. For injection moulding the code has to be taken up on both sides of the article.

6. DISCUSSION

With static and dynamic tests it has been demonstrated that plastic packages with Curvcodes can be recognized and sorted. This is an important milestone in the development of this new sorting technology. But there are still many questions. For example if it is possible to recognize packaging that are crushed and covered with dirt, the influence of humidity when waste has been stored in the rain, the influence of the speed of the conveyor belt, and finally, the possibilities to put a code in extrusion blow moulded articles.

7 Conclusion

Tests have shown that it is possible to sort packages on base of a 3D code that is taken up in relief on the package. Guidelines are needed to structure the process. Design rules are needed to reach the most effective process and results.
The next step that will be executed is sorting out coded packages from waste, fully crushed and covered with dirt.

8. ACKNOWLEDGEMENT

Acknowledgments: Filigrade for giving access to the sorting testing with the testing machine.

9. REFERENCES

[1] COMMUNICATION FROM THE COMMISSION TO THE EUROPEAN PARLIAMENT, THE COUNCIL, THE EUROPEAN ECONOMIC AND SOCIAL COMMITTEE AND THE COMMITTEE OF THE REGIONS A new Circular Economy Action Plan For a cleaner and more competitive Europe COM/2020/98 final

[2] Franz, Roland; Welle, Frank; Recycling of Post-Consumer Packaging Materials into New Food Packaging Applications—Critical Review of the European Approach and Future Perspectives; Sustainability, 14, 2022, 824; doi: 10.3390/su14020824

[3] Ruj, Biswajit; Pandey, Vivek; Jash, Priyajit; Srivastava V.K.; Sorting of plastic waste for effective recycling; Int. Journal of Applied Sciences and Engineering Research, Vol. 4, Issue 4, 2015; doi: 10.6088.ijaser.04058

[4] Seranti, Silvia; Bonifazi, Giuseppe; Techniques for separation of plastic wastes, in Use of Recycled Plastics in Eco-efficient Concrete; doi.org/10.1016/B978-0-08-102676-2.00002-5; 2019 Elsevier Ltd.

[5] Bakker EJ, PC Rem, D Hartmann, GJ Bakker (2008). Magnetic density separation; Materials and Environment; Houston, American Science Press, pp12-18

[6] Plastics, the facts 2020; An analysis of European plastics production, demand and waste data; acquired from https://plasticseurope.org/knowledge-hub/plastics-the-facts-2020/

[7] Derived from www.afvalfondsverpakkingen.nl yearly reports of waste collection in the Netherlands

[8] HolyGrail: tagging packaging for accurate sorting and high-quality recycling; Publisher(s) Ellen MacArthur Foundation, Business Sectors Waste Management and Recycling

[9] HolyGrail 2.0, acquired from www.digitalwatermarks.eu

ST-O04

Designing reusable packaging: tool development, validation, and implications

Bjorn de Koeijer[1]*, Roland ten Klooster[1], Irene Hesseling[1,2], Corné Huijben[2]

[1]*Department of Design, Production and Management, Faculty of Engineering Technology, University of Twente, Drienerlolaan 5, 7522 NB Enschede, The Netherlands*
[2]*NVC Netherlands Packaging Centre, Stationsplein 9k, 2801 AK Gouda, The Netherlands*

***Correspondence to:** Bjorn de Koeijer, Department of Design, Production and Management, Faculty of Engineering Technology, University of Twente, Drienerlolaan 5, 7522 NB Enschede, The Netherlands. E-mail: b.l.a.dekoeijer@utwente.nl*

ABSTRACT: With increasing emphasis on the environmental impact of product-packaging combinations comes a growing appreciation for systemic improvements in the way packaging acts within supply chains. One of the innovation directions at the centre of attention is reusable and refillable packaging. In various forms, these are being researched for their potential to act as a packaging concept, and developed as a market-ready alternative to one-way or single-use packaging. Both in the European Commission's Circular Economy Action Plan and packaging strategies developed by the Ellen MacArthur Foundation, reusable packaging is adopted as a valuable scenario. In this research, we set out to develop a decision support tool that can assist in developing product-packaging combinations for reuse/refill scenarios. The development of the tool is based on a designerly approach, combining literature research and interviews with industry stakeholders applying reusable packaging concepts in their product marketing, or considering to do so. Complexities in the development and application of the tool do not only come from specific product-packaging combinations, but also from the system in which these interact: retail, use, reverse logistics, and repurposing require an overhaul of their impact. The developed tool is tested by means of a multi-concept analysis, and a reflection. Eight student groups are invited to design reusable/refillable packaging concepts, for a product of their choice. Analysing the designs with the tool provides insights into the decision-making process, and a set of potential redesign considerations. With this, we assess the value of the designed concepts and the benefits of the tool for decision-making processes when developing reusable packaging concepts. The latter point is valuable for comparison to other packaging decision-making tools, and to be able to make improvements. In follow-up research, we set out to apply the decision support tool for industry-based design processes, to understand its usability for packaging designers.

Keyword: Packaging design, design tool, sustainability, reusable packaging, packaging strategy

1. INTRODUCTION

In industry, legislation, academic research, and public perception, there is increasing emphasis on the environmental impact of product-packaging combinations [1, 2]. We have established complex production-consumption systems [3] which are dominated by linear approaches, resulting in resource depletion and a surplus of waste [4, 5]. Within this, the role of packaging is often the subject of societal discussions on environmental concerns. The traditional redundancy perspective on packaging materializes in a focus on reduction or elimination, and policies aligning with that [6-8]. However, when we consider the essential position of packaging functions as facilitators of a product's ability to provide added value to a supply chain [3, 6], we can establish the potential for more systemic views and approaches towards packaging as a sustainability enabler.

Stakeholders influencing the development of product-packaging combinations aim to increase their focus on the integration of sustainability-related considerations [4, 9-11]. One of the innovation directions at the centre of attention is reusable and refillable packaging [12-17]. In various forms, these are being researched for their potential to act as a systemic packaging concept, and developed as a market-ready alternative to one-way or single-use packaging. In the European Commission's Circular Economy Action Plan [18-20], packaging strategies developed by the Ellen MacArthur Foundation [21], and national and regional plastics pacts [15, 22], reusable packaging is adopted as a valuable scenario.

However, declarations of relevance will not be sufficient to make an impact. In current development processes, sustainability-related considerations are often embedded merely as a suboptimal or late-stage differentiator in development processes, despite strategic desires and operational efforts [4, 23]. For developments to be evolutionary sustainable, a progressive transformation of economy and society is required [24]. Without the development and implementation of new business models – such as a focus on access over ownership or financial incentives addressing resources instead of labour [5, 25, 26] – product and packaging developments will not transcend trade-offs biased towards non-sustainable considerations.

This research aims to deliver a designerly approach towards a generative tool on reusable packaging, adding to current theoretical models, characterizations, and evaluations (e.g., [14, 21, 26]). This decision support tool assists in developing product-packaging combinations for reuse/refill scenarios.

2. STATE OF THE ART AND SCOPE DEFINITION

As a society, we have developed a highly efficient system in which we enable products to bridge time and distance by means of packaging, but this comes with a downside: an implicit focus on eco-efficient approaches, with a clear focus on 'eliminating packaging' [2, 6, 14]. Currently, this efficiency goes hand in hand with a significant burden on resources and increasing waste outputs [14, 27]. Critical within our current system is the single-use characteristic of much of our packaging, where products are used to satisfy a consumer's need once and thrown away right after [28]. Contrasting this, reusable packaging systems demonstrate potential environmental and economic benefits over single-use packaging systems [5, 11, 14]. Packaging systems which are focused on reusability by design [29, 30] can cut costs, adapt to individual needs, optimise operations, build brand loyalty, improve user experience, and gather intelligence [21]. In contrast, the implementation of reusable packaging systems is met with barriers and doubts as well. Examples include the actual environmental impact of reusable product-packaging combinations, value chain complexities, consumer perception, and hygiene [14, 15, 31, 32].

Reusable packaging is emerging in academic and industry focus, but it is by no means a new practice [26, 33, 34]. Historically, reusable packaging had its role in various applications [14, 16] – until recently (just a few decades ago), reusability prevailed if it was economically beneficial or sensible [3]. Only in the past decades we can observe the trend away from structural packaging reusability [14]. In packaging industry and retail environments, tangible examples of reusable packaging systems are emerging, e.g., revised deposit systems and product dispensing and refill systems ('packaging-free supermarkets') [15, 35]. The majority of classifications on reusable packaging systems build upon the reuse models framework presented by the Ellen MacArthur Foundation [21] with variations in emphasis or focus (e.g., [14, 26, 36, 37]). In addition to this, we can identify various standards, criteria, and scenarios for reusable packaging [25, 37-39]. However, these classifications and models do not structurally consider reusable packaging from a systemic perspective; a generative nature is limited, even though this is a critical point in relation to the role product-packaging combinations fulfil in supply chains [6]. Following, there is a need for designerly tools, targeting and integrating the full infrastructure, business value, and decision-making process of reusable packaging systems [25, 26].

Literature on reusable packaging demonstrates a dominant focus on business-to-business (B2B) contexts, in terms of economic and environmental impact [14, 40]. Following, this research targets business-to-consumer (B2C) as the main scope. In addition, with the established lack of generative tools, we address systemic decision-making and decision support as the key focus for development. We set out to overcome the addressed barriers, limitations, and challenges product-packaging development teams currently face.

3. TOOL DEVELOPMENT

The scope of this research covers primary (FMCG) packaging, with a focus on business-to-consumer (B2C) retail systems, since this is where we observe the major lack of current development. In addition, FMCG packaging is an important element in relation to brand equity and brand loyalty [16, 21]. The development of the tool is based on a designerly approach – essential for the development of innovative models on reusable packaging [11] – which we base on a literature research and interviews with industry stakeholders applying reusable packaging concepts in their product marketing, or considering to do so.

We consider three current packaging reuse classifications leading to build this framework on: the reusable packaging categorization by Megale Coelho et al. [14], and the reuse models frameworks by Muranko et al. [26] and the Ellen MacArthur Foundation [21]. The frameworks divide reuse models into two main categories: exclusive and sequential reuse [26]. Reusable packaging systems based on exclusive reuse are further divided into the models 'refill at home' and 'refill on-the-go' and reusable packaging systems based on sequential reuse are further divided into the models 'return from home' and 'return on-the-go' (following [21]). The different reuse models are linked to the packaging types as described by Coelho et al. [14]. Figure 1 presents the core elements of the reuse model framework.

Combining this framework with the findings and insights of six semi-structured interviews with relevant stakeholders in the FMCG sector enables the development of the reusable packaging decision support tool. This tool consists of four levels, as depicted in Figure 2. Level 1 enables users to determine the overall viability of reuse scenarios. In level 2, users explore the suitability of the four reuse models (reuse at home, refill on-the-go, return from home, and return on-the-go [21]). Level 3 determines strategic focus points for detailed development of the selected reuse model. Finally, in level 4 users explore the actual implementation from the perspectives of both the product-packaging combination as the overall system. For levels 1 and 2, the model follows a stage-gate approach [41], establishing conditional aspects to consider before continuing to levels 3 and 4.

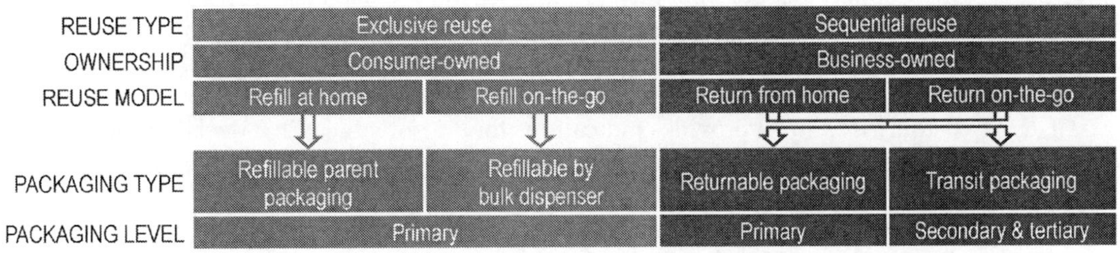

Figure 1 Combined reuse model framework (core elements, edited from [40])

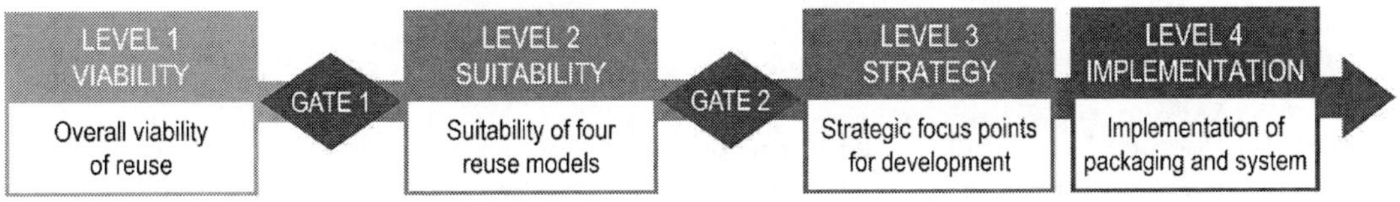

Figure 2 Level structure of the decision support tool (edited from [40])

4. ASSESSMENT AND VALIDATION

The decision support tool is designed to be applied in various kinds of B2C FMCG product-packaging development processes, for a wide range of potential stakeholders. To be able to establish an initial validation of the efficacy of the tool and the transdisciplinary nature of its application, we construct a two-fold assessment.

The first assessment is a multi-concept analysis, based on reusable product-packaging concepts developed by master students following the course Packaging Design and Management II at the University of Twente (The Netherlands). Eight student groups are invited to design reusable/refillable packaging concepts, for a branded product of their choice. With this, we can assess the value of both the designed concepts and the benefits of the decision support tool for decision-making processes, when aiming for reusable/refillable packaging concepts. It was useful to only assess level 1 of the decision support tool, since the student groups did not apply the complete tool for the development of their product-packaging concepts. The validation process is based on four perspectives by which the concepts are assessed: (1) business and brand perspective, (2) customer and consumer, (3) product and content, and (4) environment and sustainability perspective. Within each perspective, two key aspects construct a 2x2 matrix, in which the concepts are mapped. Overall, the assessment shows that the business and brand, customer and consumer, and product and content viability is reasonable to high. However, interestingly, the environment and sustainability viability is reasonably promising at best. Aligning this to the use scenario of the decision support tool, it shows the potential for assessing the conditional viability aspects of concepts – and with that: following up with level 2.

The second level of assessment is a reflection on the value of the decision support tool in relation to key factors within the realm of reusable packaging. Introspectively, we look into the key reuse factors: economics, logistics, environment, packaging design, consumer, and legislation [40]. By means of the application of the four levels of the decision support tool within product-packaging development processes, the tool provides a structured framework for the integration and alignment of these key reuse factors. The final element of the assessment builds upon the considerations by stakeholders currently applying reusable packaging concepts in their product marketing, or considering to do so. Key considerations these stakeholders face and recognize: systemic change, consumer behaviour, ease of use, retail implementation, and actual (measurable) impact [31]. Where the decision support tool does not provide clear-cut answers to these considerations, it does provide the framework to structurally implement these considerations in product-packaging development processes.

Overall, this two-fold assessment provides indications for the potential of the decision support tool in the context of development processes targeting reusable product-packaging combinations. It shows that the tool provides a good overview of the relevant decisions and considerations involved in the development process, making it suitable for use by packaging experts. Room for further improvement of the tool relates to the current high information density, and the potential use by non-packaging expert stakeholders, such as marketers [40].

5. IMPLICATIONS

With the development and initial validation of the reusable packaging decision support tool, we step into a currently underdeveloped research area. The academic implications of this research cover the alignment of theoretical insights on the potential value of reusable packaging systems and current reuse classifications with the hands-on experience of industry stakeholders applying reusable packaging concepts. With the decision support tool presented in this paper, we demonstrate the academic route to enable this alignment.

For the packaging industry at large, the implications of the decision support tool relate to its role in development processes. Where we see the challenges packaging industry stakeholders face when considering the overhaul of their packaging systems [31, 40], the decision support tool must enable product-packaging development teams to structurally consider the value of reusable packaging. When considering the three levels of sustainability integration in product packaging development processes [4], the tool primarily enables stakeholders to align the desired and achieved levels of sustainability in relation to reusable packaging concepts and systems.

However, the application of the decision support tool by no means results in an indisputable advantage for reusable packaging. Overall, reusable packaging systems have the potential to bridge time and distance with a lower environmental impact than single-use packaging systems [14, 36], but it is no one-size-fits-all solution by default. Complexities do not only come from specific product-packaging combinations, but also from the system in which these interact – retail, use, reverse logistics, and repurposing require an overhaul of their impact. For reusable packaging systems to be successful, a number of characteristics need to be aligned, such as cycle times and frequency, delivery distances, production processes, and reverse logistics [36, 37, 40, 42]. Failing to achieve a balance might result in an overall worse environmental outcome [42, 43].

An essential factor for the successful implementation of reusable packaging is the acceptance within retail environments, covering both traditional retail and e-commerce, and from the perspectives of the retailer and consumers alike. Even though research indicates customers and consumers to be willing to step away from single-use packaging [15, 36], the perception of additional efforts is a critical factor potentially hindering the acceptance of reusable packaging – a decrease in convenience might be perceived [44]. With recycling being a deeply entrenched norm for consumers, and being portrayed as a promising solution to single-use packaging waste [16, 45, 46], consumers might not actually change their purchasing behaviour unless they see a benefit [47]. Critical in this process is the perception of hygiene: customers expect reuse systems to be simple and mess-free transactions [15, 17, 35]. Since consumer acceptance of reuse systems is essential, retailers play a vital role. On the one hand, retailers might face commercial uncertainties in the transition towards reusable packaging systems [15, 48]. On the other hand, it provides valuable opportunities. Where reusable packaging could provide new ways of brand expression [14, 21, 40], this is an opportunity for supermarkets as well [15].

6. DISCUSSION AND CONCLUSIONS

This research presents a novel take on reusable packaging, with a specific focus on decision-making processes. We can identify a trend towards 'responsible packaging' [17] and packaging circularity [21], in which reusable packaging systems will play a significant role. However, when considering the current reusable packaging classifications and categorizations, we can identify a gap in relation to the practical and systematic integration of reusable packaging considerations in decision making: a limited focus on generative tools [6]. Product-packaging development processes consist of multi-stakeholder and multi-criteria challenges, for which a designerly perspective is essential [2, 4, 42]. This shapes the core of the presented decision support tool: this research presents both the preferred conditions and requirements involved in the development of reusable packaging concepts, and the translation of this into a tangible tool [40].

The decision support tool shows to provide a good overview of the decisions and considerations which are relevant in the development of reusable packaging systems, when considering the assessments and validations [40]. However, the current validations should be followed-up with additional empirical research into the application in hands-on product-packaging development processes. And with that, the iterative design process of the tool itself can be continued. An interesting route can be to explore options to step away from the current one-size-fits-all solution, and align the variety in packaging development processes with tool variations.

Where this research presents a generative approach towards reusable packaging, it does not by definition deliver 'better' solutions. Research indicates that reusable packaging has an overall lower environmental impact in comparison to single-use packaging [14, 16, 36], but this is ultimately dependent on various factors. This means that the application of the decision support tool cannot be directly linked to a reduced environmental impact by default – it will provide decision-making considerations and support, but post-development evaluations (such as LCAs [6]) will still be required [14]. This also aligns to the ways in which the tool will be implemented in packaging development practices, and the various stakeholders who will interact with it.

The decision support tool shapes a point of departure for reusable packaging development, and assists product-packaging development teams in making holistic decisions, including the aforementioned issues and challenges [40]. Reusable product-packaging development is a transdisciplinary challenge [31], where a variety of stakeholders targets a multi-criteria problem [42]. The decision support tool helps in balancing stakeholder values in relation to reusable packaging – e.g., the role of marketers in relation to other influencing stakeholders [4, 49] – going beyond the current theoretical reusable packaging classifications.

The future outlook for the decision support tool covers a number of opportunities. Firstly, as addressed, we we set out to implement and test the tool in industry-based packaging design processes, both within B2C FMCG (the current scope) and beyond. We also aim to further dive into the currently identified barriers and key considerations of industry stakeholders [31]. Finally, we see opportunities to assess the application of the tool in relation to brand-related design and marketing elements: brand expression, equity, and loyalty [37, 40], and to packaging recyclability design considerations [50], as these provide valuable design characteristics on both the product marketing and engineering dimensions of packaging development. With this, we further strengthen the decision support tool as a point of departure for reusable packaging development.

7. ACKNOWLEDGEMENTS

The authors wish to thank the student groups who provided assessment opportunities by means of their packaging concepts, and NVC Netherlands Packaging Centre and all interviewees and industry stakeholders for providing input throughout this research.

8. REFERENCES

1. Pires, A. (2021). Eco-Design and Sustainable Packaging: Challenges, Trends and Perspectives in the Food Sector. In J.R. da Costa Sanches Galvão, et al. (Eds.), Proceedings of the 1st International Conference on Water Energy Food and Sustainability (ICoWEFS 2021). Switzerland: Springer Nature Switzerland AG.
2. De Koeijer, B., I. Borgman, J. Henseler, R. ten Klooster, and J. de Lange (2021). Embedding a Sustainability Focus in Packaging Development Processes. In Y. Kishita, et al. (Eds.), EcoDesign and Sustainability II. Sustainable Production, Life Cycle Engineering and Management. Singapore: Springer Nature Singapore Pte Ltd.
3. Bruijnes, C., H. Diepenmaat, R. ten Klooster, J.P. van Soest, G. Langeveld, and V. Balk (2020). The State of Sustainable Packaging. Netherlands Institute for Sustainable Packaging (KIDV).
4. De Koeijer, B., J. de Lange, and R. Wever (2017). Desired, Perceived, and Achieved Sustainability: Trade-Offs in Strategic and Operational Packaging Development. Sustainability, 9(10), pp. 1923, DOI: 10.3390/su9101923.
5. Ellen MacArthur Foundation (2014). Towards the Circular Economy Vol. 3. Cowes, United Kingdom: Ellen MacArthur Foundation Publishing.
6. De Koeijer, B., R. Wever, and J. Henseler (2017). Realizing Product-Packaging Combinations in Circular Systems: Shaping the Research Agenda. Packaging Technology and Science, 30(8), pp. 443–460, DOI: 10.1002/pts.2219.
7. European Parliament and Council (2015). Directive 2015/720 as regards reducing the consumption of lightweight plastic carrier bags. (L 115/11).
8. European Parliament and Council (1994). Directive 94/62/EC on packaging and packaging waste.
9. Wandosell, G., M.C. Parra-Meroño, A. Alcayde, and R. Baños (2021). Green Packaging from Consumer and Business Perspectives. Sustainability, 13(3), DOI: 10.3390/su13031356.
10. Niero, M., M.Z. Hauschild, S.B. Hoffmeyer, and S.I. Olsen (2017). Combining eco-efficiency and eco-effectiveness for continuous loop beverage packaging systems: Lessons from the Carlsberg Circular Community. Journal of Industrial Ecology, 21(3), pp. 742-753, DOI: 10.1111/jiec.12554.
11. Ellen MacArthur Foundation (2017). The New Plastics Economy: Catalysing action.
12. Gardas, B.B., V.S. Narwane, and N.P. Ghongade (2021). Analyzing the Obstacles to Sustainable Packaging in the Context of Developing Economies: A DEMATEL Approach. In S.S. Muthu (Ed.), Sustainable Packaging. Environmental Footprints and Eco-design of Products and Processes. Singapore: Springer Nature Singapore Pte Ltd.
13. Steinhorst, J., and K. Beyerl (2021). First reduce and reuse, then recycle! Enabling consumers to tackle the plastic crisis – Qualitative expert interviews in Germany. Journal of Cleaner Production, 313, DOI: 10.1016/j.jclepro.2021.127782.
14. Megale Coelho, P., B. Corona, R. ten Klooster, and E. Worrell (2020). Sustainability of reusable packaging– Current situation and trends. Resources, Conservation & Recycling: X, 6(100037), DOI: 10.1016/j.rcrx.2020.100037.
15. Kramer, P., W. Meinema, and J. ter Berg (2021). Kansen om Nederlanders en Nederlandse supermarkten in beweging te krijgen om herbruikbare verpakkingen te gebruiken en introduceren. Kantar, Dutch Ministry of Infrastructure and Water Management.
16. Greenwood, S.C., S. Walker, H.M. Baird, R. Parsons, S. Mehl, T.L. Webb, A.T. Slark, A.J. Ryan, and R.H. Rothman (2021). Many Happy Returns: Combining insights from the environmental and behavioural sciences to understand what is required to make reusable packaging mainstream. Sustainable Production and Consumption, 27, pp. 1688-1702, DOI: 10.1016/j.spc.2021.03.022.
17. Mintel (2020). Global Packaging Trends 2020.
18. European Commission (2020). A new Circular Economy Action Plan - For a cleaner and more competitive Europe.
19. European Parliament and Council of the European Union (2019). Directive 2019/904 on the reduction of the impact of certain plastic products on the environment. Official Journal of the European Union, L 155/1.
20. European Commission (2018). A European Strategy for Plastics in a Circular Economy.
21. Ellen MacArthur Foundation (2019). Reuse – Rethinking Packaging.
22. Ellen MacArthur Foundation (2022). The Plastics Pact Network. https://ellenmacarthurfoundation.org/the-plastics-pact-network (accessed: March 22, 2022).

23. Kravchenko, M., D.C.A. Pigosso, and T.C. McAloone (2021). A Trade-Off Navigation Framework as a Decision Support for Conflicting Sustainability Indicators within Circular Economy Implementation in the Manufacturing Industry. Sustainability 13(1), pp. 314, DOI: 10.3390/su13010314.
24. World Commission on Environment and Development (1987). Our Common Future. Oxford, United Kingdom: Oxford University Press.
25. Kuzmina, K., S. Prendeville, D. Walker, and F. Charnley (2019). Future scenarios for fast-moving consumer goods in a circular economy. Futures, 107, pp. 74-88, DOI: 10.1016/j.futures.2018.12.001.
26. Muranko, Ż., C. Tassell, A. Zeeuw van der Laan, and M. Aurisicchio (2021). Characterisation and Environmental Value Proposition of Reuse Models for Fast-Moving Consumer Goods: Reusable Packaging and Products. Sustainability, 13(5), DOI: 10.3390/su13052609.
27. PlasticsEurope (2019). The Circular Economy for Plastics - A European Overview.
28. Zeeuw van der Laan, A., and M. Aurisicchio (2019). Archetypical consumer roles in closing the loops of resource flows for Fast-Moving Consumer Goods. Journal of Cleaner Production, 236, DOI: 10.1016/j.jclepro.2019.06.306.
29. European Parliament and Council (2008). Directive 2008/98/EC on waste and repealing certain Directives. L 312/3.
30. Ellen MacArthur Foundation (2013). Towards the Circular Economy Vol. 2. Cowes, United Kingdom: Ellen MacArthur Foundation Publishing.
31. De Koeijer, B., and M. Mulder-Nijkamp (2021). Reusable packaging. Packalicious online session.
32. Boz, Z., V. Korhonen, and C. Koelsch Sand (2020). Consumer Considerations for the Implementation of Sustainable Packaging: A Review. Sustainability, 12, pp. 2192, DOI: 10.3390/su12062192.
33. Sustainable Packaging Alliance (2002). Towards Sustainable Packaging - A Discussion Paper.
34. Lofthouse, V.A., and T.A. Bhamra (2006). Refillable Packaging Systems: Design Considerations. International Design Conference - Design 2006, pp. 1391-1398. Dubrovnik, Croatia.
35. Koetsier, D.A. (2022). A study into sustainable food chain practices, focused on food dispenser development. Industrial Design Engineering, Faculty of Engineering Technology, Department of Design, Product and Management. University of Twente (Enschede, The Netherlands).
36. Megale Coelho, P., B. Corona, and E. Worrell (2020). Reusable vs single-use packaging - A review of environmental impacts. Reloop & Zero Waste Europe.
37. PackBack (2020). Standardisation in reusable food packaging. A research exploring the potential of standardisation in primary reusable food packaging to enable and stimulate the widespread transition to reusable packaging systems in the food industry. Netherlands Institute for Sustainable Packaging (KIDV).
38. European Committee for Standardisation (CEN) (2004). EN 13429:2004 Packaging — Reuse.
39. EUROPEN (2006). Understanding the CEN Standards on Packaging and the Environment: Some Questions and Answers (4th edition).
40. Hesseling, I. (2022). From single-use to reuse: development of a decision support tool for FMCG packaging. Industrial Design Engineering, Faculty of Engineering Technology, Department of Design, Product and Management. University of Twente (Enschede, The Netherlands).
41. Cooper, R.G. (1990). Stage-gate systems: A new tool for managing new products. Business Horizons, 33(3), pp. 44-54, DOI: 10.1016/0007-6813(90)90040-I.
42. Ten Klooster, R. (2019). Food Packaging and Sustainability: A Designer's Perspective. Reference Module in Food Science, DOI: 10.1016/B978-0-08-100596-5.22645-0.
43. Lewis, H., K. Verghese, and L. Fitzpatrick (2010). Evaluating the sustainability impacts of packaging: the plastic carry bag dilemma. Packaging Technology and Science, 23(3), pp. 145-160, DOI: 10.1002/pts.886.
44. Beitzen-Heineke, E.F., N. Balta-Ozkan, and H. Reefke (2017). The prospects of zero-packaging grocery stores to improve the social and environmental impacts of the food supply chain. Journal of Cleaner Production, 140, pp. 1528-1541, DOI: 10.1016/j.jclepro.2016.09.227.
45. Van Doorn, J., and T. Kurz (2021). The warm glow of recycling can make us more wasteful. Journal of Environmental Psychology, 77, DOI: 10.1016/j.jenvp.2021.101672.
46. Thomas, C., and V. Sharp (2013). Understanding the normalisation of recycling behaviour and its implications for other pro-environmental behaviours: A review of social norms and recycling. Resources, Conservation and Recycling, 79, pp. 11-20, DOI: 10.1016/j.resconrec.2013.04.010.
47. Kunamaneni, S., S. Jassi, and D. Hoang (2019). Promoting reuse behaviour: Challenges and strategies for repeat purchase, low-involvement products. Sustainable Production and Consumption, 20, pp. 253-272, DOI: 10.1016/j.spc.2019.07.001.
48. Ubirajara Gustavo Jr., J., G. Medeiros Pereira, A.J. Bond, C.V. Viegas, and M. Borchardt (2018). Drivers, opportunities and barriers for a retailer in the pursuit of more sustainable packaging redesign. Journal of Cleaner Production, 187, pp. 18-28, DOI: 10.1016/j.jclepro.2018.03.197.
49. Ten Klooster, R., and B. De Koeijer (2016). Utilising the Potential of Design Briefs in Sustainable Packaging Development. 20th IAPRI World Conference on Packaging, pp. 200-207. Campinas, Brazil.
50. Meijer, S., B. de Koeijer, and M. Alves da Motta Filho (2021). Design for and from Recycling: Towards a designer guide for plastic packaging. 30th IAPRI member conference.

ST-O05

Investigation of the Consumer Perceptions of e-Commerce Packaging

Eduardo Molina[1]*, Laszlo Horvath[1]

[1]Department of Sustainable Biomaterials, Virginia Tech, Blacksburg, 24061, USA

***Correspondence to:** Department of Sustainable Biomaterials, Virginia Tech, Blacksburg, 24061, USA.*
E-mail: molina@vt.edu

ABSTRACT: Exponential growth of ecommerce, along with global changes in consumer expectations caused by the coronavirus pandemic, have placed a significant pressure in the packaging design processes. The objective of the research was to investigate the consumer perceptions of e-commerce packaging and identifying packaging factors that customer value the most. A total of 1,949 respondents in the United States were surveyed and their answers compared between packaging professionals and general consumers. It was identified that consumers value highly technical performance of packaging and were not interested in the experiential components of the packaging design. Packaging professionals valued the unboxing experience as a critical aspect, reflecting the differences between the industry and the consumers' expectations. The survey results provide a detailed insight of consumers' viewpoints and experiences, offering valuable information to be used in data-driven packaging development and innovation. A tool to quantitatively assess the packaging design suitability for e-commerce, based on maximizing a positive consumer experience, is proposed.

Keyword: ecommerce, consumer perceptions, packaging assessment.

1. INTRODUCTION

Retail e-commerce sales in the United State represented an estimate of 19% of the total retail sales in 2021. Although the share of total sales through e-commerce channels was increasing constantly in the last decade, a significant growth was experienced during 2020, when the Covid-19 pandemic began [1], [2]. Retail e-commerce sales refer to the sales conducted through internet services and delivered directly to the consumers, instead of the traditional retail sales which involve a consumer traveling to a brick-and-mortar location and acquiring the product.

E-commerce growth presents an opportunity for companies to reach a wider audience, but at the same time new risks and challenges arise. Research has been conducted to compare the sustainability of e-commerce packaging, as well as to compare the energy consumption of the traditional retail trade channels with e-commerce [3]–[8]. The logistics systems of either distribution channel have been widely studied and mapped, developing packaging simulation tests such as those by the International Safety Transit Association (ISTA). These studies and standardize testing methods focus exclusively on the evaluation of the suitability of packaging solutions based on performance metrics, such as the ability to deliver the products without damages.

In an effort to optimize the packaging systems, studies have also been conducted with the goal of minimizing the volume of the packages [9]. Past reviews of packaging design methodologies include extensive information regarding the different goals of the designers [10]. Although marketing goals are considered critical in the design processes, not enough information is available regarding the consumers' perceptions of e-commerce packaging.

When a package is evaluated, different elements can be assessed. Focusing on the correct elements increases the probability of success of the design.

According to Regattieri and Santarelli [11] the elements of packaging within the e-commerce supply chain can be broken down into three main categories: marketing, environment, and flow. The marketing function of a package aids in creating value and branding for the product and the company. The environment function focuses on the minimization of negative environmental impacts caused by the packaging decisions, and the flow function promotes areas such as better packaging logistics, handling, and efficiency. Within each of these three main areas of packaging, there are many packaging elements that play a role in the success or failure of a package for e-commerce. These researchers developed a questionnaire to understand perceptions of packaging elements by Italian companies and customers, allowing for a quantitative analysis of qualitative packaging elements [11]. The Regattieri and Santarelli method for translating qualitative features of a package into quantitative values through a questionnaire was adapted into a more appropriate format for this research study, which is further explained.

The goal of this study is to be able to characterize the consumer perceptions relating to e-commerce packaging. With this information, the authors proposed a tool to quantitatively assess the packaging design suitability for e-commerce, focused on how to maximize the consumer experience.

This paper is organized as follows. After the introduction, the materials and methods section explains in depth the methodology followed to conduct the survey as well as the type of information collected. The results and discussion section summarizes the responses collected and discusses the main findings. In the final section, conclusions are presented.

2. MATERIALS AND METHODS

2.1 Survey

A survey was created to develop an understanding of how consumers perceive the value of different packaging elements within the context of e-commerce. Questions 1-4 were intended to identify the demographic factors of the respondents (age, gender, ethnicity, and education). Question 5 asked about the respondents purchasing habits regarding e-commerce retail and question 6 and 7 were intended to identify and respondents that are professionals in a packaging-related field to correspondingly categorize their answers.

The survey consisted of three main sections. Each situational question asked about the importance of a fundamental packaging element related to e-commerce packages. The elements were grouped into three main areas: Packaging Functions (10 questions), Marketing (6 questions), and Environment (8 questions). The packaging functions section consisted of questions related to the performance of the packaging and its appearance upon delivery. The marketing questions focused on the aesthetic elements of the packaging and how consumers interact with them. The last section, environment, contained questions about packaging sustainability and if that would affect the consumer's perception and decisions about their purchases.

Each question was directly related to an element of the packaging design and functions. For example, one important packaging element identified was "Tamper protection". To evaluate the consumers' perception regarding tamper protection, they were asked "How important is it to you that when you receive an online order you can distinctly tell if the package has been opened or tampered with (example: noting broken seals, tape, etc.)?" A similar process was followed for each of the 22 elements identified as critical in the packaging design and engineering process. The evaluated elements are shown in Table 1, including the section to which they correspond between marketing, basic packaging functions and the environment.

Table 1 Packaging design and engineering elements evaluated in the consumers' perceptions survey per category of study.

Evaluated packaging elements		
Packaging Functions	**Marketing**	**Environment**
Appearance protection	Unboxing experience	Reusability
Handleability	Brand recognizability	Validating certification
Weight efficiency	Attractiveness	Renewability
Unboxing function	Brick-&-mortar similarity	Renewability of materials
Volume efficiency		
Containment	Novelty	Material post-use
Function protection	Doorstep discretion	Recyclability
Apportionment		
Resistance to climatic influences		
Tamper protection		

Respondents were asked to consider their past experiences with e-commerce, what they liked, and what they disliked. At the end of the survey, respondents were explicitly asked to rate the importance of each of the three main packaging areas. This allowed for a comparison between the perceived importance of each category and the assigned importance when answering each survey question. A free response question was also included; this allowed for the respondent to provide examples of a package they interacted with in the past that made an impression on them. The aggregate responses that were originally qualitative representations of importance were then translated into a quantitative format. The conversion of answer choices to numerical values is shown in Table 2.

Table 2 Survey response choices and their correlating quantitative value used in translating the data.

Survey Response Choices	**Correlating Quantitative Importance**
Extremely Important	5
Very Important	4
Moderately Important	3
Slightly Important	2
Not at all important	1

2.1.1. Distribution

This survey was distributed to the public digitally between April 16 and April 26, 2021. Researchers relied on releasing the survey on online social media sites to reach potential respondents. As an incentive, a gift card was offered and awarded to three respondents chosen randomly; this was detailed when distributing the survey. Undergraduate students shared the survey on personal social media pages and promoted it within social media groups that were associated with the packaging industry, Virginia Tech, and survey sampling on Facebook, Reddit, Instagram, and LinkedIn. This limited the breadth of the survey's data primarily to those that interact with Virginia Tech affiliated content, packaging-based content, and those who interact with the survey's creators on social media. Some individuals on the social media sites reshared the survey to a larger audience, but it is unknown how many people responded to the survey from each specific online source.

2.1.2. Bias

Many steps were taken to ensure that the data collected was accurate and as unbiased as possible. All questions began with the phrase "you receive an online order," as to remind the reader that this survey is specific to e-commerce. The consistency in phrasing allowed for respondents to easily understand the context of each question. Additionally, no duplicate data was allowed, to ensure that the data collected was accurate. This meant that the platform in which the survey was taken, Qualtrics [https://www.qualtrics.com], would identify if a respondent was answering from the same device to prevent receiving data multiple times from a single respondent. This precaution was necessary given the monetary incentive.

Both packaging-related industry professionals and consumers were asked to take the survey. Packaging industry professionals were asked to specify how long they had been in the field. The data from industry professionals and consumers have been assessed both together and separately for any noticeable trends and to investigate future opportunities for the data usage.

2.2 Packaging evaluation tool

The data from the accepted responses communicates how important each of the packaging elements are to consumers. Knowing the average importance value for each of the packaging elements referenced in the survey will allow for a package grading method to be developed. This grading is an assessment of the package's suitability for a redesign. The assessment tool can be used for a future study focusing on the redesign of a package for e-commerce using paper-based materials. Understanding if a package is comparatively in need of a redesign, is an essential step that should come prior to redesigning a package, to ensure that time, energy, and resources are appropriately allocated.

3. RESULTS AND DISCUSSION

3.1 Demographic Profile of Respondents

There was a total of 2,139 recorded responses to the survey. Controls were implemented to ensure that all responses used in the analysis were legitimate and unique. Any responses using the same name or email address were manually excluded. Any survey responses that took less than 126 seconds to complete were excluded from the analysis. It was estimated that the fastest time to fill out the survey, with an accurate understanding of the questions, was an average of three seconds per each of the 42 questions, and thus any responses that were submitted in less than that time were not counted as reliable answers. After excluding the duplicate and unreliable responses, the total number of valid respondents were 1,947.

The demographics of the respondents were compared to the population of the United States of America, as shown in Table 3. There was a higher number of responses from people ranging from 18 to 45 in comparison to their prevalence in the US population, according to the 2020 United States Population Census [12].This is likely a result of the exclusively online method used to circulate the survey. According to a report on the demographics of social media users [13], an average of 82.5% of people between 18 and 49 uses at least one social media site, while only 59% of people over the age of 50 use social media.

When asked about their purchasing patterns, 35 % of the survey respondents identify themselves as conducting online purchases delivered at home or at the place of work 2 to 3 times per month on average, 87.5% of them purchased at least once per month or more times through e-commerce channels. Figure 2 shows the frequency distribution of the purchasing patterns of the sampled population.

While filling out the survey, respondents were asked to specify whether or not they worked in the packaging industry. 50.53% of the respondents were members of the packaging industry while the other 49.56% of respondents were not. The high percentage of packaging industry professionals was a potential result of the distribution methods relying heavily on those already interested in the packaging industry. As shown in Figure 3, out of all the respondents identified as working in the packaging industry, 40.8% said to have less than three years of work experience, while 44.8% had three to six years of work experience.

Table 3 The survey sample's age demographics compared to US population age demographics.

Age Range	Survey Demographics		2020 US Census Demographics	Difference (%)
	Total	Percentage	Percentage	
18-30	843	43.79	22.45	21.34
30-45	755	39.22	25.23	13.99
45-65	277	14.39	32.78	-18.39
65+	50	2.60	19.54	-16.95

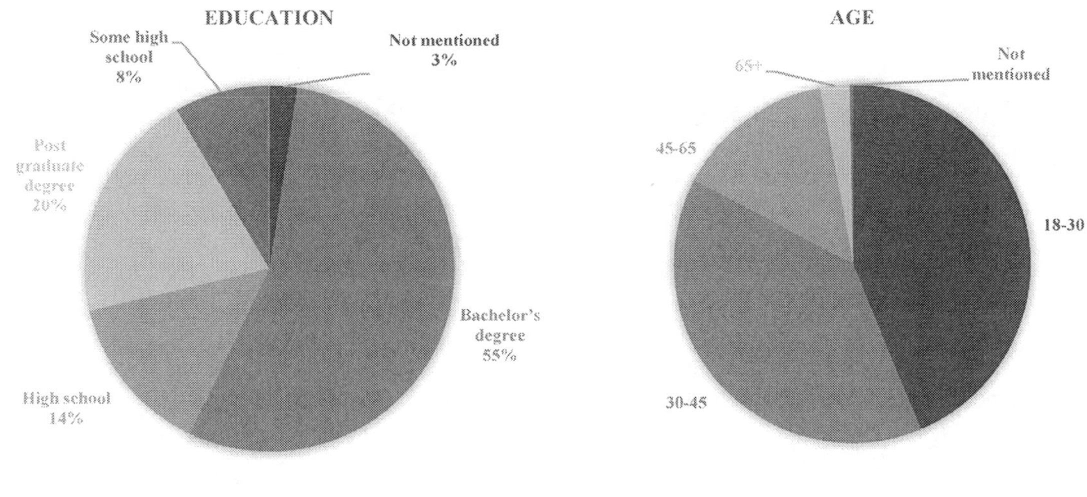

(a) (b)

Figure 1 Demographics profile of respondents participating in the survey based on (a) education, and (b) age.

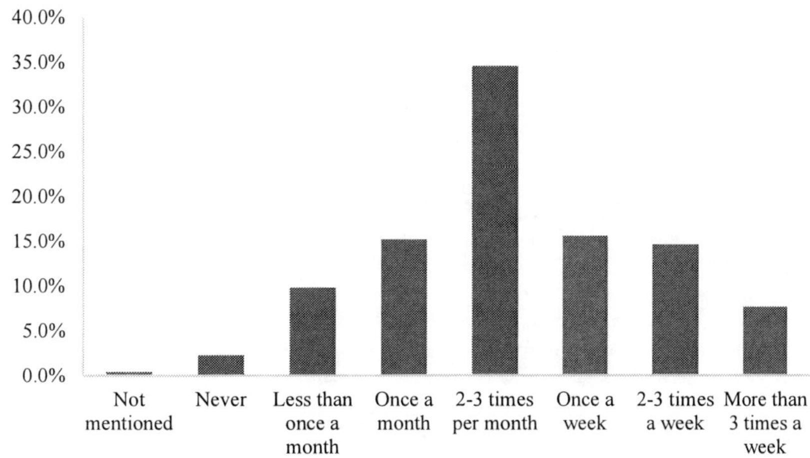

Figure 2 Frequency of e-commerce purchases conducted by the survey respondents.

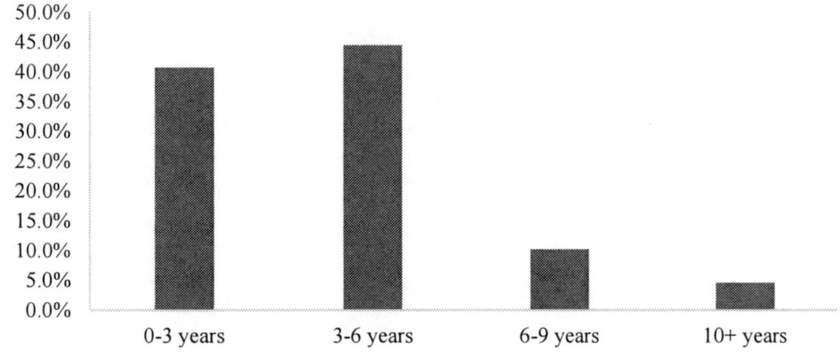

Figure 3 Years of experience in a packaging-related field from the survey respondents identified as packaging professionals.

3.2 Consumer perceptions of e-commerce packaging characteristics

Figure 4, Figure 5, and Figure 6 show the averages of the importance level assigned by the respondents to each of the areas evaluated. Figure 4 shows the responses corresponding to the ten fundamental packaging functions. Consumers evaluated “tamper protection” as the most important feature, with an average evaluation of 4.19 out of 5 possible points. This question evaluated the importance of designing packages that clearly show if the package has been opened at any stage before reaching the final consumer. On the opposite end of the spectrum consumers evaluated “appearance protection” as the least importance characteristic. For this element, consumers were asked about the importance of receiving a package that is not visibly damaged but that was appropriately protected the main product. Overall, it is clear that consumers value higher the functions directly related to the utility of a package, that is, to deliver a product safely, significantly higher than functions more closely related to perception and experience.

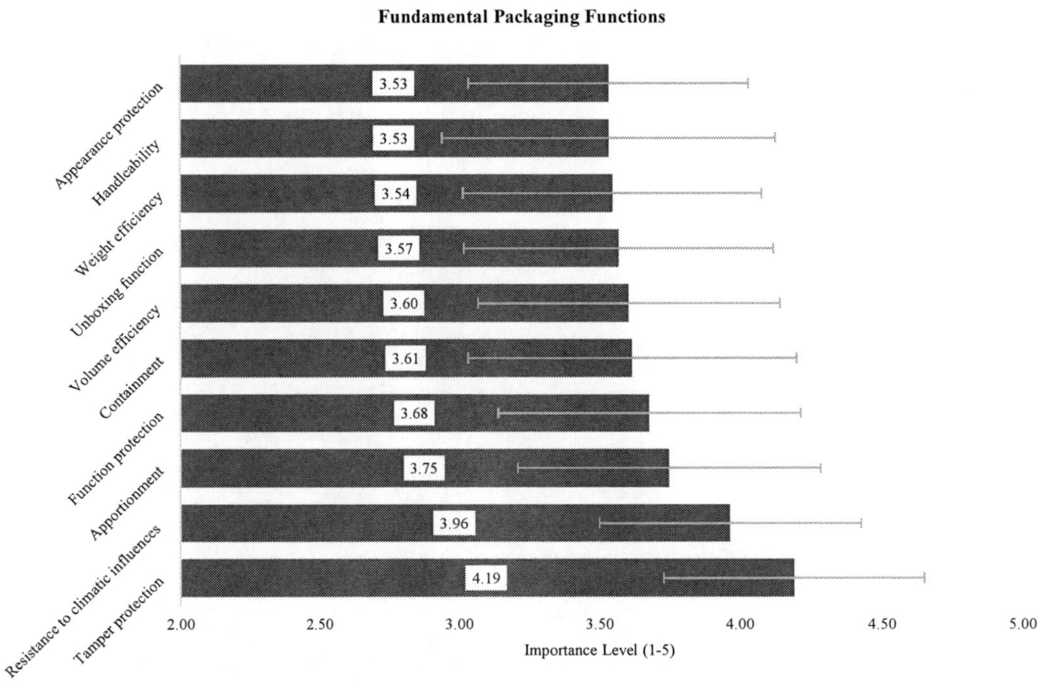

Figure 4 Average of importance assigned by the respondents to each of the elements related to fundamental packaging functions. Whiskers represent the standard deviation of the responses.

Figure 5 summarizes the importance level assigned to functions related to the marketing objectives of packaging. The highest value was assigned to “doorstep discretion” (3.79 out of 5). This element refers to the importance of concealing the products that are being delivered, be it concealment for privacy or for security reasons. Given the geographical limitation of the survey to the United States, consumers expect for their packages to be placed outside their doorsteps for a significant period of time. During this time, packages that make evident the contents could be more prone to theft or might reveal private consumer habits, none of which is desired by the survey respondents.

At the lower end of the importance spectrum, consumers minimized the relevance of factors such as “unboxing experience” and “brand recognizability”. This can be explained through the fact that consumers have already made the purchasing decision in advance of receiving the product, thus are not valuing the experiential goals of the packaging solution. Analyzing the survey responses without taking into consideration the demographics and professional experience of the respondents ignores potential significant differences in the responses. Although the "unboxing experience" was ranked the lowest among all the marketing functions, when breaking it down on whether the respondents were professionals in a packaging-related industry or not, packaging professionals valued the unboxing experience at 3.68, while regular consumers valued it at an abysmally low 2.78 out of 5. The analysis of the responses collected shows that companies could be overvaluing the customers' experience during unboxing, increasing packaging costs while not directly improving the overall perception of the packaging and the product. The alternative analysis can be supported by stating that consumers are not willing to pay a higher purchasing price for a better unboxing experience, but having one might increase their likelihood of repeating a purchase or using the same vendor, albeit unknowingly.

Consumers were asked two questions related to the unboxing section of the e-commerce purchasing process. One question asked about the importance of receiving a package that was easy to open without the need for additional tools. This question, labeled "unboxing function" was categorized as part of the basic packaging functions elements and rated 3.56, near the sample average of importance, 3.60. The second question was the "unboxing experience" previously discussed, which evaluated the importance of having a positive unboxing experience, even if it meant an increased purchasing price. Figure 7 compares the responses by age group for these two questions. It can be seen that the importance of making the process of opening a package simple is constant throughout the different age groups, while the unboxing experience (marketing function) is important for younger consumers, but not the older respondents.

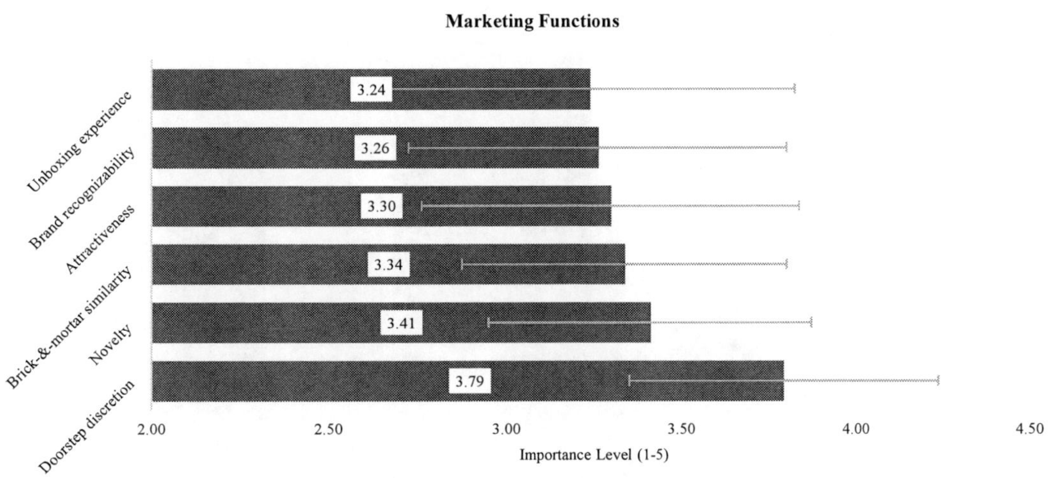

Figure 5 Average of importance assigned by the respondents to each of the elements related to marketing-related functions. Whiskers represent the standard deviation of the responses.

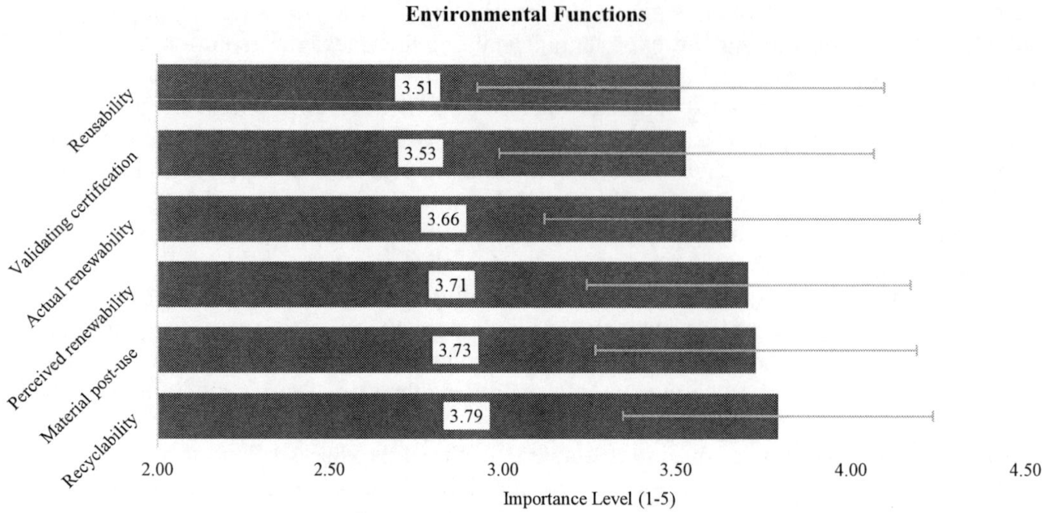

Figure 6 Average of importance assigned by the respondents to each of the elements related to environmental-related functions. Whiskers represent the standard deviation of the responses.

As an average, respondents valued the environmental elements consistently as moderately or very important, as shown in Figure 6. The highest valued element was "recyclability" of the package received, which evaluated the importance of receiving 100% curbside recyclable packages with their online purchases. The lowest valued element was the "reusability" of the packaging solution, that is, the possibility of serving a secondary purpose after receiving the product.

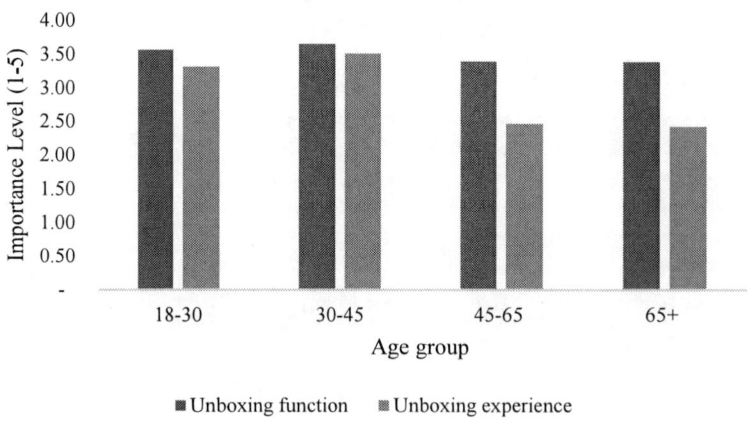

Figure 7 Average of importance for the unboxing functionality and the unboxing experience by age groups.

Overall, based on the importance assigned to each element, survey respondents considered that the fundamental packaging functions were the most important category that needed to be served when designing packaging solutions, followed by environmental aspects and lastly, marketing functions. This supports the pattern of answers followed by the individual element analysis. Respondents are interested more in packaging that fully fulfil the minimum expectations of a package than attractive solutions that will provide a nicer, more memorable experience.

Comparing the responses by those self-identified as having work experience in a packaging related field and those who are not, most of the elements were rated with a similar importance level. Table 4 shows the 9 elements with the biggest differences among the two groups of respondents. It can be seen that the three categories with higher discrepancies in the responses are "attractiveness", "brand recognizability" and "unboxing experience". All of these 3 survey questions evaluated marketing functions regarding the importance of a positive experience, such as receiving a package with nice graphics, attractive unboxing experience, and clear, recognizable branding. Those in the packaging field, being part of the brand owners, evaluate significantly higher these elements. Regular consumers seemed not to be concerned with these appearance themed elements. Further research should be conducted to fully explain the importance of these elements.

3.3 Packaging assessment tool development

An assessment tool for package suitability was developed using data from the survey and research conducted. The average importance rating of each element was extracted from the survey responses to create a tool titled "*Package Suitability for E-Commerce Redesign Assessment*. This two-part assessment tool requires at least two packages, as it comparatively evaluates a package's need for redesign. In the first part, each element of a package must be rated from 1 to 10, based on its ability to fulfill the functions as described in each category. A rating of 1 assumes not capable of fulfilling the expected function and 10 is given to packages that completely fulfills the function. In the second part, each rating is multiplied by its average weighted importance extracted from the survey data. Then, the resulting multiplications are summed, with the highest possible total being 591.7 and the lowest possible total being 59.17. The package that receives the lowest overall total is the package that is most suitable for a redesign.

Table 4 Average importance evaluation responses comparing industry participants and general consumers for a select group of packaging elements.

Average Importance Level (1-5)

	Packaging-related role or industry		
	Yes	No	Difference
Handleability	3.79	3.27	15.9%
Appearance protection	3.86	3.20	20.6%
Validating certification	3.79	3.26	16.3%
Reusability	3.77	3.24	16.4%
Novelty	3.78	3.03	24.8%
Brick-&-mortar similarity	3.70	2.97	24.6%
Attractiveness	**3.77**	**2.81**	34.2%
Brand recognizability	**3.73**	**2.78**	34.2%
Unboxing experience	**3.68**	**2.78**	32.4%

There are limited applications of this assessment tool, as the goal is to evaluate how well a package is designed to fulfill the consumer's expectations of e-commerce purchases and deliveries. It cannot be used to assess a package that hasn't experienced an e-commerce distribution system or testing to simulate e-commerce distribution.

4. CONCLUSIONS

This study investigated the consumer perceptions related to the ecommerce packaging experience. Obtaining data from a very large sample provided valuable information for packaging designers looking to optimize their designs based on the improvement of the consumer experience.

It can be concluded that consumers, when receiving e-commerce packages, value the fundamental packaging functions higher than sustainability and marketing functions. Factors such as tamper evidence and doorstep discretion where consistently assigned a higher importance by respondents. A significant difference in the responses was identified when segregating the responses by age groups as well as whether the respondent was related to the packaging industry. From the responses, it can be concluded that packaging professionals assign a higher value to the elements related to the experience, such as unboxing and package attractiveness, while the consumers valued higher those elements that focused on the functionality aspects of the packaging system.

A package assessment tool was developed in order to use the survey results as part of the packaging development workflow.

5. REFERENCES

[1] J. Young, "US ecommerce grows 14.2% in 2021," *Digital Commerce 360*, 2022. https://www.digitalcommerce360.com/article/us-ecommerce-sales/ (accessed Mar. 01, 2022).

[2] U.S. Department of Commerce, "QUARTERLY RETAIL E-COMMERCE SALES 4 th QUARTER 2021," U.S. Department of Commerce, Washington, D.C., Jan. 2022.

[3] S. Escursell, P. Llorach-Massana, and M. B. Roncero, "Sustainability in e-commerce packaging: A review," *J. Clean. Prod.*, vol. 280, p. 124314, Jan. 2021, doi: 10.1016/j.jclepro.2020.124314.

[4] H. Pålsson, F. Pettersson, and L. Winslott Hiselius, "Energy consumption in e-commerce versus conventional trade channels - Insights into packaging, the last mile, unsold products and product returns," *J. Clean. Prod.*, vol. 164, pp. 765–778, Oct. 2017, doi: 10.1016/j.jclepro.2017.06.242.

[5] H. Buldeo Rai, "The net environmental impact of online shopping, beyond the substitution bias," *J. Transp. Geogr.*, vol. 93, p. 103058, May 2021, doi: 10.1016/j.jtrangeo.2021.103058.

[6] A. Muñoz-Villamizar, J. C. Velázquez-Martínez, P. Haro, A. Ferrer, and R. Mariño, "The environmental impact of fast shipping ecommerce in inbound logistics operations: A case study in Mexico," *J. Clean. Prod.*, vol. 283, p. 125400, Feb. 2021, doi: 10.1016/j.jclepro.2020.125400.

[7] H. Buldeo Rai, C. Broekaert, S. Verlinde, and C. Macharis, "Sharing is caring: How non-financial incentives drive sustainable e-commerce delivery," *Transp. Res. Part Transp. Environ.*, vol. 93, p. 102794, Apr. 2021, doi: 10.1016/j.trd.2021.102794.

[8] K. Mommens, H. Buldeo Rai, T. van Lier, and C. Macharis, "Delivery to homes or collection points? A sustainability analysis for urban, urbanised and rural areas in Belgium," *J. Transp. Geogr.*, vol. 94, p. 103095, Jun. 2021, doi: 10.1016/j.jtrangeo.2021.103095.
[9] J. Brinker and H. I. Gündüz, "Optimization of demand-related packaging sizes using a p-median approach," *Int. J. Adv. Manuf. Technol.*, vol. 87, no. 5–8, pp. 2259–2268, Nov. 2016, doi: 10.1007/s00170-016-8630-4.
[10] A. Azzi, "Packaging Design General Framework and Research Agenda," *Packag. Technol. Sci.*, vol. 23, no. May, pp. 253–266, 2010, doi: 10.1002/pts.
[11] A. Regattieri and G. Santarelli, "The Important Role of Packaging in Operations Management," in *Operations Management*, M. Schiraldi, Ed. InTech, 2013. doi: 10.5772/54073.
[12] U. C. Bureau, "National Demographic Analysis Tables: 2020," *Census.gov*. https://www.census.gov/data/tables/2020/demo/popest/2020-demographic-analysis-tables.html (accessed Feb. 16, 2022).
[13] 1615 L. St NW, S. 800 Washington, and D. 20036 U.-419-4300 | M.-857-8562 | F.-419-4372 | M. Inquiries, "Social Media Fact Sheet," *Pew Research Center: Internet, Science & Tech*. https://www.pewresearch.org/internet/fact-sheet/social-media/ (accessed Apr. 16, 2022).

ST-P01

Bibliometric Literature Review on Sustainable Packaging

MELEKŞEN AKIN[1]*, Sadiye Peral Eyduran[2], Victoria Krauter[3]

[1]*Department of Horticulture, Iğdır University, Iğdır 76 000, TURKEY*
[2]*Department of Horticulture, Muğla Sıtkı Koçman University, Fethiye, TURKEY*
[3]*Section Packaging and Resource Management, Department Applied Life Sciences, FH Campus Wien, University of Applied Sciences, 1030 Vienna, AUSTRIA*

***Correspondence to:** Department of Horticulture, Iğdır University, Iğdır 76 000, TURKEY*
akinmeleksen@gmail.com

ABSTRACT: We conducted a bibliometric analysis on scientific documents related to sustainable packaging. The literature was extracted from Web of Science database including restrictions on document type, language, and index which resulted in 220 documents published until 2021. The bibliographic data were analyzed utilizing Bibliometrix package available on R statistical software. The first document on sustainable packaging was published in 2006, and there was an increased trend of publications over time (7.11%). The core group consisted of seven out of 98 journals with Packaging Technology and Science, followed by Journal of Cleaner Production, and Sustainability publishing the highest number of documents in the field. The United States and Italy made the largest contribution on sustainable packaging research followed by India, Brazil, China, etc. All of the publications of France and Vietnam on the topic were in collaboration with other countries, whereas Denmark, Finland and Romania had only single country publications based on corresponding author country. Collaboration network analysis was performed to describe relationships between top countries on the area, which resulted in five distinct collaboration groups. The most frequent keywords projected with the word cloud after sustainable packaging, packaging and sustainability was circular economy. There was a shift from more general to more specific research topics over time as sustainable packaging and circular economy with a trend on biodegradable and mechanical properties emerged as unique terms beside packaging. The objective of this study is to evaluate conceptual, intellectual and social characteristics defining sustainable packaging, as well as to project research trends in this area performing descriptive and retrospective bibliometric analysis.

Keyword: bibliometrix, network analysis, science mapping, sustainable packaging, thematic evolution

1. INTRODUCTION

Sustainability is not what currently describes human activity on our planet. In fact, humankind faces severe problems in multiple dimensions: ecological, economic and social [1-4]. It is to be welcomed that these topics are increasingly gaining space in public and political discourse. And a reversal of the trend is hopefully in sight. A prime example of international efforts are the 17 Sustainable Development Goals (SDGs) of the United Nations, which aim to take action for people, planet, prosperity and, beyond that, peace and partnership. Looking into these goals, it quickly becomes clear that nutrition and environment are of outstanding importance [1].

Accordingly, work is underway diligently to both determine the true impact of our food systems [3, 5, 6] and to find appropriate mitigation measures [7-9]. The interdisciplinary field of food science and technology plays a particularly important role in this endeavor, focusing on the key points of a product's journey from farm to fork as well as food quality, safety and security. However, this includes not only the food itself, but increasingly also the packaging, which can be seen as an integral part of the same and has been shown to be a key lever in terms of sustainability [10-14]. Without packaging and its basic service functions (containment, protection, communication and convenience) [15-17], our modern society as well as food systems would not be possible in their current form [10]. In order to make this system more sustainable, in recent decades a wide variety of stakeholders have set themselves the goal of focusing not only on the pure effectiveness of packaging, but increasingly also on efficiency, circularity and safety aspects. This is reflected in numerous publications, guidelines, methods, criteria and corresponding indicators, frameworks and legal texts [11, 13].

Against this background, the aim of this bibliometric review is to find out what has actually been the focus of the scientific activity in the field of sustainable packaging in recent decades, what trends there have been and what influence different journals and countries have had. Based on this, valuable conclusions can be drawn as to which areas are already very well researched and thus support the transformation towards a more sustainable future and which areas still show research potential.

2. MATERIALS AND METHODS

Bibliometrics is a literature review technique based on quantitative analysis and statistics of scientific documents. This approach provides quantitative evaluation of publications to track research progress over time and trends in a scientific community together with showing most active countries, institutions and researchers [18-20].

The most well-known two scientific literature databases are Web of Science (WoS) from Clarivate Analytics and Scopus from Elsevier [21]. We used WoS for data collection using the search string "sustainable packag*" in the topic field with restrictions on year (documents published before 2022), language (English only) and WoS index (SCI-Expanded only). The search was performed on February 13th, 2022. Bibliometric analysis was performed using Bibliometrix package and Biblioshiny interface in R studio [18, 22] to assess scientific production in sustainable packaging. Descriptive analysis on journals and countries was performed. A word cloud showing most frequent author's keywords was projected. Thematic evolution map was produced with author keywords to follow the research evolution over various time periods choosing time cutting points in accordance with annual scientific publication distribution. Collaboration network analysis was run to show distinct collaboration groups between countries [18].

3. RESULTS AND DISCUSSION

The bibliometric data were downloaded from various sources (journals, books, etc.) published until 2022. There were 200 documents consisting mainly of articles (181), followed by reviews (35) and proceeding papers (4). The first document on "sustainable packaging" was published in 2006 according to our data collection, and after this year there was an increasing annual growth rate of 7.11% until 2022 year (Fig. 1). There were sharp increases of publications in 2010 (from 2 to 9 documents), 2016 (from 5 to 15 documents) and in 2020 year (from 33 to 70 documents). A bibliometric study conducted on "food packaging material" for the 1996-2016 period showed increasing publication trend as well [23]. Another systematic review was performed on "packaging" and "marketing" for the period 1956-2019, which showed fluctuating increase over the observed years [24]. The international efforts toward sustainability [25], as well as the consumer recycling behavior and a larger awareness of the environmental impact of food packaging [26] might invigorate this increasing publication trend.

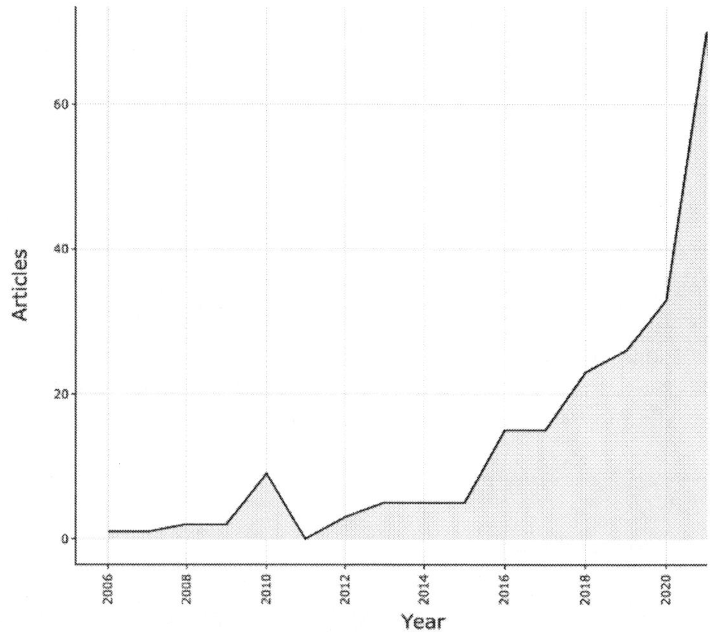

Figure 1: Annual scientific production on sustainable packaging field.

The core group consisted of seven out of 98 journals according to Bradford's Law [27] with Packaging Technology and Science (19 documents), followed by Journal of Cleaner Production (17 documents), and Sustainability (13 documents) publishing the highest number of documents in the field (Fig. 2 and 3). As mentioned above the total number of journals is 98, which is pretty high, implying that many journals started to publish "sustainable packaging" research showing the emerging popularity of this area. Nemat et al. (2019) stated that food packaging design literature on consumer recycling behavior was published in a large number of journals with various focus on topics including marketing, food, logistics, and waste management indicating that waste management, prevention and recycling are within the scope of both packaging and food research fields.

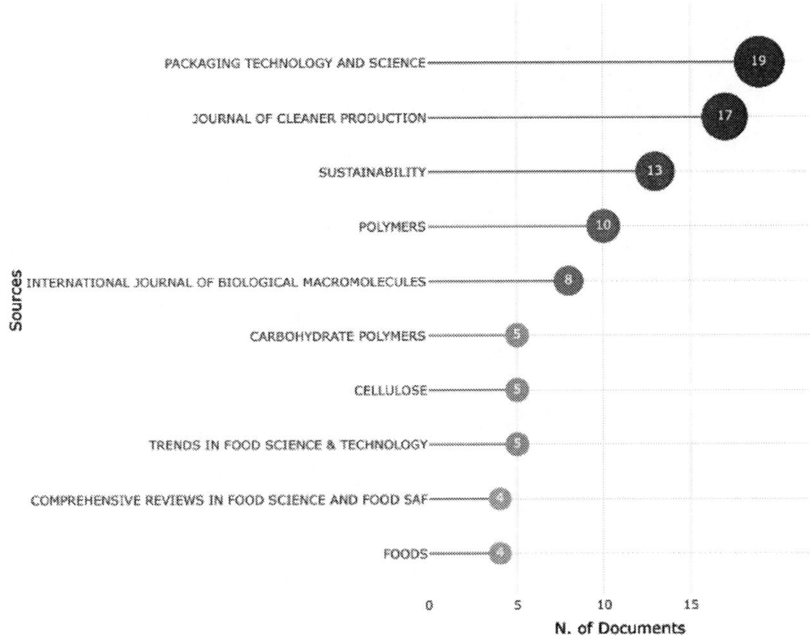

Figure 2: Most productive journals on sustainable packaging literature.

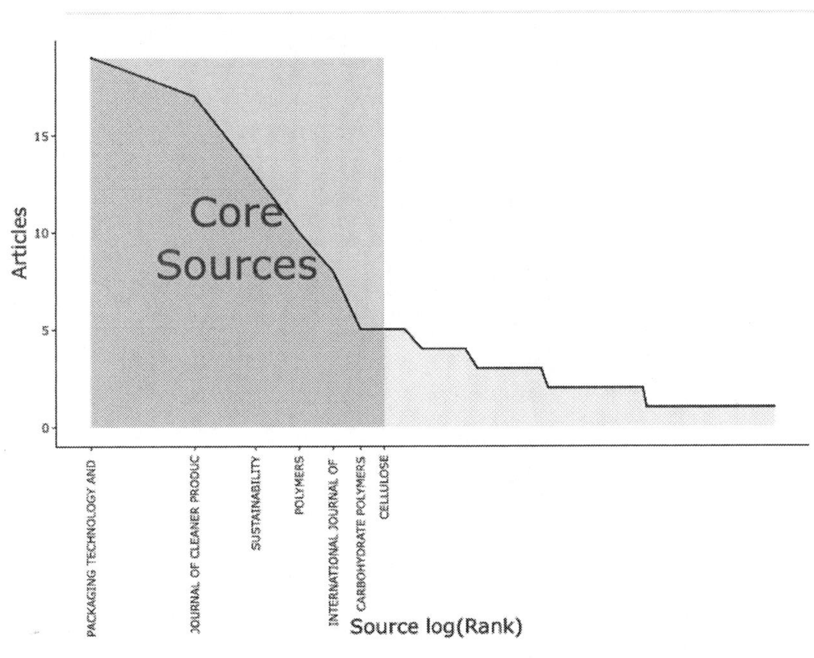

Figure 3: Core journal group based on Bradford's Law.

The United States (62 documents) and Italy (61 documents) made the largest contribution on sustainable packaging research followed by India (44 documents), Brazil (37 documents), China (36 documents), etc. There are lots of gray shaded countries on the map demonstrating no published research in sustainable packaging (Fig. 4). Rodríguez-Rojas et al. (2019) reported Italy, followed by Spain, Brazil, the United States of America (USA) and China most productive countries in the area of food packaging materials. These countries are important actors in our data collection as well. The USA was the most actively collaborating with other countries, followed by Brazil, Colombia and Italy [23]. According to our data collection, the USA mostly published single-country publications with Denmark, Finland and Romania having only single country publications based on corresponding author country, whereas all of the publications of France and Vietnam on the field were in collaboration with other countries (Fig. 5).

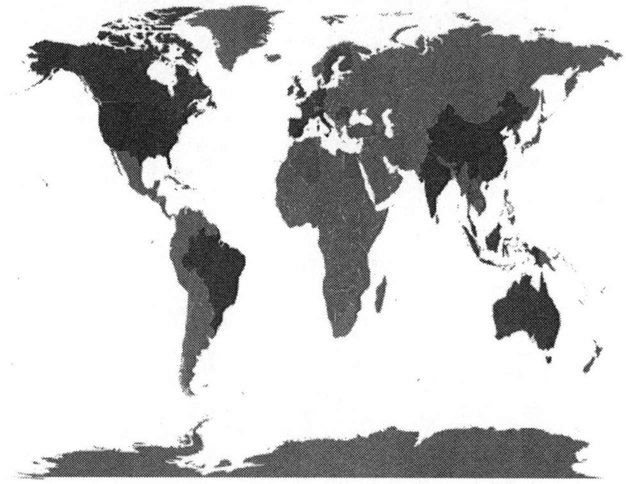

Figure 4: World map of publication frequency per country. The color concentration is proportional to publication frequency.

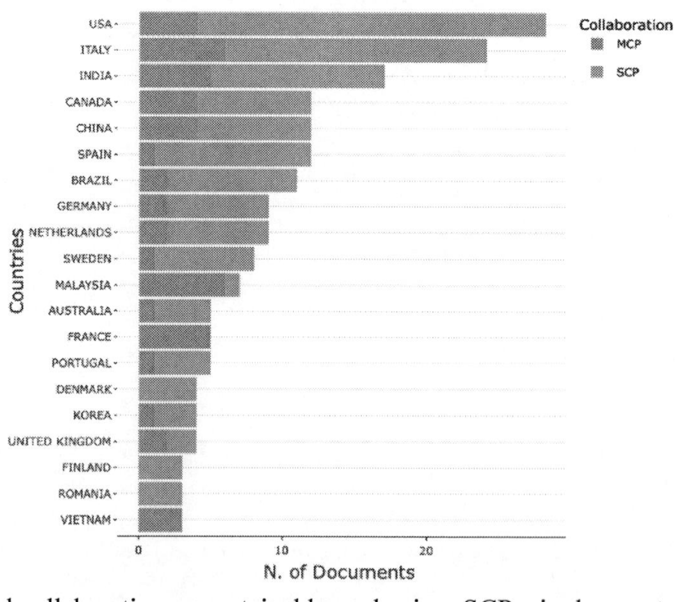

Figure 5: International collaboration on sustainable packaging. SCP: single country publication, MCP: multiple country publication

Five distinct collaboration groups of the top ranked countries were detected (Fig. 6). The size of a circle corresponds to the total publication number of one country through international collaboration. The size of the circle increase depending on published documents. The line width shows collaborative frequency, thicker line depicts closer relationship. One of these groups consists of the USA, India, Brazil, China, Canada and Korea with the USA showing the highest collaboration with the other members in the group (Fig. 6). The USA was also the most active country with respect to international collaboration on food packaging material [23]. Another distinct group is formed by Italy, Spain, France, Germany, Ireland, Netherlands, Sweden, Vietnam and Australia, with Italy being the most active member in the group and in closer relationship with Spain (Fig. 6). Italy showed a high collaboration with Spain on food packaging material, and also collaborated with Netherlands, France, Belgium, Finland, USA and Columbia [23]. Croatia and Romania are another isolated group. Turkey, Poland, Serbia and Lithuania are a distinct group which tend to collaborate together. United Kingdom, Malaysia, Indonesia and Nigeria are another subgroup, with Indonesia and Malaysia showing the closest relationship in the cluster (Fig 6).

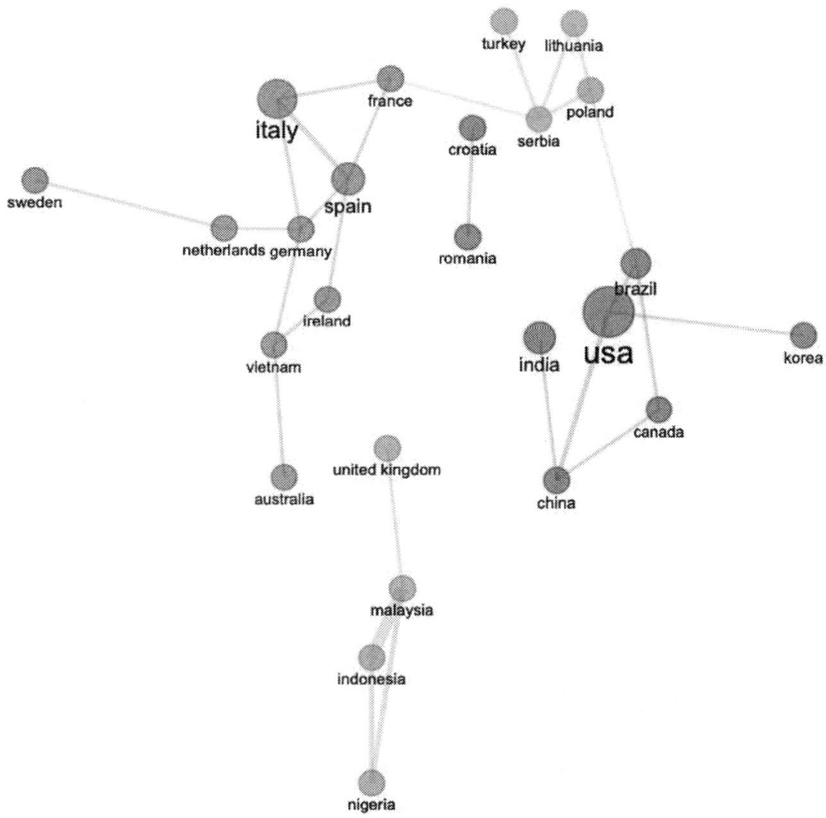

Figure 6: Country collaboration networks on sustainable packaging.

The most frequent keywords projected with the word cloud after sustainable packaging, packaging and sustainability was circular economy implying the importance of circularity in minimizing food waste and improving sustainability. Other notable keywords were mechanical properties, biodegradable, antimicrobial, barrier properties, biopolymer, shelf life, active packaging, nanocellulose, nanocomposite, recycling, etc. (Fig. 7).

Figure 7: Word cloud demonstrating most frequent author keywords.

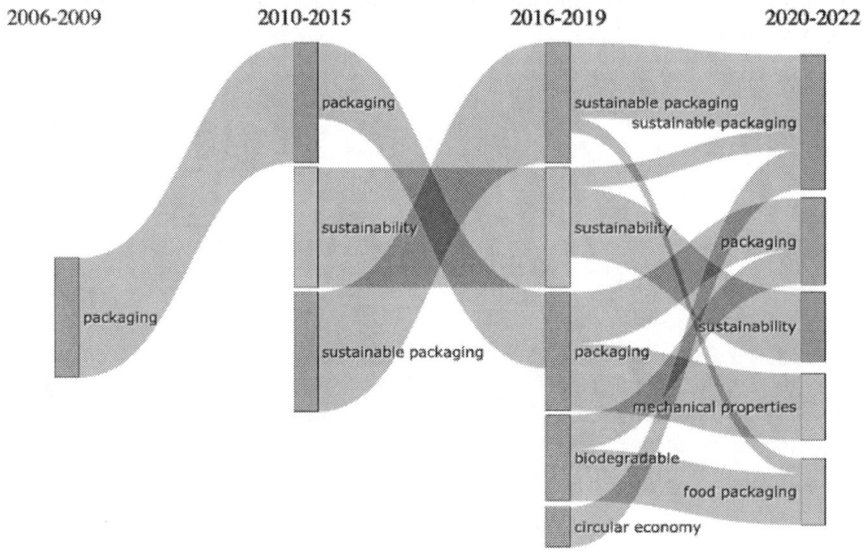

Figure 8: Evolution of research topics over time.

Packaging was the only research topic during the first time period (2006-2009) (Fig. 8) when the publication activity was low (Fig. 1). Later in the second period (2010-2015), when there was an increase in sustainable packaging publications (Fig. 1), besides packaging two new research topics emerged (sustainability and sustainable packaging). Biodegradable and circular economy showed as unique research topics during the third time period (2016-2019) along with packaging, sustainability and sustainable packaging (Fig. 8). Packaging branched to mechanical properties during the last time period (2020-2022), biodegradable merged in packaging and food packaging fields, and circular economy took place within sustainable packaging (Fig. 8). Nemat et al. (2019) showed that packaging and production research areas progressed in environmental and waste management topics over time, which is in agreement with our findings.

4. CONCLUSIONS

The first document on sustainable packaging according to our data collection was published in 2006, and there was an increased trend of publications over time (7.11%). Only seven out of 98 journals were in the core group with Packaging Technology and Science, followed by Journal of

Cleaner Production, and Sustainability being the most productive journals in the field. This high number of journals implies that many sources started to publish "sustainable packaging" research showing the emerging popularity of this area. The USA and Italy made the largest contribution on sustainable packaging research followed by India, Brazil and China. Collaboration network analysis was performed to describe relationships between top countries on the area, which resulted in five collaboration sub-groups reaffirming that research collaboration is not restricted by languages and borders. According to keyword analysis, the most frequent author keywords after sustainable packaging, packaging and sustainability was circular economy. Time evolution of research topics demonstrated that there was a shift from more general to more specific research topics starting from packaging and progressing into sustainable packaging and circular economy with a trend towards detailed characterization of packaging properties such as biodegradable and mechanical properties. Consequently, we believe that this work will be a useful source for researchers looking for international collaboration and trending topics in sustainable packaging, as well as potential journals to publish their research. Conflict of interest: The authors declare that there are no potential conflicts of interest regarding the research, authorship and publication of this manuscript.

5. ACKNOWLEDGEMENT

This article/publication is based upon work from COST Action "CA19124 Circul-a-bility" (circula-bility.org), supported by COST (European Cooperation in Science and Technology). www.cost.eu.

6. REFERENCES

1. United-Nations, Transforming Our World: The 2030 Agenda for Sustainable Development. Resolution Adopted by the General Assembly on 25 September 2015. 2015.
2. Circle-Economy, The Circularity Gap Report 2021. Circle Economy. 2021.
3. Crippa, M., et al., Food systems are responsible for a third of global anthropogenic GHG emissions. Nature Food, 2021. **2**(3): p. 198-209.
4. United-Nations, United Nations Department of Economic and Social Affairs. The Sustainable Development Goals Report 2021. United Nations. 2021.
5. Clune, S., E. Crossin, and K. Verghese, Systematic review of greenhouse gas emissions for different fresh food categories. Journal of Cleaner Production, 2017. **140**: p. 766-783.
6. Molina-Besch, K., F. Wikström, and H. Williams, The environmental impact of packaging in food supply chains—does life cycle assessment of food provide the full picture? The International Journal of Life Cycle Assessment, 2019. **24**(1): p. 37-50.
7. Garnett, T., Where are the best opportunities for reducing greenhouse gas emissions in the food system (including the food chain)? Food Policy, 2011. **36**: p. S23-S32.
8. FAO, Nutrition and food systems. A report by the High Level Panel of Experts on Food Security and Nutrition of the Committee on World Food Security, Rome. 2017.
9. Poore, J. and T. Nemecek, Reducing food's environmental impacts through producers and consumers. Science, 2018. **360**: p. 987-992.
10. Floros, J., et al., Feeding the World Today and Tomorrow: The Importance of Food Science and Technology. Comprehensive Reviews in Food Science and Food Safety, 2010. **9**(5): p. 572-599.
11. Verghese, K., H. Lewis, and L. Fitzpatrick, Packaging for Sustainability; Springer, ISBN 9780857299871. 2012.
12. Guillard, V., et al., The Next Generation of Sustainable Food Packaging to Preserve Our Environment in a Circular Economy Context. Frontiers in Nutrition, 2018. **5**.

13. Wikström, F., et al., Packaging Strategies That Save Food: A Research Agenda for 2030. Journal of Industrial Ecology, 2019. **23**(3): p. 532-540.
14. Lillford, P. and A.-M. Hermansson, Global missions and the critical needs of food science and technology. Trends in Food Science & Technology, 2021. **111**: p. 800-811.
15. Robertson, G., Food Packaging and Shelf Life; CRC Press, ISBN 9781420078459. 2009.
16. Robertson, G., Food packaging: Principles and practice, 3. ed.; CRC Press: Boca Raton, Fla., ISBN 9781439862414. 2013.
17. Soroka, W., Fundamentals of packaging technology: Fifth Edition, Inst of Packaging Professionals. 2014.
18. Aria, M. and C. Cuccurullo, bibliometrix: An R-tool for comprehensive science mapping analysis. Journal of Informetrics, 2017. **11**(4): p. 959-975.
19. Broadus, R.N., Toward a definition of "bibliometrics". Scientometrics, 1987. **12**(5): p. 373-379.
20. Crane, D., Invisible colleges: Diffusion of knowledge in scientific communities. Chicago: University of Chicago Press, 1972.
21. Forliano, C., P. De Bernardi, and D. Yahiaoui, Entrepreneurial universities: A bibliometric analysis within the business and management domains. Technological Forecasting and Social Change, 2021. **165**: p. 120522.
22. R Core Team, R: A language and environment for statistical computing. R Foundation for Statistical Computing, Vienna, Austria. URL https://www.R-project.org/. 2021.
23. Rodríguez-Rojas, A., et al., ¿What is the new about food packaging material? A bibliometric review during 1996–2016. Trends in Food Science & Technology, 2019. **85**: p. 252-261.
24. Vila-Lopez, N. and I. Küster-Boluda, A bibliometric analysis on packaging research: Towards sustainable and healthy packages. British Food Journal, 2021. **123**(2).
25. Shi, S. and J. Yin, Global research on carbon footprint: A scientometric review. Environmental Impact Assessment Review, 2021. **89**: p. 106571.
26. Nemat, B., et al., The Role of Food Packaging Design in Consumer Recycling Behavior—A Literature Review. Sustainability, 2019. **11**(16): p. 4350.
27. Bradford, S., Sources of information on specific subjects. Engineering, 1934. **137**: p. 85-86.

ST-P02

Environmental Analysis of Single-use and Multiple-use Pallet Systems for Air Shipments

Joongmin Shin* and Jay Singh

Industrial technology and Packaging, California Polytechnic State University, San Luis Obispo CA, 93407, USA

***Correspondence to:** Industrial technology and Packaging, California Polytechnic State University, San Luis Obispo CA, 93407, USA Jshin20@calpoly.edu*

ABSTRACT: Pallets, an integral component of most distribution strategies, offer a portable platform on which packaged-product can be stacked, stored, and moved using material handling equipment. Pallets can be broadly categorized as single-use or multiple-use. Single-use pallets tend to be relatively light and preferred for air shipments, but unlike the multiple-use pallets, they can't be reused, such as in a pooling system. This study undertook a comparative analysis in terms of environmental impacts based on lifecycle metrics and packaging attributes. COMPASS was used to compare six different pallets. These included plastic (HDPE and EPS composite), paper (corrugated fiberboard), metal (aluminum), and wood pallets (high density treated wood and pressed wood chips). All pallets were assumed to have an identical product capacity. To avoid redundant complexities and the overpowering of results in some transportation phases, the only packaging considered was the pallet itself. The functional unit was one pallet and truck ("Rigid 15t truck, average") and air transportation was used as the mode of distribution. An average distribution distance of 10,000 km was assumed for all pallets. The environmental impact assessment indicators included were global warming, fossil fuel use, mineral resource use, water use, freshwater eutrophication, human impact, and freshwater ecotoxicity. Overall, the single-use EPS composite pallet outperformed all other pallet types when a single one-way air shipment is assumed to be deployed from China to the US. The packaging-specific indicators used were packaging-to-product ratio and packaging-to-landfill or recycled. The EPS composite pallet, weighing 3.4 kg, was the lightest of all pallet types and had a packaging-to-product ratio of approximately 1.3%. Using the assumption of 20% of all pallet systems ending up in a landfill, the EPS composite pallet, at 0.33 kg, also outperformed all others.

Keyword: Pallets, life cycle assessment (LCA), COMPASS, Environmental impact assessment indicators, Packaging systems

1. INTRODUCTION

Every product is packaged and transported at one point during its lifetime. Packaging efficiency has drastically increased over the years as production volumes have increased exponentially. Since many products are produced in such large volumes, improving the efficiency of processes or materials can save businesses valuable resources. Most products are unitized as a palletized load. Unitization or consolidation of smaller units into a single larger unit has numerous benefits for all players in the typical supply chains of packaged products. These include improved warehousing efficiencies, increased handling and transportation productivities, enhanced safety of the packaged products and the employees, and inventory control augmentation.

Pallets are the principal interface of a unit load, and their primary role involves protecting the product from the physical stresses encountered through the logistical activities of handling, storage, and transportation. Superior quality pallets permit a higher volume of products per shipment, reduce the amount of product lost to damages or contamination, prevent worker injury from overloaded or unbalanced shipments, and reduce expenses such as fuel costs [1]. It is estimated that there are approximately five billion pallets in use worldwide [2]. Of these, approximately two billion are in use in the United States alone. It is also estimated that approximately 90% of the pallets are made of wood and the rest are made of plastic, metal, or cardboard [2]. According to a research report, the global pallets industry generated $79 billion in revenue in 2019, and is estimated to generate $110.56 billion in 2027, resulting in a compound annual growth rate (CAGR) of 5.1% from 2020 to 2027 [3].

Wooden pallets, the most used worldwide, can be classified in different ways depending on their size and construction. These include classes (stringer and block), use category (reusable and single use), entry types (two-way entry, partial four-way entry and full four-way entry), styles (single-face, double-face-nonreversible, and double-face-reversible), top deck constructions (deck board, deck board/stringer board, panel, and panel/stringer board) and bottom deck constructions (unidirectional, overlapping, perimeter, and cruciform) [4].

As compared to multiple use pallets, single use pallets tend to be relatively lighter and preferred for air shipments. Unlike the multiple-use pallets, single use pallets can't be reused, such as in a pooling system. Pooled pallets are routinely recovered, cleaned, repaired, and recirculated. Single-use pallets tend to be made from lighter weight or pressed wood, paper, and foamed or thermoformed plastic. With the ever-increasing logistics costs, especially for air freight, lighter-weight single-use pallet options are gaining momentum.

While reusable packaging formats tend to be attractive for an overall reduction in energy use and emission of greenhouse gases, the closed-loop (pooling) inventory routing is a major concern for reusable pallet systems [5-11]. This issue is enhanced for long-distance and multicontinental shipments.

Life Cycle Assessment (LCA) is the investigation and valuation of the environmental impacts of a product or service with the goal of comparing the full range of environmental damages assignable [12]. LCA attempts to predict the overall environmental burdens associated with the provision of the product and identify particularly burdensome or wasteful processes therein. This study presents the LCA results for six different pallets. These include plastic (HDPE and EPS composite), paper (corrugated fiberboard), metal (aluminum), and wood pallets (high density treated wood and pressed wood chips).

2. MATERIALS AND METHODS

The LCA is an assessment and evaluation of the input and output data and the potential environmental impacts of a product, system, or process throughout its life cycle. The widely accepted four steps in the LCA study include (1) goal and scope definition, (2) LCI, (3) life cycle impact assessment (LCIA), and (4) interpretation of the result [12]. This analysis includes many environmental impacts that are easy to overlook, such as energy consumption and atmospheric emissions over the entire life cycle. The LCAs conducted by EcoImpact-COMPASS (Ver. 3.0, TRAYAK, Mason, OH).

2.1. Goal and Scope Definition

This study aimed to understand the environmental impacts of five different pallet systems, including plastic (HDPE and EPS composite), paper (corrugated fiberboard), metal (aluminum), and wood pallets. The functional unit used for this project was one single pallet load. Preliminary analysis was performed on each pallet system carrying identical amounts of product. However, the weight of the product was overpowering the results in some of the transportation phases. In order to avoid unnecessary complexities, the pallet itself was assumed as the only packaging, and It was assumed that each pallet could protect and deliver the same product to meet the packaging requirements.

The scope of this research includes the complete process of the packaging life cycle from raw material extraction to disposal options for each plastic material, which is the cradle-to-grave perspective. However, the system boundary excludes the production, packaging, and use stage of the product, primary packaging, and secondary packaging. The geographic scope of this study is the US for the manufacturing and end-of-life (EOL) region.

2.2. Scenario Description

The life cycle scenarios of the pallet systems have been separated into four life cycle stages: Material production, manufacturing, transportation, and end of life. The material production stage includes plastic, wood, paper, and aluminum production processes. The manufacturing includes pallet formation processes such as injection molding, thermoforming, corrugated board process, etc. The transportation stage covers the transportation of packaged products from the manufacturing site to the retail locations or end-users. The same type of transportation (air-international, from China to USA) was used for all pallets, and an average distance of 10,000 km was assumed. In addition, a constant dometic transportation distance of 1678.45 km was chosen (including average distance to deliver retail stores, average distance from curb to disposal plant, average distance from curbside to material recovery facilities). The transportation distance from the package manufacturer to the point of use could vary widely for each package alternative and product combination. By choosing to model a specific package system, the study could have unknowingly chosen a system that favored one package type over any of the others. Consequently, the constant transportation distance was chosen based on the previous studies [13-16].

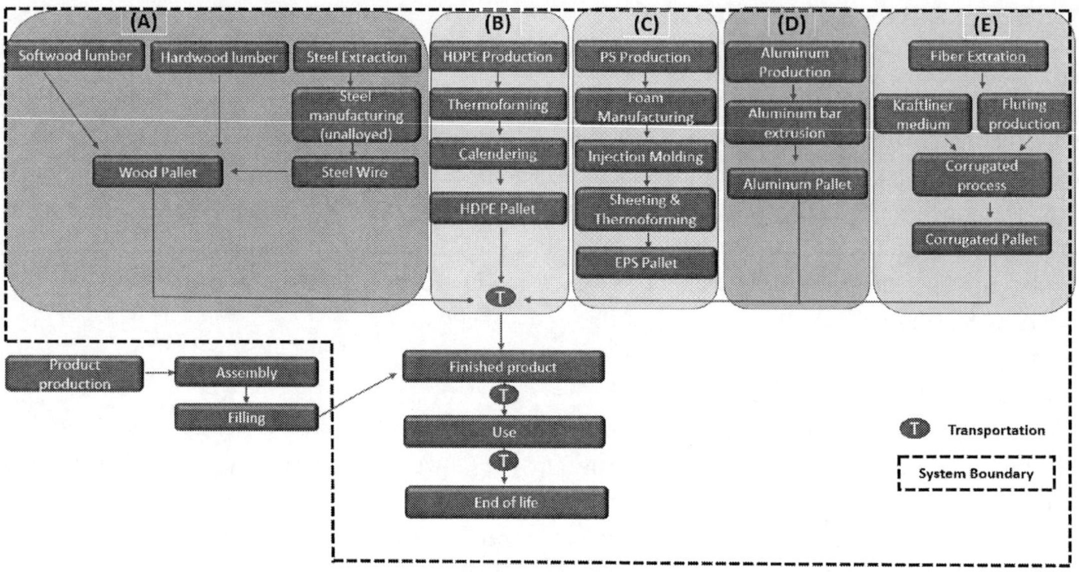

Figure 1. System boundary of 5 different pallet systems. (A) shows wood pallet. (B) shows HDPE pallet. (C) shows EPS pallet. (D) shows aluminum pallet. (E) shows Corrugated pallet.

2.3. Life cycle impact assessment methodology

Ecoinvert 3.6 database was used for most of the materials within the LCA and represented a global market. For any recycled materials within the LCA, energy and water flows have been regionalized, so they are specific to the United States region. The data sources for the various materials used within this LCA are detailed in Table 1.

Table 1. life cycle inventory for modeling the production of pallet system

Packaging System	Material	Production	Weight (g)	End of life
EPS pallet	Expanded Poly Styrene (EPS)	Injection molding, Thermoforming Plastic Sheet	3.3	Waste to Energy 20%, Landfill 10%
GMA wood pallet	Sawn hardwood, raw, air dried	No conversion	5850	Waste to Energy 20%, Landfill 80%
	Sawn softwood, raw, air dried	No conversion	7150	Waste to Energy 20%, Landfill 80%
	Steel (unalloyed)	Steel wire drawing	222.3	Waste to Energy 20%, Landfill 80%
HDPE pallet	High Density Polyethylene (HDPE)	Thermoforming with calendering	15420	Recycling 287%, Waste to Energy 12%, Landfill 60%
Aluminum pallet	Aluminum	Aluminum Bar Extrusion	16800	Waste to Energy 20%, Landfill 80%
Corrugated Pallet	Currugated board	Corruaged board process	4900	Recycling 96%, Waste to Energy 1%, Landfill 3%

2.4. Assumptions and Limitations

To conduct this LCA, some assumptions needed to be made to fill in some data gaps. The results generated through COMPASS were limited to the data provided, modeled, and assumptions made where data was absent such as proxying a material or process. For the end of life, a number of assumptions were made based on previous research and COMPASS defaults data. In case in which information was available about take-back systems, about 80% of the systems were assumed to end up in landfills. For systems made of materials such as Aluminum and Corrugated, the existing recycling data has used these materials.

3. RESULTS AND DISCUSSIONS

The radar chart in Figure 2 shows all the analyzed impacts for all pallets. For the air shipping scenario, EPS pallet outperforms most other pallet systems except for water use. Plastic pallets (such as HDPE and EPS pallet) was supposed to have a higher mineral resource impact because these are fossil fuel-based material. However, the weight of the EPS pallet is so light, and the impacts are significantly smaller. The impact in HDPE was also lower (than wood pallet) because of less weight and more recycling contents. The water consumption of EPS was the highest among all pallet systems because the conversion and sourcing process used for EPS is more water-intensive than others. The corrugated pallet is also a lighter system, although a bit more than double the weight of EPS pallet. Thus, the system also showed significantly lower overall environmental footprints. However, EPS still causes a less overall impact.

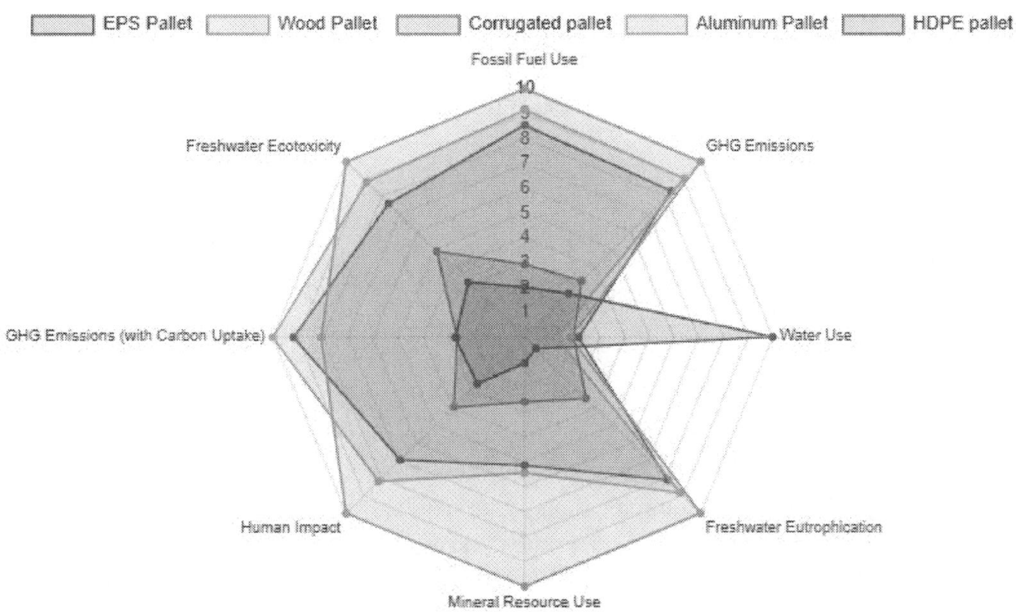

Figure 2. Overview of LCA results for air shipping scenario

The absolute values of the comparison are shown below in Table 2 .The impacts are shown for the functional units of 1 use of each pallet.

Table 2. Absolute results of life cycle impact assessment for different pallet systems

Name	EPS Pallet	Wood Pallet	Corrugated pallet	Aluminum Pallet	HDPE pallet
Fossil Fuel Use (MJ deprived)	3.51	10	4.55	9.16	8.59
GHG Emissions (kg CO2 eq.)	3.86	10	4.9	9.09	9.25
Water Use (liters)	10	2.07	3.14	1.51	2.5
Freshwater Eutrophication (kg PO_4 eq.)	2.37	10	5.21	8.91	8.28
Mineral Resource Use (kg deprived)	2.12	10	4.1	5.94	5.78
Human Impact (DALY)	3.78	10	5.89	8.35	7.8
GHG Emissions (with Carbon Uptake) (kg CO2 eq.)	4.16	8.47	4.19	9.83	10
Freshwater Ecotoxicity (CTUe)	4.28	10	7.09	8.92	7.98

The LCA results are broken into the five phases of the life cycle: material, manufacturing, transportation, end of life, and use phase. Among the 8 environmental impacts, Fossil fuel use, GHG Emissions, Freshwater Eutrophication, and Mineral resource use were significantly distinguished. Figure 3 below is shown the scale (%) of the impact comparisons with the graph. The EPS pallet, which has the lowest overall environmental impact, was used as a reference to calculate the scale. As is shown in the graph, transportation (air shipment) causes the biggest impact on all environmental potentials. Thus, material, manufacturing, and end of life have the least impact of all analyzed pallet systems. Thus, a domestic short trip scenario such as a truck or train may have different results. Additional studies in various transportation scenarios (different distances and transportation modes) are recommended for future study.

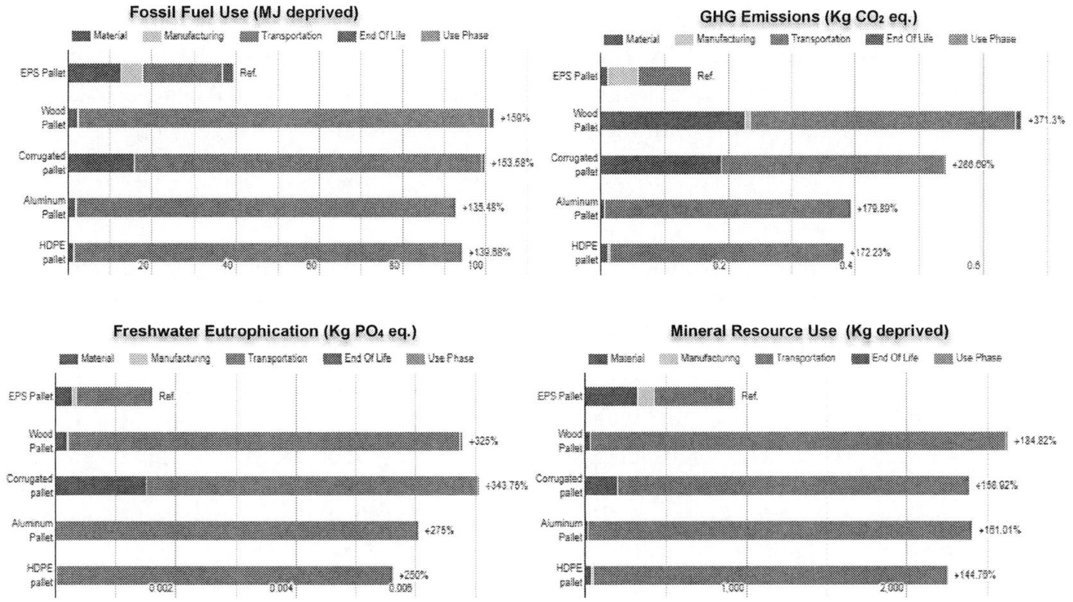

Figure 3. Results of life cycle impact assessment.

4 CONCLUSION

A life cycle study of 5 pallet systems was conducted to analyze their environmental burden. Based on the assumption that 20% of all pallet systems end up in a landfill, EPS pallets incur an overall 26 – 375% lower environmental impact than other pallet systems. The EPS composite pallet, weighing 3.4 kg, was the lightest of all pallet types and had a packaging-to-product ratio of approximately 1.3%. Although this study provides a useful implementation of the five pallet systems, it still has several imitations. First and foremost, due to the nature of a typical LCA study, several uncertainties are dealt with the analysis, including data collection uncertainty, model correctness uncertainty, and model completeness uncertainty, which may weaken the conclusion in some aspects. Moreover, due to the limited information, several assumptions such as end-of-life recycling data, number of reuse, end of life recycling data may have changed the result significantly.

5 REFERENCES

[1] Inbound Logistics. (2018). Pallets: Backbone of the Warehouse. https://www.inboundlogistics.com/cms/article/pallets-backbone-of-the-warehouse/#:~:text=Pallets%20are%20the%20primary%20interface,traveling%20through%20the%20supply%20chain. Accessed April 21, 2022

[2] Placek, M. (2022). Pallet market size worldwide 2019-2027. https://www.statista.com/statistics/1228432/global-pallet-market-size/#:~:text=Analysts%20estimate%20that%20there%20are,plastic%2C%20metal%2C%20or%20cardboard. Accessed April 2, 2022

[3] Pednekar, T & Sumant, O. (2021). Pallets Market by Type (Rackable, Nestable, Stackable, and Display), Material (Wood, Plastic via Injection Molding, Plastic via Other Methods, Corrugated Paper, and Metal), and End-user Industry (Food & Beverage, Chemical, Retail, Pharmaceutical, and Others): Global Opportunity Analysis and Industry Forecast, 2020–2027. Report. Allied Market Research. March 2021

[4] National Wooden Pallet and Container Association (NWPCA). (2014). Uniform Standard for Wood Pallets. Alexandria, VA: National Wooden Pallet and Container Association. https://cdn.ymaws.com/www.palletcentral.com/resource/collection/E8AADDDE-7CBA-4298-8341-C7F29D0C14FF/Uniform-Standard-for-Wood-Pallets-2014(REV).pdf. Accessed April 6, 2022

[5] Soysal, M. (2016). Closed-loop Inventory Routing Problem for returnable transport items. Transportation Research Part D: Transport and Environment, 48, 31-45.

[6] Stefanini, R., Borghesi, G., Ronzano, A., & Vignali, G. (2021). Plastic or glass: a new environmental assessment with a marine litter indicator for the comparison of pasteurized milk bottles. The International Journal of Life Cycle Assessment, 26(4), 767-784.

[7] Meyhoff Fry, J., Hartlin, B., Wallén, E., & Aumônier, S. (2010). Life cycle assessment of example packaging systems for milk. Waste & Resources Action Programme (WRAP).

[8] Accorsi, R., Cascini, A., Cholette, S., Manzini, R., & Mora, C. (2014). Economic and environmental assessment of reusable plastic containers: A food catering supply chain case study. International Journal of Production Economics, 152, 88-101.

[9] Gallego-Schmid, A., Mendoza, J. M. F., & Azapagic, A. (2019). Environmental impacts of takeaway food containers. Journal of Cleaner Production, 211, 417-427.

[10] Singh, J., Krasowski, A., & Singh, S. P. (2011). Life cycle inventory of HDPE bottle-based liquid milk packaging systems. Packaging Technology and Science, 24(1), 49-60.

[11] Singh, J., Shani, A. B. R., Femal, H., & Deif, A. (2016). Packaging's role in sustainability: Reusable plastic containers in the agricultural-food supply chains. In Organizing Supply Chain Processes for Sustainable Innovation in the Agri-Food Industry. Emerald Group Publishing Limited.

[12] ISO 14040 (2006): Environmental management – life cycle assessment– principles and framework. ISO 14040:2006

[13] Konstantas, A., Stamford, L., Azapagic, A., 2019. Evaluation of environmental sustainability of biscuits at the product and sectoral levels. J. Clean. Prod. 230, 1217–1228. https://doi.org/10.1016/j.jclepro.2019.05.095.

[14] Weber, C.L., Matthews, H.S., 2008. Food-miles and the relative climate impacts of food choices in the United States. Environ. Sci. Technol. 42 (10), 3508–3513. https://doi.org/10.1021/es702969f.

[15] Ploeg M.V., Mancino L., Todd J.F., Clay D.M., Scharadin, B. (2015) Where do Americans usually shop for food and how do they travel to get there? Initial findings from the National Household Food Acquisition and Purchase Survey. Food Shopping Patterns and Geographic Access to Food: comparisons and Data

[16] U.S. Department of Energy. 2008. Transportation Energy Databook 27. ORNL-6981

SP-O01

Life Cycle Assessment of Returnable Mailers used for Apparel Electronic Commerce: A Case Study in Canada

Jonghun Park[1]* and Zuha Waqar[1]

[1]Graphic Communications Management at the Creative School, Toronto Metropolitan University, Toronto, ON M5B 2K3, Canada

***Correspondence to:** Jonghun Park, Graphic Communications Management at the Creative School, Toronto Metropolitan University, 350 Victoria St., Toronto, ON M5B 2K3, Canada.*
E-mail: jaypark@ryerson.ca

ABSTRACT: E-commerce is increasingly popular in the fashion industry in Canada. Based on a case study, this research analysed the environmental sustainability of the returnable e-commerce package format distributed in Canadian apparel retail channels. The case study compared the environmental impacts of a returnable polypropylene mailer to those of two different expendable packaging formats currently used by a brand owner, such as a corrugated paperboard package and a low-density polypropylene mailer. A life cycle assessment (LCA) was conducted to examine the environmental impacts of the three packaging scenarios. The foreground data of the LCA were obtained from a brand owner and packaging suppliers in 2021. TRACI 2.1, including 10 life cycle impact categories, was used for the LCA. The comparative LCA results showed that the returnable mailer generated the lowest environmental impact for eight out of 10 impact categories when the final consumers' locations were relatively close to the brand owner or returnable packaging service provider. The expendable corrugated paperboard packaging format contributed the greatest environmental burden in all the impact categories primarily because it used a large amount of material and was relatively heavy. However, the comparative results were changed for some impact indicators depending on the final consumer's location and the number of reuse cycles for the returnable mailer. Specifically, the returnable mailer had a higher environmental impact than the expendable mailer for half of the 10 impact categories when the number of reuses for the returnable mailer was decreased to 10 cycles.

iapri Bangkok 2022

23rd World Packaging Conference
12-16 June 2022

GOLD SPONSOR:

SINGHA CORPORATION

SILVER SPONSOR:

GRAINPRO
STORING THE FUTURE

Intelgro
Intelligent

BRONZE SPONSORS:

FREE FROM FOOD ASIA BANGKOK • 2022

VIV ASIA

meat pro ASIA

8 – 10 March 2023
Bangkok, Thailand

SCGP

PROPAK ASIA

PerkinElmer
For the Better

SHIMADZU
Excellence in Science

BARA SCIENTIFIC
Solution of Success

PACK PRINT INTERNATIONAL

SUSTAINABLE EVENT BY:

carbon neutral

SUSTAINABLE EVENT

BY THAILAND CONVENTION
AND EXHIBITION BUREAU